# Plant Breeding Essentials

# Plant Breeding Essentials

Edited by **Clive Koelling**

R CALLISTO
REFERENCE

New York

Published by Callisto Reference,
106 Park Avenue, Suite 200,
New York, NY 10016, USA
www.callistoreference.com

**Plant Breeding Essentials**
Edited by Clive Koelling

International Standard Book Number: 978-1-63239-512-2 (Hardback)

Printed in the United States of America.

# Contents

# Preface

In my initial years as a student, I used to run to the library at every possible instance to grab a book and learn something new. Books were my primary source of knowledge and I would not have come such a long way without all that I learnt from them. Thus, when I was approached to edit this book; I became understandably nostalgic. It was an absolute honor to be considered worthy of guiding the current generation as well as those to come. I put all my knowledge and hard work into making this book most beneficial for its readers.

This book presents a detailed account of information on modern approaches in plant breeding. Contemporary plant breeding is regarded as a discipline whose origins lie in the science of genetics. It is considered a very intricate subject, involving the use of several integrative novel sciences and technologies which developed into business, science and art. Extraordinary growth in contemporary plant breeding has been witnessed, enriching the conventional breeding practices with accurate, effective, economical and swift breeding tools and approaches as a result of novel advancements in genomics as well as plant genetics and coupling plant "omics" accomplishments accompanied with progresses in computer science and informatics, as well as laboratory robotics. The aim of this book is to describe some of the current developments of 21st century plant breeding, elucidating new approaches, achievements, views, research efforts and perspectives in breeding of some crop species. Latest advances and comprehensive information on selected topics have been provided in this all-inclusive book which aims to improve the knowledge of the readers regarding contemporary plant breeding.

I wish to thank my publisher for supporting me at every step. I would also like to thank all the authors who have contributed their researches in this book. I hope this book will be a valuable contribution to the progress of the field.

**Editor**

# Part 1

# Breeding Approaches and Its Modelling

# Virtual Plant Breeding

Sven B. Andersen
*University of Copenhagen, Dept. Agronomy and Ecology*
*Denmark*

## 1. Introduction

Breeding better cultivars has become a highly efficient way to improve plant production for yield, quality and reduced input. Most breeding activities have been focussed into fewer and larger programs with less people engaged. Educational environments for the area have been correspondingly reduced in size, which makes it harder to provide inspiring environments to generate high level education, for the fewer people needed in industry. Still plant breeders, scientists as well as society have ample interest in widespread public understanding of the use of new as well as old technologies for improvement of our cultivated plants. This is not least to avoid future communication problems with the general public like experienced with genetically modified plants during recent years.

Many university environments dealing with plant breeding education face two main problems. The first problem is the strong tendency to privatise practical plant breeding, which moves the breeding operations away from educational institutions into private more closed environments. Without live breeding programs associated with the campus, meaningful training becomes a possibility only for the few. Advanced students dedicated to the area may enter private activities as part of their advanced education, but this is not an option for the majority. The second problem is the decreasing number of plant breeders needed in the increasingly more efficient breeding programs. With fewer jobs on the market the general interest among students is reduced and the entire educational environment diminishes. For a majority of students in biology the interest in plant breeding exists, but only on a shallow level. They may well see the point in understanding major principles of applied genetics, also to be used in basic plant science, but actual training in plant breeding theory and operations is far beyond both their time and dedication. To reach a wider public within education in the future, plant breeding may need new ways to communicate its methods and principles.

Simulated learning systems are now widely used to train special skills in complex environments. Clinical training of nurses and health care workers are time-consuming and resource-demanding, but can be well supplied by simulations in special virtual learning environments. It is possible through experimental studies to assess the effect on such simulated training of skills within health care (Tsai et al., 2008). Several more general software packages for simulation based training of complex skills are available on the internet. Several British law schools collaborate on a simulated virtual learning approach to train skills in legal transactions with the software SIMPLE (SIMPLE 2011). ISLE Interactive

Simulated Learning Environments gives as a service the construction of user defined training systems for complex and expensive equipment or environments (ISLE 2011). The software SIMWRITER from NexLearn can be customized to users' needs for simulation to train complex social skills (Nexlearn 2011). Virtual Property Manager is a simulation software developed to teach students within residential property management, developed to generate more interest among students for the area (Carswell & James 2007). The principles of efficient training with virtual learning environments have been outlined with the theory of *problem spaces* (Stefanutti & Albert 2003).

The software Supergene relates to the problem in plant breeding education that most operations related to plant breeding outside laboratories are very time-consuming and resource-demanding. Students have to devote an increasing part of their time to studies of rapidly expanding molecular and physiological issues. This leaves less time to come to understand the actual applications on plant, population and ecological levels. The program is an attempt to establish simulation based illustration of phenomena and operations in applied genetics and breeding based on very few assumptions. The program is written in Delphi version 5 and works on Windows platforms. The entire package can be downloaded from www.supergene.dk. The present interphase points towards plants, but the basic simulation process may be used for other eukaryotes with some modification.

## 2. Material and methods

### 2.1 Basics of the simulation process

The software simulates virtual genetic alleles each holding one byte of information, which means that one gene locus may form a maximum of 256 different mutations or alleles. Gene loci each holding one allele are arranged linearly into virtual chromosomes. One complete set of chromosomes for a species is simulated as a gamete, which has the ability to fuse with another gamete from the same species during a fertilization process to form a diploid zygote. The user defines number of chromosomes, length of individual chromosomes as well as total number of gene loci during creation of the species.

Diploid zygotes formed by fertilization between two gametes from the same species form the basis of individual plant simulation. Such zygotes have the ability to form gametes through meiosis. During meiosis homologous chromosomes are first copied and subsequently paired into bivalents and recombined through formation of randomly positioned chiasmata. Number of chiasmata is drawn from a Poisson distribution with an expectation to generate a recombination frequency of 1 % between gene loci. Chiasmata are positioned randomly on chromosome pairs according to a uniform distribution with assumption of no interference between double crossovers. The approximately one centimorgan between each gene locus implemented also sets the limit of resolution during simulated linkage analysis. The meiosis process includes mutagenesis of each parental gamete before formation of chiasmata and recombination. The number of mutations in a gamete is drawn from an approximate Poisson distribution adjusted to obtain expected number of mutations in each gamete desired by the user. Mutations are located randomly onto loci of chromosomes according to a uniform distribution, and mutation of one allele means that its byte content is changed to carry a new random byte.

In addition to nuclear chromosome sets each zygote also contains one gamete simulating a chloroplast genome and one gamete simulating a mitochondrial genome. During virtual meiosis these cytoplasmic genomes are copied into the new gametes after mutagenesis. The user can control the frequency of paternal or maternal inheritance of cytoplasmic organelles during the fertilization process. If transferred from both the male and the female gamete into the new zygote the mitochondrial genomes are recombined and randomly reduced to one copy. Chloroplast genomes, if inherited from both the male and the female gamete, are randomly reduced to one copy without recombination.

Successive cycles of meiosis and fertilization form the basic simulation of sexual propagation of plants in populations. Chromosome doubled haploid plants are generated through meiosis with subsequent chromosome doubling instead of fertilization. Cloning uses simple copying of a zygote without meiosis and fertilization, but with mutation. Cybridization of protoplasts to transfer cytoplasmic male sterility (*cms*) or other cytoplasm inherited factors transfers only the chromosomal genomes of the recipient. Organellar genomes are combined from both protoplast parents following the rules above for biparental inheritance. In addition to the mutation and recombination, gametes of a plant may be transformed, which means that one or more transgenic constructs (inserts) are substituted into the chromosomes instead of the ordinary gene loci. Such transgenic alleles carry one random byte value termed the "expression value" to simulate different expression levels of different inserts.

## 2.2 Major features of the user interface

Input and output for the software has been arranged in a set of material windows presenting plants, populations and groups of populations. Operation windows make it possible to perform basic breeding operations like crossing, selfing, doubled haploid generation, genetic transformation of plants and cybridization with protoplasts. Also it is possible to perform polycross, topcross and testcross to estimate combining abilities of plants and populations and use them for evaluation and selection based on offspring values. Some basic analyses like analysis of MxN crosses and diallel sets of crosses have been implemented in special windows. For markers it is possible to perform simple analysis of association between traits and markers in populations. It is possible to analyze biparental mapping populations and download all polymorphic marker scores together with trait values for subsequent quantitative trait loci (QTL) analysis. Markers found linked to genes of interest can subsequently be transferred to a special marker list in the program and used for marker assisted selection (MAS).

Other windows make it possible to generate and edit traits associated with virtual genes in the genome of the species simulated. Such definitions of a trait are methods for calculation of values of the trait for each plant, based on the alleles the plant inherited from its father and mother. The trait values for each plant or population can subsequently be used for selection of individuals or populations with desirable alleles.

For simulation of molecular markers the basic byte values, are read out directly as marker alleles (Fig. 2.). These allele values may be thought of as the migration distance on a gel. The marker allele value zero is considered a null allele that does not generate a detectable signal. Heterozygous co-dominant marker loci show both allele values, while homozygous marker loci show only one allelic value. Null alleles are not shown as a separate value when

| NR | NOR.Yield | CA.Yield | NOR.Carbohydrat | CA.Carbohydrate |
|----|-----------|----------|------------------|-----------------|
| 1  | 1022.5590 | 1568.2407 | 0.6869 | 0.6676 |
| 2  | 1716.3158 | 1845.6825 | 0.7118 | 0.6917 |
| 3  | 1672.9583 | 1974.7544 | 0.7106 | 0.6745 |
| 4  | 2176.4598 | 1790.4046 | 0.8117 | 0.6891 |
| 5  | 1544.1726 | 1571.1863 | 0.6585 | 0.6882 |
| 6  | 2449.9753 | 1633.6382 | 0.6497 | 0.6774 |
| 7  | 2079.8218 | 1835.6891 | 0.6731 | 0.6540 |

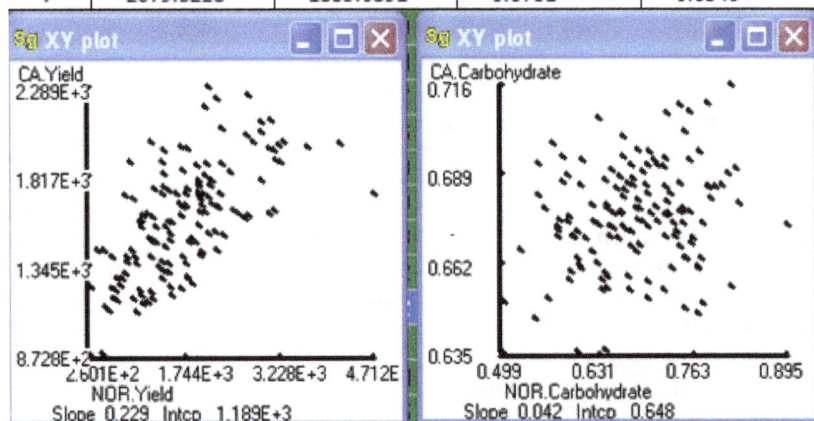

Fig. 1. Ryegrass plants after polycross evaluation showing yield of parental plants (NOR.Yield, kg/ha) and yield of their corresponding polycross offspring (CA.Yield). The offspring versus parent regression shows high heritability for yield, while carbohydrate content (per cent total carbohydrate of dry matter) show low heritability.

heterozygous, but appear as the value 0 when homozygous. The user can control the frequency of markers simulated as dominant or co-dominant. Highly co-dominant markers are like Simple Sequence Repeats (SSR), while more dominant systems behave like Amplified Fragment Length Polymorphisms (AFLP).

| F2 | NR | AthBIO2b | AthGPA1 | nga1145 |
|----|----|----------|---------|---------|
| Plant1 | 1 | 134 | 110 81 | 152 |
| Plant2 | 2 | 198 134 | 81 | 152 |
| Plant3 | 3 | 198 134 | 110 81 | 152 |
| Plant4 | 4 | 198 134 | 110 81 | 0 |
| Plant5 | 5 | 134 | 110 | 0 |
| Plant6 | 6 | 198 | 110 81 | 152 |
| Plant7 | 7 | 198 134 | 110 81 | 0 |

Fig. 2. Virtual *Arabidopsis thaliana* plants segregating for two co-dominant markers (AthBIO2b and AthGPA1), each showing two different allelic values for heterozygous and only one allelic value for homozygous genotypes. The marker nga1145 segregates dominantly with a null allele, which shows the value 0 only if homozygous.

## 2.2 Simulation of traits

The software enables user definition of traits calculated for each plant based on the allelic values inherited in its genes. Values of SIMPLE traits are calculated as a function of the allelic values in one or more genetic loci from either chromosomal or cytoplasmic genomes. Values for TRANSGENIC traits are calculated like for SIMPLE traits, but only based on expression values of transgenic inserts in the chromosomes. COMBINED traits are user defined functions of one or more SIMPLE, TRANSGENIC or COMBINED traits, which enable construction of very complex and correlated traits.

Calculation of a trait value consists of three steps: 1) calculation of a BASIC VALUE as a sum of contributions to the trait from each gene locus, 2) addition of environmental effects and 3) scaling.

$$Basic\ value = \sum_{all\ loci} Bconstant \pm (Male + Female \pm ABS(Male - Female) * dom)/2$$

Where *Bconstant* is a user defined constant to set the general level of the trait. *Male* and *Female* are the byte values from the alleles of the locus after transformation to have an approximate normal distribution with expectation zero and variation defined for each locus by the user. *dom* is a user defined factor of dominance between alleles for each locus, which can take values between 0..1. Transformation of the uniformly distributed byte values of each allele to approximately normally distributed values has the effect that alleles with extreme contribution in positive or negative direction become rare. This means that in an ordinary population, plants with extreme genotypes of the trait become rare. The user defined variation of the allelic value distribution describes different contribution of each locus to the trait in addition to the actual allelic values inherited. The ± in the summation is a sign bit, which means that the contribution from the locus to the trait is either positive or negative. This sign bit is initially set randomly by the program, but can be controlled by the user. It has been introduced to avoid unidirectional correlation between different traits affected by the same gene loci. For a TRANSGENIC trait, the basic value is *Bconstant* plus the sum of expression values from each insert of the construct. For a COMBINED trait the basic value is the value of evaluation of a function of other traits defined by the user.

Random environmental effects are added to the basic trait value to simulate non-genetic variation. A random number is drawn from a normal distribution with expectation zero and standard deviation between zero and one. The size of this environmental standard deviation is provided by the user. The random number is subsequently multiplied with the basic trait value to generate a multiplicative environmental effect before being added to the basic trait value. This type of environmental effects simulation combined with the subsequent scaling of the trait introduces a multiplicative type of gene by environment interaction.

Scaling of traits is performed to keep the trait levels within some desired bounds. A trait like yield should be limited to non-negative values. In principle, yield should be able to go to infinity, however the illusion is lost if the trait goes much above levels observed in reality. The software uses exponential scaling functions (Table 1) to introduce such limits and the user can then use the scaling constant of the trait and the variation of the genetic loci to limit the unbounded values of the trait. A logistic function is used to scale traits with both upwards and downwards bounds. With a large scaling constant this function will also

| High bound | Low bound | Trait Value |
| --- | --- | --- |
| $+\infty$ | $-\infty$ | $Const * Bvalue$ |
| $Max$ | $Min$ | $Min + (Max - Min) * \dfrac{e^{Const*Bvalue}}{1 + e^{Const*Bvalue}}$ |
| $Max$ | $-\infty$ | $Max - e^{Const*Bvalue}$ |
| $+\infty$ | $Min$ | $Min + e^{Const*Bvalue}$ |

Table 1. Functions used to scale traits with different boundaries. *Max* and *Min* are upwards and downwards limits of the trait. *Bvalue* is the basic trait value including added environmental effect, and *Const* is a user provided scaling constant.

effectively transform a basically quantitative trait into a two level qualitative trait. The scaling will introduce some types of epistasis and the degree of dominance of a trait is also affected by the scaling.

There are three traits with special functions generated automatically when a new species is initiated. "Malefertile" and "Femalefertile" determine the frequency of viable male and female gametes, respectively. The trait "Compatibile" determines the frequency of fertilization with pollen similar to the stigma in one or more gene loci using either a gametophytic or sporophytic model. All three traits are initiated with trait value 1.0, which means full fertility, but the user can make them dependent on genetic loci in the species.

The software provides a text editing window system named a "Tutorial" to write simple linear guides. Tutorials can be written by anyone to generate a sequence of windows, which can guide students through a set of operations on associated virtual plant material. Students can subsequently play the tutorial sequence and perform the operations at their own speed.

## 3. Results

### 3.1 Adaptation of virtual material

When initiating a new species the software will generate a single plant with maternal and paternal genome sets and random byte values in each gene allele. If multiplied for some generations with a high mutation and cross breeding frequency, this plant will give rise to a diversified base population. The base population, if big enough, will contain all possible alleles of all gene loci in equal frequency as long as no selection is performed.

If subsequently traits are defined to depend on different gene loci of the species then recurrent selection for different traits on the base population will lead to adapted subpopulations with changed allele frequencies. Such recurrent selection for many generations can be performed easily using a special "AutoSelect" operation window. Simultaneous index selection for several traits can be performed with a "Combined" trait. During such recurrent selection, mutation rate of the species may be elevated to introduce new alleles in the existing material. This will generate highly diversified sub populations from the original base population.

During the adaptation process of populations, trait definitions can be rescaled at any time to find suitable settings.

The original base population may be considered the undomesticated or wild material of the species. If a trait like yield, controlled by hundred or more gene loci, is defined with a lower bound of zero then single plants or lines from this population will rarely show extreme yield values (Fig. 3, generation 0). This is because of normally distributed simulation effects of allelic values, which will mostly have effects close to zero and rarely more extreme contributions. Thus without selection, it is unlikely that plants or lines with many high effect alleles for the yield trait will show up. With recurrent selection, possibly combined with an elevated mutation rate, however, domesticated material with high frequency of rare alleles contributing to yield can be generated (Fig. 3). Such domestication if performed without

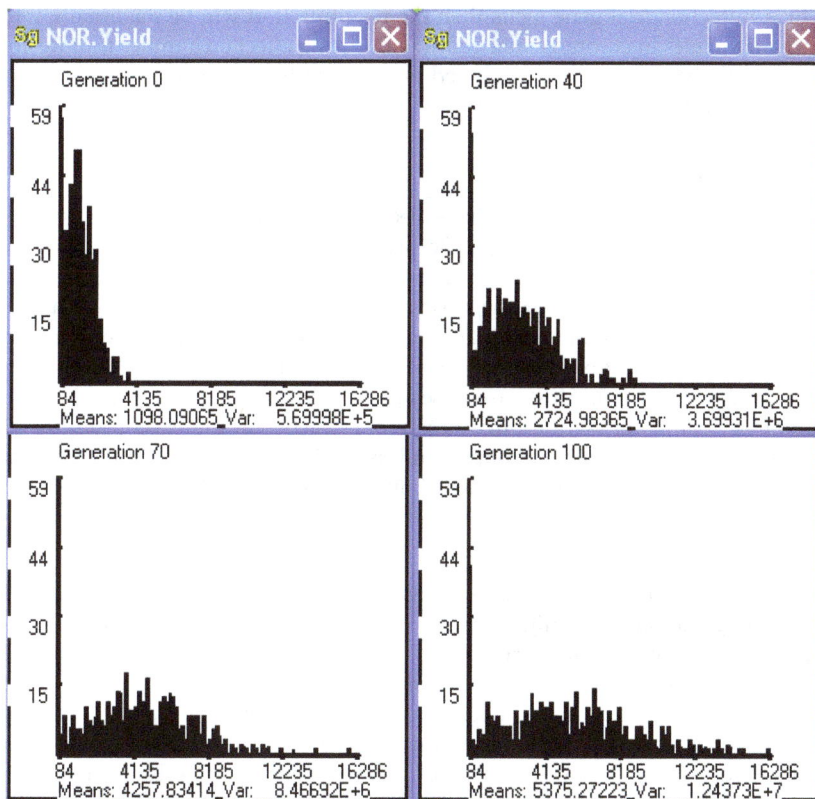

Fig. 3 Histograms illustrating adaptation of a base population through recurrent selection for a yield trait (kg seed/ha) affected by 200 different gene loci. For each generation the 20 % highest yielding plants were selected and intercrossed to obtain the next generation.

unrealistically high mutation rates, will also lead to reduced genetic diversity of the material. Of course, the domestication process may be performed for several different traits like yield, quality or disease resistance depending on different genes, either in tandem or in parallel using an index. This can generate many different domesticated or wild subpopulations. A characteristic feature of such adapted populations is that they lose their selected traits and degenerate back to the original base population if heavily mutated. This is because new alleles with extreme positive or negative effects are rare in the simulation.

Before, during or after the domestication process, genetic structure related to outbreeding tendency may be established through adjustment of the cross breeding frequency. The special trait "Compatibile" may be set to depend on one or more gene loci. Subsequent multiplication with a low degree of mutation will lead to strong selection for multiple alleles in the affecting genes and maintain low inbreeding. Also the two special traits "Malefertile" and "Femalefertile" may be set to depend on gene loci or other traits to affect male and female sterility, which will also affect genetic structure of the resulting populations. Rare mutations with strong effect on single gene traits can be selected from one or two generations of multiplication under elevated mutation frequency. If a trait named *cms* is affected by a mitochondrial gene locus then the resulting trait will be maternally inherited. Plants with rare alleles in chromosomal genes with strong effect on another trait Restorer (R) may subsequently be selected and combined with *cms* in the "Malefertile" trait, to establish a simple *cms* restoration system in the species.

## 3.2 Guided exercises

A number of tightly guided exercises have been developed based on the software. They have been used mainly to illustrate specific principles of breeding methods for students. Each exercise consists of a virtual plant material and an associated tutorial, which will guide students through the operations at their own speed. The virtual plant material for these exercises is of named type and species adapted to mimic as much as possible reality. The traits are selected to interest our students. Some of the exercises make reference to chapters in Sleper & Poehlman (2006). All the exercises have been used for two hours confrontation in groups of two students with approximately one teacher in the class room for each 10 students. After one week, student groups hand in a 4-5 pages report based on screen clips from the program embedded in text, explaining theoretical background, operations and results obtained. Reports are subsequently returned to the students with comments from teachers.

One exercise deals with the difference between qualitative and quantitative traits and introduces how traditionally qualitative traits, become quantitative, if the number of genes or the environmental effects are increased. It introduces heritability and simple selection theory. A second exercise gives an introduction to incompatibility and *cms* restoration systems in breeding. In addition to a basic introduction of these major fertility regulation systems this exercise also introduces the basic setup of hybrid cultivars. A third exercise introduces backcrossing with a dominant trait. A fourth exercise introduces linkage analysis between markers and a qualitative trait and performs marker assisted backcrossing of a recessive disease resistance into a high yielding cultivar. Students experience the characteristic complete absence of visible resistance in the material during the entire backcross operation, followed by reappearence of the resistance trait after final selfing of the backcrossed material. A fifth exercise introduces polycross for estimation of combining ability and its use to construct synthetic cultivars in highly outcrossing species. In particular this exercise discusses the number of basic clones needed to avoid the risc of inbreeding depression of the synthetic cultivar during later generations of multiplication. A sixth exercise introduces the structure of hybrid cultivars based on *cms* restoration systems their maintenance and improvement of their inbred parental lines. An existing hybrid is improved through backcrossing of dominant and recessive genes for disease resistance into its parental lines and the risc of reduced hybrid performance because of linkage drag is discussed.

It would be quite simple also with the software to generate guided exercises for population genetics to illustrate the effect of selection and cross breeding frequency on various population genetic features like deviation from Hardy-Weinberg equilibrium. Also phenomena like genetic drift of allelic frequencies during generations and the risk of losing rare alleles may be illustrated.

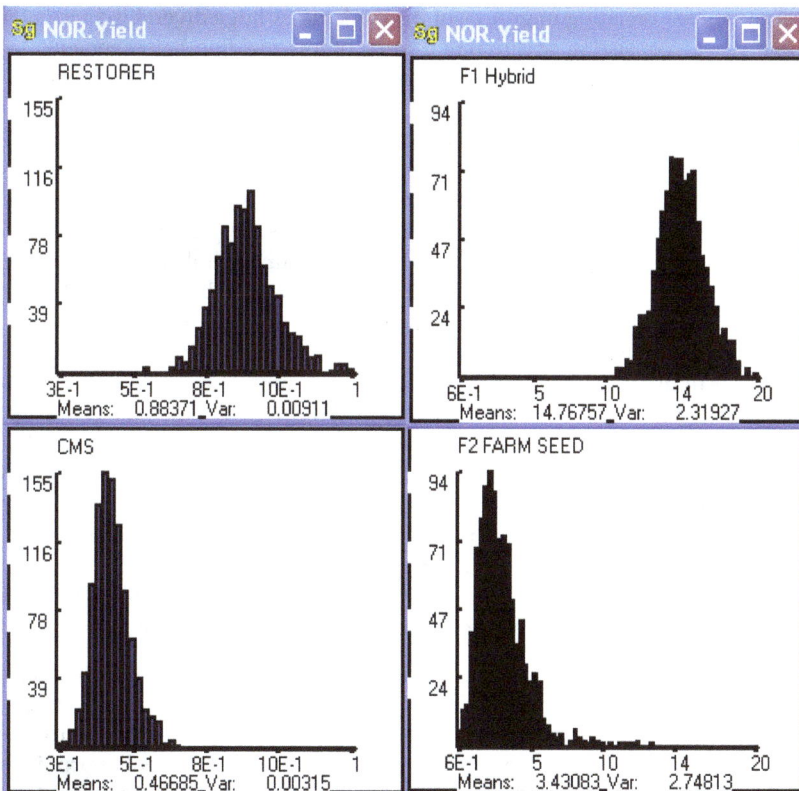

Fig. 4. Histograms showing yield levels and variation within two inbreds (Restorer and CMS), their high yielding hybrid (F1 Hybrid) and farm saved seed from the hybrid (F2 FARM SEED). Yield simulated with 200 different gene loci and approximately 500 plants per population.

A virtual *Arabidopsis thaliana* with named SSR markers from the databases has been established. It has been used in an exercise where single gene mutants for different traits are located to chromosomal areas using SSR markers. Students subsequently find markers flanking the mutation and use new markers in the interval to close map the mutation to within one centimorgan for map based cloning. In another exercise with this virtual *Arabidopsis*, students make crosses between ecotypes differing in quantitative traits. They derive mapping populations (Doubled haploids (DH), $F_2$ or Recombined inbred lines (RIL)) and analyse the mapping populations for all segregating markers with a marker download window. The operation will download scores of all markers for all plants of the mapping population together with trait measurement for each plant and a map of the markers. The downloaded file can be used directly with a supplied QTL mapping program SuperQTL to analyse for QTL of the trait.

## 3.3 Explorative learning

The simulation software has been used during several courses in molecular breeding to teach marker assisted breeding in groups of 2-3 students. A large diversified virtual material of hexaploid wheat, consisting of different groups of low yielding heterogeneous landraces and some high yielding uniform modern cultivars, was established. Students have been given this virtual genetic resource material for a two weeks working period under guidance of a teacher, but without a written tutorial. Students use various phenotypic screening and association and QTL mapping approaches to identify major and minor disease resistance genes in the landrace material followed by backcrossing into the high yielding cultivars. Some special features studied are the problem of linkage drag and associated yield loss caused by backcrossing from less adapted material. Amount of linkage drag can be studied easily using the marker system. It is possible to follow the breakdown of the linkage disequilibrium during subsequent generations to reduce the amount of introduced genetic material and markers can be used to speed up this process. Close mapping of QTL to generate diagnostic markers have been performed for subsequent selection with reduced linkage drag. The diagnostic markers have been used for pyramiding of genes for partial resistance to produce efficient disease resistance based on 2-4 different genes. Joint QTL analysis of multiple crosses have been studied to search for QTL in genetically wide materials and to study the effect of reduced size of the mapping populations. It is possible during such activities to exchange plant material between participants on different computers to generate cooperation in larger groups. The two weeks teaching periods have resulted in 25-30 pages reports composed of screen clips from the simulations, Figs and tables summarizing results, embedded in text to explain theory, ideas and results.

Table 2 shows summary of results from students' joint QTL analysis of powdery mildew infection on approximately 1000 wheat DH lines from 20 different crosses. Ten different cultivars were crossed each with cultivars Grete and Jesper and about 50 DH from each F1 hybrid were used for the joint QTL analysis with approximately 3000 polymorphic markers. The table shows the positions on the chromosomes of each QTL and the LOD values obtained. In addition, additive value from each cultivar in each QTL is shown. Negative additive values indicate alleles for reduced powdery mildew infection. The two QTL with lowest LOD of 7.1 and 6.8, respectively, were deemed false positives based on permutation analysis. This status as false positives of these two QTL was also confirmed from inspection

|  | Chrom 1 Pos 104 | Chrom 9 Pos 1436 | Chrom 13 Pos 2186 | Chrom 15 Pos2574 | Chrom 18 Pos 3000 | Chrom 20 Pos3422 |
|---|---|---|---|---|---|---|
| Ellen | -0.1 | 11.8 | -12.5 | -13.5 | 6.9 | 1.1 |
| Else | 17.4 | -3.1 | 3.5 | 9.3 | 6.3 | 0.3 |
| Grete | -5.0 | 1.4 | 1.7 | -3.3 | 7.0 | 20.0 |
| Gunhild | 10.4 | 0.2 | -18.0 | -20.2 | 12.6 | -20.4 |
| Jesper | 7.6 | -4.8 | -0.7 | 10.9 | 4.5 | 5.6 |
| Knud | -23.4 | 3.8 | 3.8 | -9.6 | -2.7 | -3.5 |
| Mads | -11.4 | 7.7 | -0.6 | -3.6 | 1.5 | -11.4 |
| Merete | 13.7 | -9.9 | -2.0 | -3.3 | 0.3 | -0.3 |
| Sigfred | -6.2 | -10.9 | 23.5 | 9.3 | -6.4 | 2.8 |
| Soren | 9.9 | 2.1 | 20.3 | 1.3 | -18.5 | 4.4 |
| Verner | -3.3 | -5.0 | -0.2 | 10.2 | -1.7 | -2.1 |
| William | -9.7 | 6.8 | -18.7 | 12.4 | -9.8 | 3.6 |
| LOD | 24.0 | 7.1 | 6.8 | 12.0 | 8.4 | 57.0 |
|  |  | False | False |  |  |  |

Table 2. LOD values and estimated additive effects of six QTL for powdery mildew infection in 12 different cultivars.

of the trait design, which did not rely on gene loci in these chromosomal areas. Simulation of the powdery mildew trait used 20 different gene loci with 20% environmental effect, however, only four of these loci showed significant functional polymorphism in the cultivar material analyzed. The cultivar Gunhild has two desirable alleles on chromosome 15 and chromosome 20, respectively. The cultivar Knud has another desirable allele on chromosome 1. The two cultivars were subsequently crossed and markers surrounding the three QTL were used to select offspring homozygous for all three alleles. One high yielding line was found among the offspring with pyramided resistance.

## 4. Discussion

A large number of simulation tools related to plant breeding can be found on the internet. Many such programs target the use and analysis of molecular data from markers or data from field experiments and simulate plant performance based on a set of underlying assumptions. Other approaches use physiological or biochemical models to predict plant behaviour and connect phenotypic performance with genes (Chapman et al. 2003, Hammer et al. 2005). Softwares like QuLine (Wang & Pfeiffer, 2007) can combine many different types of genetic data now available on the internet to simulate performance of breeding material. Although not primarily made for teaching, such programs can be extremely useful for learning though mostly for people already dedicated to the area.

The simulation for clinical training in health care and the training of focussed skills in legal management are maybe somewhat different from our needs in plant breeding education. For the plant breeding area, specific training of skills may be less important. Such skills will have to be trained anyway in actual breeding programs, should the student end up with such engagement. More important in the case of plant breeding is the communication of basic theoretical ideas and phenomena to an audience with only a part of its interest and time devoted to the area. In this situation, the most important result of such virtual plant breeding simulations is the images and associations generated in the heads of students. Stimulation of this process strongly depends on the choice of species, traits etc., which will associate positively with students intellectual background. A student with a focussed educational background in horticulture may not show much enthusiasm for a hybrid breeding program in maize, while a very similar setup named Primula or head cabbage could make a difference. In the same way, students with a highly molecular background cannot be expected to raise much interest from simulations illustrating pure traditional breeding programs. However, they may well take interest in quantitative genetics and QTL as a means to track down important genes, if wrapped in circumstances appealing to their background. Many students within plant science have interests and imaginings in genetic resources and population genetics, which can be well simulated with the present setup and used to introduce them to applied genetics. In general, however, simulations cannot substitute completely for real work with plants etc. Without hands on experience, students do not possess a background, with which the operations on the screen can generate positive associations and imaginative pictures. However, given the basic biological and ecological background, students can in a matter of hours experience complex operations and solve problems, which in real life may take several human generations and demand resources beyond imagination.

The basic setup of the present system for simulation of genetics has the advantage of being very general. The only assumptions made are the basic rules of genetics, and the byte representation of genes reduces the amount of needed computer memory. It enables more than 30 000 plants each with 3-5000 loci to be manipulated on most ordinary laptops. Still the amount of information in these genomes provides good simulation of both qualitative and quantitative traits. Like for real genetic resources the genetic diversity of populations can become exhausted if subjected to intensive inbreeding and selection. For out breeding populations, however, a quantitative trait based on 100 gene loci and a 200 plant population size will retain variation for several hundred generations of recurrent selection. Also the phenomena of heterosis and inbreeding depression are simulated very well for a trait, if its loci have been given a significant amount of dominance. As soon as ordinary populations with many heterozygous gene loci are inbred the trait shows inbreeding depression, which can be relieved through hybridization with different material. A few generations of reciprocal recurrent selection for combining ability between genetically distinct populations will efficiently build up populations showing heterosis upon hybridization. Lines derived from such reciprocally heterotic groups, generally form hybrids with good performance in the traits selected for. Such hybrids are generally quite difficult to out yield if students subsequently derive new inbreds from the original populations. Furthermore, performance of such hybrids is highly sensitive to genetic pollution from linkage drag. In general 6-7 backcrosses are needed for full recovery of hybrid performance, if one of the parentals is improved for a simple trait. In many cases, full recovery of the hybrid performance is only possible for some backcross lines, so students working in parallel, with the same starting material, can experience quite different outcomes.

Also genetic phenomena like cytoplasmic male sterility and restoration are simulated quite well. Like for real hybrid breeding the female *cms* line can only be multiplied via pollination from a similar maintainer without restoration genes. The *cms* trait can be transferred to new inbred maintainer types via simple backcrossing. If a rare allele is sought for restoration, then restorers in the breeding material will be rare. Otherwise the ordinary problems with restoration in the breeding material will arise, so only some inbreds can become maintainers. Transfer of the restoration trait to new inbreds, so they can be used as male lines for hybrids, must be done by backcrossing. Selection for the restoration ability during the backcrossing operation may be based on test crossing with a male sterile. Alternatively the genes affecting the trait may be mapped in a suitable mapping population and the linked or functional markers can be used for selection during backcrossing.

The program will also easily simulate transgenic engineered male sterility like the *barnase barstar* system (Denis et al. 1993) used in many *Brassica napus* breeding programs. One transgenic construct named barnase is used to generate the male sterility and another construct named barstar is used as restorer. A basta resistance trait, in addition, is linked to the barstar construct. This basta resistance trait can then be used to eliminate male fertile plants lacking the restorer after each multiplication. During the transformation process, different transgenic plants will receive different number of inserts and only some inserts show high expression of the construct. Like for real life breeding with these approaches, it is normally easier to backcross elite inserts into new breeding lines than to introduce them by *de novo* transformation.

Many of the features of self-pollinators appear naturally, if the material is suitably prepared. Populations of species with high self-pollination frequency, quickly lose genetic diversity, if selected without forced hybridization. Therefore, generation of basic subpopulations with different adaptation is most efficient with a high frequency of cross pollination. Subsequently, diversified populations, clearly different from the base population, are obtained through recurrent selection. Then cross pollination frequency is lowered for another ten generations to obtain heterogeneous populations of pure lines. From such heterogeneous land race like material then, pure lines can be extracted via single plant offspring. Subpopulations can be improved intensively for quantitative traits like yield and quality affected by many genes and give rice to cultivar like superior lines. Such cultivars will rarely be outperformed by material from the unselected base population or from populations of lines selected for other traits. They may, however, show heterosis upon hybridization, and new superior lines can be identified in the inbred offspring from such hybrids. Simple hybridization of high performing lines with non-adapted material to introduce new traits like resistance, in almost all cases destroys adaptation. Adaptation of the original line can be restored through 6-7 backcrosses with selection for the new trait. However, because of linkage drag, complete performance of the line is not always restored, and students working with the same initial material can experience different results.

Simulation of markers and genetic mapping is efficient with the present simple approach. In segregating offspring, students can count recombinants between traits and markers or between different markers. They can then calculate recombination frequencies, generate local maps surrounding genes for specific traits and decide, which markers to be used for selection or close mapping of the trait. Students during such operations will quite realistically face the questions related to lack of information from monomorphic markers, reduced information from dominant markers and the question of, which parental allele to use for selection. During marker-assisted backcrossing of e.g. a recessive trait, they will further face the chance of losing the linked trait because of recombination between marker and the target, if the marker is not functional. During fine mapping of a qualitative trait, students will face the challenge of, how to decide on flanking markers and subsequently how to repeatedly reduce the target area based on new markers situated between the flanking ones. In dissecting quantitative traits with QTL analysis, students can realistically experience the problems with false positives, estimation of LOD thresh holds, relative effects of different QTL and problems with linkage drag, if QTL are backcrossed. Because of the simulated trait structure, positions of mapped genes or QTL can always finally be checked with the trait structure, to see whether a gene is actually simulated in the genomic area and where exactly the gene in question is situated.

A more evolution based type of teaching of applied genetics may be possible in the future with such simulation systems. For students there is much to be learned from the process of adaptation of the virtual material to simulate well, different types of breeding material or plant genetic resources. Given the necessary time, students may generate their own virtual species with trait definitions build on information from literature and databases. Underlying genes for the traits may be approximately positioned on the genome and their major effects and interactions can be modelled based on available knowledge. Different models for a genetic phenomenon may be implemented in different student groups and their performance may be compared to identify advantages and limitations of each genetic model. Relative success of major methods of breeding like hybrid programs compared with open pollinated cultivars for different types of virtual plant material could be an example. The understanding of market performance of hybrids, compared with open pollinated cultivars is complex and includes e.g. both intellectual property rights, as well as several economical issues in addition to the genetics of the plant material. The genetic dimension of this problem in itself is often too complex for many students to invest enough efforts to understand it, based on lectures and associated literature. Simulation could be an alternative way to introduce students to the genetic dimension of this discussion and then wrap the entire teaching operation into an economic environment with simple book keeping and associated seed market situations.

Also reporting from such simulated teaching may be much developed. Future students, instead of traditional written reports with screen clips, may hand in a virtual plant material and an attached Tutorial guide, which would guide the teacher or colleague students through the process and ideas of the author. To some extent the motivation of simulated learning may not so much be the actual presentation of complex phenomena to the receiver. The activation of students to create a product, like a new high yielding disease resistant cultivar, a Tutorial guide to be seen by someone else, or simply a conventional written report with colourful screen clips, may be much more motivating. Another motivation is the random number based simulation, which assures that any student product will be unique, because two simulations never produce the same result. This effectively eliminates any sentiment of communicating with a robot.

The present implementation of Supergene does not enable simulation of polyploids. Genetics of auto-polyploidy is under implementation in a planned future Java based implementation. However, general simulation of allo-polyploidy will demand a more general simulation of evolution of genomes. Also simulation of environmental effects and gene by environmental interactions are severely restricted with this simple approach. Separate environments are not simulated and therefore general gene by environment interactions cannot be modelled. Also plant plant interactions are not simulated in the present implementation. Such plant plant interactions would be interesting for ecology. Some more complex gene by environment interactions as well as complex epistasis may be simulated using the combined trait feature of the program.

## 5. Conclusion

In spite of these and other limitations of the present implementation, it has been used with good results, both for illustration of various breeding phenomena and for more explorative teaching. The simulation has mainly been used for breeding operations too time or resource consuming to be performed in real life. This also means that teachers' personal contacts during the exercises have been prioritized. Only a minor part of the teacher contact has been used for questions regarding the program interface. The majority of questions and discussion with teachers generally have been related to scientific understanding of the breeding operations. In this way the software is efficient in raising questions from the students regarding why and how. Questions which would otherwise not surface based on reading a book chapter or listening to a lecture.

## 6. References

Carswell, A.T. & James, R.N. (2007). Virtual Property Manager: Providing a simulated learning environment in a new university program of study. *Systemics, cybernetics and Informatics*, Vol. 5, No.4, (June 2007), pp.34-40, ISSN 1690-4524

Denis, M.; Delourme, R.; Gourret, J-P.; Mariani, C. & Renard, R. (1993). Expression of Engineered nuclear male sterility in Brasscia napus. *Plant Physiology*, Vol. 101, No.4, (April 1993), pp.1295-1304, ISSN 0032-0889

ISLE (2011). http://www.biographix.com/services/pdf/ISLE.pdf

Nexlearn (2011). http://www.nexlearn.com/?q=node/6

Chapman, S.; Cooper, M.; Podlich, D. & Hammer, G. (2003). Evaluating plant breeding strategies by simulating gene action and dryland environment effects. Agronomy Journal, Vol.95, No. 1, (Jan.-Feb. 2003), pp. 99-113, ISSN 0002-1962

Hammer, G.L.; Chapman, S.; van Oosterrom, E. & Podlich, D.W. (2005). Trait physiology and crop modelling as a framework to link phenotypic complexity to underlying genetic systems: Modelling complex traits for plant improvement. *Australian Journal of Agricultural research*, Vol. 56, No.9, (2005), pp. 947-960, ISSN 0004-9409

Simple (2011). http://www.ukcle.ac.uk/projects/past-projects/simpleupdate/

Sleper, D.A. & Poehlman, J.M. (2006). *Breeding Field Crops* (Fith edition), Blackwell Publishing, ISBN-13: 978-0-8138-2428-4, 2121 State Avenue, ames, Iowa, 50014

Stefanutti L. & Albert D. (2003). Skill assessment in problem solving and simulated learning environments. *Journal of Universal Computer Science*, Vol.9, No.12, (Dec. 2003), pp.1544-1468, ISSN 0948-695X

Tsai, S-L.; Chai, S-K.; Hsieh, L-F.; Lin, S.; Taur, F-M.; Sung, W-H. & Doong J-L. (2008). The use of virtual reality computer simulation in learning Port-A Cath Injection. *Advances in Health Sciences Education*, Vol.13, No.1, (March 2008), pp. 71-87, ISSN 1382-4996

Wang, J-K. & Pfeiffer, W.H. (2007). Simulation modelling in plant breeding: principles and
applications. *Agricultural Sciences in China*, Vol. 6, No.8, (July 2007), pp.101-105,
ISSN 1671-2927

# 2

# Fundamental Cryobiology and Basic Physical, Thermodynamical and Chemical Aspects of Plant Tissue Cryopreservation

Patu Khate Zeliang and Arunava Pattanayak
*Division of Plant Breeding, ICAR Research Complex for NEH Region,*
*India*

## 1. Introduction

The greatest stability of *in vitro* plant materials with practical storage periods measured in decades can be achieved by cryogenic storage at ultra low temperatures. Liquid Nitrogen (LN) is the most common medium for cryostorage as it is relatively inexpensive and readily available (Withers, 1987; Panis and Lambardi, 2005). This process puts the cells in suspended animation where they can retain their viability indefinitely. Maintenance under these conditions effectively halts biological growth and development (Franks, 1985; Grout and Morris, 1987; Grout, 1990a; 1990b) because at below -140°C, the rates of chemical and biophysical reaction will be too slow to affect cell survival. Consequently, material that can be brought to the ultra-low temperature and recovered from it without acquiring lethal injury may be stored for extremely long periods. The challenge is to devise a protocol that allows *in vitro* plant material to be recovered from the cryogen at high viability, and without structural and functional changes (Kartha, 1997; Withers, 1987; Grout and Morris, 1987; Grout, 1990 a, 1990 b). Preservation of viability depends upon the ability to minimize the stresses of cryopreservation and protect against the damaging consequences. A variety of plant material can be used for cryopreservation including *in vitro* cultivated material, pollen, seeds, embryos, buds and meristematic tissues.

Although cryopreservation has many advantages, freezing and thawing injuries related to membrane structure and function that would result in low survival percentages are still the limiting factors (Ashmore, 1997). Prior to freezing, the cells must be treated with a cryoprotectant solution such as glycerol, dimethyl sulfoxide (DMSO), or ethylene glycol. These substances protect the cells and their membranes from damage during the freezing process. After the cells have been exposed to the freezing medium containing the cryoprotectant, they must be dehydrated so that the water inside the cells will not form ice crystals damaging the cell. To dehydrate the cells, they are cooled very slowly prior to plunging them into LN at -196° C that will maintain this constant temperature as long as there is nitrogen in the storage tank (Touchell and Dixon, 1999). When cells are to be thawed, they are warmed rapidly and the cryoprotectant solution is removed. The cells are then cultured in an incubator where they will resume their growth and development. Thawed cells have retained their viability following freezing in LN for more than half a century. The cryopreservation of plant genetic resources has two purposes: (1) preservation of genetic diversity and (2) preservation of selected varieties for the economic value of their characteristics. Maintenance of continuous culture is labour intensive. In addition, frequent subcultures can generate variants. This disadvantage can be limited by slow growth of *in vitro* cultures, whereby the frequency of subculture can be reduced. However, the principle inconvenience is the possible occurrence of somaclonal variants, incompatible with the aim of genetic resource conservation. Large differences in cell growth lag phase relate to the cryopreservation protocol, the cell source used, as well as variations in cell size and degree of vacuolation in the calli or cell suspensions used for preservation. Thus, there is a concern that cryopreservation and regrowth procedures might contribute to selection of cells with specific characteristics. In addition, tissue culture continues to play a vital role in the development of cryopreservation techniques. Thus, there is an increasing requirement to determine whether plants derived from cryopreservation are 'true to type' or not, to measure the extent of the 'normal phenotype' in cryopreserved plants, and to estimate the degree of closeness of cryopreserved plants to the 'true' parental genotype. These determinations may be achieved through the application of a range of analytical techniques to examine cryopreservation-derived changes at the phenotypic, histological, cytological, biochemical and molecular levels (Harding *et al.*, 2009).

## 2. Available plant cryopreservation protocols

Methods for the cryopreservation of a variety of *in vitro* cultured tissues such as embryogenic cells (Huang *et al.*, 1995), suspension cells (Ishikawa *et al.*, 1996), transgenic suspension cells (Cho *et al.*, 2007) have been developed. However, a general protocol for cryopreservation of plant cell cultures has not yet been developed. Also, an exact and most suitable guideline for individual specimens is not available as noted by Withers (1983). Thus, the steps involved in the process such as osmotic pre-treatment, cryoprotection, freezing, thawing and subsequent regeneration require standardization for individual species or even individual cell line (Cho *et al.*, 2007). The three most common methods are given below.

### 2.1 Classical slow-cooling/freezing protocol

The function of controlled, slow-cooling in cryopreservation is to allow cryodehydration to progress without intracellular freezing, removing water from cells to a point where their contained solutions will not form ice crystals when taken to the final cryogen temperature. Cell injuries during freezing are effectively decreased by the addition of various cryoprotectants that reduce the cell size and lower the freezing point to prevent the formation of ice crystals in the cells (Jain *et al.*, 1996). Slow-cooling rates minimize thermal stress from non-uniform temperature distribution, and cryoprotectants contribute to reduce stress by changing the microstructure of the ice formed (Gao *et al.*, 1995).

### 2.2 Encapsulation/dehydration

In this method, explants (usually meristems or embryos) are first encapsulated in alginate beads (which can contain also mineral salts and organics), thus forming "synthetic seeds" ("artificial seeds" or "synseeds"). They are then, treated with a high sucrose concentration, dried down to a moisture content of 20-30% (under airflow or using silica gel) and subsequently rapidly frozen in LN. Due to the extreme desiccation of explants, most or all freezable water is removed from cells and vitrification of internal solutes takes place during rapid exposure to LN, thus avoiding lethal intracellular ice crystallization (Engelman, 1997).

## 2.3 Vitrification

It consists of first preculturing the explants in a dilute solution of a permeating cryoprotectant (so called as loading phase), followed by a vitrification solution (so called as dehydration phase). Since the osmotic strength of the vitrification solution is very high (>8 osmol) and the duration of application fairly short, the main function is to dehydrate the sample, concentrating the permeable components and other cytoplasmic contents within the cell (Towill, 2002). Vitrification-based procedures involve removal of most or all freezable water by physical or osmotic dehydration of explants, followed by ultra-rapid-freezing which results in vitrification of intra cellular solutes, *i.e.* formation of an amorphous glassy structure without occurrence of ice crystals which are detrimental to cellular structural integrity. These techniques are less complex and do not require a programmable freezer, and are more appropriate for complex organs like embryos and shoot apices. Hence, are suited for use in any laboratory with basic facilities for tissue culture. Engelmann (2000) described seven vitrification-based procedures in use for cryopreservation: encapsulation-dehydration, vitrification, encapsulation-vitrification, desiccation, pregrowth, pregrowth-desiccation, and droplet freezing, which have been reported to be successfully used for a number of different plant species.

Slow freezing causes extracellular ice to form, thus dehydrating the cell and preventing damage by intracellular ice formation. However, if the freezing is carried out too slowly and dehydration is excessive, the cell will suffer from "solution effects" damage (Withers, 1984; Mazur, 1984). This has been ascribed to pH alterations, intracellular solute concentration, dehydration, membrane alterations and protein denaturation and intracellular ice formation at supraoptimal rates (Towill, 2002). Solution based vitrification is very time-consuming, and processing large numbers of explants is often difficult. The encapsulation procedure allows more propagules to be handled at one time, and timing is not very critical; although it usually requires several hours to attain the desired level of desiccation. If air drying is used, the duration will differ depending on the relative humidity of the atmosphere. While two step cooling systems require a suitable apparatus, vitrification procedures eliminate the necessity for expensive and sophisticated slow-cooling equipments by allowing tissues to be cryopreserved by direct immersion in LN. Explants like shoot apices have high moisture content and require specialized treatments to prevent lethal ice crystal formation during freezing. Vitrification reduces the requirement for extensive manipulation of the apices that may result otherwise in physical damage to the tissues (Towill and Jarret, 1992). However, as vitrification solutions cause extreme desiccation of shoot apices, care must be taken during washing procedures to prevent rapid deplasmolysis. To reduce cellular damage, washing solutions should be isotonic with the thawed tissues, which is usually achieved by the manipulation of the sucrose concentration in the washing medium (Touchell and Dixon, 1999). The key to high survival rates in the slow-cooling method is to carefully control the cooling procedure, whereas in the vitrification method, the cryoprotectant exposure must be carefully controlled. It is reasonable to expect that no single method will give high levels of survival for every genotype of a species after cryogenic exposure. Since a method cannot be devised for each genotype, once a useful procedure is identified the next step would be to apply it to different genotypes to determine overall utility.

## 2.4 Other protocols

The other available methods are the pre-culture/dehydration used by Dumet et al. (1993), where clumps of somatic embryos of oil palm were dissected from standard cultures and pre-grown on 0.75M sucrose for 7 days or further dehydrated under air flow and silica gel for few hours before immersing in LN. In another preculturing method given by Panis et al. (1996), the proliferating meristems of banana were pre-cultured for 2-4 weeks on MS medium with 0.3-0.5M sucrose, then excised and plunged into LN. In the droplet freezing method by Schäfer et al. (1997), apical shoot tips of potato were first incubated in DMSO, then transferred into 2.5ml droplets placed on small leaflets of aluminium foil and immersed in LN. All these techniques have been used for only a limited number of plant species.

## 3. Cryoprotective-agent

In theory, ultra low temperature, such as -196°C in LN, maintains tissue quality. In practice, the water-rich plant tissues require additional compounds to prevent ice crystal damage during cryopreservation. Different chemicals and treatments are applied to protect and to recover the plant materials during and after storage in LN. Such a chemical should sustain the viability of biological material during freeze thaw cycle and must not be toxic to tissues at the concentration required adequately to lower the freezing point.

There are two types of cryoprotectants: (i) low molecular weight compounds which penetrate the cell with ease, e.g. glycerol and DMSO, and (ii) compounds which penetrate the cell slowly e.g. sucrose, PVP, dextran etc. These compounds protect the cell surface membrane by reducing growth rate and size of ice crystals and by lowering the effective concentration of solutes in equilibrium with ice inside and outside the cell. This increases membrane permeability which aids water removal from the cell and facilitates protective dehydration during early stages of freezing. Most commonly used cryoprotective agents include glucose, DMSO and polyethelene glycol. Glucose serves as a dehydrating agent. DMSO (toxic at higher temperature) passes through the cell membrane readily and protects the cells during the process of freezing. Polyethylene glycol (PEG) gives a water stress to the cells so that ice formation is reduced and may also diminish the harmful effects of DMSO on the freeze stressed cells. Desferrioxamine, an iron-chelating agent, has also been used as a cryoprotectant (Benson et al., 1995). When establishing a cryopreservation protocol, it is also important to determine if cryoprotectants impair cell growth and development. However, published information on the toxicity of cryoprotectants prior to the cooling process is limited. In some species, cryoprotectants cause a temporary loss of semi permeability of membranes (McLellan et al., 1990), while, Pushkar et al. (1976) and Moiseyev et al. (1982) found that in the presence of low molecular weight PEGs, the activity of enzyme system was decreased. The cryoprotectant mixture of Withers and King (1980), containing DMSO (0.5M), glycerol (0.5M) and sucrose (1.0M), has been most widely used particularly in studies with rice cells and has not been found to be toxic to the rice cells or inhibit its regeneration.

## 4. Regeneration of plants after cryopreservation

Regeneration is an important criterion for most of the cryopreserved materials. Moreover, the viability rate of in vitro cultured and cryopreserved cells must be high to avoid the growing of particular types of cells (Menges and Murray, 2004). Since elite genotype selection is on the basis of both *in vitro* and *ex vitro* evaluations, the process from somatic embryo initiation to mass seedling production can be long. A consequence of this extended time period is that embryogenicity may diminish or even be irrevocably lost in a few weeks or months (Breton *et al.*, 2006). A way around this problem is to cryopreserve tissues while they are at their peak productivity, shortly after embryogenic tissue induction and when enough tissue is available (Kong and Aderkas, 2011). Thus, a relatively simple and reproducible protocol for the regeneration of plants from the cryopreserved explant is essential for long-term preservation/conservation. Plant regeneration and its frequency are dependent on various factors like: (1) genotypes, (2) age and physiological state of the culture preserved, (3) state of differentiation i.e., isolated cells or well-differentiated tissue and organs, (4) water content of cells of the explant, (5) concentration and duration of treatment with the cryoprotectant, (6) method and rate of freezing and (7) method of thawing and culture including combinations of plant growth regulators used during regeneration (Withers, 1983; Bajaj and Sala, 1991; Tsukara and Hirosawa, 1992). The pregrowth phase of plant cells is considered as very important stage for attaining successful cryopreservation. During this period, various changes may occur at the cellular level including a decrease in cell and vacuole size, changes in the flexibility and thickness of cell walls and alteration of metabolic activities (Withers, 1978). Tissue survival is mainly affected by pretreatment i.e., the longer the pretreatment, the higher the survival percentage. Embryogenicity is also affected by the temperature of the pretreatment. Lower temperatures prevent embryos from maturing and, thus extend embryogenic tissue recovery (Kong and Aderkas, 2011). After recovery from LN, explants contain living, weakened and killed cells. Undifferentiated suspensions, which consist of large vacuolated cells, are also prone to severe cryoinjury compared with embryogenic cultures and apical organs, which contain small cytoplasmic-rich meristematic cells (Wang *et al.*, 2002). In addition, suspension cells are sensitive to environmental stresses, such as dehydration, high osmotic pressure, and low temperatures. Therefore, the assessment of the condition of a specimen both quantitatively and qualitatively after the various stages in the cryopreservation procedure is one of the most important aids to the development of a freeze preservation protocol. Viability of the explants/cells after cryopreservation can be assessed by the flourescein diacetate (FDA) or triphenyltetrazolium chloride (TTC) test. The FDA gets converted to flourescein as a result of esterase activity. Cells with an intact plasma membrane fluoresce green in ultraviolet light as the larger molecules of fluorescin are unable to pass through the membrane. The TTC reduction is based on the mitochondrial respiratory efficiency of cells that converts the tetrazolium salt to insoluble formazon, which is extracted and measured spectrophotometrically.

The first report on survival of plant tissues on exposure to ultra low–temperatures was made by Sakai, when he demonstrated that very hardy mulberry (*Morus sp.*) could withstand freezing in LN after dehydration by extra organ freezing (Sakai,1956). Huang *et al.* (1995) achieved success in plant regeneration from cryopreserved suspension cells of rice. Cornejo *et al.* (1995) discovered that cryopreservation did not affect the ability of rice cells to integrate and express foreign genes. Yang *et al.* (1999) reported developing an efficient protocol for regeneration of a model rice variety Taipei 309 from long-term storage. Anther-derived rice (*O. sativa* L. sp. japonica variety) plants were obtained after cryopreservation by an encapsulation/dehydration technique. Eighty percent of the plantlets developed into normal plants after being transferred to greenhouse conditions. Histological observations showed that the origin of the plants was not modified by the cryopreservation process (Marassi *et al.*, 2006).

## 5. Variation in cryopreserved derived plant

The genome "quality" reflects its organization and structure, and the genome "flexibility" reveals the complex functionality including capability to response to intracellular and exogenous signals. The measure of tolerance of the genome to exogenous factors depends on the genome "flexibility" generating genetic variation (Skyba and Cellarova, 2009). Zhang and Hu (1999) and Moukadiri et al. (1999a) suggested that phenotypic variations seen in some of the regenerated plants were mainly due to tissue culture induced variations rather than effect of cold storage that were revealed by flow cytometric analysis (Moukadiri et al., 1999a) and randomly amplified polymorphic DNA (RAPD) markers (Moukadiri et al., 1999b). However, differences in band intensities among some but not all bands might indicate structural rearrangements in DNA caused by different types of DNA damage (Danylchenko and Sorochinsky, 2005) that might not be readily detected using the given system. The RAPD technology has previously been used successfully to detect occurrence of genetic alterations (Finkle et al., 1985; Harding, 1997; Aronen et al., 1999; Ahuja et al., 2002; Urbanova´ et al., 2005; Castillo et al., 2010), but this approach possesses limits with reproducibility, and it is currently being replaced by techniques such as Amplified Fragment Length Polymorphisms (AFLPs) and/or Simple Sequence Repeats (SSRs, microsatellites). These techniques are now being used to consider more carefully the issue of genetic fidelity after cryo-procedures, especially in the breeding of long-living conifers (Salaj et al., 2010), where genetic changes might be substantially expressed only later on, in mature trees.

Phenotypic and DNA variation among putative plant clones is termed somaclonal variation. Somaclonal variation caused by the process of tissue culture is also called tissue culture-induced variation to more specifically define the inducing environment (Kaepler et al., 2000). Somaclonal variation is a likely reflection of response to cellular stress in other situations as well. Therefore, understanding the mechanism of tissue culture variation will be useful in defining cellular mechanisms acting in the process of evolution, and in elucidating the mechanism by which plants respond to stress. Epigenetic processes are likely to play an important role in these mechanisms. Primary regenerants (R0) are often more variable than their progeny. Examples of aberrant phenotypes in regenerated plants include abnormal leaf structures and variant floral morphology (Kaeppler et al., 2000) and change in kernel color of O. rufipogon seeds in R1 plants (Zeliang et al., 2010). Qualitative mutation is frequent among tissue culture regenerants and the summation of protein assays, DNA studies and specific mutant analyses suggests that single-base changes or very small insertions/deletions are the basis of these changes (Kaeppler et al., 1998).

However, there is no convincing evidence for genetic alterations due to cryoprotectant effects in cryopreservation experiments where concentrations of protectants are relatively low, exposure time are short and reduced temperatures are likely to have an ameliorating effect. It has also been suggested that freezing damage is related, in part, to free radical effects, and that both DMSO and glycerol provide an element of protection against these agents by acting as free radical scavengers (Benson, 1990). Although one of the benefits of cryostorage is the maintenance of germplasm in a genetically secure environment, very little work has been conducted on the genetic fidelity of shoot material or plants recovered after cryostorage. Most workers relied on observations of phenotype for confirmation of stability in morphogenic cultures. Phenotypic abnormalities became more common as the number of reports of regeneration increased, suggesting that many genetic changes are not incompatible with regeneration (Withers, 1984). The climate of opinion on genetic stability in tissue culture is also changing due to the recognition of 'somaclonal variation' as an important phenomenon associated with *in vitro* works (Scowcroft and Larkin, 1983; Scowcroft, 1984). The need to screen for and to report on induced genetic variation is now widely recognized. In the case of *Manihot spp* (Withers, 1984) and *Saccharum spp* (Ulrich *et al.*, 1979), cryopreservation led to an apparent loss of totipotency. However, the reported phenomenon is different from the time-related loss of totipotency observed in long-term callus cultures. There is increasing evidence that, provided that the cryopreservation technique applied ensures the greatest possible maintenance of the integrity of the stored specimen, there will be no modification at the phenotypic, biochemical, chromosomal or molecular level due to cryopreservation (Engelmann, 1997). In a study on *Cosmos atrosanguineus* cryopreservation by Wilkinson *et al.* (2003), the use of AFLP gave no indication of any variation among the tested regenerants and any growth abnormality observed directly after regeneration was not carried over to the later growth stages. Moreover, the nature and location of cell damage appeared to be dependent on the pre-growth cryoprotectant, freezing protocol and species under study.

## 6. Current status of research and development in cryopreservation

When plant cells are cryopreserved, their plasma membranes are believed to freeze first. As the plant cell is frozen, ice crystals form in the intercellular space and eventually expand with drop in temperature. This creates an osmotic pressure difference between the inside and outside of the cell; therefore, to compensate for this difference, the cell expels water. Since the freezing outside the cell is faster than the inside, ice formation occurs first outside the cell. This formation of ice in turn reduces the water content outside the cell and causes the water inside the cell to move out thereby eventually dehydrating the cells. Temperature, as a major triggering variable in low temperature exposure, affects all structures and processes in the living matter with no exception. Skyba and Cellarova (2009), using the *Hypericum peforatum* model, studied the effect of temperature on the physical and physiology aspect of the plant after cryopreservation. They concluded that, the way in which temperature was decreased affect cell viability and choice of the explants and its seasonal rhythm affect survival rate after cryopreservation. In the case of *in vitro* cultured cells, systematic studies about cryopreservation and its applicability are yet to be determined, because even cell lines of closely related species require different parameter and the same cell line may behave differently in different laboratories. Even if a cryopreservation method has been worked out, the problems of transporting the cultures or reproducing the same method in a different laboratory remain to be solved (Dobbennack *et al.*, 2009). Further, there is no demonstrated mechanism for enhancing the survival of thawed cells. It is supposed that high sugar in culture medium lowers water content and increases endogenous sugar concentrations in cells (Matsumoto *et al.*, 1998). Sugars are known to protect membranes from desiccation events that are inherent to any preservation and added between protocol and are also known to enhance glass formation during cooling.

Cereals, especially rice and barley are two well studied model crops where a variety of information is available. Cryopreservation of rice cell suspension was first reported by Sala *et al.* (1979). Various types of *in vitro* cultures of rice, such as cell suspensions, protoplast, zygotic embryos and cultured shoot apices survive freezing in LN and retain their morphogenic potential. Rice cells suffer severe metabolic impairment after freezing in LN and show reduced uptake of glucose after freezing (Cella *et al.*, 1982) and preferentially use fructose as carbon source. A detailed study on cryopreservation of suspensions cells of Taipei309 was carried out by Lynch and Benson (1991). They suggested that successful cryopreservation depends on cryogenic technique and pre and post freezing tissue culture. In the post-freezing recovery phase, carbon source in the culture medium has been reported to be an important factor. Kuriyama *et al.* (1989) showed that viability and proliferation of thawed rice cells are depressed in presence of $NH_4^+$ ions. The effect is thought to be due to the inability of freshly thawed cells to control ionic gradients across plasma membranes. However, rice cells utilize $NH_4^+$ ions effectively once they have started recovery from cryo-injury. Freezing protocol of Withers and King (1980) has been used in majority of the studies with rice cells. Rice is normally cultured with sucrose as the sole source of nitrogen. Sucrose is rapidly hydrolyzed by actively growing rice cells to fructose and glucose, which are utilized equally. Sucrose hydrolysis is dependent on invertase activity (Amino and Tazawa, 1988), which in turn is associated with the cell wall (Schmitz and Lorz, 1990). However, long-term maintenance of freezed cells on fructose was detrimental. The direct immersion of frozen cells in liquid medium is damaging for rice suspension cells (Lynch and Benson, 1991; Huang *et al.*, 1995; Lynch *et al.*, 1995). Physiological condition of growth, cell aggregate size, embryogenicity and water content of cells has been reported to influence the cryopreservation by Lynch *et al.* (1995) and Watanabe *et al.* (1995). Their study revealed that cells from poorly cryoprotectable genotypes showed increased freezing tolerance after protoplasting (removing of cell wall).

Jain *et al.* (1996) reported a two step freezing protocol for aromatic rice varieties. Suspension cells were cryopreserved by pre-conditioning cells in mannitol, pretreatment in a cryoprotectant solution containing sucrose, DMSO, glycerol, proline and modified R2 medium, cooling to -25°C in a freezer followed by storage in LN. Plants were regenerated from freezed cells as well as from protoplasts isolated from re-established suspension cells. Cryopreservation by vitrification of rice calli has been reported by Wang *et al.* (1996). Watanabe *et al.* (1998) cryopreserved non-embryogenic rice callus cells by vitrification and found that the cell grew vigorously after cryopreservation in the same manner as untreated control and program freezed cells. Medium containing organic nitrogen (amino acids) source was found most suitable as pre-growth medium for suspension cells while inorganic nitrogen was required for successful post thawing recovery. Hu and Gou (1996) and Towill and Walters (2000) successfully cryopreserved pollen grain. Adventitious buds were also cryopreserved efficiently by Zhang and Hu (1999). Adding haemoglobin solution (ErythrogenTM) to post–thaw medium of indica rice (*O. sativa* L.) cells has been reported to enhance survival following cryopreservation (Forkan *et al.*, 2001). During rapid-freezing of rice embryogenic suspension cells, the addition of AFP-I (polar fish antifreeze protein) displayed protective action in the higher concentrated (but non vitrifying) cryoprotectants and detrimental effect in more dilute ones (Wang *et al.*, 2001). Pregrowth-desiccation of the suspension cells using sucrose and sorbitol in AA medium also gave 96-100% survival (Zhang *et al.*, 2001). In another study by Jelodar *et al.* (2001) the protoplast yields increased for re-established cell suspension cultures after cryopreservation compared to unfrozen control cultures. Direct immersion in LN of calli pre-treated with abscisic acid was found to be a fast and highly efficient freezing procedure that maintained the main characteristics of the cell populations and appeared to increase their metabolic activity (Moukadiri, 2002). Marassi *et al.* (2006) developed encapsulation-dehydration technique and successfully cryopreserved anthers with this technique.

In barley, preculturing reduced the volume of vacuoles and eliminated some osmotically sensitive mitochondria. Reduction in total water content, increase in bound water ratio and accumulation of new proteins also occurred during this step. The adaptation other than in structure and physiology improved the tolerance of cells to desiccation and freezing (Wang and Huang, 2002). Antifreeze proteins first identified in polar fishes also accumulate in freezing–tolerant overwintering cereals (Antikainen and Griffith, 1997). These proteins can lower the freezing temperature noncolligatively and inhibits the growth of ice crystals. The function of these proteins during cryopreservation of rice cells was tested by Wang and Huang (1998) and Wang *et al.* (1999). They found that at the proper concentration, antifreeze proteins may enhance viability through inhibition of ice crystallization, whereas at high concentration, they may decrease the survival rates by ice nucleation.

In a study on *Panax* (Ginseng), Mannonen *et al.* (1990) preserved *P. ginseng* cultures either in LN or under mineral oil for 6 months and compared their growth behaviour and ability to produce ginsenosides after a recovery period with the cultures maintained by frequent sub cultivation during the same period. They demonstrated that neither growth kinetics nor the degree of vacuolation that occurred during growth was affected by either storage protocol. However, some changes in secondary metabolism were found with preservation under mineral oil but not with the cryogenic method (Yoshimatsu and Shimomura, 2002).

In the case of conifers, the conventional cryopreservation method of embryogenic tissue required a few key steps: (1) pretreatment with osmotic regulators, such as sugars (usually, sucrose) or sugar alcohols (such as sorbitol and mannitol), (2) cryoprotectant treatment with DMSO, and (3) a carefully controlled slow-cooling process before the immersion in LN (Klimaszewska *et al.*, 1992; Cyr, 1999; Gale *et al.*, 2007). Currently, efforts are being made to simplify this process by exploring new methods of cryopreservation in both angiosperms and gymnosperms (Touchell *et al.*, 2002; Gale *et al.*, 2008; Popova *et al.*, 2009; Yin and Hong, 2010). DMSO was found to cause genetic alteration under some conditions by Vannini and Poli (1983) and  DNA damage and/or rearrangements in some cases viz., *Abies cephalonica* (Aronen *et al.*, 1999), *Solanum tuberosum* L. (Harding, 1997) and *Rubus grabowskii* (Castillo *et al.*, 2010) This effect of DMSO is usually explained by its effect on membrane permeability and function, thermostability of chromosome structure, or inhibition of DNA synthesis. The addition of abscisic acid to *in vitro* stock plants has been found to improve cryopreservability for cold hardy species (Ryyanen, 1998) but in general, cryopreservation procedures have been shown to be species specific.

## 7. Conclusion

Cryogenic storage is often referred to as a safe system, but this is dependent on reliable procedures and subsequent handling. For propagules from clones or desiccation-sensitive seed, uncontrolled temperature fluctuations, especially above -120° C may drastically affect viability. Hence, cryogenic storage tanks should be carefully monitored for temperature or LN level, along with proper handling, which is crucial for safe storage. It is believed that many biochemical (Stirn *et al.*, 1995), genetic (Wang *et al.*, 1992), and histological (Wang and

Huang, 1995) properties in relation to the embryogenic potential disappear rapidly during extended subculturing. Therefore, cryopreservation can be used to arrest the genetic instability that occurs by continual culture of embryogenic lines (Wang *et al.*, 2002). Techniques of controlled freezing, vitrification, encapsulation-dehydration, dormant bud preservation and combinations of these are now directly applicable with plant genotypes representing hundreds of species (Reed, 2002). Although the stresses involved in the introduction into and recovery from storage may be considerable, they can be minimized by appropriate handling and are unlikely to be genetically influential (Withers, 1984). Cryopreservation may also aid germplasm preservation of vegetatively propagated plants maintain the morphogenetic potential of cultured cells, and facilitate regeneration from young explants. The initial investment for cryopreservation is greater, but recurrent costs are minimal. As expertise in the cryopreservation of organized cultures increases, cryopreservation is likely to be the chosen long-term storage method for shoot tips and embryos because of its advantages of lower cost and greater convenience and stability. Cryopreservation techniques have been shown to be adaptable to a variety of plant tissues, but they must be tested and adapted to each new species that is tried (Pence, 1990). Although the species-specific nature of the cryostorage protocols presents problems, the overall benefits of the process argue well for cryoconservation of endangered genetic resources. The relative costs of storing cultures in the growing state and by cryopreservation will also change over time. Further studies will also increase the applicability of the procedure to other threatened species from around the world. Recent interest in the potential for turning *in vitro* related instability to advantage in plant breeding should also improve our understanding of the phenomenon and hence its control.

## 8. References

Amino, S. & Tazawa, M. (1988). Uptake and Utilization of sugars in cultured rice cells. *Plant Cell Physiol.*, Vol. 29, pp. 483-487.

Antikainen, M. & Griffith, M. (1997). Antifreeze protein accumulation in freezing tolerance cereals. *Physiol Plant*, Vol. 99, pp. 423-432.

Ahuja, S.; Mandal, BB.; Dixit, S.; Srivastava, PS. (2002). Molecular, phenotypic and biosynthetic stability in *Dioscorea floribunda* plants derived from cryopreserved shoot tips. *Plant Sci.*, Vol. 163, pp. 971–977.

Aronen, T.; Krajnakova, J.; Haggman, H. & Ryynanen, LA. (1999). Genetic fidelity of cryopreserved embryogenic cultures of open-pollinated *Abies cephalonica*. *Plant Sci.*, Vol. 142, pp.163–172.

Ashmore, SE. (1997). Status report on the development and application of *in vitro* techniques for the conservation and use of plant genetic resources. IPGRI, Italy.

Bajaj, YPS. & Sala, F. (1991). Cryopreservation of germplasm of rice. In: *Biotechnology in Agriculture and Forestry*, Y.P.S Bajaj (Ed.), Vol. 14, pp. 553-571, Springer Verlag Berlin, Heidelberg.

Benson, EE. (1990). Free radical damage in stored plant germplasm. IBPGR, Rome, Italy.

Benson, EE.; Lynch, PT. & Jones, J. (1995). The use of iron chelating agent desferrioxamine in rice cell cryopreservation: a novel approach for improving recovery. *Plant Sci. Limerick*, Vol. 110, pp. 249-258.

Breton, D.; Harvengt, L.; Trontin. JF.; Bouvet, A. & Favre, JM. (2006). Long-term subculture randomly affects morphology and subsequent maturation of early somatic embryos in maritime pine. *Plant Cell Tiss Org Cult.*, Vol. 87, pp.95–108.

Castillo, NRF.; Bassil, NV.; Wada, S. & Reed, BM. (2010). Genetic stability of cryopreserved shoot tips of *Rubus* germplasm. *In Vitro Cell Dev Biol Plant*, Vol. 46, pp. 246–256.

Cella, R.; Colombo, R.; Galli, MG.; Nielsen, E.; Rollo, F. & Sala, F. (1982). Freeze preservation of rice cells: a physiological study of freeze-thawed cells. *Plant Physiol.*, Vol. 55, pp. 279-284.

Cho, JS.; Hong, SM.; Joo, SY.; Yoo, JS. & Kim, DI. (2007). Cryopreservation of transgenic rice suspension cells producing recombinant hCTLA4Ig. *Appl Microbiol Biotechnol*, Vol. 73, pp. 1470-1476.

Cornejo, MJ.; Wong, VL.& Blechl, AE. (1995). Cryopreserved callus: a source of protoplasts for rice transformation. *Plant Cell Rep.*, Vol. 14, pp. 210–214.

Cyr, DR. (1999). Cryopreservation of embryogenic cultures of conifers and its application to clonal forestry. In: *Somatic embryogenesis in woody plants*, S.M. Jain, P.K. Gupta, R.J. Newton (Eds.), Vol. IV, pp. 239-262, Kluwer, Dordrecht.

Danylchenko, O. & Sorochinsky, B. (2005). Use of RAPD assay for the detection of mutation changes in plant DNA induced by UV-Band c-rays. *BMC Plant Biol*, Vol. 5, pp. 59.

Dobbennack, EH.; Kiesecker, H. & Schumacher, HM. (2009). Fundamental aspects and economic need of plant cell cryopreservation. *EU COST Action 871 Abstracts, Cryolett.*, Vol. 30, pp. 382, UK.

Dumet, D.; Engelmann, F.; Chabrillange, N. & Duvall, Y. (1993). Cryopreservation of oil palm (*Elaeis guinensis* Jacq.) somatic embryogenesis involving a dessication step. *Plant Cell Rep.*, Vol. 12, pp. 352-355.

Engelmann, F. (2000). Importance of cryopreservation for the conservation of plant genetic resources. In: *Cryopreservation of tropical plant germplasm*, F. Engelmann, H. Takagi (Eds.), JIRCAS. International Agricultural Series No. 8, pp. 8–20, Tsukuba, Japan.

Finkle, BJ.; Zavala, ME. & Ulrich, JM. (1985). Cryoprotective compounds in the viable freezing of plant tissues. In: *Cryopreservation of plant cells and organs*, K.K. Kartha (Ed.), pp.75–113, CRC, Boca Raton.

Forkan, AM.; Anthony, P.; Power, JB.; Davey, MR. & Lowe, KC. (2001). Effect of Erythrogen TM on post-thaw recovery of cryopreserved cell suspensions of Indica rice (*Oryza sativa* L.). *Cryolett.*, Vol. 22, pp. 367-374.

Franks, F. (1985). Biophysics and biochemistry at low temperature. Cambridge University Press, Cambridge.

Gale, S.; John, A. & Benson, EE. (2007). Cryopreservation of *Picea sitchensis* (Sitka spruce) embryogenic suspensor masses. *Cryolett.*, Vol. 28, pp.225–239.

Gale, S.;John, A.; Harding, K.& Benson, EE. (2008). Developing cryopreservation for *Picea sitchensis* (Sitka spruce) somatic embryos: a comparison of vitrification protocols. *Cryolett.*, Vol. 29, pp. 35–144.

Gao, DY.; Liu, J. & Liu, C. (1995). Prevention of osmotic injury to human spermatozoa during addition and removal of glycerol. *Biol.Reprod.*, Vol. 53, pp. 985-995.

Grout, BWW. & Morris, GJ. (1987). Freezing and cellular organization. In: *The effects of low temperatures on biological systems*, B.W.W. Grout, G.J. Morris (Eds.), pp.147-174 Edward Arnold, London.

Grout, BWW. (1990a). In vitro conservation of germplasm. In: *Plant tissue culture: application and limitations*, S.S. Bhojwani (Ed.), pp. 394-411, Elsevier, Amsterdam.

Grout, BWW. (1990b). Genetic preservation in in vitro. In: *Plant cellular and molecular biology*, H.J.J. Nijkamp, L.H.W.Van der plas, K.Van Aartrijk (Eds.), pp.13-22 Kluwer Dordrecht.

Harding, K. (1997). Stability of the ribosomal RNA genes in *Solanum tuberosum* L. plants recovered from cryopreservation. *Cryolett.*, Vol. 18, pp. 217–230.

Harding, K. (2004). Genetic integrity of cryopreserved plant cells: a review. *Cryolett.*, Vol. 25, pp. 3-22.

Harding, K.; Lynch, PT. & Johnston, JW. (2009). Epigenetic changes associated with cryopreservation of clonal crops. EU COST Action 871 Abstracts, *Cryolett.*, Vol. 30, pp.390, UK.

Hu, J. & Gou, CG. (1996). Studies on the cryopreservation (-196°C) of pollen of a restorer line in hybrid rice. *Acta Agronomica Sinica*, Vol. 22, pp.72-77.

Huang, CN.; Wang, JH.; Yan, QS. & Yan, QF. (1995). Plant regeneration from rice (*Oryza sativa* L.) embryogenic suspension cells cryopreserved by vitrification. *Plant Cell Rep.*, Vol 11, pp. 730-734.

Ishikawa, M.; Tandon, P.; Suzuki. M. & Yamaguishi-Ciampi, A. (1996). Cryopreservation of bromegrass (*Bromus inermis Leyss*) suspension cultured cells using slow prefreezing and vitrification procedures. *Plant Sci.*, Vol.120, pp. 81-88.

Jain, S.; Jain R. K. & Wu, R. (1996). A simple and efficient procedure for cryopreservation of embryogenic cells of aromatic indica rice varieties. *Plant Cell Rep.*, Vol. 15, pp. 712-717.

Jelodar, NB.; Davey, MR. & Cocking, EC. (2001). Studies on the cryopreservation of cell suspension cultures of some Iranian Indica and Japonica rice (*Oryza sativa* L.) cultivars. *Iranian Journal of Agricultural Sciences*. Vol. 32, pp. 661-670.

Kaeppler, SM.; Phillips, RL. & Olhoft, P. (1998). Molecular basis of heritable tissue culture-induced variation in plants. In: *Somaclonal Variation and Induced Mutations in Crop Improvement. Current Plant Science and Biotechnology in Agriculture*, Jain et al., (Eds.), Vol. 32. pp. 465–484, Kluwer Academic Publishers, Dordrecht, Netherlands,.

Kaeppler, SM.; Kaeppler, HF. & Rhee, Y. (2000). Epigenetic aspects of somaclonal variation in plant. *Plant Mol Biol.*, Vol. 43, pp.179-188.

Kartha, KK. (1997). Advances in cryopreservation technology of plant cells and organs. In: *Plant tissue and cell culture*, C.E. Green, D.A. Somers, W.P. Hackett, D.D.Biesboer (Eds.), pp. 447-458, Alan R. Liss. New York.

Klimaszewska, K.;Ward, C. & Cheliak, WM. (1992). Cryopreservation and plant regeneration from embryogenic cultures of larch (*Larix X eurolepis*) and black spruce (*Picea mariana*). *J Exp Bot.*, Vol. 43, pp.73–79.

Kong, L. & Aderkas, P.V. (2011). A novel method of cryopreservation without a cryoprotectant for immature somatic embryos of conifers. *Plant Cell Tiss Org Cult.*, Vol. 106, pp.115-125.

Kuriyama, A.; Watanabe, K.; Ueno, S. & Mitsuda, H. (1989). Inhibitory effect of ammonium ion on recovery of cryopreserved rice cells. *Plant Sci.*, Vol. 64, pp.231-235.

Lynch, PT. & Benson, EE. (1991). Cryopreservation, a method for maintaining the plant regeneration capability of rice cell suspension cultures In: *Proc.2nd Intl. Cong. of Rice Res.* pp. 321-331, IRRI. The Philippines.

Lynch, PT.; Benson, EE.; Jones, J.; Cocking, EC.; Power, JB.; Davey, MR.& Cassells, AC. (1995). The embryogenic potential of rice cell suspensions affects their recovery following cryogenic storage. *Euphytica,* Vol. 85, pp. 347-349.

Mannonen, L.; Toiwonen, L. & Kauppinen, V. (1990). Effects of long-term cryopreservation on growth and productivity of *Panax ginseng* and *Catharanthus roseus* cell cultures. *Plant Cell Rep.* Vol. 9. pp. 173-177.

Marassi, AM.; Scocchi, A. & Gonzalez, AM. (2006). Plant regeneration from rice anthers cryopreserved by an encapsulation/dehydration technique. *In vitro Cell Dev Bio. Plant.* Vol. 42, pp. 31-36.

Matsumoto, T.; Takahishi, C.; Sakai, A. & Nako, Y. (1998). Cryopreservation of in vitro-grown apical meristems of hybrid statice by three different procedures. *Sci Hortic.,* Vol. 76, pp.105–114.

Mazur, P. (1984). Freezing of living cells: mechanisms and implication. *The American Journal of Physiolog.,* Vol. 247, pp.125-142.

McLellan, MR.; Scrijnemakers, EWM. & Iren, FV. (1990). The responses of four cultured plant cell lines to freezing and thawing in the prescence or absence of cryoprotected mixtures.*Cryolett.,* Vol. 11, pp.189-204.

Menges, M.; Murray, JAH. (2004). Cryopreservation of transformed and wild-type *Arabidopsis* and tobacco cell suspension cultures. *Plant J.,* Vol. 37, pp. 635–644.

Moiseyev, VA.; Nardid, OA. & Belous, AM. (1982). On a possible mechanism of the protective action of cryoprotectants. *Cryolett.,* Vol. 3, pp.17.

Moukadiri, O.; Deming, J.; O'Connor, JE. & Cornejo, MJ. (1999a). Phenotype characterization of progenies of rice plants derived from cryopreserved Calli. *Plant Cell Rep.,* Vol. 18. pp. 625-632.

Moukadiri, O.; Lopes, CR. & Cornejo, MR. (1999b). Physiological and genomic variations in rice cells recovered from direct immersion and storage in liquid nitrogen. *Physologia Plantarum,* Vol. 105, pp. 442-449.

Moukadiri, O.; Connor, JE.; Cornejo, MJ. (2002). Effects of the cryopreservation procedures on recovered rice cell populations. *Cryolett.,* Vol. 23, pp. 11-20.

Panis, B.; Totté, N.; Van Nimmen, K.; Withers, LA. & Swennen, R. (1996). Cryopreservation of banana (*Musa* spp.) meristem cultures after preculture on sucrose. *Plant Sci.,* Vol. 121, pp. 95-106.

Panis, B. & Lambardi, M. (2005). Status of cryopreservation technologies in plants (crops and forest trees). Paper presented at the International workshop on, *The role of biotechnology for the characterization and conservation of crop, forestry, animal and fishery genetic resource,* 5-7 March, Turin, Italy.

Pence, VC. (1990). In vitro collection, regeneration and cryopreservation of *Brunsfelsia densifolia.* In: (Abstract) *Proceedings VIIth International Congress on Plant Tissue and Cell Culture,* pp.377.

Popova, EV.; Lee, EJ.; Wu, CH.; Hahn, EJ. & Paek, KY. (2009). A simple method for c ryopreservation of *Ginkgo biloba* callus. *Plant Cell Tiss Org Cult.,* Vol. 97, pp. 337–343.

Pushkar, NS.; Sheenberg, MG. & Oboznaya, EI. (1976). On the mechanism of cryoprotection by polyethylene oxide, *Cryobiology,* Vol. 13, pp.142-146.

Reed, BM. (2002). Implementing cryopreservation for long term germplasm preservation in vegetatively propagated species. In: *Biotechnology in Agriculture and forestry.*

*Cryopreservation of plant germplasm II*, L.E.Towill, Y.P.S.Bajaj (Eds.), Vol. 50, pp. 22-33, Springer Verlag, Berlin, Heidelberg.

Ryynen, LA. (1998). Effect of Abscicic acid, cold hardening and photoperiod on recovery of cryopreserved in vitro shoot tips of silver birch. *Cryobiology*. Vol. 36, pp. 32-29.

Sakai, A. (1956). Survival of plant tissue at super low temperatures. *Low temperature Sci.Ser.B.*, Vol. 14, pp. 17-23.

Sala, F.;Cella, R. & Rollo, F. (1979). Freeze preservation of rice cells grown in suspension culture. *Plant Physiol.*,Vol. 45, pp.170-176

Salaj, T.; Matusikova, I.; Fraterova, L.; Pirselova, B. & Salaj, J.(2010). Regeneration of embryogenic tissues of *Pinus nigra* following cryopreservation. *Plant Cell Tiss Org Cult.*, Vol. 106, pp. 55-61.

Schäfer Menuhr, A.; Schumacher, HM. & Mix-Wagner, G. (1997). Cryopreservation of cultivars – design of a method for routine application in gene banks. *Acta Hortic.*, Vol. 447,pp. 477-482.

Schmitz, U. & Lorz, H. (1990) Nutrient uptake in suspension cultures of Gramineae II. Suspension cultures of rice (*Oryza sativa* L.). *Plant Sci.*, Vol. 66, pp.95-111.

Scowcroft, WR. & Larkin, PJ. (1983). Somaclonal variation and genetic improvement of crop plants. In: *Better crops for food Ciba foundation symposium*, J. Nugent, M.O'Connor (Eds.) No.97. pp. 177-197, Pitman , London.

Scowcroft, WR. (1984). Genetic instability of in vitro cultures. *Consultant report*. Rome: IBPGR.

Skyba, M. & Cellarova, E. (2009). Fundamental aspects in cryopreservation of medicinal plants, the *Hypericum* story. EU COST Action 871 Abstracts, *Cryolett*. Vol. 30, pp. 387, UK.

Stirn, S.; Mordhorst, AP.;Fuchs, S. & Lorz, H. (1995). Molecular and biochemical markers for embryogenic potential and regenerative capacity of barley (*Hordeum vulgare* L.) cell cultures. *Plant Sci.*, Vol. 2106, pp. 195-206.

Touchell, DH. & Dixon, KW. (1999). In vitro preservation. In: *A color atlas of plant propagation and conservation*. B.G. Bowes (Ed.), pp.108-118, Manson Publishers, London UK.

Touchell, DH.; Chiang, VL. & Tsai, CJ. (2002). Cryopreservation of embryogenic cultures of *Picea mariana* (black spruce) using vitrification. *Plant Cell Rep.*, Vol. 21, pp.118–124.

Towill, LE. & Jarret, RL. (1992). Cryopreservation of sweet potato (*Ipomoea batatas* L.Lam.) shoot tips by vitrification. *Plant Cell Rep.*, Vol. 11, pp.175-178.

Towill, L E. & Walters, C. (2000). Cryopreservation of pollen. In: *Cryopreservation of tropical germplasm: Current research progress and application*, F.Engelmann, H.Takagi (Eds.), pp. 115–129, JIRCAS/IPGRI, Ibarak, Japan.

Towill, LE. (2002). Cryopreservation of plant germplasm: Introduction and some observations. In: *Biotechnology in Agriculture and forestry. Cryopreservation of plant germplasm II*, L.E. Towill, Y.P.S. Bajaj (Eds.), Vol. 50, pp.3-21, Springer Verlag Berlin, Heidelberg.

Tsukuhara, M. & Hirosawa, T. (1992). Simple dehydration treatment promotes plantlet regeneration of rice (*Oryza sativa* L.) callus. *Plant Cell Rep.*, Vol. 11, pp. 550-553.

Ulrich, JM.; Finkle, BJ.; Moore, PH. & Ginoza, H. (1979). Effects of a mixture of cryoprotectants in attaining liquid nitrogen survival of callus of a tropical plant. *Cryobiology*, Vol. 16, pp. 550-556.

Urbanova,´ M.; Kosuth, J. & Cellarova,´ E. (2005). Genetic and biochemical analysis of *Hypericum perforatum* L. plants regenerated after cryopreservation. *Plant Cell Rep.*, Vol. 25, pp.140–147.

Vannini, GL. & Poli, F. (1983). Binucleation and abnormal chromosome distribution in *Euglena gracilis* cells treated with dimethylsulphoxide. *Protoplasma*, Vol. 114, pp. 62-66.

Wang, XH.; Lazzeri, PA. & Lorz, H. (1992). Chromosomal variation in dividing protoplast derived from cell suspensions of barley (*Hordeum vulgare* L.) *Theor Appl Genet.*, Vol. 85, pp. 181-185.

Wang, JH. & Huang, CN. (1995). Histological studies on the dedifferentiation and redifferentiation pattern of barley (*Hordeum vulgare* L.) mature embryo cells. *J. Hangzhou University (Natural Science edition)*, Vol. 22, pp. 102-106.

Wang, XL.; Shu, LH.;Yuan, WJ. & Liao, LJ. (1996). Panicle culture and karyotype analysis from callus cells of a diploid wild rice, *Oryza meyeriana*. *IIRN.*, Vol. 21, pp. 7-8.

Wang, JH. & Huang, CN. (1998). Assessment of antifreeze proteins and water deficit protein during cryopreservation of *Oryza sativa* and *Dendrobium candidum* cells. Abstracts of *XVIII international congress of genetic*, pp. 186.10-15 Aug, Beijing.

Wang, JH.; Bian, HW.; Huang, CN.& Ge, JG. (1999). Antifreeze protein accumulation in freezing tolerance cereals. *Physiol Plant*, Vol. 99, pp. 423-432.

Wang, JH.; Bian, HW.; Zhang, Y X.& Cheng, HP. (2001). The dual effect of antifreeze protein on cryopreservation of rice (*Oryza sativa* L.) embryogenic suspension cells. *Cryolett.*, Vol. 22, pp. 175-182.

Wang, JH. & Huang, CN. (2002). Cryopreservation of *hordeum* (barley) In: *Biotechnology in Agriculture and forestry, Cryopreservation of plant germplasm II*. L.E. Towill, Y.P.S. Bajaj (Eds.) Vol. 50, pp. 119-135, Springer Verlag, Berlin, Heidelberg.

Wang, Q.; Gafny, R.; Sahar, N.; Sela, I.; Mawassi, M.; Tanne, E. & Perl, A. (2002). Cryopreservation of grapevine (*Vitis vinifera* L.) embryogenic cell suspensions by encapsulation-dehydration and subsequent plant regeneration. *Plant Sci.*, Vol. 162, pp. 551–558.

Watanabe, K.; Kawai, F.; Kanamori, M. (1995). Factors affecting cryoprotectability of cultured rice (*Oryza sativa* L.) cells-cell wall and cell-aggregate size. *Cryolett.*, Vol.16, pp. 147-156.

Watanabe, K.; Kuriyama, A.; Kawai, F. & Kanamori, M. (1998). Survival of rice non-embryogenic callus cells after cryopreservation in liquid nitrogen by vitrification. *Plant Biotechnology*, Vol. 15, pp. 35-37.

Wilkinson, T.; Wetten, A.; Prychid, C. & Fay, MF. (2003). Suitability of cryopreservation for the long term storage of rare and endangered plant species: a case history for *Cosmos atrosanguineus*. *Annals of Botany*, Vol. 91, pp. 65-74.

Withers, LA. (1978). A fine structural study of the freeze preservation of plant tissue cultures in the thawed state. *Protoplasma*, Vol. 94, pp. 235-247.

Withers, LA. (1983). Germplasm preservation through tissue culture: an overview. In: *Cell and Tissue culture technique for cereal crop improvement*. Inst of Genetics Academics Sinica/Int Rice Res Inst., pp. 315-341, Gordon and Breach, New York.

Withers, LA. & King, PJ. (1980). A simple freezing unit and cryopreservation method for plant cell suspensions. *Cryolett.*, Vol. 1, pp. 213-220.

Withers, LA. (1984). Germplasm conservation in vitro: present state of research and its application. In: *Crop genetic resources: conservation and evaluation,* J.H.W Holden, J.T.Williams (Eds.), IBPGR Publication.

Withers, LA. (1987). The low-temperature preservation of plant cell, tissue and organ cultures and seed for genetic conservation and improved agricultural practice. In: *The effects of low temperature on biological systems,* B.W.W.Grout, G.J.Morris (Eds.), pp. 89-409, Edward Arnold, London.

Yang, YS.; Zheng, YD.; Chen, YL. & Jian, YY. (1999). Improvement of plant regeneration from long-term cultured calluses of Taipei 309, a model rice variety in in vitro studies. *Plant Cell Tiss and Org Cult.,* Vol. 57, pp. 199-206.

Yin, MH. & Hong, SR. (2010). A simple cryopreservation protocol of *Dioscorea bulbifera* L. embryogenic calli by encapsulation–vitrification. *Plant Cell Tiss Org Cult.,* Vol. 101, pp. 349–358.

Yoshimatsu, K. & Shimomura, K. (2002). Cryopreservation of *Panax* (ginseng) In: *Biotechnology in Agriculture and forestry. Cryopreservation of plant germplasm II.* L.E.Towill, Y.P.S.Bajaj (Eds) Vol. 50, pp. 164-179, Springer Verlag Berlin, Heidelberg.

Zeliang, PK.; Pattanayak, A.; Iangrai, B.; Khongwir, E. & Sarma, BK. (2010). Fertile plant regeneration from cryopreserved calli of *Oryza rufipogon* Griff. and assessment of variation in the progeny of regenerated plants. *Plant Cell Rep.,*Vol. 29, pp.1423-1433.

Zhang, ZH. & Hu, ZL. (1999). Regenerating plants from cryopreserved adventitious buds of haploids in rice. *Wuhan University Journal of Natural Sciences,* Vol. 4, pp. 115-117.

Zhang, YX.; Wang, JH.; Bian, HW.& Zhu, MY. (2001). Pregrowth-desiccation: a simple and efficient procedure for the cryopreservation of rice (*Oryza sativa* L.) embryogenic suspension cells. *Cryolett.,* Vol. 22, pp. 221-228.

# Modelling and Simulation of Plant Breeding Strategies

Jiankang Wang

*Institute of Crop Science and CIMMYT China,*
*Chinese Academy of Agricultural Sciences (CAAS),*
*China*

## 1. Introduction

The major objective of plant breeding programs is to develop new genotypes that are genetically superior to those currently available for a specific target environment or a target population of environments (TPE). To achieve this objective, plant breeders employ a range of selection methods (Allard, 1999; Hallauer et al., 1988). Quantitative genetic theory generally provides much of the framework for the design and analysis of selection methods used within breeding programs, based on various assumptions in order to render mathematically or statistically tractable theories (Hallauer et al., 1988; Falconer and Mackay, 1996; Lynch and Walsh, 1998). Some of these assumptions can be easily tested or satisfied by certain experimental designs; others can seldom be met, such as the assumptions of no linkage and no genotype by environment (G×E) interaction. Still others, such as the presence or absence of epistasis and pleiotropy, are difficult to test. Field experiments have been conducted to compare the efficiencies from different breeding methods. However, due to the time and effort needed to conduct field experiments, the concept of modeling and prediction have always been of interest to plant breeders. Computer simulation gives breeders the opportunity to lessen the impact of these assumptions, thereby establishing more valid genetic models for use in plant breeding (Kempthone, 1988). Simulation as a tool has been applied in many special plant breeding studies that use relatively simple genetic models. A tool capable of simulating the performance of a breeding strategy for a continuum of genetic models ranging from simple to complex, embedded within a large practical breeding program including marker-assisted-selection, had not been available until recently (Wang et al., 2003; Wang and Pfeiffer, 2007). In this chapter, the principles and applications of simulation modeling in plant breeding are introduced.

## 2. Principles of plant breeding simulation

### 2.1 Available simulation tools

QU-GENE is a simulation platform for quantitative analysis of genetic models, which consists of the two-stage architecture (Podlich and Cooper, 1998). The first stage is the engine, and its role is to: (1) define the gene and environment (G×E) interaction system (i.e., all the genetic and environmental information of the simulation experiment), and (2) generate the starting population of individuals (base germplasm). The second stage encompasses the application modules, whose role is to investigate, analyze, or manipulate the starting population of individuals within the G×E system defined by the engine. The application module usually represents the operation of breeding programs.

Three application modules have been developed. QuLine, a computer program, was firstly designed in 2002-2003 for simulating CIMMYT's wheat breeding, one of the most successful wheat breeding programs in the world. QuLine can integrate enormous amounts of data from different sources, process them in many ways, and produce alternative theoretical but realistic scenarios that the breeder can draw on to make a decision. It can simulate almost all breeding activities in CIMMYT's wheat breeding program, including male master selection, female master selection, parental selection, single cross, backcross, top cross, double cross, doubled haploid, marker-assisted selection, pedigree breeding, selected bulk etc. QuLine can simulate other breeding programs for selecting inbred lines, which means all major food cereals in the world, plus basically all leguminous crops (Wang et al., 2003 and 2004).

Taking advantage of the sophisticated state of QuLine, QuHybrid was developed in 2008-2009. The major development required for QuHybrid is the implementation of test crossing. To make the testcrosses, an additional population defining all the testers was added (Zhang et al. 2011). When the testcross functionality is activated, testcrosses will be made between all families and testers. Among-family selection is conducted based on the mean performance of all testcrosses in each tested family. Breeding methods can be compared by the line *per se* and testcross genetic gains. For hybrid prediction, another population consisting of inbred lined in another heterotic group is also needed to run QyHybrid. At the end of each breeding cycle, performance of all potential $F_1$ hybrids between the final selected inbred lines and lines in the other heterotic group are predicted.

Marker assisted recurrent selection (MARS) was proposed to overcome the disadvantages when using markers in selecting complex traits. It has been commercially used for selecting complex traits in maize, sunflower and soybean breeding programs (Bernardo and Charcosset, 2006). As a result, QuMARS was developed in 2009-2010. Prediction models include best linear unbiased prediction (BLUP), and regression models. Prediction can be built on both line *per se* performance and testcross performance. Therefore, it can simulate both MARS and genomic selection (GS), starting from a single cross between two parental lines. With QuMARS, various issues using MARS or GS can be investigated. For example, how many cycles of recurrent selection are suitable? How many markers should be used? How can the breeding values of lines under development be best predicted?

## 2.2 Selected applications of modeling and simulation in plant breeding

Simulation can be used to investigate both strategic (say comparison of two breeding methods; Wang et al., 2003, 2004, 2009b) and tactical (say identification of optimum crossing and selection schemes given the gene distribution in parents; Wang et al., 2005, 2007a, 2007b, 2009a) issues in plant breeding. Two strategic and two tactical applications using QuLine are summarized in this section.

### 2.2.1 Comparison of two breeding strategies in CIMMYT's wheat breeding

The main elements of international wheat improvement program at CIMMYT have been shuttle breeding at two contrasting locations in Mexico, wide adaptation, durable rust and

Septoria resistances, international multi-environment testing, and the appropriate use of genetic variation to enhance yield gains (Rajaram et al., 1994; Rajaram, 1999). Two breeding strategies are commonly used in CIMMYT's wheat breeding programs (van Ginkel et al., 2002; Wang et al., 2003, 2004). The modified pedigree (MODPED) method begins with pedigree selection of individual plants in the $F_2$, followed by three bulk selections from $F_3$ to $F_5$, and pedigree selection in the $F_6$. In the selected bulk (SELBLK) method, spikes of selected $F_2$ plants within one cross are harvested in bulk and threshed together, resulting in one $F_3$ seed lot per cross. This selected bulk selection is also used from $F_3$ to $F_5$, whereas, pedigree selection is used only in the $F_6$. Assuming that planting intensity is similar, SELBLK uses approximately two thirds of the land allocated to MODPED, and produces smaller number of families. Therefore when SELBLK is used, fewer seed lots need to be handled at both harvest and sowing, resulting in a significant saving in time, labor, and cost. Will the two strategies result in similar genetic gain on yield and other breeding traits?

The genetic models developed accounted for epistasis, pleiotropy, and G×E. For both breeding strategies, the simulation experiment comprised of the same 1000 crosses developed from 200 parents. A total of 258 advanced lines remained following 10 generations of selection. The two strategies were each applied 500 times on 12 GE systems (Wang et al., 2003). The average adjusted genetic gain on yield across all genetic models is 5.83 for MODPED and 6.02 for SELBLK, with a difference of 3.3%. This difference is not large and, therefore, unlikely to be detected using field experiments (Singh et al., 1998). However, it can be detected through simulation, which indicates that the high level of replication (50 models by 10 runs in this experiment) is feasible with simulation and can better account for the stochastic properties from a run of a breeding strategy and the sources of experimental errors. The average adjusted gains for the two yield gene numbers 20 and 40 are 6.83 and 5.02, respectively, suggesting that genetic gain decreases with increasing yield gene number.

The number of crosses remaining after one breeding cycle is significantly different among models and strategies, but not among runs (Wang et al., 2003). The number of crosses remaining from SELBLK is always higher than that from MODPED, which means that delaying pedigree selection favors diversity. On an average, 30 more crosses were maintained in SELBLK. However, there was a crossover between the two breeding strategies. Prior to $F_5$ the number of crosses in MODPED was higher than that in SELBLK. The number of crosses became smaller in MODPED after $F_5$, when pedigree selection was applied in $F_6$. Among-family selection from $F_1$ to $F_5$ in SELBLK was equal to among-cross selection, and resulted in a greater reduction in the cross numbers for SELBLK compared to MODPED, in the early generations. In general, only a small proportion of crosses remained at the end of a breeding cycle (11.8% for MODPED and 14.8% for SELBLK); therefore, intense among-cross selection in early generations was unlikely to reduce the genetic gain. On the contrary, breeders would tend to concentrate on fewer but "higher probability" crosses. As more crosses remained in SELBLK, the population following selection from SELBLK might have a larger genetic diversity than that from MODPED. In this context also, SELBLK is superior to MODPED.

### 2.2.2 Modeling of the single backcrossing breeding strategy

Regarding the crossing strategies in CIMMYT wheat breeding, top (or three-way) crosses and double (or four-way) crosses were employed to increase the genetic variability of breeding populations in the early 1970s. By the late 1970s, double crosses were dropped due to their poor results relative to single cross, top crosses and limited backcrosses. From the 1980s onwards, all crosses onto selected $F_1$ generations were single cross, backcrosses or top crosses (van Ginkel et al., 2002). Single and top (or three-way) crosses are commonly used among adapted parental lines, while backcrosses are preferred for transferring a few useful genes from donor parents to adapted lines. In CIMMYT, the single backcrossing approach (one backcross to the adapted parent) was initially aimed at incorporating resistance to rust diseases based on multiple additive genes (Singh and Huerta-Espino, 2004). However, it soon became apparent that the single backcross approach also favored selection of genotypes with higher yield potential. The reason why single backcrossing shifts the progeny mean toward the higher side is that it favors the retention of most of the desired major additive genes from the recurrent, while simultaneously allowing the incorporation and selection of additional useful small-effect genes from the donor parents.

The breeding efficiency of this strategy compared with other crossing and selection strategies was investigated through computer simulation for many scenarios, such as the number of genes to be transferred, frequency of favorable alleles in donor and recurrent parents etc. Results indicated this breeding strategy has advantages in retaining or overtaking the adaptation of the recurrent parents and at the same time transferring most of the desired donor genes for a wide range of scenarios (Wang et al., 2009). Two times of backcrossing have advantages when the adaptation of donor parents is much lower than that of the adapted parents, and the advantage of three times of backcrossing over two times of backcrossing is minimal. We recommend the use of single backcrossing breeding strategy based on three assumptions: (1) multiple genes governing the phenotypic traits to be transferred from donor parents to adapted parents, (2) donor parents still have some favorable genes that may contribute to the improvement of adaptation in the recipient parents even under low adaptation, and (3) the conventional phenotypic selection is applied or the individual genotypes cannot be precisely identified.

### 2.2.3 Optimization of marker assisted selection (MAS)

Many breeding programs in a range of crops are using molecular markers to screen for one to several alleles of interest. The availability of an increasing number of useful molecular markers is allowing accurate selection at a greater number of loci than has been previously possible (Dekkers and Hospital, 2002; Dubcovsky, 2004). However, larger population sizes are required to ensure with reasonable certainty that an individual with the target genotype is present. Different crossing and selection strategies may require vastly different population sizes to recover a target genotype with the same certainty even when the same parents are used (Bonnett et al., 2005). Determination of the most efficient strategy has the potential to dramatically decrease the amount of resources (plants, plots, marker assays, and labor) required to combine a set of target alleles into a new genotype.

The drought-suitable lines in wheat should be semi-dwarf with long coleoptiles, resistant to multiple diseases, have good dough properties, and have productive tillers. To achieve this, nine target alleles need to be combined into one genotype (Wang et al., 2007a). Three parent lines were used: Sunstate, a commercial Australian line; HM14BS, a germplasm line combining an allele for height reduction and long coleoptiles; and Silverstar+tin, a derivative of Silverstar with a restricted tillering allele. The largest target genotype frequency was found in the Silverstar+tin/HM14BS//Sunstate topcross. The optimum MAS strategy to combine the nine target alleles from this topcross could be divided into three steps: (i) selection for Rht-B1a and Glu-B1i homozygotes, and enrichment selection of Rht8c, Cre1, and tin in top cross $F_1$, (ii) selection of homozygotes for one target allele, e.g. Rht8c, and enrich the remaining target alleles in top cross $F_2$, and (iii) selection of the target genotype with doubled haploid lines or recombination inbred lines. Enrichment of allelic frequencies in top cross $F_2$ reduced the total number of lines screened from >3500 to <600.

## 2.2.4 Design breeding with known gene information

The concept of design breeding was proposed in recent years as the fast development in molecular marker technology (Peleman and Voort, 2003; Wang et al., 2007b). Three steps are involved in design breeding. The first step is to identify the genes for breeding traits, the second step is to evaluate the allelic variation in parental lines, and the third step is to design and conduct breeding. A permanent mapping population of rice consisting of 65 non-idealized chromosome segment substitution lines (denoted as CSSL1 to CSSL65) and 82 donor parent chromosome segments (denoted as M1 to M82) was used to identify QTL with additive effects for two rice quality traits, area of chalky endosperm (ACE) and amylose content (AC), by a likelihood ratio test based on stepwise regression. These CSS lines were generated from a cross between the japonica rice variety Asominori (the background parent, denoted as P1) and the indica rice variety IR24 (the donor parent, denoted as P2) (Wan et al., 2004, 2005).

Through QTL studies, it is impossible to derive an inbred with the minimum of ACE and the maximum of AC, because QTL on segments M35, M57, and M59 have unfavorable pleiotropic effects on ACE and AC. However, the ideal inbred with relatively low ACE and high AC can be identified through simulation (Wang et al. 2007b). This designed inbred contains four segments from P2, which are, M19, M35, M57, and M60, and another genome is from the background P1. The value of ACE in this inbred is 9.2%, where the theoretical minimum ACE is 0. The value of AC is 17.73%, whereas, the theoretical maximum of AC is 22.3%. Among the 65 CSS lines, the three lines, CSSL15, CSSL29, and CSSL49, have the required target segments, therefore, can be used as the parental lines in breeding. Three possible topcrosses can be made among the three parental lines, Topcross 1: (CSSL15 × CSSL29) × CSSL49, Topcross 2: (CSSL15 × CSSL49) × CSSL29, and Topcross 3: (CSSL29 × CSSL49) × CSSL15. Different MAS schemes can be used to select the target inbred line. Here two schemes are considered, Scheme 1: 200 topcross $F_1$ (TCF$_1$) are first generated. Then 20 doubled haploid (DH) are derived from each TCF$_1$ individual. The target inbred lines are selected from the 4000 DH lines. Scheme 2: 200 TCF1 are first generated. An enhancement selection (Wang et al., 2007a), is conducted among the 200 TCF$_1$ individuals. Then 20 doubled haploid (DH) are derived from each selected TCF$_1$ individual. The target inbred lines are selected from those derived DH lines.

From 100 simulation runs, it was found that by using Scheme 1, 27 target inbred lines were selected from Topcross 1, 13 from Topcross 2, and 8 from Topcross 3. Therefore Topcross 1 had the largest probability to select the target inbred line, and should be used in breeding low ACE and AC inbred lines. The two MAS schemes resulted in significant difference in cost when genotyping for MAS. Scheme 1 required 4000 DNA samples for each topcross. On the contrary, Scheme 2 required 462 DNA samples for Topcross 1, 324 for Topcross 2, and

691 for Topcross 3. Topcross 1 combined with Scheme 2 resulted in the least DNA samples per selected line, and therefore was the best crossing and selection scheme.

## 2.3 Definition of a gene and environment system in QU-GENE

G×E system underlies the genetic and environmental model for simulation experiments. In general, information about a G×E system consists of some general information, the target population of environments (TPE) for the breeding program, traits to be selected during the breeding procedure, random environmental deviations for these traits, genes for traits, their locations on chromosomes, and their effects on traits in different environment types. Information about the population consists of the number of parents and their genotypes.

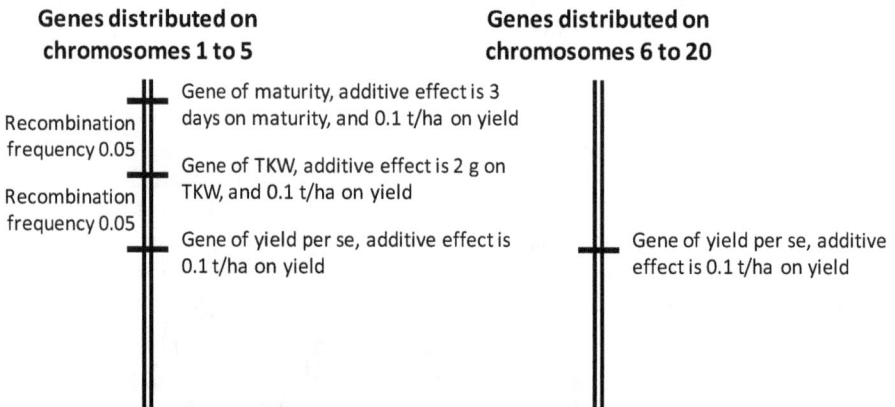

Fig. 1. A putative genetic model consisting of five genes for maturity, five genes for TKW (thousand kernel weight), and 20 gene for yield.

As a simplified example, we assume TPE of a plant breeding program only contain one environment type, and three traits are used in selection in, i.e., maturity, thousand kernel weight (TKW), and yield. A putative genetic model consists of 5 genes for maturity, five genes for TKW, and 20 genes for yield (Fig. 1). Each maturity gene has an additive effect of 3 days on maturity, and 0.1 t/ha on yield (Fig. 1). Each TKW gene has an additive effect of 2 g on TKW, and 0.1 t/ha on yield (Fig. 1). Each yield gene has an additive effect of 0.1 t/ha on yield (Fig. 1). One maturity gene, one TKW gene and one yield gene are linked on each of the first 5 chromosomes, and one yield gene is located on chromosomes 6 and 20 (Fig. 1). These information needs to be organized in certain formats in QU-GENE.

### 2.3.1 General information about a G×E system

The first part is the general information about the G×E system (Fig. 2). Number of models is specifically designed for a G×E system with random gene effects. For a G×E system with all gene effects (additive, dominance, epistasis, and pleiotropy) fixed, this parameter should be set at 1. The random effects model in a G×E system will most likely mimic the real genetic effects of a large number of genes, such as the genes for yield. With this model some genes will have relatively larger effects and others, smaller effects. The large number of G×E systems, different yield gene effects in each G×E system, and replications feasible within the simulation allow many potential realities to be compared. If one breeding methodology is superior to another for all, or most, permutations, the breeder can be confident that a superior breeding methodology has been identified that is also robust to the complexities and perturbations that may emerge, regardless of the G×E system. Random seed of random gene effects will ensure that the same gene effects will be assigned whenever the G×E system is used, so that all random gene effects are repeatable.

```
! *********************************************************************
! *   QUGENE engine input file
! *
! *********************************************************************
! *** General information on the G-E system ***

! Engine G-E output filename prefix (*.ges)
WheatModel

1                 ! Number of models
0                 ! Random seed of random gene effects
30                ! Number of genes (includes markers and qtls)
1                 ! Number of environment types
3                 ! Number of traits (not including markers)
1 1 1 0 0 0 0     ! Specify names (ETs, Trts, Genes, Alls, EPN, GPM, pop)
```

Fig. 2. General information about a gene and environment system in QU-GENE

## 2.3.2 Environment information

The TPE for a breeding program consists of a set of distinct, relatively homogenous environment types, each with a frequency of occurrence. Each environment type has its own gene action and interaction, providing the framework for defining G×E interactions (Fig. 3). Each environment type takes three rows. Row 1 is an ID number to distinguish each environment type (arranged in order and starting from 1). Row 2 gives the name of the environment type (if defined). If the indicator for environment type names is 1 (Fig. 3), a valid name must be specified for each environment type. If the indicator for environment type names is 0, the place is left blank. Row 3 specifies the frequency of occurrence in the TPE. Each frequency should be equal to or greater than 0.0, and the sum of all frequencies should be equal to 1.0. In Fig. 3, the one environment type is given the name "Obregon", with a frequency of 1.0 in TPE.

## 2.3.3 Trait information

For the purpose of simulation, the genotypic value of an individual can be calculated from the definition of gene actions in the G×E system and from its genotypic combination. However, breeders select based on the phenotypic value in the field. Therefore, the phenotypic value of a genotype in a specific environment needs to be defined from its genotypic value and associated environmental errors. The trait information will allow QuLine to define the phenotypic value from the genotypic value. Major trait information required by the QU-GENE engine is the environmental effects on traits (within-plot variance and among-plot variance) in each environment type. Either the variance or individual plant-level heritability in the broad sense needs to be specified. For heritability, the QU-GENE engine will convert the specified heritability into an estimate of environmental variance based on the provided reference population. This environmental variance is used throughout the simulation. The population structure differs from generation to generation; hence, population heritability also varies with changes in the genetic variation within the population.

Each trait takes four rows in Fig. 3. Row 1 is an ID number to distinguish each trait. Row 2 gives the name of the trait (if defined). Row 3 specifies that heritability will be defined for within plot, among-plot error will be defined as a proportion of within-plot error variance. Row 4 specified heritability or error variance in "Obregon", depending on indicators of Row 3. In Fig. 3, the within-plot heritability is 0.4 for maturity, 0.3 for TKW, and 0.2 for yield. The engine will use a reference population to calculate with-in plot error variance of each trait. The among-plot error is defined as 1.0 of the within-plot error. If indicator 2 means the error variance will be given. If more environment types are defined, same information as shown in Row 3 is needed for each environment type.

```
! ******************************************************************************
! *** Environment Type Information ***
!       Row 1: Number
!       Row 2: Name (if defined)
!       Row 3: Frequency of occurrence in TPE
! ******************************************************************************

1
Obregon
1.000

! ******************************************************************************
! *** Trait Information ***
!       Row 1: Number
!       Row 2: Name (if defined)
!       Row 3: Error Specification Type (for within,among,mixture)
!              1=heritability (spb); 2=error
!       Row 4+: Within, Among, Mixture error [each ET]
! ******************************************************************************

1
Maturity
1     1      2
0.400 1.000  0.000

2
TKW
1     1      2
0.300 1.000  0.000

3
Yield
1     1      2
0.200 1.000  0.000
```

Fig. 3. Environment and trait definitions in QU-GENE

## 2.3.4 Gene information

Gene information is the most fundamental and complicated part in defining a G×E system. It is used to generate progeny genotypes from any crossing or propagation type, and to define the genotypic value of any genotype for each trait. It consists of the location of genes on wheat chromosomes, the number of alleles for each gene locus, the number of traits affected by each gene, the genotypic effects in each defined environment type, etc. Linkage, multiple alleles, pleiotropy, epistasis, and G×E interaction are all defined in this part.

Definition of three genes located on chromosome 5, and one on chromosome 6 was shown in Fig. 4. The following parameters define each gene (including markers) (Fig. 4). Row 1 is the locus ID number to distinguish each gene (arranged in order, and starting from 1). Please note all genes should be arranged in order starting from the first chromosome or linkage group. Genes in one chromosome or linkage group should also be arranged as they appear on the chromosome. Row 2 gives the name of the gene (if defined). If the indicator for gene name is 1, a valid name must be specified, or the place is left blank. Row 3 specifies chromosome, recombination frequency, number of alleles, and number of traits the gene affects. All the genes, including markers, in the GE system are supposed to be arranged in order on the chromosomes. Recombination frequency of a gene is the crossover rate between the gene and the gene located just before it (two flanking genes). If a gene is located at the beginning of a chromosome, its recombination frequency should be set at 0.5. Row 4 specifies name of each allele (if defined). If the indicator for allele names is 1, a valid name must be specified for each allele, or the place left blank. The number of rows used to define genetic effects of the gene depends on number of traits affected, and number of environments (Fig. 4). For each affecting trait and each environment, Column 5 specifies the trait ID the gene affects. Column 6 specifies the environment ID. Column 7 specifies the three genotype to phenotype (or gene effect) types, i.e., additive (including dominance), epistasis, and QU-GENE plug-in.

```
! ***********************************************************************
!                  Columns
! CH     RF    NA NT      WT   ET  GP  EF  Gene effects
! 1      2     3  4       5    6   7   8   9+
! ******************************************************************
         Locus ID number
  13                     Locus name        Chromosome ID number, recombination with previous locus,
  Mat5                                     number of alleles at the locus, and number of traits affecting
  5   0.5000   2  2
                         1    1   1  -1    0.0 3.000 0.000    Genetic effects for all
                         3    1   1  -1    0.0 0.100 0.000    affected traits in all
  14                                                          defined environments
  TKW5
  5   0.0500   2  2
                         2    1   1  -1    0.0 2.000 0.000
                         3    1   1  -1    0.0 0.100 0.000
  15
  Yld5
  5   0.2300   2  1
                         3    1   1  -1    0.0 0.100 0.000
  16
  Yld6
  6   0.5000   2  1
                         3    1   1  -1    0.0 0.100 0.000
```

Fig. 4. Gene definition in QU-GENE.

Column 8 specifies how gene effects are stored. For additive genes, value –1 means that midpoint (m), additive (a) and dominance (d) will be specified later. This option is only available for genes with two alleles. For a gene with multiple alleles, value 0 should be used.

Value 0 means that genotypic values in the order of AA, Aa, and aa, where A-a are the two alternative alleles on the gene locus. In case of three alleles, e.g. $A_1$, $A_2$ and $A_3$ at locus A, the genotypic values are arranged in the order of $A_1A_1$, $A_1A_2$, $A_1A_3$, $A_2A_2$, $A_2A_3$, and $A_3A_3$. The order is similar for more than three alleles at a gene locus. Value 1 means that random gene effects with no dominance. In the case of two alleles A and a, genotypes AA and aa have random values AA and aa ranged from 0.0 to 1.0, but the value (Aa) of genotype Aa is at the mid-point between AA and aa. Value 2 means that random gene effects with no over-dominance. In the case of two alleles A and a, genotypes AA, Aa and aa have random values ranged from 0.0 to 1.0, but Aa is between AA and aa. Value 3 means that random gene effects with partial/over-dominance. In the case of two alleles A and a, genotypes AA, Aa and aa have independent random values ranged from 0.0 to 1.0, which will result in either partial dominance or over-dominance, depending on chance.

For epistatic genes, a number is given for the epistatic network the gene is included. Genotypic values of all possible combination in an epistatic network will be defined at a later stage, once all genes in the network have been determined. For QU-GENE plugin genes, a number is given for the plugin the gene is included. If a gene is only a marker, the trait number has to be set at 0. Trait number 0 is reserved to identify which gene locus is a marker.

### 2.3.5 Definition of starting populations

In QU-GENE, a population can be defined by gene frequency, or by genotypes. Four populations are defined in Fig. 5, and the first population "Poperror" will be used as reference to translate heritability to error variance. The other three, i.e. Pop02, Pop05, and

```
4      ! Number of populations to create
1      ! Which population to use for error estimates

1
Poperror
100
1
    0   1   1   2   1   0     0.5000

2
Pop02
20
1
    0   1   1   2   1   0     0.2000

3
Pop05
20
1
    0   1   1   2   1   0     0.5000

4
Pop08
20
1
    0   1   1   2   1   0     0.8000
```

Fig. 5. Four populations defined in QU-GENE.

Pop08, have a size of 20, and allele frequencies 0.2, 0.5, and 0.8, respectively. Each population takes 5 rows. "0" at the beginning of row 5 represents frequencies of alleles at all loci are identical. Otherwise, each locus will take a row. Pop02, Pop05, and Pop08 will be used as the starting population in breeding simulation.

## 2.4 Definition of breeding strategies in QuLine

By defining breeding strategy, QuLine translates the complicated breeding process in a way that the computer can understand and simulate. QuLine allows for several breeding strategies, which were contained in one input file, to be defined simultaneously. The program then makes the same virtual crosses for all the defined strategies at the first breeding cycle. Hence, all strategies start from the same point (the same initial population, the same crosses and the same genotype and environment system), allowing appropriate comparison. A breeding strategy in QuLine is defined as all the crossing, seed propagation, and selection activities in an entire breeding cycle. For illustration, two breeding strategies, denoted by I-M, and II-M, are described in Fig. 6. Strategy I-M is similar to modified pedigree and bulk, where pedigree is used two times in $F_2$ and $F_5$ generations. Strategy II-M is similar to selected bulk, where pedigree is used only once in $F_5$ generation (Wang et al. 2003).

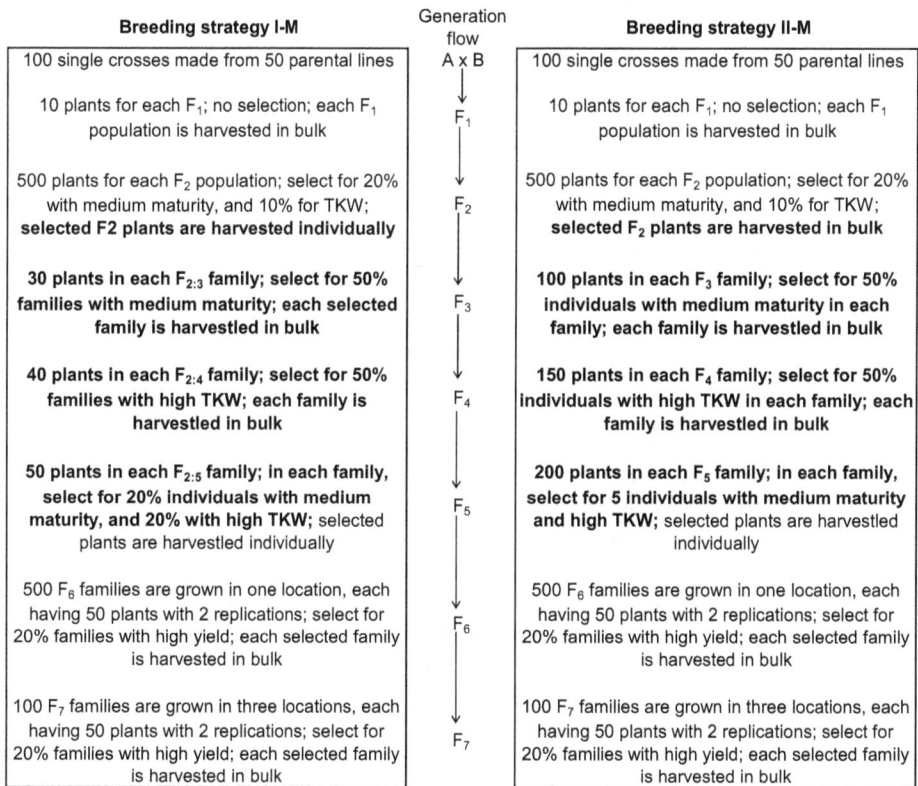

| Breeding strategy I-M | Generation flow | Breeding strategy II-M |
|---|---|---|
| 100 single crosses made from 50 parental lines | A x B | 100 single crosses made from 50 parental lines |
| 10 plants for each $F_1$; no selection; each $F_1$ population is harvested in bulk | $F_1$ | 10 plants for each $F_1$; no selection; each $F_1$ population is harvested in bulk |
| 500 plants for each $F_2$ population; select for 20% with medium maturity, and 10% for TKW; **selected F2 plants are harvested individually** | $F_2$ | 500 plants for each $F_2$ population; select for 20% with medium maturity, and 10% for TKW; **selected $F_2$ plants are harvested in bulk** |
| **30 plants in each $F_{2:3}$ family; select for 50% families with medium maturity; each selected family is harvestled in bulk** | $F_3$ | **100 plants in each $F_3$ family; select for 50% individuals with medium maturity in each family; each family is harvestled in bulk** |
| **40 plants in each $F_{2:4}$ family; select for 50% families with high TKW; each family is harvestled in bulk** | $F_4$ | **150 plants in each $F_4$ family; select for 50% individuals with high TKW in each family; each family is harvestled in bulk** |
| **50 plants in each $F_{2:5}$ family; in each family, select for 20% individuals with medium maturity, and 20% with high TKW;** selected plants are harvestled individually | $F_5$ | **200 plants in each $F_5$ family; in each family, select for 5 individuals with medium maturity and high TKW;** selected plants are harvestled individually |
| 500 $F_6$ families are grown in one location, each having 50 plants with 2 replications; select for 20% families with high yield; each selected family is harvested in bulk | $F_6$ | 500 $F_6$ families are grown in one location, each having 50 plants with 2 replications; select for 20% families with high yield; each selected family is harvested in bulk |
| 100 $F_7$ families are grown in three locations, each having 50 plants with 2 replications; select for 20% families with high yield; each selected family is harvested in bulk | $F_7$ | 100 $F_7$ families are grown in three locations, each having 50 plants with 2 replications; select for 20% families with high yield; each selected family is harvested in bulk |

Fig. 6. Planting and selection details in two plant breeding strategies I-M and II-M. Major difference between the two strategies was highlighted in bold.

### 2.4.1 General simulation information

Generation information specifies the number of strategies to be simulated or compared, number of simulation runs, number of breeding cycles, number of crosses to be made at the beginning of each breeding cycle, indicator for crossing block update, and indicators for outputting simulation results. Indicator 0 for crossing block update means that only the final selected lines will be used as the parents for next breeding cycle. The parents in the current crossing block will not be considered for crossing in the following cycles. Indicator 1 means that for the next cycle, some parents come from the current crossing block, and some from the final selected lines. A breeding cycle begins with crossing and ends at the generation when the selected advanced lines are returned to the crossing block, as new parents.

### 2.4.2 Number of generations and number of selection rounds in each generation

In the breeding program in Fig. 6, the best advanced lines developed from the $F_7$ generation will be returned to the crossing block to be used for new crosses. Therefore, the number of generations in one breeding cycle is seven for both strategies (Figs. 6, 7 and 8). The crossing block (viewed as $F_0$) and the seven generations need to be defined in QuLine. The parameters to define a generation consist of the number of selection rounds in the generation, an indicator for seed source (explained later), and the planting and selection details for each selection round. Most generations in plant breeding programs have just one selection round, but some generations may have more than one selection round (Wang et al. 2003). More rounds of selection also allow the selection on traits measured by seeds instead of plants grown in the field. All generations in Strategies I-M and II-M have one round (Figs. 7 and 8).

### 2.4.3 Seed propagation type for each selection round

The seed propagation type describes how the selected plants in a retained family, from the previous selection round or generation, are propagated, to generate the seed for the current selection round or generation. There are nine options for seed propagation, presented here in the order of increasing genetic diversity ($F_1$ excluded): (i) clone (asexual reproduction), (ii) DH (doubled haploid), (iii) self (self-pollination), (iv) single cross (single cross between two parents), (v) backcross (back crossed to one of the two parents), (vi) topcross (crossed to a third parent, also known as three-way cross), (vii) doublecross (crossed between two F1s), (viii) random (random mating among the selected plants in a family), and (ix) no selfing (random mating but self-pollination is eliminated). The seed for $F_1$ is derived from crossing among the parents in the initial population (or crossing block). QuLine randomly determines the female and the male parents for each cross from a defined initial population, or alternately, one may select some preferred parents from the crossing block. The selection criteria used to identify such preferred parents (grouped here as the male and female master lists) can be defined in terms of among-family and within-family" selection (see below for details) within the crossing block (referred to as $F_0$ generation). By using the parameter of seed propagation type, most, if not all methods of seed propagation in self-pollinated crops can be simulated in QuLine. Three seed propagation types are used in defining Strategies I-M, and II-M, which were clone, singlecross (only used for $F_1$ generation) and selfing (Figs. 7 and 8).

### 2.4.4 Generation advance method for each selection round

The generation advance method describes how the selected plants within a family are harvested. There are two options for this parameter: pedigree (the selected plants within a family are harvested individually, therefore each selected plant will result in a distinct family in the next generation), and bulk (the selected plants in a family are harvested in bulk, resulting in just one family in the next generation). This parameter and the seed propagation type allow QuLine to simulate not only the traditional breeding methods such as pedigree breeding and bulk population breeding, but also many combinations of different breeding methods. The bulk generation advance method will not change the number of families in the following generation if among-family selection is not applied in the current generation, whereas the pedigree method increases the number of families rapidly if among-family selection intensity is weak, and several plants are selected within each retained family. For a generation with more than one selection round, the generation advance method for the first selection round can be either pedigree or bulk. The subsequent selection rounds are used to determine which families derived from the first selection round will advance to the next generation. In the majority of cases, bulk generation advance is the preferred option for the subsequent selection rounds. It can be seen from Fig. 7 that pedigree is used in $F_2$ and $F_5$, and bulk is used in the other generations in Strategy I-M. In comparison, pedigree is used only in $F_5$ in Strategy II-M.

### 2.4.5 Field experimental design for each selection round

The parameters used to define the virtual field experimental design in each selection round include the number of replications for each family, the number of individual plants in each replication, the number of test locations, and the environment type for each test location (Figs. 7 and 8). Each environment type defined in the genotype and environment system has its own gene action and gene interaction, which provides the framework for defining the genotype by environment interaction. Therefore, by defining the target population of environments as a mixture of environment types, genotype by environment interactions are defined as a component of the genetic architecture of a trait.

### 2.4.6 Among-family selection and within-family selection for each selection round

Three traits have been defined before and now can be used in selection. There are two levels of selection in plant breeding, among-family and within-in family. The definition of these two types of selections is essentially the same: the number of traits to be selected is followed by the definition of each trait (Wang et al., 2004). Apart from the trait code there are two parameters that define a trait used in the selection, selection mode and selected amount. Selected amount can be a proportion of the number of families, individuals in selection, a threshold value or a specified number. The four options for defining selected proportions are (i) T (top), where the individuals or families with highest phenotypic values for the trait of interest will be selected; (ii) B (bottom), where the individuals or families with the lowest phenotypic values will be selected; (iii) M (middle), where individuals or families with medium trait phenotypic values will be selected; and (iv) R (random), where individuals or families will be randomly selected. The two options for defining threshold selection are (i) TV (top value), where the individuals or families whose phenotypic values are higher than the threshold will be selected; and (ii) BV (bottom value), where the individuals or families whose phenotypic values are lower than the threshold will be selected. The three options for defining number selection are (i) TN (top number), where a specified number of the individuals or families with highest phenotypic values for the trait of interest will be selected; (ii) BN (bottom number), where a specified number of the individuals or families with lowest phenotypic values for the trait of interest will be selected; and (iii) RN (random number), where a specified number of the individuals or families will be selected randomly. Independent culling is used if multiple traits are considered for among-family or within-family selection. If there is no among-family or within-family selection for a specific selection round, the number of selected traits is noted as 0. The traits for both among-family and within-family selections can be the same or different, as is the case for selected proportions. The traits for selection may also differ from generation to generation with the selected amounts for traits.

```
!*******************General information for the simulation experiment*********************************
!NumStr NumRun NumCyc NumCro CBUpdate OutGES OutPOP OutHIS OutROG OutCOE OutVar Cross   RMtimes PopSize
 2      5      10     100    0        0      0      0      0      0      0      random  0       0


!*******************Information for selection strategies to be simulated*****************************
!StrategyNumber StrategyName  NumGenerations
 1              StrategyI-M    7

!NR SS  GT     PT          GA         RP  PS   NL  ET...                                  Row 1
!                                              AT  (ID  SP   SM)...                        Row 2
!                                              WT  (ID  SP   SM)...                        Row 3
 1  0   CB     clone       bulk       1   1    1   1
                                              0
                                              0
 1  0   F1     singlecross bulk       1   10   1   1
                                              0
                                              0
 1  0   F2     self        pedigree   1   500  1   1
                                              0
                                              2   1  M  0.20   2  T  0.10
 1  0   F3     self        bulk       1   30   1   1
                                              1   1  M  0.50
                                              0
 1  0   F4     self        bulk       1   30   1   1
                                              1   2  T  0.50
                                              0
 1  0   F5     self        pedigree   1   50   1   1
                                              0
                                              2   1  M  0.20   2  T  0.20
 1  0   F6     self        bulk       2   50   1   1
                                              1   3  T  0.20
                                              0
 1  0   F7     self        bulk       2   50   3   1  1  1
                                              1   3  T  0.20
                                              0
```

Fig. 7. General simulation information and definition of strategy I-M in QuLine

Taking $F_2$ generation of Strategy I-M as an example, no among-family selection is conducted, but two traits are used for within-family selection, i.e. maturity (ID=1), and TKW (ID=2). Selection mode is M for maturity, and selected amount is 0.2, indicating 20% of the 500 F2 individuals (i.e. 100) with medium maturity will be first selected. Selection mode is T for TKW, and selected amount is 0.1, indicating 10% of the 100 retained $F_2$ individuals (i.e. 10) with highest TKW will be selected. The ten final selected $F_2$ individuals will be harvested individually, as "pedigree" is defined as the generation advance method (Fig. 7). For comparison, two other strategies were defined, where the selection mode is B for maturity, denoted by I-B and II-B. Other selection details are the same as those in I-M and II-B, respectively.

```
!******************Information for selection strategies to be simulated******************************
!StrategyNumber StrategyName  NumGenerations
 2              StrategyII-M    7

!NR SS  GT    PT          GA        RP  PS  NL  ET...                              Row 1
!                                          AT (ID  SP  SM)...                      Row 2
!                                          WT (ID  SP  SM)...                      Row 3
 1  0   CB    clone       bulk      1   1   1   1
                                            0
                                            0
 1  0   F1    singlecross bulk      1   10  1   1
                                            0
                                            0
 1  0   F2    self        bulk      1   500 1   1
                                            0
                                            2   1   M   0.20    2   T   0.10
 1  0   F3    self        bulk      1   50  1   1
                                            0
                                            1   1   M   0.50
 1  0   F4    self        bulk      1   50  1   1
                                            0
                                            1   2   T   0.50
 1  0   F5    self        pedigree  1   200 1   1
                                            0
                                            2   1   M   0.20    2   T 0.125
 1  0   F6    self        bulk      2   50  1   1
                                            1   3   T   0.20
                                            0
 1  0   F7    self        bulk      2   50  3   1   1   1
                                            1   3   T   0.20
                                            0
```

Fig. 8. Definition of strategy II-M in QuLine

## 2.5 Simulation experimental design

A G×E system called "WheatModel" (Figs. 2, 3, and 4), three starting populations called Pop02, Pop05, and Pop08 (Fig. 5), and four breeding strategies called I-M, II-M, I-B, and II-B (Figs. 7 and 8) have been defined previously. A total of 12 simulation experiments are designed (Table 1). Each experiment was repeated for 1000 times, and mean across the 100 times will be used to compare the efficiency of different strategies.

| Experiment | G×E system | Population | Breeding strategy |
|---|---|---|---|
| 1 | WheatModel | Pop02 | Strategy I-M |
| 2 | WheatModel | Pop02 | Strategy II-M |
| 3 | WheatModel | Pop05 | Strategy I-M |
| 4 | WheatModel | Pop05 | Strategy II-M |
| 5 | WheatModel | Pop08 | Strategy I-M |
| 6 | WheatModel | Pop08 | Strategy II-M |
| 7 | WheatModel | Pop02 | Strategy I-B |
| 8 | WheatModel | Pop02 | Strategy II-B |
| 9 | WheatModel | Pop05 | Strategy I-B |
| 10 | WheatModel | Pop05 | Strategy II-B |
| 11 | WheatModel | Pop08 | Strategy I-B |
| 12 | WheatModel | Pop08 | Strategy II-B |

Table 1. Designing a simulation experiment

## 3. Explanation of simulation results

Various kinds of information can be output by setting appropriate outputting indicators (Fig. 7). These information includes genetic variance, correlation among traits for each environment, correlation among environment for each trait, number of crosses retained after each round of selection, mean genotypic values, percentage of fixed genes for all traits and the percentage of fixed genes for each trait, gene frequency, Hamming distance, selection history, number of families, number of individual plants in each generation for each simulated strategy, etc. Not all outputs are required in any simulations.

### 3.1 Genetic gains from different breeding strategies

Table 2 clearly indicated that the genetic gain on yield from Strategy II was either equal to or higher than the genetic gain from Strategy I. For the starting population Pop02, yield is 4.20 t/ha in the parental population (Table 2). When families and individuals with medium maturity are selected in breeding (i.e. I-M and II-M), Strategy I increased yield to 8.35 t/ha after 10 cycles, and Strategy II to 8.44 t/ha. This is 1.08% higher than the yield from Strategy I. When short maturity is selected (i.e. I-B and II-B), Strategy I increased yield to 7.77 t/ha after 10 cycles, and Strategy II to 7.80 t/ha that is 0.34% higher than the yield from Strategy I. The difference between medium and short maturity selections is caused by the pleiotropic effects of maturity genes on yield (Figs. 1 and 4).

For the starting population Pop05, yield is 6.00 t/ha in the parental population (Table 2). When families and individuals with medium maturity are selected in breeding, Strategy I increased yield to 8.95 t/ha after 10 cycles and Strategy II increase it to 8.96 t/ha. When short maturity is selected, Strategy I increased yield to 8.20 t/ha after 10 cycles and Strategy II increased it to 8.26 t/ha. Difference of genetic gains from the two strategies is minor. For the starting population Pop08, yield is 7.80 t/ha in the parental population (Table 2). When families and individuals with medium maturity are selected in breeding, both strategies increased yield to 9.00 t/ha after 10 cycles. When short maturity is selected, Strategy I increased yield to 8.72 t/ha after 10 cycles and Strategy II increases it to 8.86 t/ha. That is 1.54% higher than the yield from Strategy I.

| Cycle | Pop02 | | | | Pop05 | | | | Pop08 | | | |
|-------|------|------|------|------|------|------|------|------|------|------|------|------|
|       | I-M  | II-M | I-B  | II-B | I-M  | II-M | I-B  | II-B | I-M  | II-M | I-B  | II-B |
| 0     | 4.20 | 4.20 | 4.20 | 4.20 | 6.00 | 6.00 | 6.00 | 6.00 | 7.80 | 7.80 | 7.80 | 7.80 |
| 1     | 5.04 | 5.05 | 4.95 | 4.92 | 7.06 | 7.08 | 6.90 | 6.92 | 8.50 | 8.49 | 8.40 | 8.40 |
| 2     | 5.80 | 5.84 | 5.62 | 5.60 | 7.67 | 7.70 | 7.39 | 7.42 | 8.77 | 8.76 | 8.58 | 8.62 |
| 3     | 6.45 | 6.52 | 6.21 | 6.21 | 8.11 | 8.14 | 7.73 | 7.76 | 8.90 | 8.90 | 8.66 | 8.72 |
| 4     | 6.97 | 7.04 | 6.67 | 6.68 | 8.42 | 8.45 | 7.94 | 7.97 | 8.96 | 8.96 | 8.70 | 8.78 |
| 5     | 7.38 | 7.45 | 7.03 | 7.04 | 8.64 | 8.66 | 8.07 | 8.10 | 8.99 | 8.98 | 8.71 | 8.81 |
| 6     | 7.71 | 7.78 | 7.30 | 7.31 | 8.78 | 8.79 | 8.14 | 8.17 | 9.00 | 8.99 | 8.72 | 8.83 |
| 7     | 7.96 | 8.04 | 7.50 | 7.52 | 8.87 | 8.88 | 8.18 | 8.21 | 9.00 | 9.00 | 8.72 | 8.84 |
| 8     | 8.14 | 8.23 | 7.64 | 7.66 | 8.92 | 8.93 | 8.19 | 8.24 | 9.00 | 9.00 | 8.72 | 8.85 |
| 9     | 8.27 | 8.35 | 7.72 | 7.75 | 8.94 | 8.95 | 8.20 | 8.25 | 9.00 | 9.00 | 8.72 | 8.86 |
| 10    | 8.35 | 8.44 | 7.77 | 7.80 | 8.95 | 8.96 | 8.20 | 8.26 | 9.00 | 9.00 | 8.72 | 8.86 |

Table 2. Genetic gains on yield from four breeding strategies and three starting populations

In simulation, genotypic value of an individual plant (denoted as F for fitness) in each environment type is calculated from the genetic effects defined in the G×E system. The adjusted genotypic value is define as

$$F_{ad} = \frac{F - TG_l}{TG_h - TG_l} \times 100 ,$$

where $TG_l$ and $TG_h$ are the genotypic values for the two extreme target genotypes with the lowest and the highest trait values in the G×E system, respectively. This standardization is useful specifically when diverse G×E systems are used to compare the performances of different breeding strategies. The adjusted genetic gains on the three traits were shown in Fig. 9 form medium maturity selection and in Fig. 10 for bottom maturity selection.

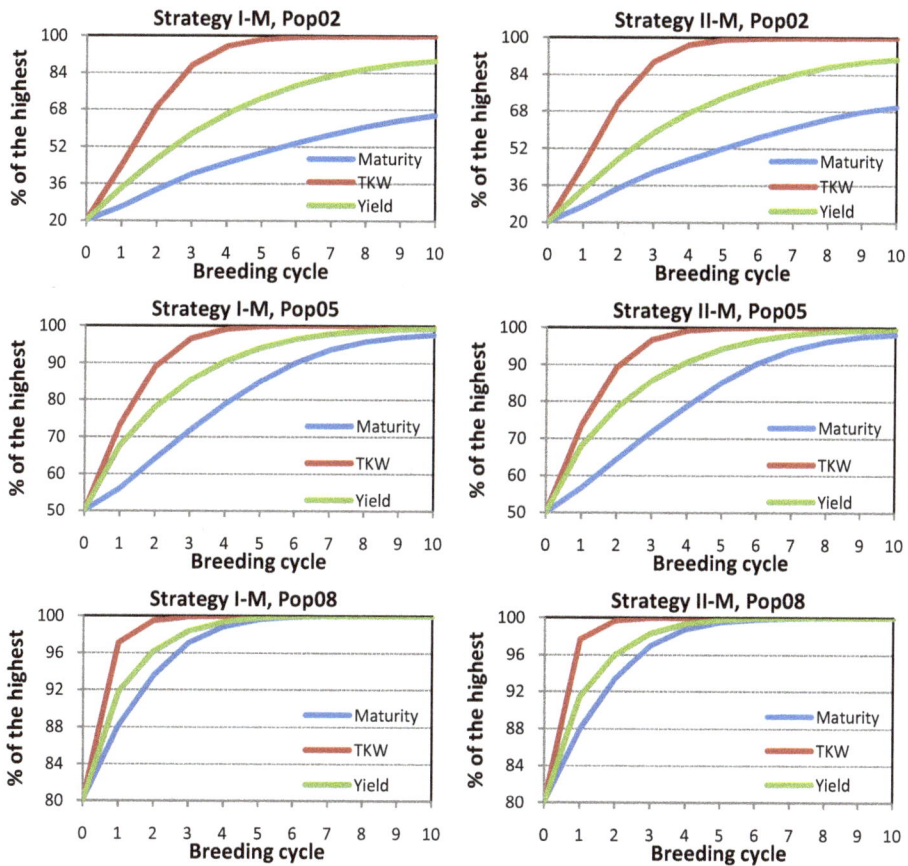

Fig. 9. Adjusted genetic gains from breeding strategies I-M and II-M

When medium maturity is selected (Fig. 9), TKW reaches to the highest value after 5 breeding cycles for Pop02, after 4 cycles for Pop05, and after 2 cycles for Pop08, for both strategies. TKW genes have pleiotropic effects on yield in definition (Figs. 1 and 4), and

TKW and yield were both selected for top performance (Figs. 6, 7 and 8). The selection on TKW and yield both helps increase the frequency of favourable TKW alleles. If there is no correlation between maturity and yield, maturity should keep unchanged. The increase in maturity is due to the selection for high yield. From the genetic model defined in Figs. 1 and 4, the longer the maturity, the higher the yield. Therefore, the selection for high yield retained the alleles of long maturity.

In practice, the breeders may want to select for short maturity cultivars. When short maturity is selected (Fig. 10), there is no much difference for TKW. For Pop02 and Pop05, both strategies reduced maturity. For Pop08, strategy I reduced maturity, but strategy II increased maturity slowly, indicating strategy II may result in less selection intensity on maturity.

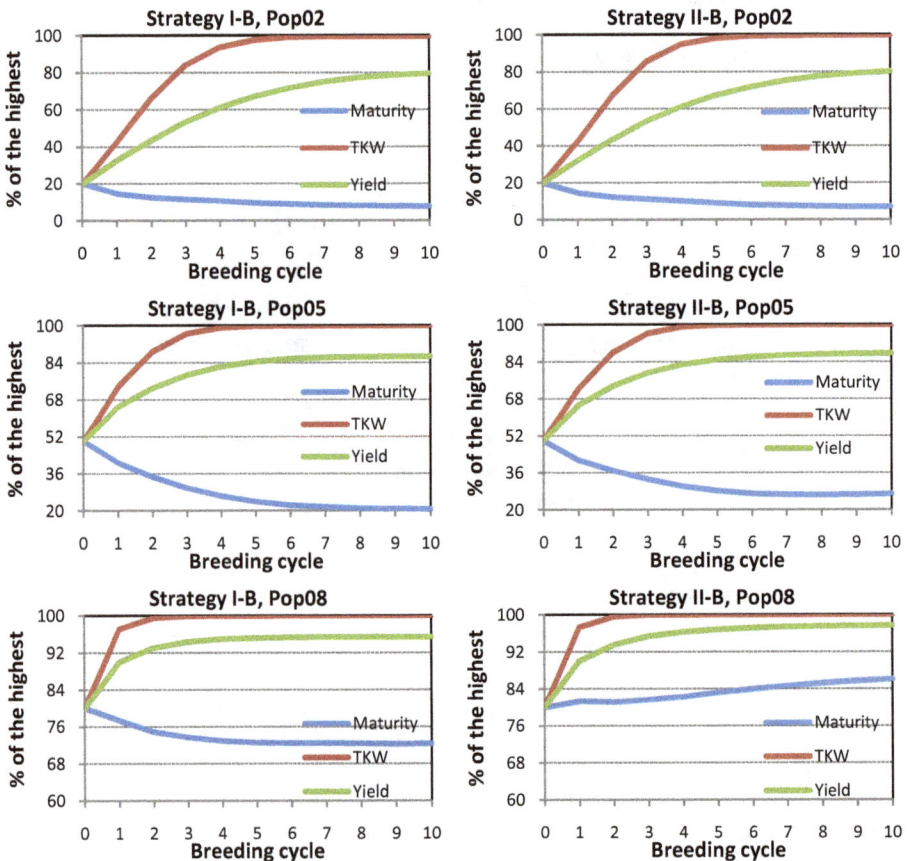

Fig. 10. Adjusted genetic gains from breeding strategies I-B and II-B

## 3.2 Cost and benefit analysis

Previous results showed that the genetic gain on yield from Strategy II was either equal to or higher than the genetic gain from Strategy I (Table 2, Figs. 9 and 10). How much cost will be needed to run each strategy? For this purpose, we compared the number of families and individual plants to be grown in the two strategies (Table 3). Less families means less seed lots to be prepared by labor for planting, and less individuals means less land to be used. In one breeding cycles, the number of families generated from strategy II is 43.14% of the number of families generated from strategy I. The number of plants to be grown in strategy II is 85.41% of the grown plant number in strategy I. Therefore when strategy II is used, fewer seed lots need to be handled at both harvest and sowing and less land is used, resulting in a significant saving in time, labor and cost.

| Gene-ration | Families before selection | | Plants before selection | | Families after selection | | Plants after selection | |
|---|---|---|---|---|---|---|---|---|
| | Strategy I | Strategy II | Strategy I | Strategy II | Strategy I | Strategy II | Strategy I | Strategy II |
| F1 | 100 | 100 | 1000 | 1000 | 100 | 100 | 1000 | 1000 |
| F2 | 100 | 100 | 50000 | 50000 | 1000 | 100 | 1000 | 1000 |
| F3 | 1000 | 100 | 30000 | 5000 | 500 | 100 | 15000 | 2500 |
| F4 | 500 | 100 | 15000 | 5000 | 250 | 100 | 7500 | 2500 |
| F5 | 250 | 100 | 12500 | 20000 | 500 | 500 | 500 | 500 |
| F6 | 500 | 500 | 50000 | 50000 | 100 | 100 | 10000 | 10000 |
| F7 | 100 | 100 | 30000 | 30000 | 20 | 20 | 6000 | 6000 |
| Total | 2550 | 1100 | 188500 | 161000 | | | | |

Table 3. Families and individual plants to be grown in strategies I and II

The simulation results (Tables 2 and 3; Figs. 9 and 10) clearly indicated that strategy II resulted in similar genetic gain on yield, but was more cost-effective compared with strategy I. Strategy I is called MODPED and II is called SELBLK in CIMMYT's wheat breeding. By applying bulk, we may not know which $F_2$, $F_3$ or $F_4$ individual derives which final fixed line, but parental lines deriving each fixed line are still known, which provides the most important information for the next cycle of breeding.

## 4. Conclusion

Conventional plant breeding largely depends on phenotypic selection and breeder's experience; therefore, the breeding efficiency is low and the predictions are inaccurate. Along with the fast development in molecular biology and biotechnology, a large amount of biological data are available for genetic studies of important breeding traits in plants, which in turn allows the conduction of genotypic selection in the breeding process. However, gene information has not been effectively used in crop improvement because of the lack of appropriate tools. The simulation approach can utilize the vast and diverse genetic information, predict the cross performance and compare different selection methods. Thus, the best performing crosses and effective breeding strategies can be identified. On the basis of the results from simulation experiments, breeders can optimize their breeding methodology and greatly improve the breeding efficiency.

On the other hand, a great amount of studies on QTL mapping have been conducted for various traits in plants and animals in recent years (Dekkers and Hospital, 2002; Peleman and Voort, 2003; Wang et al., 2005, 2007b, 2009). As the number of published genes and QTLs for various traits continues to increase, the challenge for plant breeders is to determine how to best utilize this multitude of information for the improvement of crop performance. Breeding simulation allows the definition of complicated genetic models consisting of multiple alleles, pleiotropy, epistasis, and genes by environment interaction that provides a useful tool to breeders, who can efficiently use the wide spectrum of genetic data and information available. This approach will be very helpful when the breeders want to compare breeding efficiencies from different selection strategies, to predict the cross performance with known gene information, and to investigate the efficient use of identified QTLs in conventional breeding.

## 5. Acknowledgments

Development of QuLine was originally funded by the Grains Research and Development Corporation (GRDC) of Australia (2000-2004). The continuous development of QuLine and the development of QuHybrid and QuMARS are funded by GCP (2005 to now) and HarvestPlus (2006 to now) Challenge Programs of CGIAR. Simulation tools described in this chapter are avaiable from www.uq.edu.au/lcafs/qugene/.

## 6. References

Allard, R.W. (1999). *Principles of plant breeding*, 2nd Edition, John Wiley & Sons, ISBN 0-471-02309-4, New York, USA

Bernardo, R. & Charcosset, A. (2006). Usefulness of gene information in marker-assisted recurrent selection: A simulation appraisal. *Crop Science*, Vol.46, No.2, (March 2006), pp. 614–621

Bonnett, D.G.; Rebetzke, G.J. & Spielmeyer, W. (2005). Strategies for efficient implementation of molecular markers in wheat breeding. *Molecular Breeding*, Vol.15, No.1, (January 2005), pp. 75-85

Dekkers, J.C.M. & Hospital, F. (2002). The use of molecular genetics in the improvement of agricultural populations. *Nature Review Genetics*, Vol.3, No.1, (January 2002), pp. 22-32

Dubcovsky, J. (2004). Marker-assisted selection in public breeding programs: the wheat experience. *Crop Science*, Vol.44, No.6, (December 2004), pp. 1895-1898

Falconer, D.S. & Mackay, T.F.C. (1996). Introduction to quantitative genetics, 4th edition, Longman, ISBN 0582-24302-5, Essenx, England

Hallauer, A.R.; Carena, M.J. & Miranda Filho, J.B. (1988). *Quantitative Genetics in Maize Breeding*, 2nd Edition, Springer, ISBN 978-1-4419-0765-3, New York, USA

Kempthorne, O. (1988). An overview of the field of quantitative genetics, In: *Proceedings of the 2nd International Conference on Quantitative Genetics*, B.S. Weir, E.J. Eisen, M.M. Goodman, G. Namkoong (eds), 47-56, Sinauer Associates, Sunderland, Massachusetts, USA

Lynch, M. & Walsh, B. (1998). *Genetics and analysis of quantitative genetics*, Sinauer Associates, ISBN 0-87893-481-2, Sunderland, Massachusetts, USA

Peleman, J.D. & Voort, J.R. (2003). Breeding by design. *Trends in Plant Science*, Vol.8, No.7, (July 2003), pp. 330-334

Podlich, D.W. & Cooper, M. (1998). QU-GENE: a platform for quantitative analysis of genetic models. *Bioinformatics*, Vol.14, No.7, (August 1998), pp. 632-653

Rajaram, S.; van Ginkel, M. & Fischer, R.A. (1994). CIMMYT's wheat breeding mega-environments (ME), *Proceedings of the 8th International Wheat Genetics Symposium*, pp. 1101-1106, China Agricultural Scientech, Beijing, China

Rajaram, S. (1999). Historical aspects and future challenges of an international wheat program. In: *Septoria and Stagonospora Diseases of Cereals: A Compilation of Global Research*, M. van Ginkel, A. McNab, and J. Krupinsky (eds.), 1-17, CIMMYT, D.F., Mexico

Singh, R. P.; Rajaram, S.; Miranda, A.; Huerta-Espino J. & Autrique, E. (1998). Comparison of two crossing and four selection schemes for yield, yield traits, and slow rusting resistance to leaf rust in wheat. *Euphytica*, Vol.100, No. 1, (May 1998), pp35-43

Singh, R.P. & Huerta-Espino, J. (2004). The use of "single-backcross selected-bulk" breeding approach for transferring minor genes based rust resistance into adapted cultivars, *Proceedings of 54th Australian Cereal Chemistry Conference and 11th Wheat Breeders Assembly*, Canberra, Australia, September21-24, 2004

van Ginkel, M.; Trethowan, R.; Ammar, K.; Wang, J. & Lillemo, M. (2002). Guide to bread wheat breeding at CIMMYT (Revised edition). *Wheat Program Special Report*, No.5, (Februrary 2002), CIMMYT, ISSN 0187-7787, D.F., Mexico

Wan, X.; Wan, J.; Su, C.; Wang, C.; Shen, W.; Li, J.; Wang, H.; Jiang, L.; Liu, S.; Chen, L.; Yasui, H. & Yoshimura, A. (2004). QTL detection for eating quality of cooked rice in a population of chromosome segment substitution lines. *Theoretical and Applied Genetics*, Vol.110, No.1, (December 2004), pp. 71-79

Wan, X.; Wan, J.; Weng, J.; Jiang, L.; Bi, J.; Wang, C. & Zhai, H. (2005). Stability of QTLs for rice grain dimension and endosperm chalkiness characteristics across eight environments. *Theoretical and Applied Genetics*, Vol.110, No.7, (May 2005), pp.1334-1346

Wang, J.; van Ginkel, M.; Podlich, D.; Ye, G.; Trethowan. R.; Pfeiffer, W.; DeLacy, I.H.; Cooper, M. & Rajaram, S. (2003). Comparison of two breeding strategies by computer simulation. *Crop Science* Vol.43, No.5, (September 2003), pp. 1764-1773

Wang, J.; van Ginkel, M.; Trethowan, R.; Ye, G.; DeLacy, I.H.; Podlich, D. & Cooper, M. (2004). Simulating the effects of dominance and epistasis on selecting response in the CIMMYT wheat breeding program using QuLine. *Crop Science* Vol.44, No.6, (November 2004), pp. 2006-2018

Wang, J.; Eagles, H.A.; Trethowan, R. & van Ginkel, M. (2005). Using computer simulation of the selection process and known gene information to assist in parental selection in wheat quality breeding. *Australian Journal of Agricultural Research*, Vol.56, No.5, (May 2005), pp. 465-473

Wang, J. & Pfeiffer, W. H. (2007). Simulation approach and its applications in plant breeding. *Agricultural Sciences in China*, Vol.6, No.8, (August 2007), pp. 908-921

Wang, J.; Chapman, S.C.; Bonnett, D.B.; Rebetzke, G.J. & Crouch, J. (2007a). Application of population genetic theory and simulation models to efficiently pyramid multiple genes via marker-assisted selection. *Crop Science*, Vol. 47, No.2, (March 2007), pp. 580-588

Wang, J.; Wan, X.; Li, H.; Pfeiffer, W.; Crouch, J. & Wan, J. (2007b). Application of identified QTL-marker associations in rice quality improvement through a design breeding approach. *Theoretical and Applied Genetics*, Vol.115, No.1, (June 2007), pp. 87-100

Wang, J.; Chapman, S. C.; Bonnett, D. G. & Rebetzke, G. J. (2009a). Simultaneous selection of major and minor genes: use of QTL to increase selection efficiency of coleoptile length of wheat (*Triticum aestivum* L.). *Theoretical and Applied Genetics*, Vol.119, No.1, (June 2009), pp. 65-74

Wang, J.; Singh, R.P.; Braun, H.-J. & Pfeiffer, W.H. (2009b). Investigating the efficiency of the single backcrossing breeding strategy through computer simulation. *Theoretical and Applied Genetics*, Vol.118, No.4, (february 2009), pp. 683-694

# Part 2

# Cytological Technologies

# Haploids and Doubled Haploids in Plant Breeding

Jana Murovec and Borut Bohanec
*University of Ljubljana, Biotechnical Faculty*
*Slovenia*

## 1. Introduction

Haploids are plants (sporophytes) that contain a gametic chromosome number (n). They can originate spontaneously in nature or as a result of various induction techniques. Spontaneous development of haploid plants has been known since 1922, when Blakeslee first described this phenomenon in *Datura stramonium* (Blakeslee et al., 1922); this was subsequently followed by similar reports in tobacco (*Nicotiana tabacum*), wheat (*Triticum aestivum*) and several other species (Forster et al., 2007). However, spontaneous occurrence is a rare event and therefore of limited practical value. The potential of haploidy for plant breeding arose in 1964 with the achievement of haploid embryo formation from *in vitro* culture of *Datura* anthers (Guha and Maheshwari, 1964, 1966), which was followed by successful *in vitro* haploid production in tobacco (Nitsch and Nitsch, 1969). Many attempts have been made since then, resulting in published protocols for over 250 plant species belonging to almost all families of the plant kingdom (reviewed in Maluszynski et al., 2003). In fact, under optimal conditions, doubled haploids (DH) have been routinely used in breeding for several decades, although their common use is still limited to selected species. There are several reasons for this. These might be categorized as *biological*, based on plant status (annual, biannual, perennial, authogamous, allogamous, vegetativelly propagated) and flower morphology or *technical*, which are the result of the feasibility and efficiency of DH induction protocol. Induction protocols substantially vary, in fact, not only among species but also among genotypes of the same species.

## 2. Production of haploids and doubled haploids

Haploids produced from diploid species (2n=2x), known as monoploids, contain only one set of chromosomes in the sporophytic phase (2n=x). They are smaller and exhibit a lower plant vigor compared to donor plants and are sterile due to the inability of their chromosomes to pair during meiosis. In order to propagate them through seed and to include them in breeding programs, their fertility has to be restored with spontaneous or induced chromosome doubling. The obtained DHs are homozygous at all loci and can represent a new variety (self-pollinated crops) or parental inbred line for the production of hybrid varieties (cross-pollinated crops). In fact, cross pollinated species often express a high degree of inbreeding depression. For these species, the induction process *per se* can serve not only as a fast method for the production of homozygous lines but also as a selection tool for the elimination of genotypes expressing strong inbreeding depression. Selection can be expected for traits caused by recessive deleterious genes that are associated with vegetative growth. Traits associated with flower fertility might not be related and should be eliminated by recurrent selection among DH lines.

The production of pure lines using doubled haploids has several advantages over conventional methods. Using DH production systems, homozygosity is achieved in one generation, eliminating the need for several generations of self-pollination. The time saving is substantial, particularly in biennial crops and in crops with a long juvenile period. For self incompatible species, dioecious species and species that suffer from inbreeding depression due to self-pollination, haploidy may be the only way to develop inbred lines.

The induction of DH lines in dioecious plants, in which sex is determined by a regulating gene, has an additional advantage. Such a case is well studied in asparagus, in which sex dimorphism is determined by a dominant gene *M*. Female plants are homozygous for the recessive alleles (*mm*), while male plants are heterozygous (*Mm*). Androgenically produced DH lines are therefore female (*mm*) or 'supermale' (*MM*). An advantage of supermales is that, when used as the pollinating line, all hybrid progeny are male.

Haploids from polyploid species have more than one set of chromosomes and are polyhaploids; for example dihaploids (2n=2x) from tetraploid potato (*Solanum tuberosum* ssp. *tuberosum*, 2n=4x), trihaploids (2n=3x) from heksaploid kiwifruit (*Actinidia deliciosa*, 2n=6x) etc. Dihaploids and trihaploids are not homozygous like doubled haploids, because they contain more than one set of chromosomes. They cannot be used as true-breeding lines but they enable the breeding of polyploid species at the diploid level and crossings with related cultivated or wild diploid species carrying genes of interest.

The main factors affecting haploid induction and subsequent regeneration of embryos are:

- genotype of the donor plants,
- physiological condition of donor plants (i. e. growth at lower temperature and high illumination),
- developmental stage of gametes, microspores and ovules,
- pre-treatment (i. e. cold treatment of inflorescences prior to culture, hot treatment of cultured microspores)
- composition of the culture medium (including culture on "starvation" medium low with carbohydrates and/or macro elements followed by transfer to normal regeneration medium specific to the species),
- physical factors during tissue culture (light, temperature).

## 3. Haploid techniques

### 3.1 Induction of maternal haploids

### 3.1.1 *In situ* induction of maternal haploids

*In situ* induction of maternal haploids can be initiated by pollination with pollen of the same species (e.g., maize), pollination with irradiated pollen, pollination with pollen of a wild relative (e.g., barley, potato) or unrelated species (e.g., wheat). Pollination can be followed by fertilization of the egg cell and development of a hybrid embryo, in which paternal chromosome elimination occurs in early embryogenesis or fertilization of the egg cell does not occur, and the development of the haploid embryo is triggered by pollination of polar nuclei and the development of endosperm.

### Pollination with pollen of the same species

Maternal haploid induction in **maize** (*Zea mays* L.) is a result of legitimate crossing within one species with selected inducing genotypes (line, single cross or population). It results in a majority of regular hybrid embryos and a smaller proportion of haploid maternal embryos with normal triploid endosperms. The first recognized inducer line was the genetic strain Stock 6, with an haploid induction rate of up to 2.3% (Coe, 1959), which was subsequently improved by hybridization and further selection. Today, modern haploid inducing lines display high induction rates of 8 to 10% (Geiger & Gordillo, 2009). They are routinely used in commercial DH-line breeding programs due to their high effectiveness and lower genotype dependence. In contrast to other induction techniques, no *in vitro* culture is needed, since kernels containing haploid embryos display a normal germination rate and lead to viable haploid seedlings. Haploid embryos can be selected early in the breeding process, based on morphological and physiological markers.

Pollination with **irradiated pollen** is another possibility for inducing the formation of maternal haploids using intra-specific pollination. Embryo development is stimulated by pollen germination on the stigma and growth of the pollen tube within the style, although irradiated pollen is unable to fertilize the egg cell. It has been used successfully in several species (Table 1).

| Species | Reference |
| --- | --- |
| apple | Zhang & Lespinasse, 1991; Hofer & Lespinasse, 1996; De Witte & Keulemans, 1994 |
| blackberry | Naess et al., 1998 |
| carnation | Sato et al., 2000 |
| cucumber | Przyborowski & Niemirowicz-Szczytt, 1994; Faris et al., 1999; Faris & Niemirowicz-Szczytt, 1999; Claveria et al., 2005 |
| European plum | Peixe et al., 2000 |
| kiwifruit | Pandey et. al., 1990; Chalak & Legave, 1997; Musial & Przywara, 1998, 1999 |
| mandarin | Froelicher et al., 2007; Aleza et al., 2009 |
| melon | Sauton & Dumas de Vaulx, 1987; Cuny et al., 1993; Lotfi et al., 2003 |
| onion | Dore & Marie, 1993 |
| pear | Bouvier et al., 1993 |
| petunia | Raquin, 1985 |
| rose | Meynet et al., 1994 |
| species of the genus *Nicotiana* | Pandey, 1980; Pandey & Phung, 1982 |
| squash | Kurtar et al. 2002 |
| sunflower | Todorova et al. 1997 |
| sweet cherry | Höfer & Grafe, 2003 |
| watermelon | Sari et al., 1994 |

Table 1. Induction of haploid plants by pollination with irradiated pollen

The production of maternal haploids stimulated by irradiated pollen requires efficient emasculation, which has in some cases been shown to limit its use because the method is too laborious. To overcome such an obstacle in onion, for instance, only male sterile donor plants were used as donor plants, but such lines, possessing cytoplasmically inherited male sterility, are of very limited practical use. Apart from the factors affecting haploid production already mentioned, the dose of irradiation is the main factor controlling *in situ* haploid production. At lower doses, the generative nucleus is partly damaged and therefore maintains its capacity to fertilize the egg cell. It results in large numbers of obtained embryos but all of hybrid origin and abnormal (mutant) phenotype. An increase in the irradiation dose causes a decrease in the total number of developed embryos but the obtained regenerants are mostly of haploid origin.

For most plant species, *in vitro* embryo rescue is necessary to recover haploid plants. The collection of mature seeds has only been reported for kiwifruit (Pandey et al., 1990; Chalak & Legave, 1997), onion (Dore & Marie, 1993), mandarin (Froelicher et al., 2007) and species of the genus *Nicotiana* (Pandey & Phung, 1982). Even for the aforementioned species, *in vitro* germination of seeds enhanced the recovery of haploid plants.

**Wide hybridization**

Wide crossing between species has been shown to be a very effective method for haploid induction and has been used successfully in several cultivated species. It exploits haploidy from the female gametic line and involves both inter-specific and inter-generic pollinations. The fertilization of polar nuclei and production of functional endosperm can trigger the parthenogenetic development of haploid embryos, which mature normally and are propagated through seeds (e.g., potato). In other cases, fertilization of ovules is followed by paternal chromosome elimination in hybrid embryos. The endosperms are absent or poorly developed, so embryo rescue and further *in vitro* culture of embryos are needed (e.g., barley).

In **barley,** haploid production is the result of wide hybridization between cultivated barley (*Hordeum vulgare*, 2n=2x=14) as the female and wild *H. bulbosum* (2n=2x=14) as the male. After fertilization, a hybrid embryo containing the chromosomes of both parents is produced. During early embryogenesis, chromosomes of the wild relative are preferentially eliminated from the cells of developing embryo, leading to the formation of a haploid embryo, which is due to the failure of endosperm development. A haploid embryo is later extracted and grown *in vitro*. The 'bulbosum' method was the first haploid induction method to produce large numbers of haploids across most genotypes and quickly entered into breeding programs. Pollination with maize pollen could also be used for the production of haploid barley plants, but at lower frequencies.

Paternal chromosome elimination has also been observed after interspecific crosses between **wheat** (*Triticum aestivum*) and maize. After pollination, a hybrid embryo between wheat and maize develops but, in the further process, the maize chromosomes are eliminated so that haploid wheat plantlets can be obtained. Such haploid wheat embryos usually cannot develop further when left on the plant, because the endosperm fails to develop in such seeds. By applying growth regulator 2,4-dichlorophenoxyacetic acid *in planta*, embryo growth is maintained to the stage suitable for embryo isolation and further *in vitro* culture. The maize chromosome elimination system in wheat enables the production of large numbers of haploids from any genotype. Pollination with maize is also effective for inducing haploid embryos in several other cereals, such as barley, triticale (x *Triticosecale*), rye (*Secale cereale*) and oats (*Avena sativa*) (Wędzony, 2009). Similar processes of paternal chromosome elimination occur after the pollination of wheat with wild barley (*H. bulbosum*), sorghum (*Sorghum bicolour* L. Moench) and pearl millet (*Pennisetum glaucum* L.R.Br.; Inagaki, 2003). In contrast to maize and pearl millet pollination, pollination with *H. bulbosum* is strongly influenced by the maternal genotype.

Haploid production in cultivated **potato** (*Solanum tuberosum* L. ssp. *tuberosum*, 2n=4x) can be achieved by inter-generic pollination with selected haploid inducer clones of *S. phureja* (2n=2x). The tetraploid female *S. tuberosum* produces an embryo sac containing one egg cell and two endosperm nuclei, all with the genetic constitution n=2x, while the diploid pollinator *S. phurea* produces two sperms of the genetic constitution n=x or 2x. After pollination, dihaploid (2n=2x) embryos can develop from un-fertilized egg cells, which are supported by a 6x endosperm formed by the fusion of polar nuclei with both reduced sperm cells. The frequency of dihaploid seeds is low; they have to be selected from hybrid seeds containing 3x or 4x embryos developed from egg cells (n=2x) fertilized with haploid (n=x) or diploid (n=2x) sperm cells. (Maine, 2003). Dihaploid potatoes can be used for breeding purposes, including alien germplasm introgression or selection at the diploid level, but such plants are not homozygous. Haploids have a significant role in potato breeding programs of quite a few companies, since they enable interspecific hybridization, which would not otherwise be possible due to differences in ploidy levels and endosperm balance numbers. The gene pool of potato can be broadened and certain valuable traits, such as disease resistance characters from the wild solanaceous species, can be more efficiently introgressed into cultivated potato (Rokka, 2009).

### 3.1.2 *In vitro* induction of maternal haploids - gynogenesis

*In vitro* induction of maternal haploids, so-called gynogenesis, is another pathway to the production of haploid embryos exclusively from a female gametophyte. It can be achieved with the *in vitro* culture of various un-pollinated flower parts, such as ovules, placenta attached ovules, ovaries or whole flower buds. Although gynogenetic regenerants show higher genetic stability and a lower rate of albino plants compared to androgenetic ones, gynogenesis is used mainly in plants in which other induction techniques, such as androgenesis and the pollination methods above described, have failed. Gynogenic induction using un-pollinated flower parts has been successful in several species, such as onion, sugar beet, cucumber, squash, gerbera, sunflower, wheat, barley etc. (for a detailed list and protocols overview, see Bohanec, 2009 and Chen et al., 2011) but its application in breeding is mainly restricted to onion and sugar beet.

The success of the method and its efficiency is greatly influenced by several biotic and abiotic factors. The genotype of donor plants, combined with growth conditions, is the crucial factor. In onion, for example, pronounced differences in embryo yields have been recorded among accessions and among plants within accessions. The average frequencies of induced embryos (calculated from ovaries possessing 6 ovules) varied between 0% in non-responding accessions to 18.6-22.6% in extremely responsive accessions, with individual donor plants producing up to 51.7% embryos. The high haploid production frequency was tested in two consecutive years and showed to be stable over years (Bohanec & Jakše, 1999).

Induction rates were even higher in preselected onion genotypes, achieving frequencies of 196.5%embryos from a doubled haploid line (Javornik et al., 1998) or 82.2% for an inbred line (Bohanec, 2003).

Fig. 1. Production of onion haploid plants with *in vitro* gynogenesis. (A) *In vitro* culture of un-pollinated flower buds on BDS medium (Dunstan and Short, 1977) supplemented with 500 mg/l myo-inositol, 200 mg/l proline, 2 mg/l BAP, 2 mg/l 2,4-D, 100 g/l sucrose and 7 g/l agar; (B) germination of haploid embryos after 60 to 180 days in culture; (C) elongation of haploid plantlets and (D) acclimatization of haploid plants in the greenhouse.

Developmental stage of gametes, the pre-treatment of flower buds prior to inoculation, *in vitro* culture media and culture conditions are other factors affecting the embryogenic response of gametes in culture. The female gametophyte is usually immature at inoculation and, in contrast to androgenesis, its development continues during *in vitro* culture, leading to a mature embryo sac (Musial et al., 2005). Mature embryo sacs contain several haploid cells theoretically capable of forming haploid embryos, such as the egg cell, synergids, antipodal cells and non-fused polar nuclei. However, under optimal conditions, the egg cells in most gynogenetic responsive species undergo sporophytic development (haploid parthenogenesis) (Bohanec, 2009). They can develop into haploid plants directly, avoiding the risk of gametoclonal variation, or through an intermediate callus phase.

Media components, mainly the type and concentration of carbohydrates and plant growth regulators, play an important role in reprogramming haploid cells from gametophytic to the sporophytic pathway. The requirements are species and genotype dependent and no universal protocol for *in vitro* gynogenesis exists.

## 3.2 Induction of paternal haploids - androgenesis

Androgenesis is the process of induction and regeneration of haploids and double haploids originating from male gametic cells. Due to its high effectiveness and applicability in numerous plant species, it has outstanding potential for plant breeding and commercial exploitation of DH. It is well established for plant breeding, genetic studies and/or induced mutations of many plant species, including barley, wheat, maize, rice, triticale, rye, tobacco, rapeseed, other plants from *Brassica* and other genera (for protocols, see Maluszynski et al., 2003). Its major drawbacks are high genotype dependency within species and the recalcitrance of some important agricultural species, such as woody plants, leguminous plants and the model plant *Arabidopsis thaliana*. The method relies on the ability of microspores and immature pollen grains to convert their developmental pathway from gametophytic (leading to mature pollen grain) to sporophytic, resulting in cell division at a haploid level followed by formation of calluses or embryos.

Androgenesis can be induced with *in vitro* culture of immature anthers, a technically simple method consisting of surface sterilization of pre-treated flower buds and subsequent excision of anthers under aseptic conditions. The anthers are inoculated and cultured *in vitro* on solid, semi-solid or liquid mediums or two-phase systems (liquid medium overlaying an agar-solidified medium). Anther culture was the first discovered haploid inducing technique of which efficiency was sufficient for plant breeding purposes (Maluszynski et al., 2003). It is still widely used, although isolated microspore culture is an improved alternative. During isolation of microspores, the anther wall tissues are removed, thus preventing interference of maternal sporophytic tissue during pollen embryogenesis and regeneration from somatic tissue. Moreover, basic research of haploid embryogenesis can be performed directly at the cellular, physiological, biochemical and molecular levels.

Androgenesis, like other haploid inducing techniques, is influenced by several biotic and abiotic factors. The developmental stage of male gametes at the time of anther or microspore isolation, in combination with suitable stress treatments, are the main factors determining the androgenetic response. It can be triggered within a relatively wide developmental window around the first pollen mitosis, when uninucleate microspores divide asymmetrically resulting in a generative cell embedded in a vegetive cytoplasm (Touraev et al., 1997; Maraschin et al., 2005).

The application of suitable physiochemical factors promotes a stress response, which arrests the microspores or young pollen grains in their gametophytic pathway. Their development is triggered through embryogenesis by promoting cell divisions and the formation of multicellular structures contained by the exine wall. Finally, the embryo-like structures are released from the exine wall (Maraschin et al., 2005). The most widely used triggering factors are temperature pre-treatment, sucrose and nitrogen starvation and osmotic stress. Depending on the plant species and genotype, temperature stress can be applied by subjecting excised flower buds, whole inflorescences or excised anthers to low (barley, wheat, maize, rice, triticale, rye) or high (rapeseed, *Brassica* species, tobacco, wheat) temperatures for several hours or days. As demonstrated in rapeseed and tobacco, different triggering factors can promote embryogenesis from microspores or immature pollen cells at different developmental stages. In rapeseed, early binucleate pollen grains can be converted to the embryologic pathway by applying a heat shock treatment at 32°C, while late binucleate pollen needs an extra heat shock treatment at 41°C (Maraschin et al., 2005). In tobacco, a heat shock treatment is effective in triggering unicellular microspores but not in triggering immature bicellular pollen grains, which successfully start embryogenesis after sucrose and nitrogen starvation (Touraev et al., 1997).

Several other triggering factors such as irradiation, colchicine, auxin and water stress are also used for reprogramming microspores, but to a limited extent. The androgenetic response can be enhanced by *in vivo* pre-treatments of donor plants with nitrogen starvation, short days and low temperature culture conditions.

In addition to stress treatments, the majority of studies have focused on culture media constituents. In general, the concentration of salts in the culture media is lower compared to micropropagation media, but there is no general rule. Several commonly used recipes of macro and micro elements are often used, such as potato-2 (Ying, 1986), W-14, (Jia et al., 1994), NLN (Lichter, 1982), A2 medium (Touraev et al., 1996) and others. The same media are often used for systematically diverse taxa, for instance NLN medium developed for *Brassica napus* was also efficient in *Apiaceae* (Ferrie et al., 2011), while A2 medium developed

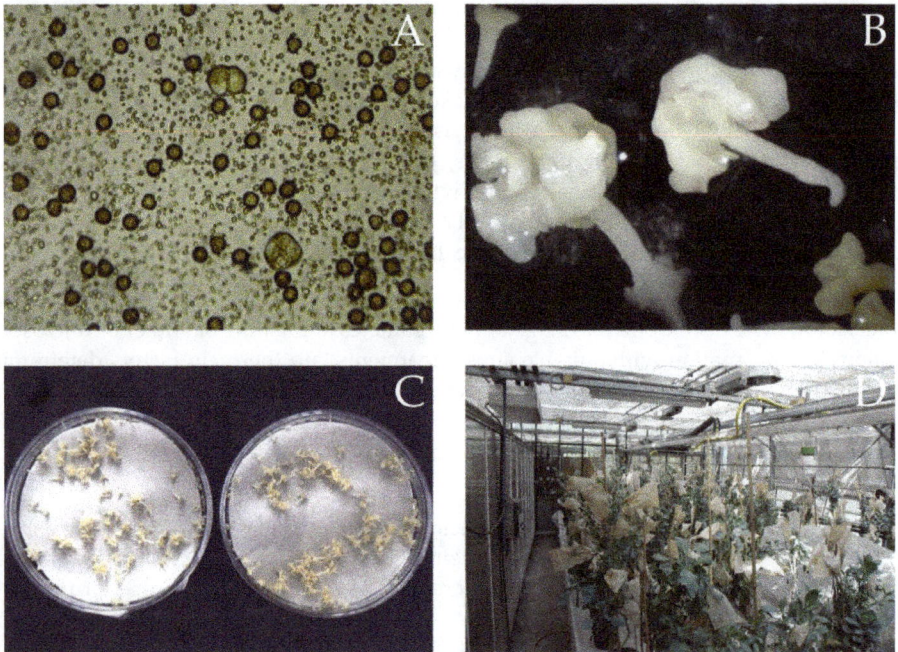

Fig. 2. Microspore culture of cabbage: (A) first divisions of microspores in NLN medium, (B) regenerated embryos, (C) embryos at desiccation treatment needed for regrowth, (D) selfing of DH lines

for tobacco was also optimal for wheat. Occasionally, even an increased concentration of micronutrients might be of high value, for instance elevated copper concentration reduced albinism in cereals (Jacquard et al., 2009). Choices and concentrations of carbohydrates are often essential. The most commonly used carbohydrate is sucrose, particularly in microspore media, and is added in high concentrations (i.e., 13%), while substitution of sucrose by maltose (Hunter et al., 1988) has been an important innovation, first discovered for barley anther culture. Although not always required, plant growth regulators might be essential. The influence of all groups of growth regulators has been tested, with positive effects of polyamines being among the latest studied.

Under optimal *in vitro* culturing conditions, androgenetic plants are regenerated from embryo-like structures (direct microspore embryogenesis) or from microsporial callus cells (organogenesis). Direct embryogenesis is preferred, since regeneration through the callus stage might induce undesired gametoclonal variation and might also cause albinism.

## 4. Identification of (doubled) haploids: ploidy level determination and homozygosity testing

After a successful haploid induction and the regeneration procedure, evaluation of regenerants is needed to distinguish between desired haploids (or spontaneously doubled haploids) and redundant heterozygous diploids. The haploid inducing techniques presented here differ in haploid induction rates and in the type of undesired regenerants which can be obtained. In this regard, isolated microspore culture is superior to other techniques because the filtering of microspore suspensions during isolation for the most part prevents diploid plant residuals from entering into the *in vitro* culture and later the regeneration of heterozygotes. Moreover, a relatively high proportion of induced haploids spontaneously double their chromosome number, leading to regeneration of homozygous doubled haploids. The phenomenon of spontaneous diploidization of microspore derived embryos has been studied in more detail in barley. The study revealed that, in microspores pre-treated with mannitol, chromosome doubling is caused by nuclear fusion after the first nuclear division (Kasha et al., 2001). In contrast to microspore culture and androgenesis in general, gynogenic regeneration leads to predominantly haploid regenerants in the majority of species.

During the production of homozygous lines, various undesired heterozygous plantlets can be obtained. In anther culture and *in vitro* gynogenesis, such plants can be regenerated from the somatic tissue of inoculated plant organs such as anther wall cells, somatic cells of flower buds, ovaries or ovules. Moreover, heterozygous hybrids, produced after wide hybridization or after pollination with irradiated pollen, are another class of regenerants that could negatively affect breeding progress, if not discarded. Reliable and fast selection of regenerants is therefore necessary before further employment of putative haploids and doubled haploids.

Several direct and indirect approaches are available for determining the ploidy level of regenerated plants. Indirect approaches are based on comparisons between regenerated and donor plants in terms of plant morphology (plant height, leaf dimensions and flower morphology), plant vigor and fertility, number of chloroplasts and their size in stomatal guard cells. They are fairly unreliable and subject to environmental effects but do not require costly equipment. Direct methods for ploidy determination are more robust and reliable and include conventional cytological techniques, such as counting the chromosome number in root tip cells (for a protocol, see Maluszynska, 2003) and measurement of DNA content using flow cytometry (for a protocol, see Bohanec, 2003). The latter provides a rapid and simple option for large-scale ploidy determination as early as in the *in vitro* culturing phase. It also enables detection of mixoploid regenerants (having cells with different ploidy) and the determination of their proportion.

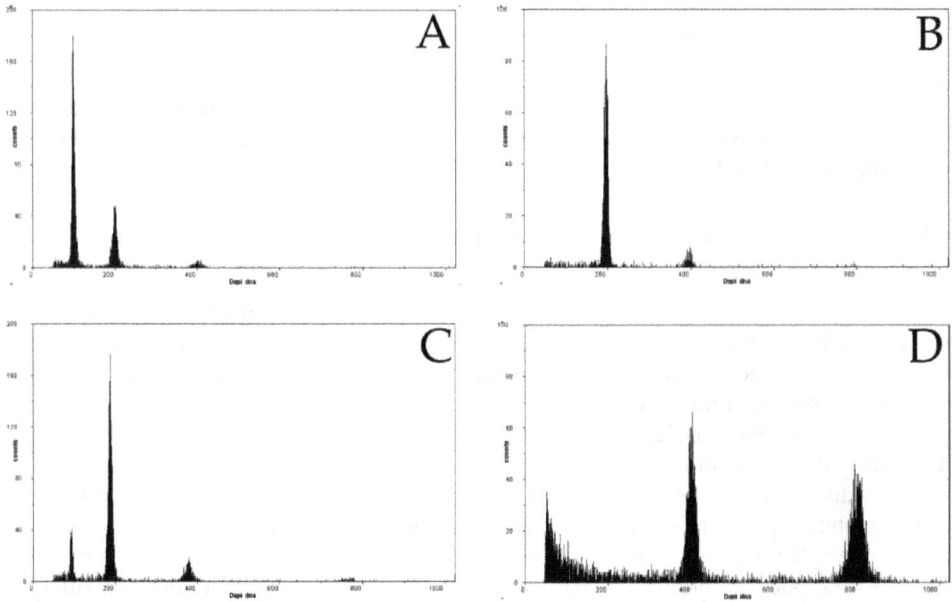

Fig. 3. Determination of ploidy using flow cytometry according to position of peaks representing size of nuclei as determined for microspore derived regenerants of rocket, *Eruca sativa*; (A) haploid, (B) diploid, (C) mixoploid and (D) tetraploid. Note: position of the first peak on the left determines the ploidy, the rest are G2 nuclear stages, G1 and G2 in mixoploids or endoreduplicated nuclei.

Regeneration of diploid plants is not always caused by unwanted adventitious regeneration from somatic cells or germination of hybrid embryos. Spontaneously doubled haploids may also occur, thus eliminating the need for chromosome doubling (see next section). Several markers can be used for assessing the origin of diploids, depending on their availability for a particular plant species. In the past, evaluation of regenerants mainly relied on phenotypic markers, progeny testing after self-pollination and isozyme analysis. Nowadays, DNA molecular markers, such as AFLP (Amplified Fragment Length Polymorphism), RAPD (Random Amplified Polymorphic DNA), SCAR (Sequence Characterized Amplified Regions) or SST (Simple Sequence Repeat), are commonly used for homozygosity testing and assessment of plant origin. There is a considerable difference in interpretation between dominant or co-dominant electrophoretic profiles. Co-dominant molecular markers, as well as isozyme markers, have the advantage that a single locus, when heterozygous in donor plants, might be used for homozygosity determination. In contrast, a more complex profile is analyzed with dominant markers. In such a case, bands missing from the donor profile indirectly indicate homozygosity.

A fast and reliable haploid identification method is needed for large scale production of DHs. Morphological markers expressed at the embryo, seed or early seedling stages are preferentially used. In **maize**, the most efficient haploid identification marker is the 'red crown' kernel trait, which causes deep pigmentation of the aleurone layer in the crown region (endosperm) and scutellum (embryo tissue) (Geiger & Gordillo, 2009). In a haploid inducing cross, the marker should be homozygous recessive in the female parent and homozygous dominant in the pollinator inducer line. After pollination, kernels with a red aleurone crown (resulting from regular fertilization of polar nuclei) containing a non-pigmented scutellum are visually selected from the hybrid kernel of regular fertilization with both aleurone and scutellum pigmented. A similar approach is used in **potato,** in which selection is based on a homozygous dominant color marker gene carried by the pollinator line (Maine, 2003). The purple spot embryo marker shows up on seeds whose embryos possess a genome from the pollinator. Those hybrid seeds are discarded, while spotless dihaploid seeds are included in breeding process. Selection can be repeated at the seedling stage, when a purple nodal band can be detected on the hybrid's stem. In the case of both maize and potato selectable markers, it is not possible to distinguish hybrid seeds resulting from unintentional self-pollination of donor plants. Selection has to be supplemented with other morphological or molecular markers.

## 5. Chromosome doubling

Following regeneration, haploid plants obtained from either anther or ovule culture may grow normally under *in vitro* conditions or can even be acclimatized to form vital mature plants. Such plants often express reduced vigor but in some crops such as onion, even haploid plants might grow vigorously. At the flowering stage, haploid plants form inflorescences with evident malformations. Due to the absence of one set of homologous chromosomes, meiosis cannot occur, so there is no seed set. Duplication of the chromosome complement is therefore necessary.

As described above, in pollen derived plants, spontaneous duplication of chromosomes may occur in cultures, often in a sufficient proportion, thus eliminating the need for chemically induced doubling. Spontaneously doubled plants are sometimes preferred because of the fear that the duplication process might induce undesired mutations. Mechanisms of spontaneous doubling differ, with nuclear fusion being the most common cause. As first described by Sunderland et al. (1974), synchronous division of two or more nuclei in early stages of embryo development might develop a common spindle. The nuclear fusion theory is supported by the frequent occurrence of a small proportion of triploid regenerants. Nuclear fusions might be associated with delayed cell wall formation, which, as reviewed by Kasha (2005), is typical of cereals. Other mechanisms such as endomitosis are another possible mechanism that is currently less understood. For maternally derived haploid plants, the rate of spontaneous doubling is often much lower or entirely absent, so a doubling procedure is essential.

Various methods have been applied over several decades and are still in development. The most frequently used application is treatment with anti-microtubule drugs, such as colchicine (originally extracted from autumn crocus *Colchicum autumnale*), which inhibits microtubule polymerization by binding to tubulin. Although colchicine is highly toxic, used at a millimolar concentration and known to be more efficient in animal than in plant tissues, it is still the most widely used doubling agent. Other options are oryzalin, amiprophos-methyl (APM), trifluralin and pronamide, all of which are used as herbicides and are effective in micromolar concentrations. Anti-microtubule drugs might be applied at various stages of androgenesis, such as being incorporated into microspore pretreatment media. Colchicine application on anther culture medium, for instance, showed a significant increase in embryo formation and green plant regeneration in wheat (Islam, 2010). More often, duplication treatments are applied after regeneration at either embryo, shoot or plantlet level. Similarly, treatments of gynogenically derived embryos with colchine have also been found to be appropriate (Jakše et al., 2003). The treatment of plants at later developmental stages has the advantage that only already tested haploid regenerants are treated either *in vitro* (for instance at the shoot culture stage) or *in vivo* following acclimatization.

The concentration and duration of treatments must be always determined in relation to two effects: the percentage of doubled plants and the percentage of survival. Optimization treatments often require large experimental units (such as 300 explants per treatment) due to the substantial variation of response. High doses/durations can lead to tetraploidization.

Treatment with nitrogen oxide ($N_2O$), which was developed for maize seedlings (Kato & Geiger, 2002) is a special case. Plantlets are treated at a high pressure of 600 kPa for two days at the six-leaf stage, in which plants develop flower primordia. The mechanism of action was studied by Kitamura et al. (2009) in *Lillium* and depolymerization of microtubules was found to be the cause.

Chemical treatment might be avoided by using *in vitro* adventitious somatic regeneration, which itself frequently leads to increased ploidy. Such an approach was efficient in onion (Alan et al., 2007). The method has two advantages: the first being that no potentially damaging chemicals are used in the process and the second that regenerants do not for the most part show a mixoploid character. Up to 100% doubling efficiency in relation to individual line treatment can be achieved using this method (Jakše et al., 2010).

## 6. Applications of doubled haploids in plant breeding

The induction and regeneration of haploids followed by spontaneous or induced doubling of chromosomes are widely used techniques in advanced breeding programs of several agricultural species. They have been successfully used for commercial cultivar production of species such as asparagus, barley, *Brassica juncea*, eggplant, melon, pepper, rapeseed, rice, tobacco, triticale, wheat and more than 290 varieties have already been released (http://www.scri.ac.uk/assoc/COST851/Default.htm). Using DH technology, completely homozygous plants can be established in one generation thus saving several generations of selfing comparing to conventional methods, by which also only partial homozygosity is obtained.

Another feature that should be considered is the breeding strategy. Within the breeding process, DH lines can be induced as soon as from $F_1$ generation (note that gametes on $F_1$ plants represent the $F_2$ generation), although some breeders prefer to induce DH lines from later generations. Induction in the $F_2$ generation was proposed as an option because lines originated from $F_3$ generation gametes had passed through another recombination cycle. However, Choo et al. (1982), comparing DH and single seed descent methods showed that there was no difference in the sample of recombinants.

The role of DH in the breeding process largely depends on the plant mode of reproduction. In self-pollinated species, they can represent final cultivars or they can be used as parental lines in hybrid production or test-crosses of cross-pollinated species. The basic breeding scheme in self-pollinated species starts with crossings of desired genotypes, leading to hybrids containing chromosome sets of both parents. During gamete formation, recombinations enable new gene combinations, which are fixed in the process of doubled haploid production. Doubled haploids thus represent recombinant products of parental genomes in a completely homozygous state. They can be propagated as true breeding lines, facilitating large-scale testing of agronomic performance over the years. Due to complete homozygosity, the efficiency of selection for both qualitative and quantitative characters is increased since recessive alleles are fixed in one generation and directly expressed. Additionally, doubled haploids can be used in a recurrent selection scheme in which superior doubled haploids of one cycle represent parents for hybridization for the next cycle. Several cycles of crossing, doubled haploid production and selection are performed and gradual improvement of lines is expected due to the alternation of recombination and selection.

Similarly as with self-pollinated species, the use of doubled haploids in cross-pollinated species improves selection efficiency and can be used at any or each cycle of recurrent selection. Cross-pollinated species are known to possess numerous deleterious recessive alleles that are not expressed in heterozygous states. They are gradually fixed during self-pollination, causing inbreeding depression and difficulties in producing homozygous lines during conventional breeding. Doubled haploid technology helps to overcome these problems through the rapid fixation of genes in one generation and early elimination of deleterious alleles from populations. The recovered recombination products thus represent more viable combinations of genes. Their complete homozygosity enables true breeding and stable field performance over generations of progeny, although the complete lack of heterozygosity and heterogeneity in varieties is thought to be more vulnerable to environmental changes and altered cropping systems.

It should be noted that, following chromosome doubling, DH plants are normally selfed for maintenance and for further multiplication. In cross-pollinated species with strongly expressed self-incompatibility, various techniques are used to overcome the incompatibility reaction. For instance in *Brassicas*, bud pollination is enhanced by treatment in a $CO_2$ enriched atmosphere (Nakanishi & Hinata, 1973) or by application of gibberelic acid, sodium chloride, urea or ammonium sulphate on stigmas (Sun et al., 2005). Alternatively, DH lines might be clonally propagated, in which case micropropagation is often the best choice.

Mutation breeding is another area of plant improvement for which doubled haploid techniques can help to accelerate the process. Homozygosity of regenerants and true breeding propagation enables the fixation of mutations in the first generation after mutagenic treatment. All mutated traits are immediately expressed, allowing screening for both recessive and dominant mutants in the first generation without the need for self-pollination. The first option is, that mutagenic treatment is applied to dormant seeds that, on germination and flowering, produce M1 gametes, which are used as donor material for haploid culture. The second option relies on mutagenic treatment of haploid cells *in vitro*. The mutagenic agent is usually applied soon after microspore isolation at the uninucleate stage, before the first nuclear division in order to avoid heterozygosity and chimerism caused by spontaneous diploidization through nuclear fusion. *In vitro* mutagenic treatment can be followed by *in vitro* selection of desired traits, such as disease and herbicide resistance.

## 7. Novel approaches combining DHs and molecular genetics

A simplified scheme for backcrossing has been proposed (Forster et al., 2007), aimed at shortening the period needed for the introduction of a particular trait from donor to recipient germplasm. According to the scheme, DHs are produced from the BC1 generation. Segregation of parental chromosomes into the filial generation is followed by molecular markers to identify lines with only recipient chromosomes. The gene of interest should thus be introduced into the recipient chromosome by a random crossing over event in the $BC_1$ generation.

A protocol for "reverse breeding" was proposed by Wijnker et al. (2007). According to this invention, superior hybrid genotypes are first identified among the segregating population. Using genetic transformation, a gene for induced suppression of meiotic recombination is then introduced, and several DH lines are produced. Segregation of chromosomes is followed by chromosome specific molecular markers and a final combination of two lines represents complementary sets constituting the original heterozygous superior hybrid.

Both methods described above are at present predominantly theoretical and one of their obvious limitations is the number of chromosomes of a particular species. As described by Dirks (2009), the probability of finding two lines with a complete set of homologous chromosomes sharply decreases with the number of chromosomes of a particular species. For instance, the probability of identifying such complementary lines is 1 in 47 for *Arabidopsis* (n=x=5) but as high as 1 in 532 in a species with n=x=12.

Genetic transformation at haploid level has been studied in several ways. The most common approach has been for haploid plants to be transformed using established transformation methods. To give just one example, Chauhan et al. (2011) transformed haploid bread wheat with an HVA1 gene to obtain drought tolerance. Chromosome doubling thus enabled stable fixation of the integrated gene, and this feature was tested for 14 generations. Another approach has been for haploid cells themselves, mainly microspores, to be targets of transformation prior to haploid induction. Touraev et al. biolistically transformed tobacco microspores, induced maturation and pollinated to achieve transformed progeny. Eudes and Chugh (2008) and Chugh et al. (2009) transformed triticale microspores using the coupling of cell-penetrating peptides with plasmid DNA and regenerated haploid transformed plants.

A completely novel approach for haploid induction has recently been developed by the genetic engineering of the centromeric region (Ravi & Chan, 2010). Centromeres are chromosomal regions in which DNA sequences serve as binding sites for kinetochore proteins, on which spindle microtubules bind during mitosis and meiosis. In this novel method, a kinetochore protein (Cenh3) was first disabled by mutation and the altered version was then inserted by genetic transformation. In such plants, this novel CENH3 protein is also disabled but only to such an extent that its chromosome segregating function is maintained, while defective kinetochores cause elimination of this chromosomal set during mitotic divisions in zygotic cells. To achieve haploid induction, therefore, the method requires inactivation of the endogenous *CENH3* gene by mutation or RNAi interference and the insertion of an additional gene coding for the *CENH3* variant. This method has some resemblance to the genome elimination described previously in wide crosses or in the case of maize intra-specific crosses and potentially allows its use in any plant species. The authors claim that another feature of this system is that the 'inducer line' (line with the altered centromeric gene) can be used to induce either maternal or paternal haploids by crossing the mutant with female or male wild-type plants. The procedure has so far been demonstrated in *Arabidopsis thaliana*, causing up to 50% of the $F_1$ progeny to be haploid. At the same time, this protocol is the first demonstration of haploid induction in this model species, which has been recalcitrant to all haploid induction protocols available so far. Attempts to test this procedure are currently ongoing in other species.

Protocols involving genetic engineering in agricultural applications have given rise to opposition in several countries, thus limiting their availability in breeding programs. It should be noted that, in the case of the presented 'transgenic inducer technology', the final haploid line would not possess any transgenic elements, because the chromosomes of the inducer line are outcompeted by a wild-type parent. It remains to be resolved how such a process will be regulated under GMO legislation. At least at the EU level, legal regulation/deregulation of such new techniques is already under discussion.

## 8. Conclusion

Doubled haploidy is and will continue to be a very efficient tool for the production of completely homozygous lines from heterozygous donor plants in a single step. Since the first discovery of haploid plants in 1920 and in particular after the discovery of *in vitro* androgenesis in 1964, techniques have been gradually developed and constantly improved. The method has already been used in breeding programs for several decades and is currently the method of choice in all species for which the technique is sufficiently elaborated. Species for which well-established protocols exist predominantly belong to field crops or vegetables, but the technique is gradually also being developed for other plant species, including fruit and ornamental plants and other perennials.

It should be mentioned that, in addition to breeding, haploids and doubled haploids have been extensively used in genetic studies, such as gene mapping, marker/trait association studies, location of QTLs, genomics and as targets for transformations.

Furthermore the haploid induction technique can nowadays be efficiently combined with several other plant biotechnological techniques, enabling several novel breeding achievements, such as improved mutation breeding, backcrossing, hybrid breeding and genetic transformation.

## 9. References

Alan, A.R.; Lim, W.; Mutschler, M.A. & Earle, E.D. (2007). Complementary strategies for ploidy manipulations in gynogenic onion (*Allium cepa* L.). *Plant Science*, Vol. 173, No. 1, (July 2007), pp. 25-31, ISSN 0168-9452

Aleza, P.; Juarez, J.; Hernandez, M.; Pina, J.A.; Ollitrault, P. & Navarro, L. (2009). Recovery and characterization of a *Citrus clementina* Hort. ex Tan. 'Clemenules' haploid plant selected to establish the reference whole *Citrus* genome sequence. *BMC Plant Biology*, Vol. 9, (August 2009), pp. 110, ISSN 1471-2229

Blakeslee, A.F.; Belling, J.; Farnham, M.E. & Bergner, A.D. (1922). A haploid mutant in the Jimson weed, *Datura stramonium*. *Science*, Vol. 55, pp. 646-647, ISSN 0036-8075

Bohanec, B. & Jakše, M. (1999). Variations in gynogenic response among long-day onion (*Allium cepa* L.) accessions. *Plant Cell Reports*, Vol. 18, No. 9, (May 1999), pp. 737-742, ISSN 0721-7714

Bohanec, B. (2003). Ploidy determination using flow cytometry. In: *Doubled Haploid Production in Crop Plants: A Manual*, Maluszynski, M., Kasha, K.J., Forster, B.P. & Szarejko, I., pp. 397-403, Kluwer Academic Publishers, ISBN 1-4020-1544-5, Dordrecht

Bohanec, B. (2009). Doubled Haploids via Gynogenesis. In: *Advances in Haploid Production in Higher Plants*, Touraev, A., Forster, B.P., & Jain, S.M., pp. 35-46, Springer Science + Business Media B.V, ISBN 978-1-4020-8853-7

Bouvier, L.; Zhang, Y.-X. & Lespinasse, Y. (1993). Two methods of haploidization in pear, *Pyrus communis* L.: greenhouse seedling selection and I situ parthenogenesis induced by irradiated pollen. *Theoretical and Applied Genetics*, Vol. 87, No. 1-2, (October 1993), pp. 229-232, ISSN 0040-5752

Chalak, L. & Legave, J.M. (1997). Effects of pollination by irradiated pollen in Hayward kiwifruit and spontaneous doubling of induced parthenogenetic trihaploids. *Scientia Horticulturae*, Vol. 68, No. 1-3, (March 1997), pp. 83-93, ISSN 0304-4238

Chauhan, H. & Khurana, P. (2011). Use of doubled haploid technology for development of stable drought tolerant bread wheat (*Triticum aestivum* L.) transgenics. *Plant Biotechnology Journal*, Vol. 9, No. 3, (April 2011), pp. 408-417, ISSN 1467-7652

Chen, J.F.; Cui, L.; Malik, A.A. & Mbira, K.G. (2011). *In vitro* haploid and dihaploid production via unfertilized ovule culture. *Plant Cell and Tissue Culture*, Vol. 104, No. 3, (March 2011), pp. 311-319, ISSN 1573-5044

Choo, T.M.; Reinbergs, E. & Park S.J. (1982). Comparison of frequency distributions of doubled haploid and single seed descent lines in barley. *Theoretical and Applied Genetics*, Vol. 61, No. 3, (September 1982), pp. 215-218, ISSN 0040-5752

Chugh, A.; Amundsen, E. & Eudes, F. (2009). Translocation of cell-penetrating peptides and delivery of their cargoes in triticale microspores. *Plant Cell Reports*, Vol. 28, No. 5, (May 2009), pp. 801-810, ISSN 0721-7714

Claveria, E.; Garcia-Mas, J. & Dolcet-Sanjuan, R. (2005). Optimization of cucumber doubled haploid line production using *in vitro* rescue of *in vivo* induced parthenogenic embryos. *Journal of the American Society for Horticultural Science*, Vol. 130, No. 3, (July 2005), pp. 555-560, ISSN 0003-1062

Coe, E. H. (1959). A line of maize with high haploid frequency. *American Naturalist*, Vol. 93, pp. 381-382, ISSN 0003-0147

Cuny, F.; Grotte, M.; Dumas De Vaulx, R. & Rieu, A. (1993). Effects of gamma irradiation of pollen on parthenogenetic haploid production in muskmelon (*Cucumis melo* L.). *Environmental and Experimental Botany*, Vol. 33, No. 2, (April 1993), pp. 301-312, ISSN 0098-8472

De Witte, K. & Keulemans, J. (1994). Restrictions of the efficiency of haploid plant production in apple cultivar Idared, through parthenogenesis in situ. *Euphytica*, Vol. 77, No. 1-2, (February 1994), pp. 141-146, ISSN 0014-2336

Dirks, R.; van Dun, K.; de Snoo., C.B.; van den Berg, M.; Lelivelt, C.L.; Voermans, W.; Woudenberg, L.; de Wit, J.P.; Reinink, K.; Schut, J.W.; van der Zeeuw, E.; Vogelaar, A.; Freymark, G.; Gutteling, E.W.; Keppel, M.N.; van Drongelen, P.; Kieny, M.; Ellul, P.; Touraev, A.; Ma, H.; de Jong, H. & Wijnker, E. (2009). Reverse breeding: a novel breeding approach based on engineered meiosis. *Plant Biotechnology Journal,* Vol. 7, No. 9, (December 2009), pp. 837–845, ISSN 1467-7652

Dore, C. & Marie, F. (1993). Production of gynogenetic plants of onion (*Allium cepa* L.) after crossing with irradiated pollen. *Plant Breeding,* Vol. 111, No. 2, (September 1993), pp. 142-147, ISSN 0179-9541

Dunstan, D.I. & Short, K.C. (1977). Improved growth of tissue cultures of the onion, *Allium cepa. Physiologia Plantarum,* Vol. 41, No. 1, (September 1977), pp. 70-72, ISSN 0031-9317

Eudes, F. & Chugh, A. (2008). Nanocarrier based plant transfection and transduction. Patent WO/2008/148223

Faris, N.M. & Niemirowicz-Szczytt, K. (1999). Cucumber (*Cucumis sativus* L.) embryo development in situ after pollination with irradiated pollen. *Acta Biologica Cracoviensia Series Botanica,* Vol. 41, pp. 111-118, ISSN 0001-5296

Faris, N.M.; Nikolova, V. & Niemirowicz-Szczytt, K. (1999). The effect of gamma irradiation dose on cucumber (*Cucumis sativus* L.) haploid embryo production. Acta *Physiologiae Plantarum,* Vol. 21, No. 4, (December 1999), pp. 391-396, ISSN 0137-5881

Ferrie, A.M.R. & Caswell, K.L. (2011). Isolated microspore culture techniques and recent progress for haploid and doubled haploid plant production. *Plant Cell, Tissue and Organ Culture,* Vol. 104, No. 3, (March 2011), pp. 301-309, ISSN 0167-6857

Forster, B.P.; Heberle-Bors, E.; Kasha, K.J. & Touraev, A. (2007). The resurgence of haploids in higher plants. *Trends in Plant Science,* Vol. 12, No. 8, (July 2007), pp. 368-375, ISSN 1360-1385

Froelicher, Y.; Bassene, J.B.; Jedidi-Neji, E.; Dambier, D.; Morillon, R.; Bernardini, G.; Costantino, G. & Ollitrault, P. (2007). Induced parthenogenesis in mandarin for haploid production: induction procedures and genetic analysis of plantlets. *Plant Cell Reports,* Vol. 26, No. 7, (July 2007), pp. 937-944, ISSN 0721-7714

Geiger, H.H. & Gordillo, G.A. (2009). Doubled haploids in hybrid maize breeding. Maydica, Vol. 54, (January 2009), pp. 485-499, ISSN 0025-6153

Guha, S. & Maheshwari, S.C. (1964). *In vitro* production of embryos from anthers of *Datura. Nature,* Vol. 204, No. 4957, (October 1964), pp. 497, ISSN 0028-0836

Guha, S. & Maheshwari, S.C. (1966). Cell division and differentiation of embryos in the pollen grains of *Datura in vitro. Nature,* Vol. 212, pp. 97–98, ISSN 0028-0836

Höfer, M. & Grafe, Ch. (2003). Induction of doubled haploids in sweet cherry (*Prunus avium* L.). *Euphytica,* Vol. 130, No. 2, (March 2003), pp. 191-197, ISSN 0014-2336

Höfer, M. & Lespinasse, Y. (1996). Haploidy in apple. In: *In Vitro Haploid Production in Higher Plants Vol. 3: Important Selected Plants,* Jain, S.M., Sopory, S.K. & Veilleux, R.E., pp 261-276, Kluwer Academic Publishers, ISBN: 9780792335795, Dordrecht

Hunter, C.P. (1987). Patent Number: EP 972007737

Inagaki, M.N. (2003). Doubled haploid production in wheat through wide hybridization. In: *Doubled Haploid Production in Crop Plants: A Manual,* Maluszynski, M., Kasha, K.J., Forster, B.P. & Szarejko, I., pp. 53–58, Kluwer Academic Publishers, ISBN 1-4020-1544-5, Dordrecht

Islam, S.M.S. (2010). The effect of colchicine pretreatment on isolated microspore culture of wheat (*Triticum aestivum* L.). *Australian Journal of Crop Science,* Vol. 4, No. 9, (November 2010) pp. 660-665, ISSN 1835-2693

Jacquard, C.; Nolin, F.; Hecart, C.; Grauda, D.; Rashal, I.; Dhondt-Cordelier, S.; Sangwan, R.S.; Devaux, P.; Mazeyrat-Gourbeyre, F. & Clement, C. (2009). Microspore embryogenesis and programmed cell death in barley: effects of copper on albinism in recalcitrant cultivars. *Plant Cell Reports*, Vol. 28, No. 9, (September 2009), pp. 1329-1339, ISSN 0721-7714

Jakše, M.; Havey, M.J.& Bohanec, B. (2003). Chromosome doubling procedures of onion (*Allium cepa* L.) gynogenic embryos. *Plant Cell Reports*, Vol. 21, No. 9, (June 2003), pp. 905-910, ISSN 0721-7714

Jakše, M.; Hirschegger, P.; Bohanec, B. & Havey, M.J. (2010). Evaluation of Gynogenic Responsiveness and Pollen Viability of Selfed Doubled Haploid Onion Lines and Chromosome Doubling via Somatic Regeneration. *Journal of the American Society for Horticultural Science*, Vol. 135, No. 1, (January 2010), pp. 67-73, ISSN 0003-1062

Javornik, B.; Bohanec, B. & Campion, B. (1998). Second cycle gynogenesis in onion, *Allium cepa* L, and genetic analysis of the plants. *Plant Breeding*, Vol. 117, No. 3, (July 1988), pp.275-278, ISSN 0179-9541

Jia, X.; Zhuang, J.; Hu, S.; Ye, C. & Nie, D. (1994). Establishment and application of the medium of anther culture of intergeneric hybrids of *Triticum aestivum* × *Triticum agropyron*. *Scientia Agricultura Sinica*, Vol. 27, pp. 83-87, ISSN 0578-1752

Kasha, K,J.; Hu, T.C.; Oro, R.; Simion, E. & Shim, Y.S. (2001). Nuclear fusion leads to chromosome doubling during mannitol pretreatment of barley (*Hordeum vulgare* L.) microspores. *Journal of Experimental Botany*, Vol. 52, No. 359, (June 2001), pp. 1227-1238, ISSN 0022-0957

Kasha, K.J. (2005). Chromosome doubling and recovery of doubled haploid plants. In: *Biotechnology in Agriculture and Forestry Vol. 56, Haploids in crop improvement II*, Palmer, C.E., Keller, W.A. & Kasha, K.J., pp. 123-152, Spirnger-Verlag, ISBN 3-540-22224-3, Berlin Heildelberg

Kato, A. & Geiger, H.H. (2002). Chromosome doubling of haploid maize seedlings using nitrous oxide gas at the flower primordial stage. *Plant Breeding*, Vol. 121, (October 2002), pp. 370-377, ISSN 0179-9541

Kitamura, S.; Akutsu, M. & Okazaki, K. (2009). Mechanism of action of nitrous oxide gas applied as a polyploidizing agent during meiosis in lilies. *Sexual Plant Reproduction*, Vol. 22, No. 1, (March 2009), pp. 9-14, ISSN 0934-0882

Kurtar, E.S. & Balkaya, A. (2010). Production of in vitro haploid plants from in situ induced haploid embryos in winter squash (*Cucurbita maxima* Duchesne ex Lam.) via irradiated pollen. *Plant Cell, Tissue and Organ Culture*, Vol. 102, No. 3, (September 2010), pp. 267-277, ISSN 0167-6857

Kurtar, E.S.; Sari, N. & Abak, K. (2002). Obtention of haploid embryos and plants through irradiated pollen technique in squash (*Cucurbita pepo* L.). *Euphytica*, Vol. 127, No. 3, (October 2002), pp. 335-344, ISSN 0014-2336

Lichter, R. (1982). Induction of haploid plants from isolated pollen of *Brassica napus*. *Zeitschrift fur Pflanzenphysiologie*, Vol. 105, No. 5, pp. 427-434, ISSN 0044-328X

Lotfi, M.; Alan, A.R.; Henning, M.J.; Jahn, M.M. & Earle, E.D. (2003). Production of haploid and doubled haploid plants of melon (*Cucumis melo* L.) for use in breeding for multiple virus resistance. *Plant Cell Reports*, Vol. 21, No. 11, (July 2003), pp. 1121-1128, ISSN 0721-7714

Maine, M.J. De. (2003). Potato haploid technologies. In: *Doubled Haploid Production in Crop Plants: A Manual*, Maluszynski, M., Kasha, K,J., Forster, B.P. & Szarejko, I., pp. 241-247, Kluwer Academic Publishers, ISBN 1-4020-1544-5, Dordrecht

Maluszynska, J. (2003). Cytogenetic tests for ploidy level analyses – chromosome counting. In: *Doubled Haploid Production in Crop Plants: A Manual*, Maluszynski, M., Kasha,

K.J., Forster, B.P. & Szarejko, I., pp. 391-395, Kluwer Academic Publishers, ISBN 1-4020-1544-5, Dordrecht

Maluszynski, M.; Kasha, K.J. & Szarejko, I. (2003). Published doubled haploid protocols in plant species. In: *Doubled Haploid Production in Crop Plants: A Manual*, Maluszynski, M., Kasha, K.J., Forster, B.P. & Szarejko, I., pp. 309–335, Kluwer Academic Publishers, ISBN 1-4020-1544-5, Dordrecht

Maluszynski, M.; Kasha, K.J.; Forster, B.P. & Szarejko, I. (Eds.). (2003.) *Doubled Haploid Production in Crop Plants: A Manual*, Kluwer Academic Publishers, ISBN 1-4020-1544-5, Dordrecht

Maraschin, S.F.; de Priester, W.; Spaink, H.P. & Wang, M., (2005). Androgenic switch: an example of plant embryogenesis from the male gametophyte perspective. *Journal of Experimental Botany*, Vol. 56, No. 417, (July 2005), pp. 1711-1726, ISSN 0022-0957

Meyneta, J.; Barradea, R.; Duclosa, A. & Siadousb, R. (=1994). Dihaploid plants of roses (*Rosa x hybrid*a, cv 'Sonia') obtained by parthenogenesis induced using irradiated pollen and in vitro culture of immarure seeds. *Agronomie*, Vol. 14, No. 3, (1994), pp. 169-175, ISSN 0249-5627

Musial, K. & Przywara, L. (1998). Influence of irradiated pollen on embryo and endosperm development in kiwifruit. *Annals of Botany*, Vol. 82, No. 6, (December 1998), pp. 747-756, ISSN 0305-7364

Musial, K. & Przywara, L. (1999). Endosperm response to pollen irradiation in kiwifruit. *Sexual Plant Reproduction*, Vol. 12, No. 2, (June 1999), pp. 110-117, ISSN 0934-0882

Musial, K.; Bohanec, B.; Jakše, M. & Przywara, L. (2005). The development of onion (*Allium cepa* L.) embryo sacs in vitro and gynogenesis induction in relation to flower size. *In Vitro Cellular and Developmental Biology-Plant*, Vol. 41, No. 4 (July-August 2005), pp. 446-452, ISSN 1054-5476

Naess, S.K.; Swartz, J.H. & Bauchan, G.R. (1998). Ploidy reduction in blackberry. *Euphytica*, Vol. 99, No. 1, (January 1998), pp. 57-73, ISSN 0014-2336

Nakanishi, T. & Hinata, K. (1973). Effective time for $CO_2$ gas treatment in overcoming self-incompatibility in *Brassica*. *Plant and Cell Physiology*, Vol. 14, No. 5, (October 1973), pp. 873-879, ISSN 0032-0781

Nitsch, J.P. & Nitsch, C. (1969). Haploid plants from pollen grains. *Science*, Vol. 163, No. 3862, (January 1969), pp. 85-87, ISSN 0036-8075

Pandey, K.K. & Phung, M. (1982). 'Hertwig effect' in plants: induced parthenogenesis through the use of irradiated pollen. *Theoretical and Applied Genetics*, Vol. 62, No. 4, pp. 295-300, ISSN 0040-5752

Pandey, K.K. (1980). Parthenogenetic diploidy and egg transformation induced by irradiated pollen in *Nicotiana*. *New Zeland Journal of Botany*, Vol. 18, No. 2, pp. 203-207, ISSN 0028-825X

Pandey, K.K.; Przywara, L. & Sanders, P.M. (1990). Induced parthenogenesis in kiwifruit (*Actinidia deliciosa*) through the use of lethally irradiated pollen. *Euphytica*, Vol. 51, No. 1, (November 1990), pp. 1-9, ISSN 0014-2336

Peixe, A.; Campos, M.D.; Cavaleiro, C.; Barroso, J. & Pais, M.S. (2000). Gamma-irradiated pollen induces the formation of 2n endosperm and abnormal embryo development in Euroean plum (*Prunus domestica* L., cv. 'Rainha Claudia Verde'. *Scientia Horticulturae*, Vol. 86, No. 4, (December 2000), pp. 267-278, ISSN 0304-4238

Przyborowski, J. & Niemirowicz-Szczytt, K. (1994). Main factors affecting cucumber (*Cucumis sativus* L.) haploid embryo development and haploid plant characteristics. *Plant Breeding*, Vol. 112, No. 1, (January 1994), pp. 70-75, ISSN 0179-9541

Raquin, C. (1985). Induction of haploid plants by *in vitro* culture of *Petunia* ovaries pollinated with irradiated pollen. *Zeitung Pflanzenzucht*, Vol. 94, pp. 166-169

Ravi, M. & Chan, S.W. (2010) Haploid plants produced by centromere-mediated genome elimination. *Nature*, Vol. 464, (March 2010), pp. 615–618, ISSN 0028-0836

Rokka, V.M. (2009). Potato Haploids and Breeding. In: *Advances in Haploid Production in Higher Plants*, Touraev, A., Forster, B.P., & Jain, S.M., pp. 199-208, Springer Science + Business Media B.V, ISBN 978-1-4020-8853-7

Sari, N.; Abak, K.; Pitrat, M.; Rode, J.C. & Dumas de Vaulx, R. (1994). Induction of parthenogenetic haploid embryos after pollination by irradiated pollen in watermelon. HortScience, Vol. 29, No. 10, (October 1994), pp. 1189-1190, ISSN 0018-5345

Sato, S.; Katoh, N.; Yoshida H.; Iwai, S. & Hagimori M. (2000). Production of doubled haploid plants of carnation (*Dianthus caryophyllus* L.) by pseudofertilized ovule culture. *Scientia Horticulturae*, Vol. 83, No. 3-4, (March 2000), pp. 301-310, ISSN 0304-4238

Sauton, A. & Dumas de Vaulx, R. (1987). Obtention de plantes haploids chez le melon (Cucumis melo L.) par gynogenese induite par pollen irradie. *Agronomie*, Vol. 7, pp. 141-148, ISSN 0249-5627

Sunderland, N. (1974). Anther culture as a means of haploid production. In: *Haploids in higher plants, advances and potential*, Kasha K.J., pp. 91–122, University of Guelph, Guelph

Todorova, M.; Ivanov, P.; Shindrova, P.; Christov, M. & Ivanova, I. (1997). Doubled haploid production of sunflower (*Helianthus annuus* L.) through irradiated pollen-induced parthenogenesis. *Euphytica*, Vol. 97, No. 3, (November 1997), pp. 249-254, ISSN 0014-2336

Touraev, A.; Indrianto, A.; Wratschko, I.; Vicente, O. & Heberle-Bors, E. (1996) Efficient microspore embryogenesis in wheat (*Triticum aestivum* L.) induced by starvation at high temperatures. *Sexual Plant Reproduction*, Vol. 9, No. 4, (July 1996), pp. 209–215, ISSN 0934-0882

Touraev, A.; Stoger, E.; Voronin, V. & Heberle-Bors, E. (1997). Plant male germ line transformation. *Plant Journal*, Vol. 12, No. 4, (February 1997), pp. 949-956, ISSN 0960-7412

Wędzony, M.; Forster, B.P.; Žur, I.; Golemiec, E.; Szechyńska-Hebda, M.; Dubas, E. & Gołębiowska, G. (2009). Progress in doubled haploid technology in higher plants. In: *Advances in Haploid Production in Higher Plants*, Touraev, A., Forster, B.P., & Jain, S.M., pp. 1-34, Springer Science + Business Media B.V, ISBN 978-1-4020-8853-7

Wijnker, E.; Vogelaar, A.; Dirks, R.; van Dun, K.; de Snoo, B.; van den Berg, M.; Lelivelt, C; de Jong, H. & Chunting, L. (2007). Reverse breeding: Reproduction of F1 hybrids by RNAi-induced asynaptic meiosis. *Chromosome Research*, Vol. 15, No. 2, (August 2007), pp. 87-88, ISSN 0967-3849

Ying, C. (1986) Anther and pollen culture of rice. In: *Haploids of higher plants in vitro*, Hu, H. & Yang, H.Y., pp. 3-25, China Academic Publishers and Springer- Verlag, ISBN 3-540-16003-5

Zhang, Y.X. & Lespinasse, Y. (1991). Pollination with gamma-irradiated pollen and development of fruits, seeds and parthenogenetic plants in apple. *Euphytica*, Vol. 54, No. 1, (April 1991), pp. 101-109, ISSN 0014-2336

# 5

# Chromosome Substitution Lines: Concept, Development and Utilization in the Genetic Improvement of Upland Cotton

Sukumar Saha[1] et al.[*]
*[1]United States Department of Agriculture-Agriculture Research Service,
Crop Science Research Laboratory,
USA*

## 1. Introduction

Cotton is the most important natural fiber source for the textile industry world-wide. It is also an alternative of the man-made petroleum-based "synthetic fibers" providing an advantage for a sustainable environment. Cotton is formed by developing seed of several *Gossypium* species, which are mainly grown as an important cash crop in more than 70 countries including USA, India, China and Uzbekistan (Smith and Coyle, 1997). Although cotton plants are best known as the renewable source of textile materials for clothing, the fiber, seed and plants have many other uses, including home insulation to save energy, protein-rich seed-derived feed for animals, cottonseed oil as a foodstuff for humans, and as a source of mulch and biomass (Cotton Incorporated, 2010, http://cottontoday.cottoninc.com/sustainability-about/responsible-economic-development, verified on October 14, 2011). This brings significant humanitarian and economic benefits. For example, scientists are exploring genetic means to better harness its highly nutritious seed for food and feed (Sunilkumar et al., 2006).

Cotton is facing some serious challenges in production and marketing, such as competition from synthetic fibers, large year-to-year variability in yield and fiber qualities, and recent changes in textile technologies, for which the optimization of cotton requires altered fiber quality characteristics. Genetic solutions to these and other challenges require adequate genetic variation to be present in the breeding germplasm. In cotton, however, the genetic diversity available in the breeding gene pool is narrow, and is recognized as a cause of yield stagnation, declining fiber quality, and increasing genetic vulnerability to biotic and abiotic stresses in worldwide cotton production.

* David M. Stelly[2], Dwaine A. Raska[2], Jixiang Wu[3], Johnie N. Jenkins[1], Jack C. McCarty[1], Abdusalom Makamov[4], V. Gotmare[5], Ibrokhim Y. Abdurakhmonov[4] and B.T. Campbell[6]
[1]United States Department of Agriculture-Agriculture Research Service, Crop Science Research Laboratory, USA
[2]Department of Soil and Crop Sciences, Texas A&M University, USA
[3]Plant Science Department, South Dakota State University, USA
[4]Center of Genomic Technologies, Institute of Genetics and Plant Experimental Biology,
Academy of Sciences of Uzbekistan, Uzbekistan
[5]Division of Crop Improvement, Central Institute for Cotton Research, India
[6]USDA-ARS Coastal Plains Soil, Water, and Plant Research Center, USA

Cotton breeding programs were extremely successful over several decades by increasing the frequency of beneficial alleles for important traits at many loci, which generated high yielding lines with superior agronomic qualities. As a consequence, breeders continued making crosses among closely related high-yielding cultivars. In fact, most Upland cotton breeding programs rely primarily on crosses among closely related elite domesticated genotypes with high yield and superior fiber qualities, and on reselection from existing cultivars (Esbroeck and Bowman, 1998).

The rate of change of USA cotton yield has steadily declined since 1985. The absolute cotton yield reached at a disturbing rate (3.3% annual rate) by 1998, demanding the immediate need for genetic improvement (Paterson et al., 2004). The history on the genetic improvement of Upland cotton suggested that some of the favorable alleles were fixed in the elite breeding gene pool and that significant additional improvements would require introgressing genes from outside of the pool (Bowman et al., 1996). Exotic unadapted species have contributed beneficial alleles for improving agronomically valuable traits in many crop species (Tanksley and McCough, 1997). Although it is well known that exotic germplasm contains potentially valuable genes, the exotic gene pools of *Gossypium hirsutum* L. have been under-characterized and under utilized.

There are several impediments in conventional methods of interspecific introgression in cotton: 1) complex antagonistic relationships among important traits; 2) cytogenetic differences among the species due to different ploidy levels, meiotic affinity and chromosomal structural differences including translocations and inversions; 3) "linkage drag effects" leading to poor agronomic qualities; 4) reduced recombination; 4) loss of alien genetic materials in early generations; 5) sterility in the hybrids; 6) complex genetic interactions such as Muller-Dobzhansky complexes and 7) distorted segregation (Endrizzi et al., 1985). These kinds of difficulties are most severe in crosses of *G. hirsutum* with diploid species, but most of them also apply significantly to the crosses of the "primary gene pool" which includes all of the other natural tetraploid (2n=52) cotton species *G. barbadense*, *G. tomentosum*, *G. mustelinum* and *G. darwinii* along with *G. hirsutum*.

As part of the primary gene pool, these four tetraploid species are especially accessible reservoirs of important genes for pest and disease resistance, and for improved agronomic and fiber traits. However, the effects of beneficial genes from wild unadapted germplasm are often obscured by other genes that affect the trait negatively in the wild species. Only upon appropriate genetic analysis, the presence of such valuable genes can be detected. Though single genes can significantly affect individual traits, they typically involve complex direct and indirect molecular interactions with other genes. Thus, genes often affect multiple traits, and traits are generally determined by multiple genes. Furthermore, the potential utility and value of specific alien genes is usually compromised by co-inheritance of closely linked genes that have deleterious agricultural effects on productivity. Rates of recombination with alien chromosomes can be reduced considerably, especially in certain segments, therein greatly raising the likelihood of undesirable linkages. To physically separate the beneficial and undesirable alien genes by breeding can thus be extremely difficult. Several generations of crosses with special breeding strategies are needed to eliminate such undesirable traits and fix the selected few desirable traits in interspecific introgression to improve cultivars.

The success of introgressing useful genetic variation from unadapted species into Upland cotton depends on several factors: 1) the breadth of diversity that can be accessed; 2) the speed and efficacy with which the useful alleles can be transferred, given various biological constraints; and 3) whether useful variation can be transferred without the deleterious traits. An alternative approach to conventional pedigreed or population-based interspecific introgression is to use alien chromosome substitution, which entails a modified pedigreed backross breeding approach (Stelly et al., 2005). A similar and potentially complementary approach is marker-assisted chromosome segment substitution, as first demonstrated in tomato "introgression lines (ILs)". Categorically, these methods trace back to the classical study on quantitative trait analysis of wheat "backcross inbed lines (BILs)" by the famous quantitative geneticist/breeder Robert Allard (Wehrhahn and Allard, 1965).

All five of the 52-chromosome allotetraploid *Gossypium* species, including the four species most amenable to interspecific introgression into Upland cotton (common name for the widely grown middle-staple *G. hirsutum* cultivars), are suggested to have arisen from a common ancient polyploidization event (Wendell, 1989). Cytological observations of hybrids and comparative molecular mapping indicate that synteny and colinearity are mostly conserved among the five tetraploid species, and that there are few major cytostructural barriers to interspecific introgression among them. However, genetic limitations are manifested by the significant level of "$F_2$ breakdown" that is commonly observed after hybridization of Upland cottons with the other species due to extensive genetic incompatibility between these species at the whole-genome level (Beasley and Brown 1942; Reinisch et al. 1994). As Allard's work on wheat would suggest, backcross-inbreeding can be used to circumvent some of these issues. We have begun to exploit a similar approach by chromosome and chromosome segment substitution (CS) lines, using a modified form of backcross-inbred line development and subsequent quantitative genetic evaluations.

We released a set of 17 disomic alien chromosome substitution *G. barbadense* (CS-B) lines through hypoaneuploid-based backcrossing in a near-isogenic genetic background of Texas Marker-1 (TM-1) line (Stelly et al., 2005). In a series of complementary quantitative analyses, the chromosomal effects on agronomic and fiber properties were documented using these CS-B lines in our previous studies (Saha et al., 2004, 2006, 2008, 2010, 2011a, Jenkins et al., 2006, 2007). We also showed that the chromosomal substitution lines constitute important breeding resources, increasing the genetic diversity available in Upland cotton (Jenkins et al., 2006; 2007). The objective of this paper is to provide a summarized report on the concept, development and utilization of CS lines from our previous studies in genetic analysis and germplasm improvement of Upland cotton.

## 2. Cytogenetic resources of cotton

Cotton was one of the first crop plants to which Mendelian principles of genetics were applied (Balls, 1906; Shoemaker, 1908). Cytogenetic analyses revealed that the cultivated species *G. hirsutum* and *G. barbadense* and several other New World allotetraploids contained 52 chromosome, whereas most other diploid species contained just 26 chromosomes. Furthermore, they indicated that all of 52-chromosome species were allotetraploids with two pairs of meiotically independent genomes, AADD. In that, the A and D components are very similar to the A and D diploid genomes of certain extant 26-chromosome species that occur naturally in Africa and the New World (Endrizzi, 1984, 1985; Wendel 1989).

Reciprocal translocations were developed for *G. hirsutum* in order to identify and meiotically detect each of the 26 different chromosomes in translocation heterozygotes, which allowed assignment of the individual chromosomes to the A and D subgenomes. The chromosomes of A genome and D genome of *G. hirsutum* were designated as chromosomes 1-13, and 14-26, respectively (Menzel and Brown, 1978; Brown, 1980; Brown et al., 1981).

Representative types of the original primary monosomic, monotelodisomic and tertiary monosomic reciprocal translocation (NTN) have been or are being backcrossed to TM-1, a highly inbred line from Deltapine 14 that serves as a standard reference for genetic and cytogenetic research (Fig. 1; Kohel et al., 1970). Stocks in the current Cotton Cytogenetics Collection are thus nearly isogenic to TM-1 and each other (Fig. 2). Monosomic stocks have

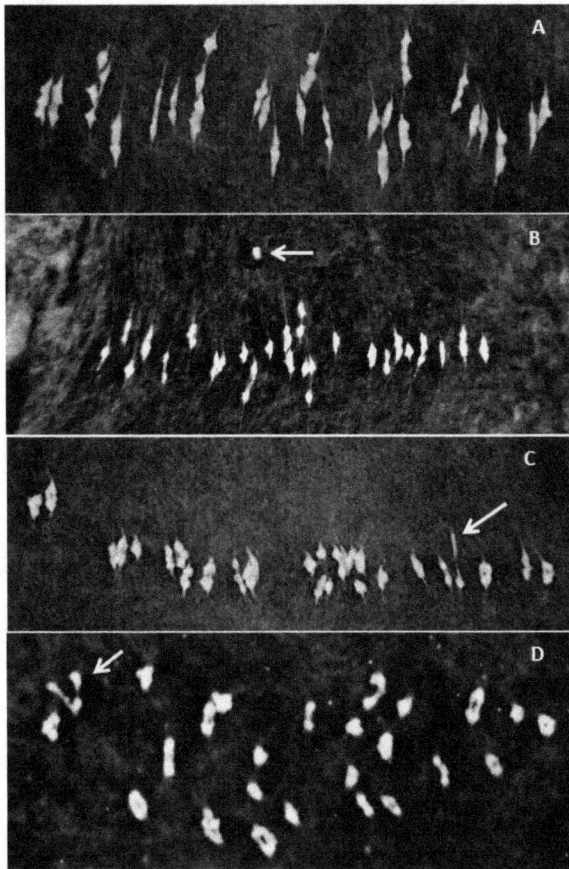

Fig. 1. Cytological identification of the aneuploid lines. (A) Metaphase showing 26II pairs of chromosomes in a normal plant (2n); (B) Metaphase stage from a monosomic plant (2n-1); (C) Metaphase stage from a monotelodisomic plant (2n-1/2) and (D) metaphase from a heterozygous translocation plant. The arrow shows the monosomic and telosomic chromosome and translocated chromosomes respectively.

been used extensively to locate and develop markers for specific chromosomes (Endrizzi, 1963; Kohel, 1978; White and Endrizzi, 1965; Guo et al., 2008; Gutierrez et al., 2009). Monosomic stocks have also been used to identify the chromosomes of translocations (Endrizzi, 1985). Monotelodisomic plants have been used to locate markers to a specific arm of a chromosome (Endrizzi and Kohel, 1966; Endrizzi and Taylor, 1968; Endrizzi and Bray, 1980) and to map the distance between a centromere and a marker locus (Endrizzi et al., 1985). The Cotton Cytogenetic Collection currently includes chromosome-deficient stocks of *G. hirsutum*, mainly primary monosomics (missing one chromosome) for 15 of the 26 chromosomes, and monotelodisomics (missing most of one chromosome arm) for 31 of the 52 chromosome arms, including at least four chromosomes not represented among the monosomic stocks. To achieve additional cytogenetic coverage of the genome, new monosomics and monotelodisomics have been identified, and tertiary monosomics have been synthesized from intercrosses between monosomics and related translocations, and repeated backcrossing to the tertiary monosomics. These chromosome-deficient stocks of *G. hirsutum* have been used as recurrent parents in the development of CS lines.

Through a Cooperative (Co-op) Agreement between TAMU and USDA-ARS, Mississippi (MS), all core monosomic and monotelodisomic cotton plants are developed at Texas A&M University (College Station, Texas), where they become part of the Cotton Cytogenetics Collection; the Collection and most developmental stocks involving *G. barbadense* 3-79 chromatin substitution are duplicated in the greenhouse at USDA/ARS (Mississippi State, MS). This Co-op Agreement enables a safety backup to the Collection, provides some resources for maintenance and development of the Collection, places important research resources at USDA/MS, and fosters multi-dimensional collaborations. These increase capability in use of the cytogenetic stocks for basic genetics, genomics and applied breeding studies.

## 3. Development of chromosome substitution lines

The detailed method of chromosome substitution line development was discussed by Stelly et al. (2005). Each alien species chromosome substitution line development involves four specific stages: (1) development of the respective TM-1-like hypoaneuploid stock, (2) use of the cytogenetic stock as a recurrent seed parent in a recurrent backcrossing program to create a monosomic or monotelodisomic $F_1$ substitution stock, followed by (3) inbreeding with the TM-1-like hypoaneuploid stock to recover a euploid disomic substitution line and (4) confirmation of the cytogenetic and genetic constitution of the disomic lines by cytological analysis and chromosome-specific SSR markers (Figs. 1, 2, 3, and 4).

The strategy in developing these lines follows principles of cytogenetic behavior, transmission and inheritance as reported by Endrizzi et al. (1985). The strategy is based on the principle of differential transmission rate between mega- versus micro-gametophytes, i.e., between the "ovule" ("seed" or "female" parent) versus the pollen ("male") parent. It is believed that transmission of hypoaneuploidy through the ovule parent is common (up to 50%) for most chromosomes or chromosome arms in monosomic and monotelodisomic conditions, but that transmission through cotton pollen is rare to nil for all whole-chromosomes and most large-segment deletions (e.g., telosomes).

**Normal**

**"Monosomic" (4)**          **"Telosomic" ("4-short")**          **"Telosomic" (4-long)**

| Isogenic or Quasi-isogenic | | | | | | | | |
|---|---|---|---|---|---|---|---|---|
| Chr. (A) | Primary monosome | Telosomes Short | Telosomes Long | Others* | Chr. (A) | Primary monosome | Telosomes Short | Telosomes Long | Others* |
| 1 | v | v | v | | 14 | | | v | v |
| 2 | v | v | v | | 15 | | | v | v |
| 3 | v | v | v | | 16 | v | v | v | v |
| 4 | v | v | v | v | 17 | v | v | v | v |
| 5 | | | v | v | 18 | v | v | v | |
| 6 | v | v | v | v | 19 | | | | v |
| 7 | v | v | v | v | 20 | v | v | v | |
| 8 | | | v | v | 21 | v | | | v |
| 9 | v | v | v | v | 22 | | v | v | |
| 10 | v | v | v | v | 23 | v | | v | |
| 11 | v | v | v | v | 24 | | | | |
| 12 | v | v | v | v | 25 | v | | v | |
| 13 | | | | v | 26 | v | v | | |

CS Line Development

*G. barbadense* (3-79)          *G. tomentosum*          *G. mustelinum*

| CS-B lines (17 released lines) | CS-T lines (in development) | CS-M lines (in development) |

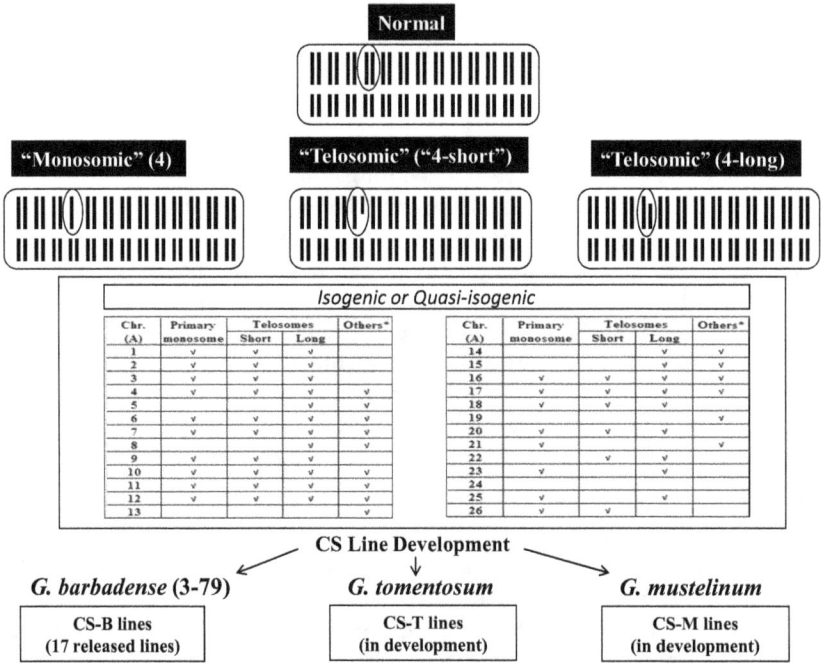

Fig. 2. Available cytogenetic deficient stocks for use in chromosome substitution line development.

In each hypoaneuploid $F_1$ stock, the targeted alien chromosome or segment is hemizygous and thus lacks a homologous partner to undergo meiotic pairing and recombination. By avoiding recombination, the alien chromosomes or segments remain intact when they are transmitted to the gametes and next generation (Endrizzi et al. 1963, 1984, 1985). Even marker-assisted selection based on large number of markers could not guarantee transfer of all alien loci, especially in high recombination regions, which tend to be gene rich (Zhang et al., 2008; Lacape et al., 2009). This is the clear difference between backcross derived materials in Upland cotton from *G. barbadense* versus interspecific chromosome substitution lines. Previous studies showed that the actual number of recombinants obtained were significantly fewer than the expected number in an interspecific backcross program of Upland cotton (Rhyne, 1958) and also limited genetic transmission in advanced generation of the interspecific hybrids (Lacape et al., 2009).

Chromosome substitution lines were developed as follows. The initial cross was made between a chromosomally normal (euploid) *G. barbadense* 3-79 line, as male, and a cytogenetic derivative of *G. hirsutum* TM-1 that was deficient for one copy of a specific chromosome ("monosomic" plant) or chromosome arm ("monotelodisomic" plant), as female. Due to their chromosomal constitution, primary monosomic, monotelodisomic, tertiary monosomic and euploid plants were isolated form diagnostic metaphase I meiotic configurations (25 II + I; 25 II + Ii; 24 II + III; and 26 II, respectively) in microsporocytes ("pollen mother cells", Fig. 1). Interspecific $F_1$ progeny were screened phenotypically, cytogenetically (metaphase-I chromosome analysis of microsporocytes) and in some instances by a molecular marker analysis [loss of heterozygosity (LOH)], to identify a plant deficient for the chromosome or chromosome arm. The identified interspecific aneuploid $(BC_0F_1)$ plant was subsequently used as male parent in backcrosses onto the recurrent aneuploid TM-1 plant. As in the preceding $(BC_0F_1)$ generation, an aneuploid $BC_1F_1$ progeny was selected based on metaphase I analysis, as in the $BC_0F_1$ generation. This procedure was repeated until the fifth backcross. The selected $BC_5F_1$ hypyoaneuploid was selfed to recover a euploid plant $(BC_5F_1S_1\ 26II)$. Euploids were again selected phenotypically, cytogenetically and, in some cases, by a molecular marker analysis. Finally, one euploid $BC_5F_1$ plant was selfed to establish each euploid $BC_5S_1$ substitution line. In each CS-B line, a pair of chromosomes (or chromosome arm segments) of *G. hirsutum* inbred TM-1 was replaced by the respective pair from *G. barbadense* doubled-haploid line 3-79 lines.

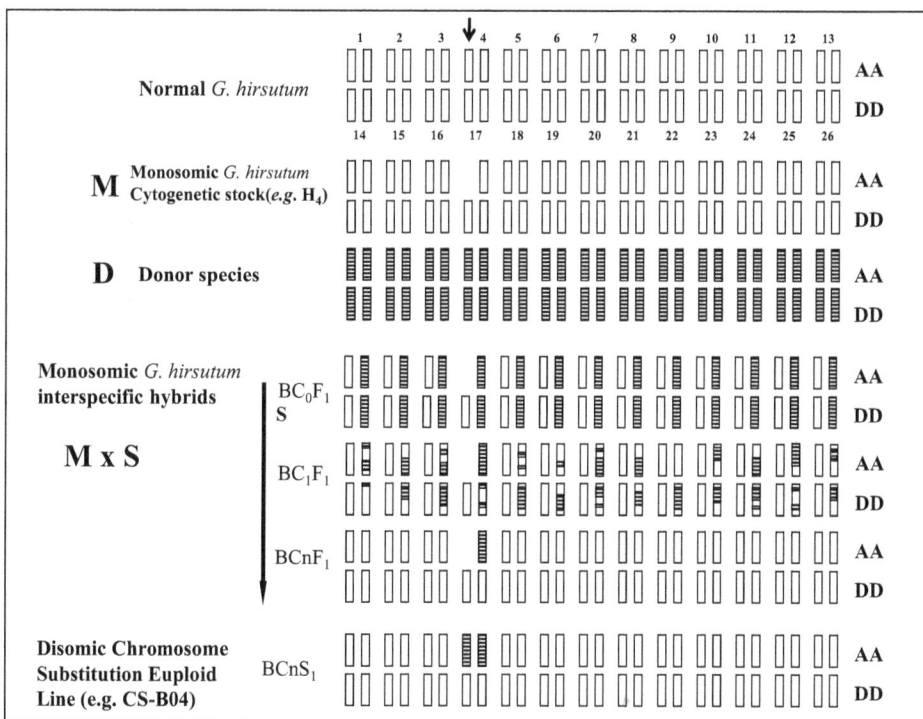

Fig. 3. The overall strategy on the development of a chromosome substitution line. The diagrammatic picture showed chromosome of two species in different colors.

As mentioned above, seventeen CS-B lines were developed using the above procedure and released for use in Upland cotton improvement (Stelly et al., 2005). We designated the line with the number of the specific substituted chromosome or chromosome segment (sh=short arm, Lo=long arm). We also designated NTN to the lines derived from the cross with translocation stocks carrying two segments of the substituted chromosomes from the donor alien species. We dubbed the 3-79 chromosome substitution lines as "CS-B" lines, e.g. CS-B01 for chromosome-1, to reflect the species of origin, i.e., *G. barbadense*. Future series involving other species will be named analogously, e.g., "CS-T" for *G. tomentosum* chromosomes or chromosome segment(s), and CS-M for *G. mustelinum*. Additional CS-B, -T and -M lines are in development; several of the most advanced are now being investigated with regard to effects on fiber and agronomic traits.

We could not confirm the genetic identity of all of the CS lines based on molecular markers at the release time because very few chromosome specific molecular markers were available at that time in the public domain. Recently, in addition to the cytological analysis, we undertook an assignment to confirm the genetic identity of the CS lines using chromosome specific SSR markers (Fig. 4). We used a slightly modified protocol of our previous studies (Guo et al., 2008; Gutierrez et al., 2009). It is expected that a CS line will have the allele of the donor alien species specific to the substituted chromosome or chromosome segment and will miss the TM-1, the recurrent parent, allele specific to the locus of the substituted chromosome or chromosome segment in molecular results.

Most marker results were concordant, but several were inconclusive or discordant (Fig. 4). For the latter, we are investigating further to discern the underlying basis, which is likely an infrequent type(s) of event, and could be a procedural and/or biological in nature. It is interesting that most of these CS lines had different agronomic and fiber properties than TM-1 suggesting that these lines carried genetic materials from some other sources than TM-1, possibly the donor parent, *G. barbadense* (Saha et al., 2004, 2006, 2008, 2010 and 2011a).

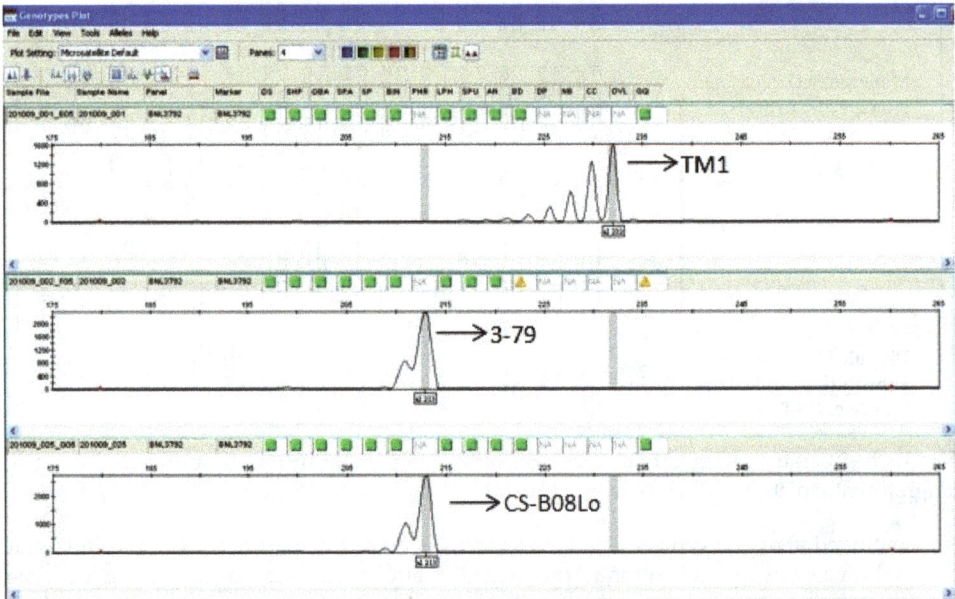

Fig. 4. Electrophoregram results by ABI3130 *xl* capillary electrophoresis showing CS-B08Lo line has only the same allele of 3-79 for BNL3792 SSR marker specific to the long arm of chromosome eight, but missing the allele of TM-1. Accordingly the molecular results confirmed that CS-B08Lo line has the substituted the long arm of chromosome eight from 3-79.

## 4. Development of different genetic resources using the CS lines

### 4.1 Chromosome substitution lines

In each CS-B line, a pair of chromosomes (or chromosome arms) of G. *hirsutum* inbred TM-1 was replaced by the respective pair from G. *barbadense* doubled-haploid line 3-79 (Fig. 5). We also developed several chromosome substitution lines from G. *tomentosum* species (CS-T) and for some new chromosomes of G. *barbadense*. Currently, we are evaluating these lines for fiber and agronomic traits. These substitution lines are nearly isogenic to the common parent TM-1 for 25 chromosome pairs, as well as to each other, for 24 chromosome pairs. The comparative analysis of such unique genetic materials greatly empowers the detection of genetic effects of novel alleles by specific alien chromosomes associated with quantitative traits.

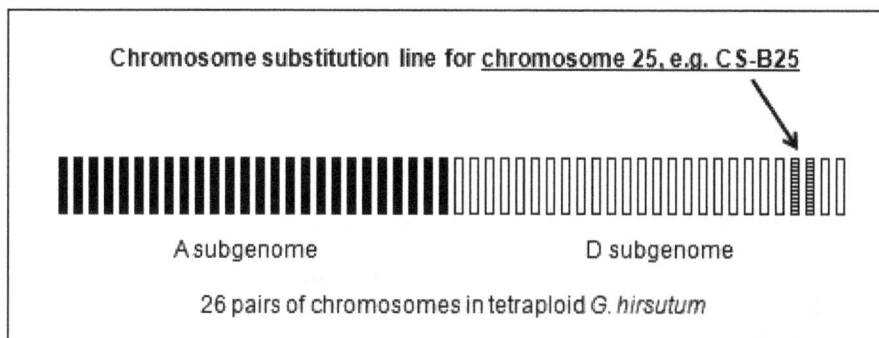

Fig. 5. A diagrammatic picture showing different chromosomes in a CS-B line genome.

### 4.2 Chromosome specific recombinant inbred line development (CS-RILS)

The CS-RILs were developed from crosses of a CS-B line (e.g. CS-B17) of G. *barbadense* chromosome substitution lines with their common recurrent parent, inbred G. *hirsutum* TM-1 (Fig. 6, Saha et al., 2011b). An individual plant of the $F_2$ population from this cross was maintained by selfing via single seed decent method until $F_6$ generation. It is expected that the CS-RIL population gets stabilized in a near homozygous condition by the $F_6$ generation. The CS-RIL populations are currently being evaluated in field trials for fiber and agronomic traits. This CS-RIL population will be very useful for high-resolution mapping of QTLs. Using molecular markers in such a RIL will facilitate detection and cloning of QTLs.

### 4.3 Crosses among the CS lines to create different chromosome specific genetic resources

The near-isogenic nature of the substitution lines to the common parent TM-1 provides a unique opportunity to use specific mating designs among the CS lines of interest to create different chromosome specific combinations of genetic resources (Fig. 7). Using a mating design within the same alien species of two CS lines of interest or between two alien species for the same substituted chromosome or different substituted chromosome, we will be able to create a unique source of genetic resources useful as genomic tools for germplasm improvement. For example, a cross between the same chromosome substitution lines of two

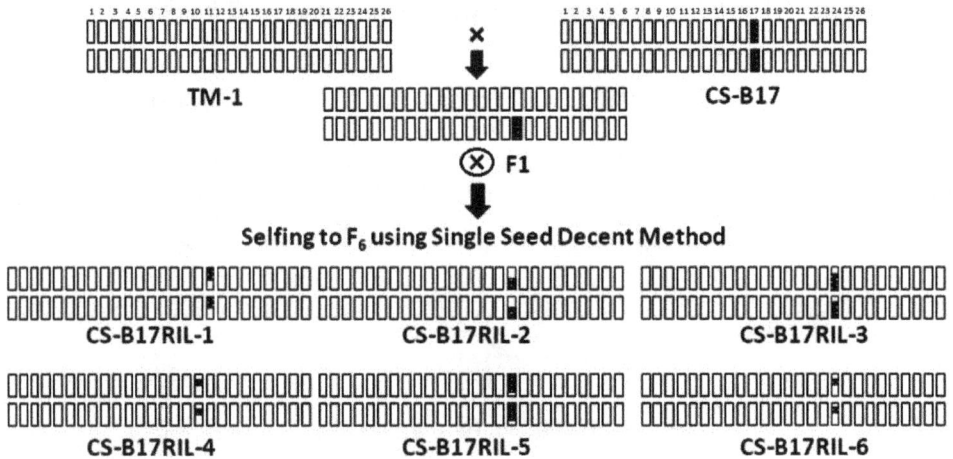

Fig. 6. The overall strategy on the development of a chromosome specific recombinant inbred line (CS-RIL) population.

different species will provide the opportunity to create a tri-species $F_1$ plant where the substituted chromosomes will be a heterozygous chromosomes from two donor alien species, while the other 25 chromosome pairs will be from the recurrent species of *G. hirsutum*. Selfing such a $F_1$ plant will create the $F_2$ progenies segregating primarily for the genes located on the substituted heterozygous chromosomes from two donor species. Such a RIL will help to unveil epistasis, a difficult gene interaction to measure, due to interactions of the alleles from the two donor alien species with the alleles from other chromosomes of Upland cotton. Also, the crosses between two different substitution lines from the same donor alien species can provide unique opportunity to combine beneficial alleles from two alien substituted chromosomes of our interest (Fig. 7). Such genetic materials can be used as valuable tools for high-resolution genetic mapping and targeted chromosome specific introgression of valuable traits from wild and unadapted species.

## 5. Discovery of alien chromosomal effects on important traits using the CS lines

We have used several types of family structure, experimental design and statistical analysis to discover the alien chromosomal effects on important traits using the CS lines (Saha et al., 2004, 2006, 2008, 2010, 2011a; Jenkins et al., 2006, 2007; McCarty et al., 2006; Wu et al., 2006). Studies have enabled the discovery of some important chromosomal effects on agronomic and fiber quality traits, and point toward additional opportunities, as more CS lines are developed.

Each substitution line is nearly isogenic to the common parent TM-1 for 25 chromosome pairs; pairs of non-homologous CS lines are nearly isogenic to each other for 24 chromosome pairs. Given n=26, each CS contains about 4% of the *G. barbadense* genome in a common *G. hirsutum* background (~96%). The near-isogenic nature of CS lines with TM-1 and each other renders them extremely amenable to dissection of quantitative traits of interest because they

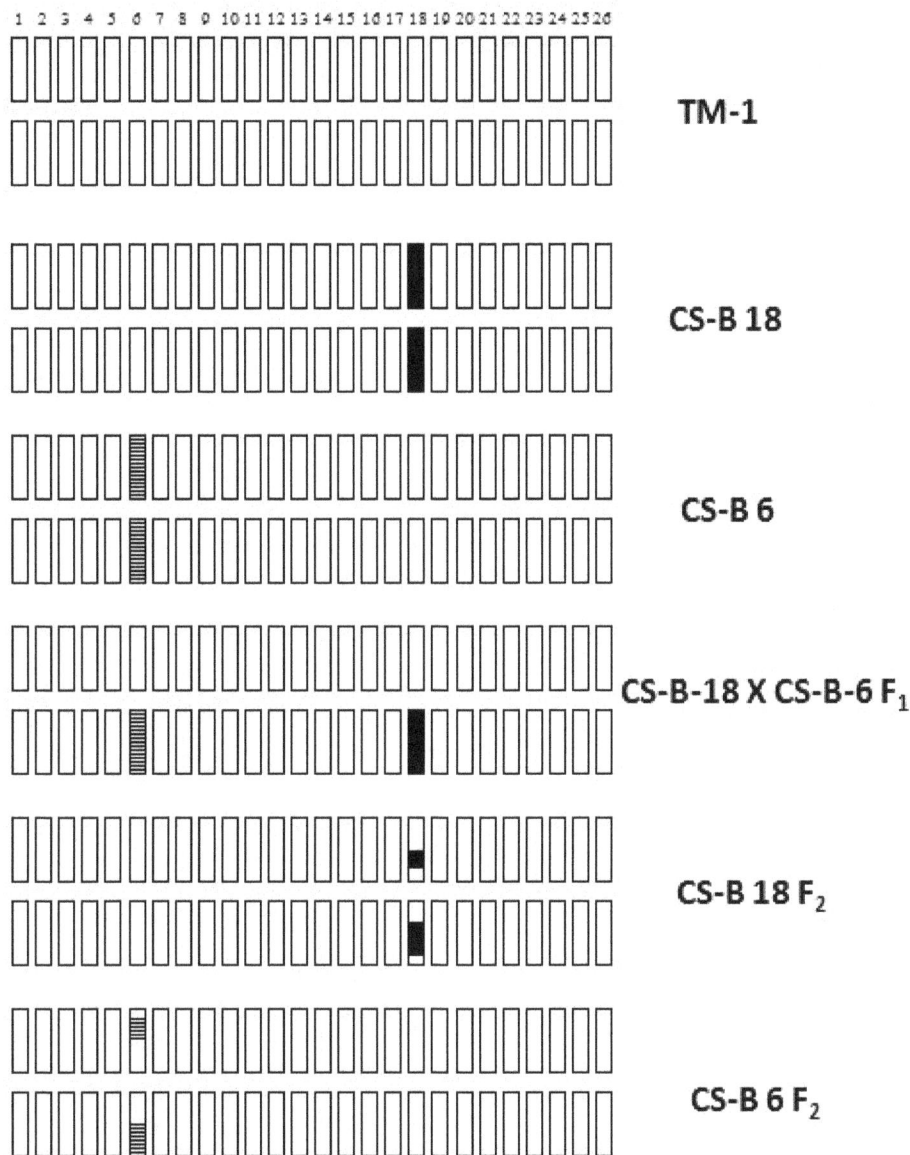

Fig. 7. Diagrammatic figure showing some representative materials developed from crossing different CS lines.

largely eliminate genetic "noise" from 96% of the genome. This greatly facilitates the detection of genetic effects of novel alleles by specific alien chromosomes on quantitative traits. We demonstrated that interspecific chromosome substitution is among the most powerful means of introgression and steps toward identification of chromosomal association with a QTL in cotton.

As a matter of thoroughness, let us note that many of the *G. hirsutum* aneuploids used to develop CS lines were first discovered in non-TM-1 lines; they were backcrossed multiple times to TM-1 to create TM-1-like hypoaneuploids, which served as recurrent parents for CS line development. Through repeatedly backcrossing, it is nonetheless likely that small residual pieces of the original aneuploid remain in the recurrent parents with TM-1 background used to create the CS lines. These could conceivably affect chromosomal associations of some traits. Similarly, note that residual genetic contributions from the donor, e.g., *G. barbadense* '3-79' could also affect CS lines, because 5-6 backcrosses is unlikely to eliminate 100% of alien germplasm in the non-substituted portion (~96%) of the genome. Thus, there is also a possibility that the observed genetic effects could have been due to some unlinked residual effect of the donor (3-79) genome, i.e., independent of the substituted chromosome or chromosome arm (Saha et al., 2006). For statistical quantitative genetic calculations, however, we have considered that the differences in any trait among the lines exclusively due to the contribution of the alien species substituted chromosome on the assumption of isogenic nature of non-substituted chromosomes.

## 5.1 Direct comparison of the CS lines

The comparative analysis of CS-B lines in a uniform genetic background has provided an opportunity to detect net genetic effects of agronomic and fiber quality traits from all genes on the specific substituted chromosome or chromosome arms (Saha et al., 2004). Empirically, one can predict that single cotton chromosome contain an average of 1000-4000 genes. The mode of development of these CS-B lines uses hemizygosity to preclude recombination during introgression, so all genes within the alien chromosome or chromosome segment are transferred into Upland cotton. The CS-B line evaluation for QTL localization does not require segregating populations, so it differs from traditional QTL mapping, and offers certain statistical advantages.

A mixed linear model of genotype with genotype by environment (G × E) interaction model was used in the analysis of results (Saha et al., 2004). The results showed that the genotypic effects greatly exceeded G × E interaction effects for all quantitative traits. The residual variance accounted for less than 20% contribution to the phenotypic variance for all traits except elongation, suggesting that the genotypic effects or G × E interaction effects were readily detected. CS-B25 had significantly lower micronaire than TM-1 (Fig 8). The detailed results of this research were presented in Saha et al. (2004). Seven CS-B lines (2, 6, 16, 18, 5sh, 22Lo, 22sh) had greater lint percentage than 3-79 or TM-1. Micronaire for substituted chromosomes 17 and 25 was lower than TM-1 (Fig. 8). Substituted chromosome 17 line had a significantly greater fiber elongation than either TM-1 or 3-79. Fibers of substituted chromosomes 2 and 25 were significantly stronger than TM-1. The results showed that backcrossed chromosome substitution lines had both positive and negative net effects on various fiber traits. Moreover, they clearly showed the potential of some of the backcrossed chromosome substitution lines from comparative analysis for improving Upland cotton germplasm.

## 5.2 Comparison of CS lines and their F₂ hybrids

The quantitative genetic analysis of CS lines can be extended by collective analysis of the lines and various hybrid generations, which allows partitioning of source effects. We initiated

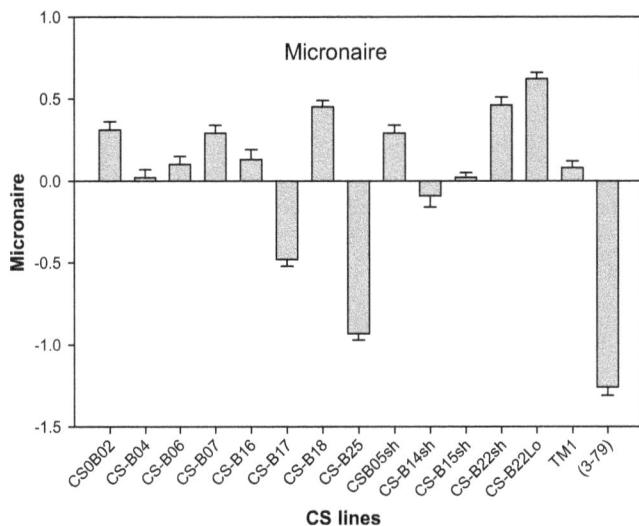

Fig. 8. The comparative analysis of CS-B lines showed CS-B25 has reduced micronaire compared to all other CS-B and TM-1 lines.

such analyses by characterizing the chromosomal association of important agronomic and fiber traits by comparative analysis of CS-B lines in a TM-1 background along with their parental lines and their $F_2$ hybrids (Saha et al., 2006; Saha et al., 2008). Given the isogenic nature of the whole-chromosome CS-B lines, the expected relative genetic complexity for single-locus effects would be approximately inversely proportional to the haploid chromosome number, $1/n$. For multilocus interactions, the reductions would be geometric and thus much more extreme. In CS-B $F_2$'s, segregation is largely to completely eliminated for 25 of the 26 chromosomes (about 96% of genome), rendering CS-B $F_2$ analyses relatively free of the extensive numbers and kinds of interchromosomal interactions that arise in a conventional interspecific $F_2$ population. The CS-B $F_2$'s thus provided an opportunity to discern effects of alleles in homozygous versus heterozygous conditions, on a chromosome-by chromosome or arm-by-arm basis. Multiple comparisons such as this can be used to determine if a substituted chromosome(s) or a chromosome arm(s) is associated with a quantitative trait of interest.

The analyses were facilitated by use of an additive dominance (AD) genetic model with G × E interaction for genetic analysis of the parental lines and their $F_2$ hybrids (Zhu 1994, Tang et al. 1996). Whereas additive and additive epistatic effects were confounded in the comparison of homozygous CS-B and parental lines in Saha et al. (2004), application of the AD model to $F_2$ generations enabled dissection of genetic effects into additive and dominance effects. In this model, additive genetic effect reveal an estimate of general combining ability due to the presence of specific alien chromosome or chromosome arm of 3-79 in TM-1 background. Thus, evaluating the additive effects may help cotton researchers choose an appropriate CS-B line as a source of good general combining ability in cultivar development. Heterozygous dominance effects can be considered as an estimate of specific combining ability (SCA) of parents in specific hybrid combinations. The dominance effects include homozygous and heterozygous effects which are related to inbreeding depression and heterosis (Jenkins et al., 2006).

The genetic background of the CS lines and the comparative analysis among the lines and their hybrids provided an opportunity to dissect the effect of each chromosome or chromosome segment under four different conditions: (i) the homozygous condition in the TM-1 genetic background, totally lacking the alien 3-79 chromosome or chromosome segment, ii)the homozygous condition of euploid CS lines in the TM-1 genetic background, iii) the average heterozygous condition in an $F_2$ generation segregating for the TM-1 and 3-79 alleles in the specific substituted chromosome or chromosome arm and iv) the homozygous condition in the 3-79 genetic background. Additive and dominance genetic effects were detected when $F_2$ hybrids and parents were evaluated (Saha et al., 2006, 2008). Additive effects were predicted based on the genetic effects of CS lines in TM-1 genetic background, TM-1 and 3-79 lines. On the other hand, dominance effects were predicted based on the results of the homozygous condition (euploid CS lines) in the TM-1 genetic background, TM-1, and 3-79 and heterozygous condition in the hybrids between CS lines and TM-1 (Saha et al., 2006; Saha et al., 2008). For example, results revealed that the alien chromosome 2 and 25 had significant net positive additive effects on fiber strength compared to TM-1, suggesting that these chromosomes carry genes for improving fiber strength and the alien chromosomes 17 and 25, respectively, had negative additive and homozygous dominance effects on micronaire, indicating that these chromosomes carry genes potentially useful for manipulation of micronaire (Saha et al., 2006, 2008).

## 5.3 Comparison of parental lines with their $F_2$ and $F_3$ hybrids

Six CS-B lines (CS-B14sh, CS-B16, CS-B17, CS-B22sh, CS-B22Lo, and CS-B25) and TM-1 (the recurrent parent) were crossed in a half diallele mating design and agronomic and fiber properties of the parental lines were compared with their $F_2$ and $F_3$ hybrids (Saha et al., 2010, 2011a). By applying the additive-dominance model, we were able to partition genetic effects into additive, dominance, additive and additive interaction effects for each of the substituted chromosome or chromosome arm (Saha et al., 2010; Saha et al., 2011a). An extended AD model including additive-by-additive epistatic effects (ADAA model) was used to estimate variance components and to predict genetic effects (Wu et al. 2006). The additive-dominance model provided means to dissect further the genetic effects into additive, dominance, additive and additive interaction effects for each of the substituted chromosome or chromosome arm in these studies (Saha et al., 2010, 2011a). We also used the method of Patterson (1939) to test the significance of the difference between the genetic effects for two lines based on the standard error of the difference between two effects. We used a one-tailed $t$-test to estimate the significance of variance components, and a two-tailed $t$-test for estimating the genetic effects. A significant difference in genetic effects between a specific CS-B line and TM-1 was considered a genetic effect attributable to the specific substituted chromosome or chromosome arm. Assuming uniform genetic background of the CS-B lines, the comparative analysis of the double-heterozygote combinations (CS-B × CS-B $F_1$) versus their respective single heterozygotes from the cross of the respective CS-B line and TM-1 with their $F_1$, $F_2$ and $F_3$ progenies revealed that epistatic effects between the genes in the chromosomes strongly affected most of the fiber quality and agronomic traits.

A summary of important genetic effects is presented in this section from these studies (Saha et al., 2010 and 2011a). CS-B14sh and CS-B25 had the highest Upper half means (UHM), and both were significantly longer than TM-1. The average UHM was higher in the hybrids of

CS-B17 × CS-B22sh and CS-B22sh × CS-B22Lo than their respective parents, which seems to indicate dominance effects of the alien allele(s) caused the hybrid vigor. Although fiber strength of the donor, 3-79, greatly exceeds that of the recurrent parent, TM-1, we observed that only one of the six substitution lines, CS-B14sh, had fiber significantly stronger than TM-1. The micronaire values of CS-B17 and CS-B25 were significantly lower than all other CS-B lines and TM-1. The $F_1$, $F_2$ and $F_3$ hybrids between these two lines, CS-B17 × CS-B25, also had exceptionally low micronaire relative to the hybrids of other CS-B lines, revealing the respective 3-79 chromosomes carry genes that can reduce micronaire, a positive breeding value. We observed positive additive effects greater than TM-1 for CS-B14sh and CS-B25 for fiber length, strength, and uniformity. Almost all (>98%) CS-B lines and their crosses had additive-by-additive epistatic effects on fiber quality traits. Additive-by-additive interaction effects for uniformity ratio were higher for most of the CS-B17 hybrids and lower for CS-B22sh hybrids than their respective CS-B parents, suggesting opposite epistatic interaction between the alleles in homozygous versus heterozygous condition for this trait. CS-B16 and CS-B17 had negative additive-by-additive interaction effects for fiber strength when carrying the only alien chromosome in homozygous condition or heterozygous condition for the respective single alien chromosome, however, when the hybrid between CS-B16 and CS-B17 carrying both of the alien chromosomes in heterozygous condition showed high positive additive-by-additive interaction effect for fiber strength implying that the epistatic interaction between the alleles in heterozygous condition of the two substituted chromosomes caused the increased genetic effect in fiber strength.

CS-B16, CS-B22sh, and CS-B22Lo lines had higher additive genetic effects on lint percentage compared to TM-1 (Saha et al., 2010). All of the lines, except CS-B17, had opposite dominance effects on seedcotton yield in heterozygous versus homozygous conditions specific to the substituted chromosomes. CS-B14sh, CS-B22Lo and CS-B25 lines had higher dominance effects on both seed cotton and lint yield under homozygous condition compared to TM-1. The majority of the hybrids had positive additive and additive epistatic effects. Hybrids of CS-B16 x CS-B22sh had the highest additive and additive epistatic effect on boll weight among all others thus the epistatic interaction of the genes located on these two substituted chromosome likely responsible for this effect. We documented in this research that epistasis made a substantial contribution to each of the complex quantitative trait loci (QTLs) showing the effect of masking of alleles at one gene locus by an allele at another locus under homozygous and heterozygous condition for the same specific chromosome on complex agronomic traits. We detected for the first time in this research many cryptic alleles located on several 3-79 substituted chromosomes or chromosome arms of 3-79 (*G. barbadense*), commonly associated with poor agronomic qualities including yield, had the potential to improve agronomic traits including seed cotton and lint yield in TM-1 (*G. hirsutum*).

### 5.4 Evaluation of chromosome specific RILs to study genetic effects

Two different CS-RILs populations were developed from crosses of CS-B05sh and CS-B17 *G. barbadense* chromosome substitution lines with their common recurrent parent, inbred *G. hirsutum* TM-1 (Saha et al., 2011b). Each population included 50 CS-RILs, which were used in field trials with commercial varieties DP393 and PHY370 WR in four environments at Mississippi State University, MS. A randomized complete block design with four replications was used in each environment. Four agronomic and five fiber traits were evaluated for these two CS-RIL populations. Mean comparisons among different lines were used subject to ANOVA analysis with least significance difference (LSD) at probability level of 0.05. The collective variation in each CS-RIL population was assessed by cluster analysis, using the Mahalanobis distance and Wald method, where all data were standardized with a variance of one and mean of zero. The detail results of this research will soon be presented in a separate paper.

Both CS-RIL populations showed significant genetic diversity for all of the traits being investigated. One RIL (CS-B05sh RIL5) had stronger fiber than DP393 and two RILs (RIL 5 and 18) had stronger fiber than PHY 370 WR. Sixty percent of CS-B17 RILs had lower micronaire than PHY 370 WR and 46% of CS-B17 RILs had lower micronaire than DP393. Boll weights were higher in 54% and 64% of CS-B17 RILs than DP393 and PHY 370WR, respectively. No RILs in the CS-B17 RIL population had higher lint percentage, seed cotton yield, or lint yield than each of two commercial cultivars. Cluster analysis results showed that the two commercial cultivars were in the same group and quiet distinct from CS-B17 RIL population. However, some of the CS-B05sh RILs were close to DP393 suggesting their potential to improve fiber traits. Results showed that some of these RILs have genetic potential to improve some agronomic and fiber traits in Upland cotton. This research provided a scope for additional resolution in genetic mapping and for the targeted use of exotic germplasm to improve fiber quality in a cotton breeding program.

## 5.5 Top crosses with improved cultivars

In commercial breeding programs, the use of unadapted germplasm has often been restricted to the introgression of traits that have already been identified, are simply inherited and offer a significantly advantageous near-term cost-benefit ratio. However, the breeding value of any new allele for a quantitative trait is dependent on the genetic backgrounds in which it is evaluated. Without introgression, it is difficult to identify the merits of alien genotypes or alleles in applied breeding programs (Jordan et al., 2011). We employed the classic AD genetic model to evaluate the merit of CS-B lines by crossing 13 CS-B lines with five elite cultivars from different cotton breeding companies (Jenkins et al., 2006, 2007). When these lines are crossed to commercial cultivars, these dominance effects included homozygous effects for cultivars and, heterozygous effects between CS lines and cultivars and TM-1 and cultivars respectively (Jenkins et al., 2006, 2007). The CS lines can be extensively utilized as "probes" in the detection of favorable alleles associated with traits of importance in the specific chromosome or chromosomes arm to improve inbred lines (Wu et al., 2010). Our results provided valuable genetic information to help breeders in the improvement of several traits of interest. Over the years different seed companies developed their improved lines by selecting desirable alleles located in different chromosomes. Discoveries of chromosomes with favorable alleles in different chromosomes of improved cultivars will provide the opportunity to combine beneficial alleles in a crossing program by selecting desirable cultivar lines. Thus we provided for the first time a genetic tool to uncover the desirable allele located in different chromosome of improved cultivars for the traits of interest (Jenkins et al., 2006, 2007). For example, chromosomes 2 and 25 from DP90, PSC355, and FM966 carried alleles for increased additive genetic effect in fiber strength. It is most likely that the use of these lines as parents in crossing program will influence fiber strength in cotton breeding program. We observed that both arms 22sh and 22Lo in FM966 had additive effects associated with increased lint percentage; chromosome 25 and arms 5sh, 14sh, and 22Lo in FM 966 were associated with improved additive effects for boll weight. Chromosome 16 in crosses with SG747 and FM966 was associated with improved additive effects for lint yield. Chromosome 25, and arms 15sh, 22sh, and 22Lo in crosses with FM966 showed with improved additive effects for fiber length. Chromosomes 2 and 25 in FM966 were associated with improved additive effects for fiber strength. Thus, cultivar FM966 has many favorable genetic factors of agronomic and fiber traits associated with different chromosomes or chromosome arms of the CS-B lines. This is especially significant considering that neitherTM-1 or 3-79, the parents of CS-B lines, is considered a line with high breeding values and thus not popular among cotton breeders.

Our studies suggested that CS-B lines could be used as tester stocks to reveal different sources of beneficial alleles, thereby, providing a tool to combine beneficial alleles for improving a fiber quality trait. Thus these CS-B lines open a new paradigm in cotton breeding programs to combine the favorable alleles located in different or same chromosomes by crossing the desirable improved cultivars.

### 5.6 CS lines in the study of developmental genetics

Genes act differentially at different stages of a plant life cycle, where their interactions in a network determine the phenotypes of complex traits. Therefore, discovering genetic changes in time-specific traits at different developmental stages during a growing season has recently become an important issue (McCarty et al., 2006; Wu et al., 2009). Cotton plants grow in an important time-specific manner to produce squares, flowers, bolls, and yields. However, most of the breeding research is based of the data at harvesting period. We used CS lines for time-specific genetic variance components determination using a mixed linear model approach, where phenotypic values observed at time $t$ were conditioned on the events occurred at time $t$-1, revealing the new genetic variations arising at several time intervals during cotton plant growth (McCarty et al., 2006; Wu et al., 2009). In our studies, we used CS-B lines to detect chromosomal associations with the flowering pattern (McCarty et al., 2006), and plant height, number of nodes, and internode length (Wu et al., 2009), major contributors to cotton yield, at different stages in the primary flowering season. We recorded the data every week in the primary growing season. We discovered that the short arm of chromosome 5 of 3-79 in TM-1 background exhibited a positive genetic association with flowering number during this primary flowering time (McCarty et al., 2006). There was no additive genetic effect detected for flower number increase at the initial stage, but additive genetic effects were detected during later stages for flower number increase. On the contrary, large dominance effects were detected for flower number increase for the first two weeks compared to the last two weeks in primary flowering period. We also discovered that the additive effects played a major role for plant height, number of nodes, and internode length at different developmental stages after initial flowering (Wu et al., 2009). We observed that CS-B16 and CS-B14sh had consistently lower additive effects for plant height across different growth stages implying that chromosome 16 and the short arm of chromosome 14 of 3-79 in TM-1 genetic background were associated with shorter plants. On the contrary CS-B26Lo consistently had greater additive effects across all of the growth stages among the CS-B lines suggesting the long arm of chromosome 26 in TM-1 genetic background was associated with taller plants in the primary flowering stages (Wu et al.,

2009). Correlation analysis showed that plant height had significant additive correlations with number of nodes at six different developmental stages suggesting that plant height had common influences with number of nodes during growth after initial flowering in cotton.

## 6. Conclusion

We have developed, released and partially analyzed several backcrossed *G. barbadense* chromosome or chromosome arm substitution (CS-B) lines. More recently, we also have developed analogous chromosome substitution lines involving *G. tomentosum*, which we have dubbed "CS-T" lines. In current research, some of the CS-T lines are being characterized, while others are still being synthesized or increased. Moreover, we are similarly developing monosomic or monotelodisomic chromosome substitution lines from *G. mustelinum*, which are to be named "CS-M" lines.

The experimental results show that the CS lines are useful from at least four different perspectives: 1) to improve genetic diversity for important traits in Upland cotton, 2) to discover the untapped potential of the novel alleles from the other tetraploid species, 3) to understand the ramification of epistasis in complex agronomic and fiber traits and 4) to identify chromosomal locations of important fiber and agronomic traits. Therefore, CS lines have been recommended for many cotton breeding programs worldwide and are being widely used in the cotton research community and by breeders. For instance, application of CS-B lines in improvement of commercialized cultivars of Uzbekistan is in progress within the frame of international collaborative projects.

We presented here a brief review on non-traditional method of interspecific germplasm introgression via the use of alien chromosome substitution lines in cotton. We showed that chromosome substitution line can be a useful tool for additional resolution in genetic mapping and for the targeted exploitation of exotic germplasm to improve fiber quality and agronomic traits in a cotton breeding program.

## 7. Acknowledgements

We thank Ms. Lillie Hendrix and Dr. Russell Hayes for helping in field and Greenhouse research. We acknowledge partial support from the following sources: Texas AgriLife Research, Cotton Inc., Texas State Support Committee, and Texas Dept. Agriculture Food & Fiber Research Grant Program. Joint publication of USDA/ARS, Mississippi Agricultural and Forestry Experiment Station, approved for publication as Journal Article No. J-12079 of the Mississippi Agricultural and Forestry Experiment Station. We thank the Office of International Research Programs (OIRP) of United States Department of Agriculture (USDA) for continual funding of our collaborative research on cotton germplasm characterization. We acknowledge Civilian Research Development Foundation (CRDF), USA for project coordination and Academy of Sciences of Uzbekistan for their continual in-house support of the research efforts.

Disclaimer: Mention of trademark or proprietary product does not constitute a guarantee or warranty of the product by the United States Department of Agriculture and does not imply its approval to the exclusion of other products that may also be suitable.

# 8. References

Balls, W. (1906). Studies in Egyptian cotton, In: Yearbook of the Khedival Agricultural Society, pp. 29-89, Cairo, Egypt

Beasley, J. & Brown, M. (1942). Asynaptic Gossypium plants and their polyploids. *Journal of Agricultural Research*, Vol.65, No. 9, (November 1942), pp. 421–427, ISBN 0-12-017623-8.

Bowman, D.; May, O. & Calhoun, D. (1996). Genetic base of upland cotton cultivars released between 1970 and 1990. *Crop Science*, Vol.36, No.3, (May-June 1996), pp. 577–581, ISSN 1435-0653.

Brown, M.; Menzel, M.; Hasenkampf, C. & Nagi, S. (1981). Chromosome configuration and orientations in 58 heterozygous translocations in Gossypium hirsutum. *Journal of Heredity*, Vol.72, No.3, (May 1981), pp. 161-168, ISSN 1465-7333.

Brown, M. (1980). Identification of the chromosomes of Gossypium hirsutum L. by means of Translocations. *Journal of Heredity*, Vol.71, No.4, (July 1980), pp. 266-274, ISSN 1465-7333.

Endrizzi, J.; Turcotte, E. & Kohel, R. (1985). Genetics, cytology and evolution of Gossypium. *Advances in Genetics*, Vol.23, December, pp. 271-375, ISSN 0065-2660.

Endrizzi, J.; Turcotte, E. & Kohel, R. (1984). Quantitative genetics, cytology and cytogenetics. In: *Cotton*, R.J. Kohel and C.F. Lewis (Eds), 81-129, American Society of Agronomy, ISBN 089118077X, Madison, WI.

Endrizzi, J. & Bray, R. (1980). Cytogenetics of disomics, monotelo- and monoiso-disomics and $Ml_1st_1$ mutants of chromosome 4 of cotton. *Genetics*, Vol.94, No.4, (April 1980), pp. 979-988, ISSN 1943-2631.

Endrizzi, J. & Taylor, T. (1968). Cytogenetic studies of $NL_{c1yg}2R2$ marker genes and chromosome deficiencies in cotton. *Genetics Research*, Vol.12, No.3, (March 1968), pp. 295-304, ISSN 0016-6723.

Endrizzi, J. & Kohel, R. (1966). Use of telosomes in mapping three chromosomes in cotton. *Genetics*, Vol.54, No.2, (August 1966), pp. 535-550, ISSN 1943-2631.

Endrizzi, J. (1963). Genetic analysis of six primary monosomes and one tertiary monosome in Gossypium hirsutum. *Genetics*, Vol.48, No.12, (December 1963), pp. 1625–1633, ISSN 1943-2631.

Esbroeck, G. & Bowman, D. (1998). Cotton germplasm diversity and its importance to cultivar development. *Journal of Cotton Science*, Vol.2, No.3, (Jul-Aug-Sep 1998), pp. 121-129, ISSN 1523-6919.

Guo, Y.; Saha, S.; Yu, J.; Jenkins, J.; Kohel, R.; Scheffler, B. & Stelly, D. (2008). BAC-derived SSR chromosome locations in cotton. *Euphytica*, Vol.161, No.3, (June 2008), pp. 361-370, ISSN 1573-5060.

Gutierrez, O.; Stelly, D.; Saha, S.; Jenkins, J.; McCarty, J.; Raska, D. & Scheffler, B. (2009). Integrative placement and orientation of non-redundant SSR loci in cotton linkage group by deficiency analysis. *Molecular Breeding*, Vol.23, No.4, (May 2009), pp. 693-707, ISSN 1572-9788.

Jenkins, J.; Wu, J.; McCarty, J.; Saha, S.; Gutierrez, O.; Hayes, R. & Stelly, D. (2007). Genetic effects of thirteen Gossypium barbadense L. chromosome substitution lines in topcrosses with Upland cotton cultivars: II Fiber quality traits. *Crop Science*, Vol.47, No.2, (March-April 2007), pp. 561-570, ISSN 1435-0653.

Jenkins, J.; Wu, J.; McCarty, J.; Saha, S.; Gutierrez, O.; Hayes, R. & Stelly, D. (2006). Genetic effects of thirteen *Gossypium barbadense* L. chromosome substitution lines with Upland cotton cultivars: I. Yield and yield component. *Crop Science*, Vol.46, No.3, (May-June 2006), pp. 1169-1178, ISSN 1435-0653.

Jordan, D.; Mace, E.; Cruickshank, A.; Hunt, C. & Henzell, R. (2011). Exploring and exploiting variation from unadapted Sorghum germplasm in a breeding program. *Crop Science*, Vol.51, No.4, (July-August 2011), pp. 1444-1457, ISSN 1435-0653.

Kohel, R. (1978). Monosomic analysis of cotton mutants. *Journal of Heredity*, Vol.69, No.4, (July 1978), pp. 275-276, ISSN 1465-7333.

Kohel, R.; Richmond, T. & Lewis, C. (1970). Texas marker 1. Description of a genetic standard for *Gossypium hirsutum* L. *Crop Science*, Vol.10, No.6, (November-December 1970), pp. 670-671, ISSN 1435-0653.

Lacape Jean-Marc; Jacobs, J.; Arioli, T.; Derijcker, R.; Forestier-Chiron, N.; Llewellyn, D.; Jean, J.; Thomas, E. & Viot, C. (2009). A new interspecific, Gossypium hirsutum × G. barbadense, RIL population: towards a unified consensus linkage map of tetraploid cotton. *Theoretical and Applied Genetics*, Vol.119, No.2, (July 2009), pp. 281-292, ISSN 1432-2242.

McCarty, J.; Wu, J.; Saha, S.; Jenkins, J. & Hayes, R. (2006). Effects of Chromosome 5sh from *Gossypium barbadense* L. on flower production in *G. hirsutum* L. *Euphytica*, Vol.152, No.1, (November 2006), pp. 99-107, ISSN 1573-5060.

Menzel, M. & Brown, M. (1978). Reciprocal chromosome translocations in *Gossypium hirsutum. Journal of Heredity*, Vol.69, No.6, (November 1978), pp. 383-390. ISSN 1465-7333.

Patterson, D. (1939). Statistical technique in agricultural research. McGraw Hill Book Company, ISBN-13: 978-1406771640, New York and London.

Paterson, A.; Boman, R.; Brown, S.; Chee, P.; Gannaway, J.; Gingle, A.; May, O. & Smith, C. (2004). Reducing the genetic vulnerability of cotton. *Crop Science*, Vol.44, No.6, (November-December 2004), pp. 1900-1901, ISSN 1435-0653.

Reinisch, A.; Dong, J.; Brubaker, C.; Stelly, D.; Wendel, J. & Paterson, A. (1994). A detailed RFLP map of cotton, *Gossypium hirsutum* × *Gossypium barbadense*: chromosome organization and evolution in a disomic polyploid genome. *Genetics*, Vol.138, No.3, (November 1994), pp. 829-847, ISSN 1943-2631.

Rhyne, C. (1958). Linkage studies in *Gossypium*. I. Altered recombination in allotetraploid *G. hirsutum* L. following linkage group transference from related diploid species. *Genetics*, Vol.43, No.5, (September 1958), pp. 822-834, ISSN 1943-2631.

Saha, S.; Wu, J.; Jenkins, J.; McCarty, J.; Hayes, R. & Stelly, D. (2011a). Delineation of interspecific epistasis on fiber quality traits in *Gossypium hirsutum* by ADAA analysis of intermated *G. barbadense* chromosome substitution lines. *Theoretical and Applied Genetics*, Vol.122, No.7, (May 2011), pp. 1351-1361, ISSN 1432-2242.

Saha, S.; Wu, J.; Jenkins, J.; McCarty, J.; Stelly, D. & Campbell, B. (2011b). Evaluation of chromosome specific RI lines for improved fiber traits, *Proceedings of the Beltwide Cotton Conference*, ISBN 0000000, Atlanta, Georgia, USA, January 2011.

Saha, S.; Wu, J.; Jenkins, J.; McCarty, J.; Hayes, R. & Stelly, D. (2010). Genetic dissection of chromosome substitution lines discovered novel alleles in *Gossypium barbadense* L. with potential for improving agronomic traits including yield. *Theoretical and Applied Genetics*, Vol.120, No.6, (April 2010), pp. 1193-1205, ISSN 1432-2242.

Saha, S.; Jenkins, J.; Wu, J.; McCarty, J. & Stelly, D. (2008). Genetic analysis of agronomic and fiber traits using four interspecific chromosome substitution lines in cotton. *Plant Breeding*, Vol.127, No.6, (December 2008), pp. 612-618, ISSN 1439-0523.

Saha, S.; Jenkins, J.; Wu, J.; McCarty, J.; Gutierrez, O.; Percy, R.; Cantrell, R. & Stelly, D. (2006). Effect of chromosome specific introgression in Upland cotton on fiber and agronomic traits. *Genetics*, Vol.172, No.3, (March 2006), pp. 1927-1938, ISSN 1943-2631.

Saha, S.; Wu, J.; Jenkins, J.; McCarty, J.; Stelly, D.; Percy, R.; Raska, D. & Gutierrez, O. (2004). Effect of chromosome substitutions from *Gossypium barbadense* L. 3-79 into G. *hirsutum* L. TM-1 on agronomic and fiber traits. *Journal of Cotton Science*, Vol.8, No.3, (Jul-Aug-Sep 2004), pp. 162-169, ISSN 1439-0523.

Shoemaker, D. (1908). A study of leaf characters in cotton hybrids. *Report of American Breeders Association*, Vol.5, December pp. 116-119.

Smith, C. & Coyle, G. (1997). Association of fiber quality parameters and within boll yield components in Upland cotton. *Crop Science*, Vol.37, No.6, (November-December 1997), pp. 1775-1779, ISSN 1435-0653.

Stelly, D.; Saha, S.; Raska, D.; Jenkins, J.; McCarty, J. & Gutierrez, O. (2005). Registration of 17 Upland (*Gossypium hirsutum*) germplasm lines disomic for different G. *barbadense* chromosome or arm substitutions. *Crop Science*, Vol.45, No.6, (November-December 2005), pp. 2663-2665, ISSN 1435-0653.

Sunilkumar, G.; Campbell, L.M.; Puckhaber, L.; Stipanovic, R.D. &Rathore K. S. (2006). Engineering cottonseed for use in human nutrition by tissue-specific reduction of toxic gossypol. *Proc Natl Acad Sci USA*. Vol. 103, No. 48, (November 2006), pp. 18054-18059, ISSN 0027-8424.

Tang, B.; Jenkins, J.; Watson, C.; McCarty, J. & Creech, R. (1996). Evaluation of Genetic variances, heritabilities, and correlations for yield and fiber traits among cotton F2 hybrid populations. *Euphytica*, Vol.91, No.3, (January 2006), pp. 315–322, ISSN 1573-5060.

Tanksley, S. & McCough, S. (1997). Seed banks and molecular maps: Unlocking genetic potential from the wild. *Science*, Vol.7, No.5329, (August 1997), pp. 1063–1066, ISSN 1095-9203.

Wehrhahn , C., and Allard, R.W. 1965. The detection and the measurement of the effects of individual genesinvolved in the inheritance of a quantitative character in wheat. Genetics 51:109-119. ISSN 1943-2631.

Wendel, J. (1989). New World tetraploid cotton contains Old World cytoplasm. *Proceedings of the National Academy of Science*, Vol.86, No.11, (June 1989), pp. 4132-4136 s10709-010-9507-3 (online), ISSN 1091-6490.

White, T. & Endrizzi, J. (1965). Tests for the association of marker loci with chromosomes in *Gossypium hirsutum* L. by the use of aneuploids. *Genetics*, Vol.51, No.4, (April 1965), pp. 605-612, ISSN 1943-2631.

Wu, J.; McCarty, J.C.; Saha, S.; Jenkins, J. & Hayes, R. (2009). Genetic changes in plant growth and their associations with chromosomes from *Gossypium barbadence* L. in G. *hirsutum* L. *Genetica*, Vol.137, No.1, (September 2009), pp. 57-66, ISSN 1573-6857.

Wu, J.; Jenkins, J.; McCarty, J.; Saha, S & Percy, R. (2008). Genetic association of lint yield with its components in cotton chromosome substitution Lines. *Euphytica*, Vol.164, No.1, (November 2008), pp. 199-207, ISSN 1573-5060.

Wu, J.; Jenkins, J.; McCarty, J. & Saha, S. (2010). Genetic effects of individual chromosomes in cotton cultivars detected by using chromosome substitution lines as genetic probes. *Genetica*, Vol.138, No.11-12, (December 2010), pp. 1171-1179, ISSN 1573-6857.

Wu, J.; Jenkins, J.; McCarty, J.; Saha, S. & Stelly, D. (2006). An additive dominance model to determine chromosomal effects in chromosome substitution lines and other germplasms. *Theoretical and Applied Genetics*, Vol.112, No.3, (February 2006), pp. 391-399, ISSN 1432-2242.

Zhang, Y.; Lin, Z.; Xia, Q.; Zhang, M. & Zhang, X. (2008). Characteristics and analysis of simple sequence repeats in the cotton genome based on a linkage map constructed from a BC1 population between *Gossypium hirsutum* and *G. barbadense*. *Genome*, Vol.51, No.7, (July 2008), pp. 534–546, ISSN 1480-3321.

Zhu, J. (1994). General genetic models and new analysis methods for quantitative traits. *Journal of Zhejianiang Agricultural University*, Vol.20, No.6, (November-December 1994), pp.551-559, ISSN 1008-9209.

# 6

# Use of 2n Gametes in Plant Breeding

A. Dewitte[1], K. Van Laere[2] and J. Van Huylenbroeck[2]
*[1]KATHO Catholic University College of Southwest Flanders,*
*Department of Health Care and Biotechnology,*
*[2]Institute for Agricultural and Fisheries Research (ILVO),*
*Plant Sciences Unit,*
*Belgium*

## 1. Introduction

Genome doubling (polyploidization) has played a major role in the evolution and diversification of the plant kingdom and is regarded as an important mechanism of speciation and adaptation in plants (Otto & Whitton, 2000). The term ploidy refers to the number of basic chromosome sets (represented by 'x') present in a somatic plant cell (2n) or gamete (n). Scaling whole sets of chromosomes up or down is a powerful and commonly applied strategy to produce altered genotypes for breeding purposes.

Polyploids differ from their diploid progenitors in morphological, ecological, physiological and cytological characteristics. Their wider geographical distribution reflects the variety of their ecological tolerances (Carputo et al., 2003; Chen, 2007; Knight et al., 2005; Otto & Whitton, 2000; Soltis & Soltis, 2000; Thompson et al., 2004). Polyploids have breeding value as they can have broader and thicker leaves, larger flowers, longer internodes, fewer stems per plant, fewer inflorescences, higher vegetative yields, more compact plants, greater tolerance to environmental stress, higher (additive) resistance to several diseases, increased seed weight but fewer seeds or higher chlorphyll production. Doubling the chromosome number can rescue and stabilise interspecific hybrids that would otherwise show a high degree of sterility due to failure in meiosis. Furthermore, polyploidy might result in the development of sterile cultivars (e.g. triploids), loss of self-incompatibility, flowering time variation, changes in growth vigour or shifts in reproductive system (e.g. to asexual reproduction). Apomixis, an extreme form of reproductive modification, is commonly associated with polyploidy. Negative side effects of polyploidization might be infertility (which can also be a desired trait), brittle wood, watery fruit, stunting and malformation (Andruezza & Siddiqi, 2008; Baert et al., 1992; Barringer, 2007; Bretagnolle & Thompson, 1995; Briggs & Walters, 1997; Chen, 2007; Chahal & Gosal, 2002; Comai, 2005; Contreras et al., 2007; De Nettancourt, 1997; Eeckhaut et al., 2004; Grant, 1981; Gu et al., 2005; Kermani et al., 2003; Lamote et al., 2002; Otto & Whitton, 2000; Stebbins, 1971; Thomas, 1993; Van Huylenbroeck et al., 2000; Van Laere et al., 2011).

Several phenotypic characteristics have been used as an indirect measure of ploidy level. Polyploids frequently have larger pollen sizes, stomata sizes, and seeds than diploids but have slower developmental rates. Further, the number of chloroplasts in guard cells, leaf area index, stomatal density, or pollen fertility might be related to the ploidy level (Aryavand et al., 2003; Kessel & Rowe, 1975; Mishra, 1997; Otto & Whitton, 2000; Vandenhout et al., 1995; Zlesak, 2009).

In general, polyploidy is accompanied by genome-wide changes in gene expression and epigenetic modifications. The genotypic and phenotypic differences are caused mainly by the increased cell size, gene dosage effect, allelic diversity (level of heterozygosity), gene silencing and genetic or epigenetic interactions (Andruezza & Siddiqi, 2008; Jovtchev et al., 2006; Kondorosi et al., 2000; Leitch & Bennett, 1997; Levin, 1983; Lewis, 1980; Mittelsten Scheid et al., 1996; Osborn et al., 2003; Pikaard, 1999).

Polyploids can be induced by two mechanisms. First, mitosis spindle inhibitors such as colchicine, oryzalin and trifluralin can be applied to create artificial (mitotic) polyploids (for a review see Dhooghe et al., 2011). Second, gametes with somatic chromosome numbers, also referred to as 2n gametes or (numerically) unreduced gametes are considered to be the driving force behind the formation of polyploids in nature (Bretagnolle & Thompson, 1995; Otto & Whitton, 2000).

Harlan & De Wet (1975) showed that almost all plant species produce 2n gametes in some frequencies and argued the importance of 2n gametes in the origin of polyploids. Although efforts to synthesize polyploids through the use of 2n gametes were performed much earlier (e.g. Skiebe 1958), it was assumed that 2n gametes occur only rarely and with little contribution to the origin of polyploids (Ramanna & Jacobsen, 2003). The use of 2n gametes in plant breeding, resulting in the establishment of sexual or meiotic polyploids, has been very useful for crop improvement (Ramanna & Jacobsen, 2003). These gametes combine the genetic effects of raised ploidy level with meiotic recombination, which makes them an attractive alternative for mitotic chromosome doubling. 2N gametes are an effective and efficient way to transmit genetic diversity (allelic variation) to cultivated forms, including both valuable qualitative and quantitative traits (Peloquin et al., 1999). More recently plant breeders have become interested in the practical use of 2n gametes in breeding programmes due to the new tools available for 2n gamete manipulation and insights into the genetic background of their formation.

This chapter presents a review of the recent advances in the practical breeding applications of 2n gametes. It addresses tools for detection, induction, and enrichment of 2n pollen, as well as the potential to engineer 2n gamete production in agricultural crops.

## 2. 2n gametes: mechanisms and genetic background

2n gametes originate from meiotic aberrations. In a normal meiosis, one mother cell (2n) divides in 4 (n) daughter cells (tetrad formation). Aberrations during chromosome pairing, spindle formation (parallel spindle, tripolar spindle, fused spindle, sequential spindle or lack of spindle) or cytokinesis might result in the formation of triads, dyads or monads (Bretagnolle & Thompson, 1995; Dewitte et al., 2010c; Taschetto & Pagliarini, 2003). This finally results in pollen grains with the same DNA content (2n in the case of dyads or triads) or a doubled DNA content (4n in the case of monads) compared to the somatic 2n plant cell. Premeiotic and postmeiotic chromosome doubling and cytomixis have also been proposed as possible mechanisms for the production of 2n gametes (Bastiaanssen et al., 1998; Falistocco et al., 1995; Ghaffari, 2006; Lelley et al., 1987; Singhal & Kumar, 2008).

Two main categories of 2n gamete formation have been described: first division restitution (FDR), and second division restitution (SDR) (Bretagnolle & Thompson, 1995; Ramanna & Jacobsen, 2003; Zhou et al., 2008). In FDR, the pairing and/or the separation of the homologous chromosomes at meiosis I does not occur (univalent formation) or occurs only at low frequencies, while the second division occurs normally with the two sister chromatids of each chromosome moving to opposite poles. With the exception of cross-over segments, the resulting FDR gametes retain all parental chromosomes. In SDR, the pairing and the separation of the homologous chromosomes during meiosis I occurs normally (bivalent formation). In meiosis II, the centromeres of the half-bivalents divide, but the chromatids do not migrate to the poles. Finally, SDR gametes contain only half of the parental chromosomes as in normal gametes (random combinations), but each of these chromosomes is present twice. Besides these two main categories, several other meiotic aberrations exist where the final chromosome constitution in the microspores is equivalent to the FDR or SDR pathways as described above (Bretagnolle & Thompson, 1995). Less frequently reported are indeterminate meiotic restitution (IMR) and post meiotic restitution (PMR). IMR has been detected in lily (Lim et al., 2001), and shows characteristics similar to both SDR and FDR. PMR, where chromosomes duplicate after meiosis, was observed by Bastiaanssen et al. (1998) in potato. Genomic *in situ* hybridisation (GISH) has made detection of the restitution mechanism possible by monitoring the meiosis of hybrids and identifying the chromosomes of individual genomes in the sexual polyploid progeny.

Examples in maize (*Zea mays*; Rhoades & Dempsey, 1966), potato (*Solanum tuberosum*; Mok & Peloquin, 1975b; Veilleux, 1985; Watanabe & Peloquin, 1989), red clover (*Trifolium pratense*; Parrot & Smith, 1986), rye (*Secale cereale*; Lelley et al., 1987), alfalfa (*Medicago sativa*; Barcaccia et al., 2000; Ortiz & Peloquin, 1991) and banana (*Musa*; Ortiz, 1997) have shown the complexity of the genetic base of 2n gamete formation. Often, one major locus is responsible for 2n gamete formation while several other genes controls its frequency. Some authors tried to map 2n gamete production or developed molecular markers associated with 2n gamete production (e.g. Barcaccia et al., 2000, 2003; Zhang et al., 2007). However, research on the model plant thale cress (*Arabidopsis thaliana*) has led to many recent advances in elucidating the molecular mechanisms as well as the first genes in which mutations result in the production of viable 2n gametes (for a review see Brownfield & Köhler, 2011). d'Erfurth et al. (2008) were the first to successfully isolate and characterize a gene involved in 2n gamete production. The *Arabidopsis thaliana Parallel Spindle1* (*AtPS1*) gene is involved in abnormal orientation of spindles at meiosis II, which controls diploid 2n gamete formation in *Arabidopsis thaliana*. Parallel, fused or tripolar spindles are different phenotypic expressions of this gene. A similar male-specific dyad-triad phenotype was observed and characterised in another mutant called *jason* (De Storme & Geelen, 2011; Erilova et al., 2009). Mutants of two other genes, OSD and TAM (CYCA1;2), were shown to omit the second meiotic division in both male and female sporogenesis at high frequency resulting in the formation of both 2n pollen and egg cells (d'Erfurth et al., 2009, 2010).

## 3. Sources of 2n gametes

Three important sources of 2n gametes are commonly reported (Bretagnolle & Thompson, 1995), beginning with interspecific hybrids. In many genera interspecific or intergeneric hybrids have produced 2n gametes e.g. lily (*Lilium*; Barba-Gonzalez et al., 2005a; Lim et al.,

2001; Lim et al., 2004), Peruvian lily (*Alstroemeria;* Ramanna et al., 2003), wheat (*Triticum;* Xu & Joppa, 1995; Xu & Joppa, 2000; Zhang et al., 2010), *Impatiens* (Stephens, 1998), *Citrus* (Chen et al., 2008), rose of sharon (*Hibiscus;* Van Laere et al., 2009), clover (*Trifolium;* Meredith et al., 1995), and the *Festuca/Lolium* complex (Gymer & Whittington, 1973; Morgan et al., 1995). Frequently, these interspecific hybrids show abnormal or absent chromosome pairing and the presence of univalents, lagging chromosomes and chromosome bridges (Islam & Shepherd, 1980; Del Bosco et al., 1999; Trojak-goluch & Berbeć, 2003). Interspecific hybrids usually share two important features. First, both 2n eggs and 2n pollen are produced simultaneously by the same hybrid. Second, neither the two parents of the F$_1$ hybrids nor their (F$_2$) sexual polyploid progenies can produce 2n gametes in any notable frequencies (Ramanna & Jacobsen, 2003).

Meiotic mutants are a second source of 2n gametes. A mutation in a gene active in meiosis might disturb during spindle formation or cytokinesis, resulting in 2n gametes. As different genes are active within the micro- and macrosporogenesis, 2n pollen can be formed independent from 2n egg cells and vice versa. Meiotic mutants have been described in potato (Jongedijk et al., 1991; Peloquin et al., 1999; Ramanna, 1983), red clover (Parrot & Smith, 1984), alfalfa (Barcaccia et al., 2003), wheat (Jauhar, 2003; Roberts et al., 1999) and *Arabidopsis* (d' Erfurth et al., 2008; Yang et al., 1999), among others.

A third source are odd polyploids. For instance, crosses with triploids revealed that euploid gametes of triploids can be 1x, 2x or 3x (Husband, 2004; Van Huylenbroeck et al., 2005). Although these 2x gametes are not exactly 2n gametes (they are 3x gametes), they result in higher ploidy levels of the progeny and mostly act as a bridge between diploids and tetraploids (Köhler et al., 2010).

## 4. Practical use of 2n gametes in plant breeding

### 4.1 Detection of 2n gametes

2n gametes must be correctly identified when used in ploidy breeding programmes. Most detection techniques focus on pollen, since it is more convenient to isolate than egg cells. 2N pollen can be detected in four ways (Bretagnolle & Thompson, 1995): pollen size measurements, flow cytometric detection of pollen DNA content, analysis of the microsporogenesis and ploidy analysis of the progeny. In Table 1, an updated overview is given of the different methods used in several crops since Bretagnolle & Thompson's 1995 review. Only the last two of the above-mentioned techniques (ploidy and macrosporogenesis analysis) can be used to detect 2n egg formation. The frequency of 2n egg formation has often been estimated after crosses between 2x x 4x plants. A cross between a diploid seed parent and tetraploid pollen parent will only result in good tetraploid seed when 2n egg cells are present (Conicella et al., 1991; De Haan et al., 1992; Erazzu & Camadro, 2007; Estrada-Luna et al., 2004; Jongedijk, 1987; Lamote et al., 2002; Ortiz & Peloquin, 1991; Veronesi et al., 1986; Van Laere et al., 2009; Werner & Peloquin, 1987). In *Triticum-Aegilops* hybrids, Zhang et al. (2007a, 2010) used the selfed seedset as a good indication for the formation of 2n gametes.

An easy and commonly used method to screen for 2n pollen is searching for large pollen within a population. Large pollen has frequently been attributed to 2n pollen in many genera. This association is caused by the positive correlation between DNA content and cell

| PS | FC | AP | MS | Crop | Reference |
|----|----|----|----|------|-----------|
| \<br>**Detection method** | | | | **Crop** | **Reference** |
| x |  |  | x | *Achillea borealis* | Ramsey, 2007 |
|  | x |  | x | *Actidinia spp.* | Yan et al., 1997 |
| x |  | x | x | *Alstroemeria spp.* | Ramanna et al., 2003 |
| x |  |  |  | *Anthoxanthum alpinum* | Bretagnolle, 2001 |
| x | x | x | x | *Begonia spp.* | Dewitte et al., 2009 |
|  |  |  | x | *Brachiaria spp.* | Gallo et al., 2007 |
|  | x |  |  | *Brassica napus* | Pan et al., 2004 |
| x |  | x |  | Cacti hybrids | Tel-Zur et al., 2003 |
|  |  | x | x | *Capsicum annuum* | Yan et al., 2000 |
| x | x |  |  | *Cupressus spp.* | Pichot & El Maataoui, 2000 |
| x | x | x |  | *Diospyros kaki* | Sugiura et al., 2000 |
| x |  |  | x | *Diospyros spp.* | Xu et al., 2008 |
|  | x | x | x | *Durum wheat* | Jauhar, 2003 |
| x | x | x | x | *Hibiscus spp.* | Van Laere et al., 2009 |
| x |  |  | x | *Ipomoea batatas* | Becerra Lopez-Lavalle & Orjeda, 2002 |
|  | x | x | x | *Lilium spp.* | Lim et al., 2004 |
| x | x | x |  | *Lilium spp.* | Akutsu et al., 2007 |
| x |  | x |  | *Lilium spp.* | Barba-Gonzalez et al., 2004 |
| x |  | x | x | *Lotus tenuis* | Negri & Lemmi, 1998 |
| x |  |  |  | *Musa spp.* | Ortiz, 1997; Ssebulita et al., 2008 |
|  |  |  | x | *Paspalum spp.* | Pagliarini et al., 1999 |
|  |  |  | x | *Pfaffia spp.* | Taschetto & Pagliarini, 2003 |
|  |  |  | x | *Populus tomentosa* | Zhang & Kang, 2010 |
| x |  |  | x | *Populus hybrid* | Wang et al., 2010 |
| x |  |  | x | *Rhododendron spp.* | Jones & Ranney, 2009 |
| x |  |  | x | *Rosa hybrida* | Crespel et al., 2006 |
| x |  | x | x | *Rosa hybrida* | El Mokadem et al., 2002a |
|  | x |  |  | *Rosa rugosa* | Roberts, 2007 |
| x |  |  |  | *Rosa spp.* | Zlesak, 2009 |
| x |  | x |  | *Trifolium pratense* | Simioni et al., 2006 |
| x | x | x |  | *Tulipa spp.* | Okazaki et al., 2005 |
| x |  |  | x | *Ziziphus jujube* | Xue et al., 2011 |

Table 1. Overview of studies on the detection of 2n pollen since Bretagnolle & Thompson's 1995 review. FC = Flow cytometry on pollen; PS = Pollen size measurements; AP = Ploidy analysis of progeny; MS = Microsporogenesis analysis

volume which in turn influences pollen diameter. In crops as Japanese persimmons (*Diospyros kaki*), banana, rose (*Rosa*) and sweet potato (*Ipomoea batatas*), the diameter of the 2n pollen was approximately 30% larger than that of the n pollen (Becerra Lopez-Lavalle & Orjeda, 2002; Crespel et al., 2006; Ortiz, 1997; Sugiura et al., 2000). In Chinese jujube (*Ziziphus jujube*) 2n pollen was more than 1.5 times larger compared to n pollen (Xue et al., 2011). The presence of large 2n pollengrains results in a bimodal pollengrain size distribution instead of a normal distribution. Although the size distribution of normal and giant pollen grains show some overlap, a threshold value of the pollen grain size is often used to select individuals that produce 2n gametes (e.g. Crespel et al., 2006; Ortiz, 1997; Sugiura et al., 2000). The presence of large pollen only indicates the presence of 2n pollen but does not present proof of doubled DNA content. Another disadvantage of this screening technique is the broad overlap in size distribution between small and large pollen in some genera such as grasses. In these cases, the frequency of 2n pollen based on pollen size only is difficult to determine (Jansen & Den Nijs, 1993). A bimodal size distribution, on the other hand, can also be related to a population of small unviable and normal pollengrains as was observed in *Hibiscus* (Van Laere et al., 2009). Other methods are necessary to confirm the association between large and 2n pollen, and supplementary evaluation of pollen viability is necessary for breeding purposes. Besides pollen size also pollen shape (spherical instead of ellipsoidal) has been associated with the ploidy level, which simplifies the determination of 2n pollen (Akutsu et al., 2007; Dewitte et al., 2009; Ramanna et al., 2003).

A more stringent method is the direct quantification of nuclear pollen DNA using flow cytometry. To determine the DNA content of pollen, nuclei have to be released from pollen and purified from the pollen wall. Several enzymatical, chemical, mechanical and osmotical methods for nuclear isolation of pollen have been used in different plants, e.g. maize, *Plumbago zeylanica*, lily hybrids, tobacco (*Nicotiana tabacum*) and rape (*Brassica napus*) (Dupuis et al., 1987; Pan et al., 2004; Russel, 1991; Van Tuyl et al., 1989; Xu et al., 2002; Zhang et al., 1992). The presence of a complex outer exine layer on the pollen surface is the main obstacle in releasing the nuclei from pollen. This outer layer is a biopolymer that is highly resistant to enzymatic breakdown and hydrolytic decomposition in strong acid or alkaline media (Bohne et al., 2003). The isolation of nuclei from pollen is often difficult and the efficiency of nuclear isolation techniques must be investigated each time when a new genus is explored.

Flow cytometric analysis compares the DNA content of pollen nuclei to the DNA content of somatic leaf tissue. Pollen nuclei are expected to have only half of the DNA content (1C) compared to nuclei from somatic cells (2C) of the same plant. Consequently, 2n pollen have a nuclear DNA content equal to somatic cells. This is true for species as *Brassica napus* or *Triticum aestivum* (Pan et al., 2004). However, several reports on crops from the genus *Begonia* (Dewitte et al., 2009), *Lilium* (Van Tuyl et al., 1989), *Cupressus* (Pichot & Maâtaoui, 2000), *Hibiscus* (Van Laere et al., 2009) and *Rosa* (Roberts et al., 2007) have shown that pollen nuclei can be at the G2 phase of cell division, and have a temporate doubled DNA content (2C) which is equal to the DNA content of somatic cells. In *Begonia* for instance, analysis of normal (binucleate) pollen results in 2 peaks in a flow cytometric histogram at the 1C (vegetative nuclei) and 2C level (generative nuclei). If 2n pollen are present, a peak at the 4C level of the 2n generative nuclei can be observed (Dewitte et al., 2009).

Although flow cytometric screening of 2n pollen can be used routinely in breeding programmes, sample preparation for flow cytometric screening generally requires more time than for microscopic evaluation. Therefore, the use of flow cytometry can be limited to genotypes which produce pollen grains with highly variable sizes to confirm the occurrence of 2n pollen grains. Some quick physical techniques to isolate pollen nuclei, such as bead beating (Roberts, 2007) or chopping of pollen grains (Pichot & Maâtaoui, 2000; Sugiura et al., 2000; Van Laere et al., 2009; Van Tuyl et al., 1989), may speed up flow cytometric screening of 2n pollen.

Moreover, Dewitte et al. (2006, 2009) developed a nuclear isolation protocol which only releases nuclei from germination tubes. As a result, flow cytometric analysis is based only on viable pollen grains and no supplementary evaluation of pollen viability is necessary.

The presence of 2n pollen is associated with the occurrence of monads, dyads or triads during microsporogenesis, except when 2n gamete formation is the result of pre- or postmeiotic restitution. Analysis of microsporogenesis may therefore provide an alternative method to confirm the presence of 2n pollen but this method does not provide any information about pollen viability. Moreover, what is observed in the meiocytes is not necessarily reflected in the mature pollen (Dewitte et al., 2010a), since the production of n and 2n pollen also depends on balanced chromosome segregation during meiosis and further maturation steps after meiosis. Scoring the meiotic products is thus not the best method to determine the frequency of the final 2n pollen. Although these observations give no information about the viability of 2n pollen, they provide insight in the mechanisms (FDR, SDR) behind 2n gamete formation. The identification of the mechanisms behind the formation of 2n gametes is however complex, because different cytological mechanisms may operate within one individual. The use of molecular cytological techniques (genomic in situ hybridization, GISH or fluorescent in situ hybridization, FISH) or marker analysis (such as amplified fragment length polymorphism, AFLP) on meiocytes or polyploid progeny may provide more accurate or additional information on the mechanisms behind 2n gamete formation (Barba-Gonzalez et al., 2005b; Crespel et al., 2002; Lim et al., 2001). Molecular cytological approaches have been successfully used in the case of allopolyploids, where the constituent genomes can be clearly discriminated. This includes the unequivocal identification of not only genomes and individual alien chromosomes but also recombinant segments in the sexual polyploid progenies. Through DNA in situ hybridisation, genomes of allopolyploids can be more critically assigned and intergenomic translocations and recombinations can be detected such as in *Gasteria-Aloe* hybrids (Takahashi et al., 1997), *Alstroemeria* species (Ramanna et al. 2003) and *Lilium* species (Barba-Gonzalez et al., 2005; Karlov et al., 1999; Lim et al., 2001). As such, GISH can also be used to discover the mechanism of 2n gamete formation (Karlov et al., 1999).

Finally, ploidy analysis of the progeny (usually using flow cytometry) can reveal the presence of 2n gametes in parent plants. Progeny analysis has the advantage of indicating the existence of viable 2n pollen in parent plants, but 2n pollen in parent plants may remain unnoticed because of preferential pairing between normal gametes. This technique has the disadvantage of being very time consuming with no guarantee of information about the production frequency in the parent plant (Bretagnolle & Thompson, 1995) due to possible differences in pollen viability, germination speed or pollen tube growth between n and 2n pollen.

Although different techniques can be used to detect 2n pollen, frequencies should be considered carefully because they depend on the method used (Dewitte et al., 2009). Moreover, frequencies may vary in different populations, within a certain genotype and during time (season) due to environmental fluctuations (Bani-Aameur et al., 1992; Crespel et al., 2006; McCoy, 1982; Ortiz et al., 1998; Ortiz & Vulsteke, 1995; Parrott & Smith, 1984; Pécrix et al., 2011; Veilleux et al., 1982). In Table 2, some recently published data on 2n gamete frequency in different (hybrid and non hybrid) populations are given. In general, the proportion between normal and 2n gametes over an at random sampled population (the mean population frequency, Fm) is low, close to 1%. However, the proportion of plants producing 2n gametes in a population (population frequency, Fp) as well as 2n gametes formation within an individual genotype (individual frequency, Fi) can be very variable. In several reports, Fp is higher than 10% and in some genera as *Ipomoea*, poplar (*Populus*) or *Pfaffia*, Fp values are noticed higher than 80%. A similar conclusion can be drawn from Fi, where values are reported from less than 1% up to 100% in *Begonia* (Dewitte et al., 2009).

As a general rule, it appears that approximately 0.1 to 2.0% of the gametes in a nonhybrid population are expected to be unreduced (Ramsey, 2007). However, the mean frequency of 2n gametes found in studies of hybrids (27.52%) was nearly a 50-fold greater than in nonhybrids (0.56%) (Ramsey and Schemske, 1998). Hence, the chance to find individuals that produce 2n gametes increases when hybrids are investigated.

| Crop | No of genotypes | Fp[1] (%) | Fi[2] (%) | Fm[1] (%) | Reference |
|------|------|------|------|------|------|
| *Achillea borealis* | 50-75 | 34.7-50.0 | 15.8 | 0.03-0.54 | Ramsey, 2007 |
| *Anthoxanthum alpinum* | 31-78 | 9-12.8 | 39.5 | 0.8-1.3 | Bretagnolle, 2001 |
| *Begonia spp.* | 70 | 14.3 | 100 | - | Dewitte et al., 2009 |
| *Hibiscus spp.* | 15 | 13.3 | 10 | - | Van Laere et al., 2009 |
| *Ipomoea batatas* | 64 | 86 | 84.2 | - | Becerra Lopez-Lavalle & Orjeda, 2002 |
| *Lolium perenne* | 154 | 9.7 | - | - | Lamote et al., 2002 |
| *Musa spp.* | 156 | 14-56 | - | - | Ortiz, 1997 |
| *Paspalum spp.* | 112 | 46.4 | - | - | Pagliarini et al., 1999 |
| *Pfaffia spp.* | 14 | 100 | 18.9 | - | Taschetto & Pagliarini, 2003 |
| *Populus tomentosa* | 224 | 97.3 | 21.9 | 1.8-7.5 | Zhang et al., 2007b |
| *Rosa spp.* | 53 | 26.4-50.9 | 9.6 | 1-2 | Crespel et al., 2006 |
| *Solanum okadai* | 118 | 20.3 | 5 | | Camadro et al., 2008 |

Table 2. Examples of 2n gamete frequencies in different crops (since Bretagnolle & Thompson's 1995 review). Fp: population frequency; Fi: individual frequency; Fm: mean population frequency. [1]Two values represent a frequency interval in which the value is dependent on the population or season. [2]The highest reported value is represented

## 4.2 Induction of 2n gametes

To date, the major drawback to use 2n gametes in plant breeding is that only a minority of genotypes regularly produce 2n gametes. Identifying such genotypes in the breeding stock thus requires a great deal of screening. Additionally, not many superior genotypes produce these gametes. Recently different attempts, with variable success, were made to induce these gametes in any genotype of interest. An overview of the methods used to induce 2n gametes is given in Table 3.

| Induction method | Crop | Reference |
| --- | --- | --- |
| N₂O | *Begonia spp.* | Dewitte et al., 2010 |
| | *Lilium spp.* | Akutsu et al. 2007; Barba-Gonzales et al. 2006; Kitamura et al., 2008; Sato et al., 2010 |
| | *Tulipa spp.* | Okazaki et al. 2005 |
| Trifluralin | *Begonia spp.* | Dewitte et al., 2010 |
| | *Zea Mays* | Kato, 1999 |
| Colchicine | Chinese cabbage | Zhong et al., 2010 |
| | *Lilium spp.* | Wu et al., 2007 |
| | *Populus* hybrid | Li et al., 2008 |
| | *Strelitzia reginae* | Xiao et al., 2007 |
| Temperature | *Lilium spp.* | Lokker et al., 2004 |
| | *Rosa spp.* | Pécrix et al., 2011 |
| EMS | *Arabidopsis thaliana* | De Storme et al., 2007; De Storme & Geelen, 2011 |
| GA₃ | *Carthamus tinctorious* | Kumar & Srivastava, 2009 |

Table 3. Methods used to induce 2n gamete formation

N₂O treatments have been reported to be an effective way to induce 2n gametes (both 2n pollen and egg cells) in tulip (*Tulipa*), *Lilium* and *Begonia* (Akutsu et al., 2007; Barba-Gonzales et al., 2006; Dewitte et al., 2010b; Okazaki et al., 2005). N₂O is suitable for treating organs inside tissues as the gas simply permeates the tissue, thereby protecting the tissues from harmful after-effects as soon as the gas is released (Kato & Geiger, 2002; Östergren, 1954). N₂O treatments are performed in a pressure tolerant vessel at 6 bar for 24 or 48h on bulbs (when microsporogenesis occurs inside the bulb) or flower buds at the stage of meiosis. In the latter case, the stage of meiosis can be estimated by the size of the flower bud (Akutsu et al., 2007; Barba-Gonzalez et al., 2006; Dewitte et al., 2010b). Consequently, bud size has been used as a criterium to start N₂O treatments. Akutsu et al. (2007) showed that effects were optimal when treatments started during pollen mother cell progression to metaphase I. By applying this technique, male sterile hybrids may produce fertile 2n gametes after treatments, but the efficiency of the treatment seems to be genotype specific (Barba-Gonzalez et al., 2006; Dewitte et al., 2010b). In *Begonia* for example, viable 2n pollen could be induced in the male sterile hybrid *B. schmidtiana* x *B.cucullata* in 14 of the 49 treated flowers, while no 2n pollen were induced in *B. fischeri* with a similar number of flowers treated.

The small number of studies on $N_2O$ induction of 2n gametes calls for additional testing in other genera. Detailed studies on the exact mechanism of $N_2O$ mediated chromosome doubling during meiosis are also limited, but it has been suggested that $N_2O$ disrupts the spindle mechanism in both mitosis (Dvorak et al., 1973; Kato & Birchler, 2006; Kihari & Tsunewaki, 1960; Östergren, 1954) and meiosis (Akutsu et al., 2007; Barba-Gonzalez et al., 2006; Okazaki et al., 2005;). Consequently, aneuploidy is frequently reported after treatments. In *Lilium*, Barba-Gonzalez et al. (2006) showed that $N_2O$ fumigation produced mainly first division restitution (FDR) gametes, indicating a disruption in meiosis I. Furthermore, Kitamura et al. (2009) showed in *Lilium* that microtubules were depolymerised during metaphase I, which prevented chromosomes from moving to the poles.

Dewitte et al. (2010b) tried to induce 2n pollen by submerging flower buds of *Begonia* in a trifluralin solution. Their treatments resulted in a disturbed meiosis and finally in 4n gametes but no progeny with raised ploidy level could be obtained after crosses with these gametes. Another way to increase the ploidy level in pollen was achieved in maize by spraying tassels with a trifluralin solution before flowering. As such, the generative nucleus was mitotically arrested and viable bicellular pollen was obtained (Kato, 1999). In some genera, flower buds were treated with colchicine (Table 2). These treatments resulted in the induction of both 2n pollen and 2n egg cells, and polyploid progeny was established after crosses with the treated plants.

Some authors tried to induce 2n gametes by manipulation of the temperature. Lokker et al. (2004) exposed four complete sterile lily genotypes to heat shock treatments and observed that three of the four genotypes became fertile by the production of viable 2n gametes. Also Pécrix et al. (2011) observed a production of 2n gametes up to 24.5% in *Rosa spp.* through an exposition of a high temperature gradient, but the range of efficient temperatures is narrow and should be applied during early meiosis. The disturbed meiosis resulted in the production of dyads and triads which mainly resulted from spindle misorientations in meiosis II. This supports the hypothesis that polyploidization events could have occurred in adverse conditions and may be of importance during species evolution.

Interspecific hybridisation is another way to induce 2n gametes. The occurrence of 2n gametes has been reported frequently in interspecific hybrids (see above). For example, in a breeding program between the tetraploids *Hibiscus syriacus* and *Hibiscus paramutabilis*, 2 of the 5 $F_2$ hybrids produced between 6 and 10% 2n pollen. These hexaploid $F_2$ hybrids were all raised in ploidy level compared to the tetraploid $F_1$ hybrids as a result of 2n egg cells in the $F_1$ population (Van Laere et al., 2009). However, even in interspecific hybrids, 2n gametes are rather exception than rule and many hybrids may be screened to detect 2n gamete producing genotypes. In *Lilium*, only 12 of the 708 (1.2%) OA $F_1$ hybrids produce 2n gametes on a regular basis, while the other seedlings were sterile (Barba-Gonzales et al., 2004).

Different attempts were made to increase the frequency of 2n pollen in genotypes which produce a low number of 2n pollen. Specific efforts have been performed using temperature variation, genetic selection, velocity sedimentation or pollen sieving (Table 4). For example, by using genetic selection, Negri & Lemmi (1998) were able to increase the frequency 2n pollen (Fi) in *Lotus tenuis* from 0-13% in the natural populations to 47.5-77.6% in the selected individuals. A further increase in frequency with a factor 1.5 to 3 could be obtained in a warm chamber.

| Increase method | Crop | Reference |
|---|---|---|
| Temperature | *Begonia spp.* | Dewitte, 2010a |
| | Loquat tree | Wang et al., 2003 |
| | *Lotus tenuis* | Negri & Lemmi, 1998 |
| | *Prunus persica* | Ye et al., 2010 |
| | *Prunus spp.* | Zhang & Liu, 1998 |
| | *Rosa spp.* | Crespel et al., 2006 |
| | *Solanum phureja* | McHale, 1983; Veilleux & Lauer, 1981; |
| | | Werner & Peloquin, 1987 |
| Velocity sedimentation | *Solanum spp.* | Simon & Sanford, 1990 |
| Pollen sieving | *Cyclamen persicon* | Takamura & Miyajima, 2002 |
| | *Diospyros kaki* | Sugiura et al., 2000 |
| | *Lilium spp.* | Okazaki et al., 2005 |
| | *Solanum spp.* | Eijlander, 1988 |
| Genetic selection | *Lotus tenuis* | Negri & Lemmi, 1998 |
| | *Medicago sativa* | Tavoletti et al., 1991b; Calderini & |
| | | Mariani, 1997; Mariani et al., 2000 |
| | *Solanum* | Ortiz & Peloquin, 1992 |
| | *Trifolium pratense* | Parrot & Smith, 1986 |

Table 4. Overview of methods used to increase the frequency of 2n gametes

## 4.3 Use of molecular tools to engineer 2n gamete formation

The isolation of genes involved in 2n gamete production opens the way for new strategies in plant breeding programmes. More specifically, techniques that involve knockdown of RNA levels, such as RNA interference (RNAi), virus induced gene silencing (VIGS) or mutagenesis of the encoding gene using techniques such as site-directed mutagenesis could be used to knock down the level of specific proteins which play a role in the unreduced gamete formation (Brownfield & Köhler, 2010). The translation of this knowledge from the model plant *Arabidopsis* into plant breeders work still has to start. However, the vision of crop generation via designed gametes is becoming realistic.

Potential strategies to mutate genes active in meiosis and induce 2n gamete producing genotypes also include general mutagenesis strategies using chemicals such as ethyl methane sulphonate (EMS) (used in *Arabidopsis* to create 2n gametes; De Storme et al., 2007), random insertional mutagenesis or irradiation of seeds or buds (Shu-Ye.& Srinivasan, 2010).

## 5. Role of 2n gametes in plant breeding

2N gametes have already been used to create new cultivars at higher ploidy levels as well as creating a bridge to transfer desirable genes from wild diploid species into the cultivated polyploid gene pool (Carputo et al., 2000 Peloquin & Ortiz, 1992). Although 2n gametes have been documented in several genera, they have been extensively investigated in potato (Den Nijs & Peloquin, 1977; Mok & Peloquin, 1975a; Peloquin et al., 1999), rose (Crespel et al., 2002; Crespel et al., 2006; El Mokadem et al., 2002a; El Mokadem et al., 2002b;), lily (Barba-Gonzalez et al., 2005a, 2005b; Lim et al., 2001a; Lim et al., 2004), and alfalfa (Tavoletti et al., 1991a; Tavoletti et al., 1991b; Barcaccia et al., 2003). The 2n gametes have in turn contributed to these crops' breeding programmes. For example, ploidy manipulations have been used in potato (*Solanum tuberosum*) breeding for many decades. Cultivated potatoes are tetraploid but most wild species are diploid. Haploidisation techniques can create dihaploids from cultivated potatoes. Via a series of hybridisations between selected dihaploids and 2x species, desirable agronomic traits from wild and closely related species can be captured. These dihaploids (producing 2n gametes) can then be introduced to tetraploids via interploidy crosses between 2x and 4x plants. The most successful breeding scheme for potatoes involves obtaining 4x progeny from 4x-2x crosses, where the 2x parent forms 2n pollen via the meiotic mutant *ps* (Peloquin et al., 1999). Several ploidy series have been developed in potato by using haploidisation and sexual polyploidization techniques, ranging from the monoploid to the hexaploid level (Carputo & Barone, 2005).

Other examples of the usefulness of 2n gametes for crop improvement in plant breeding have been demonstrated in *Alstroemeria,* carnation (*Dianthus*), primrose (*Primula*), *Triticum*, cassava (*Manihot*), blueberry (*Vaccinium*), cocksfoot grasses (*Dactylis*), *Lotus* and *Trifolium* (Carroll & Borrill, 1965; Hahn et al., 1990; Hayashi et al., 2009; Lyrene et al., 2003; Negri & Veronesi, 1989; Nimura et al., 2008; Parrot & Smith, 1984; Ramanna et al., 2003; Zhang et al., 2010), among others. 2N gametes also occur after haploidisation, which is useful for scaling the ploidy level upward again without artificial ploidy doubling (e.g. El Mokadem et al., 2002b; Nelson et al., 2009).

One advantage of 2n gametes is one-step triploid generation. Breeding for triploids may become an aim in itself. In *Citrus* for example, seedlessness is one of the most important characteristics in mandarin oranges. *Citrus* triploid plants can be recovered by sexual hybridisation of diploid plants as a consequence of the formation of 2n gametes at low frequency (Aleza et al., 2010). If breeding for triploids is not the aim of the breeding programme, further crosses are required to achieve the tetraploid or higher ploidy levels. For this purpose, the production of 2n gametes should be inherited from the diploid to the triploid plant. Several examples have been described where triploid plants produce 2n gametes, and often these are the only viable source of gametes. Crosses with these triploids resulted in tetraploid progeny, and even in this tetraploid progeny, 2n gamete production was observed (Brown, 1988; Dewitte et al., 2010d; Dweikat & Lyrene, 1988; Mok et al., 1975).

Plant breeding seeks to maximize the heterozygosity in the generations used for commercial production, and also polyploid induction will only contribute to plant improvement if substantial heterozygosity can be incorporated (Sanford, 1983). Higher heterozygosity can be achieved at the polyploid rather than at the diploid level. This is because polyploids have a greater probability of possessing three or more different alleles at the same locus whereas diploids only have two (Bingham, 1980; Lewis, 1980; Sanford, 1983). The heterozygosity present in the polyploid depends on the heterozygosity present in the parent plants and the mechanism behind 2n gamete formation. In general, FDR pollen are more important in producing heterozygous hybrids because of the highly heterozygous 2n gametes formed (Bretagnolle & Thompson, 1995). In FDR, each gamete is, except for recombinant fragments, identical to the somatic cell and thus contains a high level of heterozygosity. In SDR, chromosome assortment is random (as in normal gametes) which results in a very heterogeneous population of gametes, but with a lower heterozygosity within one cell. For breeding purposes, 2n gametes of the FDR type are more advantageous than those obtained by SDR for transferring parental heterozygosity (Barcaccia et al., 2000; Barcaccia et al., 2003), although Hutten et al. (1994) could not confirm this FDR superiority for all agronomic characteristics investigated in potato. By using molecular techniques as AFLP, random amplification of polymorphic DNA (RAPD) or isozyme analysis, the heterozygosity transmitted through 2n gametes can be calculated. Several reports on *Solanum spp.* indicate that FDR gametes transmit roughly 70-80% of the parental heterozygosity, whereas this is only about 30-40% for SDR (Barone et al., 1995; Douches & Quiros, 1988; Werner & Peloquin, 1991a; Werner & Peloquin, 1991b). These values were also reported in other crops: roses, ryegrass, *Begonia* and *Vaccinium darrowi* (Chen et al., 1997; Crespel et al., 2002; Dewitte et al., 2010a; Vorsa and Rowland, 1997; Qu & Hancock, 1995). In general FDR is more than twice as effective in transmitting heterozygosity as SDR (Peloquin et al., 2008).

Ramanna & Jacobsen (2003) report that sexual polyploids have been much more useful for crop improvement than mitotic polyploids. Beuselinck et al. (2003) has shown that 2n gametes from *Lotus glaber* may aid intercrossing with *Lotus corniculatus* to produce progeny with a greater heterotic advantage than progeny obtained from the colchicine induced tetraploid (4x) *Lotus glaber*. In alfalfa, McCoy & Rowe (1986) showed better breeding value of 2n gametes from the diploids than n gametes from chromosome doubled tetraploids. Lim et al. (2001) has shown that recombinant chromosomes were present in 2n gametes in contrast to 2x gametes from mitotically doubled plants. Hence, 2n gametes have great potential to selectively introgress certain characteristics. The genetic consequences of 2n gametes indicate that sexual polyploidization results in greater variability, fitness and heterozygosity than does somatic doubling (Carputo et al., 2003).

One potential strategy to maximize heterozygosity in polyploids is analytic breeding: breeding for superior genotypes at the diploid level followed through sexual hybridisation using 2n gametes. The potential of analytical breeding to maximize heterozygosity and intergenomic recombination has been shown in *Lilium* using GISH. In traditional methods, mitotically doubled tetraploids are crossed with a diploid to produce triploids. However, in the tetraploid no recombination is expected to occur due to autosyndetic pairing. This results in a narrow selection of cultivars from mitotic polyploidization. In contrast, analytic breeding strategies allow intergenomic recombination to produce triploid *Lilium* varieties (Khan et al., 2009; Khan et al., 2010). Other examples of analytical breeding for crop improvement are vegetatively propagated species such as potato, sweet potato, cassava, among roots and tubers, and plantain/banana (Ortiz, 2002).

## 6. Conclusions

The exploitation of 2n gametes creates a plethora of opportunities for practical breeding. In general, several reports show that a) 2n gametes are mostly easy to detect, although this depends on the method used, b) 2n pollen may transmit a high level of heterozygosity (genetic variation) to the progeny, c) the ability to produce 2n gametes may be passed on to the progeny and d) 2n gamete production can be induced by artificial techniques. Besides, recent progress in identifying the genes and molecular mechanisms involved in 2n gamete production offers breeders new opportunities to design new tools and techniques to engineer 2n gamete production in specific crops and plants. The increasing reports and knowledge about the practical use of 2n gametes in several crops and their genetic consequences show that 2n gametes are recognized as a very important tool in plant breeding. With ongoing research at the molecular level and research towards efficient methods to induce 2n gametes, the importance of 2n gametes for plant breeding is sure to increase.

## 7. References

Akutsu, M.; Kitamura, S.; Toda, R.; Miyajima, I. & Okazaki, K. (2007). Production of 2n pollen of Asiatic hybrid lilies by nitrous oxide treatments. *Euphytica*, Vol.155, No.1-2, (May 2007), pp.143-152, ISSN 0014-2336

Aleza, P.; Juárez, J. ; Cuenca, J.; Ollitrault, P. & Navarro, L. (2010). Recovery of citrus triploid hybrids by embryo rescue and flow cytometry from 2x x 2x sexual hybridisation and its application to extensive breeding programs. *Plant Cell Reports*, Vol.29, No.9, (September 2010), pp.1023-1034, ISSN 0721-7714

Andruezza, S. & Siddiqi, I. (2008). Spindle positioning, meiotic nonreduction, and polyploidy in plants. *PLOS Genetics*, Vol. 4, No.11, (November 2008), pp. 1-2, ISSN 1553-7404

Aryavand, A.; Ehdaie, B.; Tran, B. & Waines, JG. (2003). Stomatal frequencies and size differentiate ploidy levels in *Aegilops neglecta*. *Genetic Resources and Crop Evolution*, Vol.50, No.2, (March 2003), pp.175-182, ISSN 0925-9864

Baert, J.; Reheul, D.; Van Bockstaele, E. & De Loof, A. (1992). A rapid method by flow cytometry for estimating persistence of tetraploid perennial ryegrass in pasture mixtures with diploid perennial ryegrass. *Biologia Plantarum*, Vol. 34, No.5-6, (September 1992), pp. 381-385, ISSN 0006-3134

Bani-Aameur, F.; Lauer, FI. & Veilleux, RE. (1992). Frequency of 2n pollen in diploid hybrids between *Solanum phureja* Juz. & Buk. and *Solanum chacoense* Bitt. *Potato Research*, Vol. 35, No.2, (June 1992), pp. 161-172, ISSN 0014-3065

Barba-Gonzalez, R.; Lim, K-B.; Ramanna, MS.; Visser, RGF. & Van Tuyl, JM. (2005a). Occurrence of 2n gametes in the F1 hybrids of Oriental x Asiatic lilies *(Lilium)*: Relevance to intergenomic recombination and backcrossing. *Euphytica*, Vol. 143, No.1-2, (January 2005), pp.67-73, ISSN 0014-2336

Barba-Gonzalez, R.; Ramanna, MS.; Visser, RGF. & Van Tuyl, JM. (2005b). Intergenomic recombination in F1 lily hybrids *(Lilium)* and its significance for genetic variation in the BC1 progenies as revealed by GISH and FISH. *Genome*, Vol.48, No.5, (October 2005), pp.884-894, ISSN 0831-2796

Barba-Gonzalez, R.; Lokker, AC.; Lim, K-B.; Ramanna, MS. & Van Tuyl, JM. (2004). Use of 2n gametes for the production of sexual polyploids from sterile Oriental x Asiatic hybrids of lilies *(Lilium)*. *Theoretical and Applied Genetics*, Vol.109, No.6, (October 2004), pp.1125-1132, ISSN 0040-5752

Barba-Gonzalez, R.; Miller, CT.; Ramanna, MS. & Van Tuyl, JM. (2006). Nitrous oxide N₂O induces 2n gametes in sterile F1 hybrids of Oriental x Asiatic lilies *(Lilium)* and leads to intergenomic recombination. *Euphytica*, Vol.148, No.3, (April 2006), pp.303-309, ISSN 0014-2336

Barcaccia, G.; Albertini, E.; Rosellini, D.; Tavoletti, S. & Veronesi, F. (2000). Inheritance and mapping of 2n egg production in diploid alfalfa. *Genome*, Vol.43, No.3, (June 2000), pp.528-537, ISSN 0831-2796

Barcaccia, G.; Tavoletti, S.; Mariani, A. & Veronesi, F. (2003). Occurrence, inheritance and use of reproductive mutants in alfalfa improvement. *Euphytica*, Vol.133, No.1, (July 2003), pp.37-56, ISSN 0014-2336

Barone, A.; Gebhardt, C. & Frusciante, L. (1995). Heterozygosity in 2n gametes of potato evaluated by RFLP markers. *Theoretical and Applied Genetics*, Vol.91, No.1, (July 1995), pp.98-104, ISSN 0040-5752

Barringer, BC. (2007). Polyploidy and self-fertilisation in flowering plants. *American Journal of Botany* ,Vol.94, No.9, (September 2007), pp.1527-1533, ISSN 0002-9122

Bastiaanssen, HJM.; Van Den Berg, PMM.; Lindhout, P.; Jacobsen, E. & Ramanna, MS. (1998). Postmeiotic restitution in 2n-egg formation of diploid potato. *Heredity*, Vol.81, No.1, (June 1998), pp.20-27, ISSN 0018-067X

Becerra Lopez-Lavalle, LA. & Orjeda, G. (2002). Occurrence and cytological mechanism of 2n pollen formation in a tetraploid accession of *Ipomoea batatas* (sweet potato). *Journal of Heredity*, Vol.93, No.3, (May 2002), pp.185-192, ISSN 0022-1503

Beuselinck, PR.; Steiner, JJ. & Rim, YW. (2003). Morphological comparison of progeny derived from 4x-2x and 4x-4x hybridizations of *Lotus glaber* Mill. and *L. corniculatus* L. *Crop Science*, Vol.43, No.5, (September 2003), pp.1741-1746, ISSN 0011-183X

Bingham, ET. (1980). Maximizing heterozygosity in autoploids. In: *Polyploidy, biological relevance (Vol 13)*, Lewis, WH., (ed.), 471-489, Plenum Press, ISBN 9780306403583, New York

Bohne, G.; Richter, E.; Woehlecke, H.; Ehwald, R. (2003). Diffusion barriers of tripartite sporopollenin microcapsules prepared from pine pollen. *Annals of Botany*, Vol.92, No.2, (August 2003), pp.289-297, ISSN 0305-7364

Bretagnolle, F. & Thompson, JD. (1995). Gametes with the somatic chromosome number: mechanisms of their formation and role in the evolution of autoploid plants. *New Phytologist*, Vol.129, No.1, (Januari 2005), pp.1-22, ISSN 0028-646X

Bretagnolle, F. (2001). Pollen production and spontaneous polyploidization in diploid populations of *Anthoxanthum alpinum*. *Biological Journal of the Linnean Society*, Vol.72, No.2, (Februari 2001), pp.241-247, ISSN 0024-4066

Briggs, D. & Walters, SM. (1997). *Plant Variation and Evolution*. Cambridge University Press, ISBN 9780521452953, Cambridge, UK

Brown, CR. (1988). Characteristics of 2n pollen producing triploid hybrids between *Solanum stoloniferum* and cultivated diploid potatoes. *American Journal of Potato Research*, Vol.65, No.2, (February 1988), pp.75-84, ISSN 1099-209X.

Brownfield, L. & Köhler, C. (2011). Unreduced gamete formation in plants: mechanisms and prospects. *Journal of Experimental Botany*, Vol.62, No.2, (March 2011), pp. 1659-1968, ISSN 0022-0957

Calderini, O. & Mariani, A. (1997). Increasing 2n gamete production in diploid alfalfa by cycles of phenotypic recurrent selection. *Euphytica*, Vol.93, No.1, (January 1997), pp. 113-118, ISSN 0014-2336

Camadro, EL.; Saffarano, SK.; Espinillo, JC.; Castro, M. & Simon, PW. (2008). Cytological mechanisms of 2n pollen formation in the wild potato *Solanum okadae* and pollen-pistil relations with the common potato, *Solanum tuberosum* L. *Genetic Resources and Crop Evolution*, Vol.55, No.3, (May 2008), pp.471–477, ISSN 0925-9864

Carputo, D. & Barone, A. (2005). Ploidy level manipulations in potato through sexual hybridisation. *Annals of Applied Biology*, Vol.146, No.1, (January 2005), pp.71-79, ISSN 0003-4746

Carputo, D.; Barone, A. & Frusciante, L. (2000). 2N gametes in the potato: essential ingredients for breeding and germplasm transfer. *Theoretical and Applied Genetics*, Vol.101, No.5-6, (October 2000), pp.805-813, ISSN 0040-5752

Carputo, D.; Frusciante, L. & Peloquin, SJ. (2003). The role of 2n gametes and endosperm balance number in the origin and evolution of polyploids in the tuber-bearing Solanums. *Genetics*, Vol.163, No.1, (January 2003), pp.287-294, ISSN 0016-6731

Carroll, CP. & Borrill, M. (1965). Tetraploid hybrids from crosses between diploid and tetraploid *Dactylis* and their significance. *Genetica*, Vol.36, No.1, (December 1965), pp.65-82, ISSN 0016-6707

Chahal, GS. & Gosal, SS. (2002). *Principles and procedures of plant breeding: biotechnological and conventional approaches*. Alpha Science International Ltd, ISBN 9788173193743, Pangbourne, UK

Chen, C.; Lyon, MT.; O'Malley, D.; Federici, CT.; Gmitter, J.; Grosser, JW.; Chaparro, JX.; Roose, ML. & Gmitter Jr., FG. (2008). Origin and frequency of 2n gametes in *Citrus sinensis* x *Poncirus trifoliate* and their reciprocal crosses. *Plant Science*, Vol.174, No.1, (January 2008), pp.1-8, ISSN 0168-9452

Chen, C.; Sleper, DA.; Chao, S.; Johal, GS. & West, CP. (1997). RFLP detection of 2n pollen formation by first and second division restitution in perennial ryegrass. *Crop Science*, Vol.37, No.1, (January 1997), pp.76-80, ISSN 0011-183X

Chen, ZJ. (2007). Genetic and epigenetic mechanisms for gene expression and phenotypic variation in plant polyploids. *Annual Review of Plant Biology*, Vol.58, (June 2007), pp. 377-406, ISSN 1543-5008

Comai, L. (2005). The advantages and disadvantages of being polyploid. *Nature Reviews Genetics*, Vol.6, No.11, (November 2005), pp.836-846, ISSN 1471-0056

Conicella, C.; Barone, A.; Del Giudice, A.; Frusciante, L. & Monti, LM. (1991). Cytological evidences of SDR-FDR mixture in the formation of 2n eggs in a potato diploid clone. *Theoretical and Applied Genetics*, Vol.81, No.1, (January 1991), pp.59-63, ISSN 0040-5752

Contreras, RN.; Ranney, TG. & Tallury, SP. (2007). Reproductive behavior of diploid and allotetraploid *Rhododendron* L. 'fragrant affinity'. *HortScience*, Vol.42, No.1, (February 2007), pp.31-34, ISSN 0018-5345

Crespel, L.; Gudin, S.; Meynet, J.; Zhang, D. (2002). AFLP-based estimation of 2n gametophytic heterozygosity in two parthenogenetically derived dihaploids of *Rosa hybrida* L. *Theoretical and Applied Genetics*, Vol. 104, No.2-3, (February 2002), pp. 451-456, ISSN 0040-5752

Crespel, L.; Ricci, SC. & Gudin, S. (2006). The production of 2n pollen in rose. *Euphytica*, Vol.151, No.2, (September 2006), pp.155-164, ISSN 0014-2336.

D'Erfurth, I.; Jolivet, S.; Froger, N.; Catrice, O.; Novatchkova, M. & Mercier, R. (2009). Turning meiosis in mitosis. *PLOS Biology*, Vol.7, No.6, (June 2009), e1000124, ISSN 1544-9173

D'Erfurth, I.; Jolivet, S.; Froger, N.; Catrice, O.; Novatchkova, M.; Simon, M.; Jenczewski, E. & Mercier, R. (2008). Mutations in *AtPS1* (*Arabidopsis thaliana* Parallel Spindle 1)

lead to the production of diploid pollen grains. *PLOS Genetics*, Vol.4, No.11, (November 2008), e1000274, ISSN 1553-7404

D'Erfurth, I.; Cromer, L.; Jolivet, S.; Girard, C.; Horlow, C.; Sun, YJ.; To, JPC.; Berchowitz, LE., Copenhaver, GP. & Mercier, R. (2010). The cyclin-A CYCA1;2/TAM is required for the meiosis I to meiosis II transition and cooperates with OSD1 for the prophase to first meiotic division transition. *PLoS Genetics*, Vol.6, No.6, (June 2010), e100989, ISSN 1553-7404

De Haan, A.; Maceira, NO.; Lumaret, R. & Delay, J. (1992). Production of 2n gametes in diploid subspecies of *Dactylis glomerata* L. 2. Occurrence and frequency of 2n eggs. *Annals of Botany*, Vol. 69, No.4, (April 1992), pp.345-350, ISSN 0305-7364

De Nettancourt, D. (1997). Incompatibility in angiosperms. *Sexual Plant Reproduction*, Vol.10, No.4, (August 1997), pp. 185-199, ISSN 0934-0882

De Storme, N. & Geelen, D. (2011). The Arabidopsis mutant jason produces unreduced FDR male gametes through a parallel/fused spindle mechanism in meiosis II. *Plant Physiology*, Vol.155, No.3, (March 2011), pp.1403-1415, ISSN 0032-0889

De Storme, N.; Van Labbeke, M-C. & Geelen, D. (2007). Formation of unreduced pollen in *Arabidopsis thaliana*. *Communications in Agricultural and Applied Biological Sciences*, Vol.72, No.1, pp.159-162, ISSN 1379-1176

Del Bosco, SF.; Tusa, N. & Conicella, C. (1999). Microsporogenesis in a *Citrus* interspecific tetraploid somatic hybrid and its fusion parents. *Heredity*, Vol.83, No.4, (October 1999), pp.373-377, ISSN 0018-067X

Den Nijs, TPM. & Peloquin, SJ. (1977). 2N gametes in potato species and their function in sexual polyploidization. *Euphytica*, Vol.26, No.3, (December 1977), pp.585-600, ISSN 0014-2336.

Dewitte, A.; Eeckhaut, T.; Van Huylenbroeck, J.; Van Bockstaele, E. (2006). Flow cytometric detection of unreduced pollen in *Begonia*. *Acta Horticulturae*, Vol.714, (September 2006), pp.107-112, ISSN 0567-7572

Dewitte, A. (2010a). Exploitation of 2n pollen to create genetic variation in the genus Begonia, PhD thesis. Faculty of Bioscience engineering, Ghent University, ISBN 9789059893993

Dewitte, A.; Eeckhaut, T.; Van Huylenbroeck, J. & Van Bockstaele, E. (2010b). Induction of unreduced pollen by trifluralin and N$_2$O treatments. *Euphytica*, Vol.171, No.2, (October 2010), pp.283-293, ISSN 0014-2336

Dewitte, A.; Eeckhaut, T.; Van Huylenbroeck, J. & Van Bockstaele, E. (2010c). Meiotic aberrations during 2n pollen formation in *Begonia*. *Heredity*, Vol.104, No.2, (February 2010), pp. 215-223, ISSN 0018-067X

Dewitte, A.; Eeckhaut, T.; Van Huylenbroeck, J. & Van Bockstaele, E. (2009). Occurrence of viable unreduced pollen in a *Begonia* collection. *Euphytica*, Vol.168, No.1, (February 2009), pp.81-94, ISSN 0014-2336

Dewitte, A.; Van Laere, K.; Van Huylenbroeck, J. & Van Bockstaele, E. (2010d). Inheritance of 2n pollen formation in an F1 and F2 population of *Begonia* hybrids. *Acta Horticulturae*, Vol.855, (February 2010), pp.95-100, ISSN 0567-7572

Dhooghe, E.; Van Laere, K.; Eeckhaut, T.; Leus, L. & Van Huylenbroeck, J. (2011). Mitotic chromosome doubling of plant tissues in vitro. *Plant Cell Tissue and Organ Culture*, Vol.104, No.3, (March 2011), pp.359-373, ISSN 0176-6857

Douches, DS. & Quiros, CF. (1988). Genetic recombination in a diploid synaptic mutant and a *Solanum tuberosum* x *S. chacoense* diploid hybrid. *Heredity*, Vol.60; No.2, (April 1988), pp.183-191, ISSN 0018-067X

Dupuis, I.; Roeckel, P.; Matthys-Rochon, E. & Dumas, C. (1987). Procedure to isolate viable sperm cells from corn (*Zea mays* L.) pollen grains. *Plant Physiology*, Vol.85, No.4, (december 1987), pp.876-878, ISSN 0032-0889.

Dvorak, J.; Harvey, BL.; Coulman, BE. (1973). The use of nitrous oxide for producing eupolyploids and aneuploids in wheat and barley. *Canadian Journal of Genetics and Cytology*, Vol.15, No.1, (March 1973), pp.205-214, ISSN 0008-4093

Dweikat, IM. & Lyrene, PM. (1988). Production and viability of unreduced gametes in triploid interspecific blueberry hybrids. *Theoretical and Applied Genetics*, Vol.76, No.4, (October 1988), pp.555-559, ISSN 0040-5752

Eeckhaut, TGR.; Werbrouck, SPO.; Leus, LWH.; Van Bockstaele, EJ. & Debergh PC (2004). Chemically induced polyploidization in *Spathiphyllum wallisii* Regel through somatic embryogenesis. *Plant Cell Tissue and Organ Culture*, Vol.78, No.3, (September 2004), pp.241-246, ISSN 0176-6857

Eijlander, R. (1988). Manipulation of the 2n gametes frequencies in *Solanum* pollen. *Euphytica*, Vol.39, No.3, (December 1988), pp.45-50, ISSN 0014-2336

El Mokadem, H.; Crespel, L.; Meynet, J. & Gudin, S. (2002a). The occurrence of 2n pollen and the origin of sexual polyploids in dihaploid roses (*Rosa hybrid* L.). *Euphytica*, Vol.125, No.2, (May 2002), pp.169-177, ISSN 0014-2336

El Mokadem, H.; Meynet, J.; Crespel, L. (2002b). The occurrence of 2n eggs in the dihaploids derived from *Rosa hybrid* L. *Euphytica*, Vol.124, No.3, (April 2002), pp.327-332, ISSN 0014-2336

Erazzu, LE. & Camadro, EL. (2007). Direct and indirect detection of 2n eggs in hybrid diploid families derived from haploid tbr x wild species crosses. *Euphytica*, Vol.155, No.1-2, (May 2007), pp. 57-62, ISSN 0014-2336

Erilova, A.; Brownfield, L.; Exner, V.; Rosa, M.; Twell, D., Scheid, OM., Hennig, L. & Kohler, C. (2009). Imprinting of the Polycomb Group gene MEDEA serves as a ploidy sensor in *Arabidopsis*. PLOS Genetics, Vol.5, No.9, (September 2009), e1000663, ISSN 1553-7404

Estrada-luna, AA. ; Garcia-Aguilar, M. & Vielle-Caldaza, JP. (2004). Female reproductive development and pollen tube growth in diploid genotypes of *Solanum cardiophyllum* Lindl. *Sexual Plant Reproduction*, Vol.17, No.3, (September 2004), pp.117-124, ISSN 0934-0882

Falistocco, E.; Tosti, N. & Falcinelli, M. (1995). Cytomixis in pollen mother cells of diploid *Dactylis*, one of the origins of 2n gametes. *Journal of Heredity*, Vol.86, No.6, pp.448-453, ISSN 0022-1503

Gallo, P.; Micheletti, P.; Boldrini, K.; Risso-Pascotto, C.; Pagliarini, M. & Valle, C. (2007). 2N gamete formation in the genus *Brachiaria* (Poaceae: Paniceae). *Euphytica*, Vol.154, No. 1-2, (March 2007), pp.255-260, ISSN 0014-2336

Ghaffari, S.M. (2006). Occurrence of diploid and polyploid microspores in *Sorghum bicolour* (Poaceae) is the result of cytomixis. *African Journal of Biotechnology*, Vol.5, No.16, (August 2006), pp. 1450-1453, ISSN 1684-5315

Grant, V. (1981). *Plant speciation*. Columbia University press, ISBN 9780231051125, New York

Gu, XF.; Yang, AF.; Meng, H.; Zhang, JR. (2005). In vitro induction of tetraploid plants from diploid *Zizyphus jujuba* Mill. cv. Zhanhua. *Plant Cell Reports*, Vol.24, No.11, (December 2005), pp. 671-676, ISSN 0721-7714

Gymer, PT. & Whittington, WJ. (1973). Hybrids between *Lolium perenne* L and *Festuca pratensis* Huds. *New Phytologist*, Vol.72, No.2, (March 1973), pp.411- 424, ISSN 0028-646X.

Hahn, SK.; Bai, KV.; Asiedu, R. (1990). Tetraploids, triploids and 2n pollen from diploid interspecific crosses with cassava. *Theoretical and Applied Genetics*, Vol.79, No.4, (April 1990), pp. 433-439, ISSN 0040-5752

Harlan, J. & De Wet, J. (1975). On Ö. Winge and a prayer: the origins of polyploidy. *The Botanical Review*, Vol.41, No.4, (October 2004), pp. 361-390, ISSN 0006-8101

Hayashi, M.; Kato, J.; Ohashi, H. & Masahiro, M. (2009). Unreduced 3x gamete formation of allotriploid hybrid derived from the cross of *Primula denticulata* (4x) × *P. rosea* (2x) as a causal factor for producing pentaploid hybrids in the backcross with pollen of tetraploid *P. Denticulata*. *Euphytica*, Vol. 169, No.1, (September 2009), pp.123-131, ISSN 0014-2336

Husband, BC. (2004). The role of triploid hybrids in the evolutionary dynamics of mixed-ploidy populations. *Biological Journal of the Linnean Society*, Vol.82, No.4, (August 2004), pp.537–546, ISSN 0024-4066

Hutten, RCB.; Schippers, MGM.; Hermsen, JGT. & Ramanna, MS. (1994). Comparative performance of FDR and SDR progenies from reciprocal 4x-2x crosses in potato. *Theoretical and Applied Genetics*, Vol.89, No.5, (November 1994), pp.545-550, ISSN 0040-5752

Islam, AKMR. & Sheperd, KW. (1980). Meiotic restitution in wheat-barley hybrid. *Chromosoma*, Vol.79, No.3, (July 1980), pp.363-372, ISSN 0009-5915

Jansen, RC. & Den Nijs, APM (1993). A statistical mixture model for estimating the proportion of unreduced pollen grains in perennial ryegrass (*Lolium perenne* L.) via the size of pollen grains. *Euphytica*, Vol.70, No.3, (January 1993), pp.205-215, ISSN 0014-2336

Jauhar, PP. (2003). Formation of 2n gametes in durum wheat haploids: sexual polyploidization. *Euphytica*, Vol.133, No.1, (July 2003), pp.81-94, ISSN 0014-2336

Jones, JR. & Ranney, TG. (2009). Fertility of neoppolyploid *Rhododendron* and occurrence of unreduced gametes in triploid cultivars. *Journal of the American Rhododendron Society*, Vol.63, No.3, pp.131-135, ISSN 0003-0821

Jongedijk, E. (1987). A rapid methyl salicylate clearing technique for routine phase contrast observations on female meiosis in *solanum*. *Journal of Microscopy*, Vol.146, pp.157-162, ISSN 1365-2818

Jongedijk, E.; Ramanna, MS.; Sawor, Z. & Hermsen, JGT. (1991). Formation of first division restitution (FDR) 2n-megaspores through pseudohomotypic division in *ds-1* (desynapsis) mutants of diploid potato: routine production of tetraploid progeny from 2xFDR x 2xFDR crosses. *Theoretical and Applied Genetics*, Vol.82, No.5, (October 1991), pp.645-656, ISSN 0040-5752

Jovtchev, G.; Schubert, V.; Meister, A.; Barow, M. & Schubert, I. (2006). Nuclear DNA content and nuclear and cell volume are positively correlated in angiosperms. *Cytogenet.ic and Genome Research*, Vol.114, No.1, (May 2006), pp.77-82, ISSN 1424-8581

Karlov, GI.; Khrustaleva, LI., Lim, KB. & Van Tuyl, JM. (1999). Homoeologous recombination in 2n-gamete producing interspecific hybrids of *Lilium* (Liliaceae) studied by genomic *in situ* hybridisation (GISH). *Genome*, Vol.42, No.4, (August 1999), pp.681-686, ISSN 0831-2796

Kato, A. (1999). Induction of bicellular pollen by trifluralin treatment and occurrence of triploids and aneuploids after fertilization in maize. *Genome*, Vol.42, No.1, (February 1999), pp.154-157, ISSN 0831-2796

Kato, A. & Geiger, HH. (2002). Chromosome doubling of haploid maize seedling using nitrous oxide gas at the flower primordial stage. *Plant Breeding*, Vol.121, No.5, (October 2002), pp.370-377, ISSN 0179-9541

Kato, A. & Birchler, JA. (2006). Induction of tetraploid derivates of maize inbred lines by nitrous oxide gas treatment. *Journal of heredity*, Vol.97, No.1, (January 2006), pp.39-44, ISSN 0022-1503

Kermani, MJ.; Sarasan, V.; Roberts, AV.; Yokoya, K.; Wentworth, J. & Sieber, VK. (2003). Oryzalin-induced chromosome doubling in *Rosa* and its effect on plant morphology and pollen viability. *Theoretical and Applied Genetics*, Vol.107, No.7, (November 2003), pp.1195-1200, ISSN 0040-5752

Kessel, R. & Rowe, PR. (1975). Production of intraspecific aneuploids in the genus *Solanum*. Triploids produced from tetraploid-diploid crosses in potato. *Euphytica*, Vol.24, No.2, (June 1975), pp.65-75, ISSN 0014-2336

Khan, N.; Barba-Gonzalez, R.; Ramanna, MS.; Arens, P.; Visser, RGF. & Van Tuyl, JM. (2010). Relevance for unilateral and bilateral sexual polyploidization in relation to intergenomic recombination and introgression in *Lilium* species hybrids. *Euphytica*, Vol.171, No.2, (January 2010), pp.157-173, ISSN 0014-2336

Khan, N.; Zhou, S.; Ramanna, MS.; Arens, P.; Herrera, J.; Visser, RGF. & Van Tuyl, JM.(2009). Potential for analytic breeding in allopolyploids: an illustration from Longiflorum x Asiatic hybrid lilies (*Lilium*). *Euphytica*, Vol.166, No.3, (April 2009), pp.399-409, ISSN 0014-2336

Kihari, H. & Tsunewaki, K. (1960). Production of polyploidy wheat by nitrous oxide. *Proceedings of the Japan Academy*, Vol.36, No.10, pp.658-663, ISSN 0021-4280

Kitamura, S.; Akutsu, M. & Okazaki, K. (2009). Mechanisms of action of nitrous oxide gas applied as a polyploidizing agent during meiosis in lilies. *Sexual Plant Reproduction*, Vol.22, No.1, (March 2009), pp.9-14, ISSN 0934-0882

Knight, CA.; Molinari, NA. & Petrov, DA. (2005). The large genome constraint hypothesis: evolution, ecology and phenotype. *Annals of Botany*, Vol.95, No.1, (January 2005), pp.177-190, ISSN 0305-7364

Köhler, C.; Mittelsten Scheid, O. & Erilova, A. (2010). The impact of the triploid block on the origin and evolution of polyploid plants. *Trends in Genetics*, Vol.26, No.3, pp.142-148, (March 2010), ISSN 0168-9525.

Kondoresi, E.; Roudier, F.& Gendreau, E. (2000). Plant cell-size control: growing by ploidy? *Current Opinion in Plant Biology*, Vol.3, No.6, (December 2000), pp. 488-492, ISSN 1359-5266

Kumar, G. & Srivastava, P. (2009). Gibberellic acid-induced pollen mortality and abnormal microsporogenesis in safflower. Cytologia, Vol.74, No.2, (Februari 2009), pp.171-176, ISSN 0011-4545

Lamote, V.; Baert, J.; Roldan-Ruiz, I.; De Loose, M. & Van Bockstaele, E. (2002). Tracing of 2n egg occurence in perennial ryegrass (*Lolium perenne* L.) using interploidy crosses. *Euphytica*, Vol.123, No.2, (January 2002), pp.159-164, ISSN 0014-2336.

Leitch, I. & Benett, M. (1997). Polyploidy in angiosperms. *Trends in Plant Science*, Vol.2, No.12, (December 1997), pp.470-476, ISSN 1360-1385

Lelley, T.; Mahmoud, AA. & Lein, V. (1987). Genetics and cytology of unreduced gametes in cultivated rye (*Secale cereale* L.). *Genome*, Vol.29, No.4, (August 1987), pp.635-638, ISSN 0831-2796

Levin, D. (1983). Polyploidy and novelty in flowering plants. *The American Naturalist*, Vol.122, No.1, (July 1983), pp.1-25, ISSN 0003-0147

Lewis, W. (1980). Polyploidy in species populations. In: *Polyploidy, biological relevance (Vol 13)*, Lewis, WH., (ed.), 103-144, Plenum Press, ISBN 9780306403583, New York

Li, YH.; Kang, XY.; Wang, SD.; Zhang, ZH. & Chen, HW. (2008). Triploid induction in *Populus alba* x *P. glandulosa* by chromosome doubling of female gametes. *Silvae Genetica*, Vol.57, No.1, pp.37-40, ISSN: 00375349

Lim, K.; Ramanna, S.; De Jong, J.; Jacobsen, E. & Van Tuyl, J. (2001). Indeterminate meiotic restitution (IMR): a novel type of meiotic restitution mechanism detected in interspecific lily hybrids by GISH. *Theoretical and Applied Genetics*, Vol.103, No.2-3, (August 2001), pp.219-230, ISSN 0040-5752

Lim, K.; Shen, T.; Barba-Gonzalez, R.; Ramanna, M. & Van Tuyl, J. (2004). Occurrence of SDR 2n-gametes in *Lilium* hybrids. *Breeding Science*, Vol.54, No.1, pp.13-18, ISSN 1344-7610

Lokker, AC.; Barba-Gonzalez, R.; Lim, K-B.; Ramanna, MS. & Van Tuyl, JM. (2004). Genotypic and environmental variation in production of 2n gametes of oriental x Asiatic lily hybrids. *Acta Horticulturae*, Vol.673, (May 2005), pp. 453-456, ISSN 0567-7572

Lyrene, P.; Vorsa, N. & Ballington, J. (2003). Polyploidy and sexual polyploidization in the genus *Vaccinium*. *Euphytica*, Vol.133, No.1, (July 2003), pp.27-36, ISSN 0014-2336

Mariani, A.; Campanoni, P.; Gianì, S. & Breviario, D. (2000). Meiotic mutants of *Medicago sativa* show altered levels of alpha- and beta-tubulin. *Genome*, Vol.43, No.1, (February 2000), pp.166-71, ISSN 0831-2796

McCoy, TJ. (1982). Inheritance of 2n pollen formation in diploid alfalfa (*Medicago sativa* L.). *Canadian Journal of Genetics and Cytology*, Vol.24, No.3, (June 1982), pp.315-323, ISSN 0008-4093

McCoy, TJ. & Rowe, DE. (1986). Single cross alfalfa (*Medicago sativa* L.) hybrids produced via 2n gametes and somatic chromosome doubling: experimental and theoretical comparisons. *Theoretical and Applied Genetics*, Vol.72, No.1, (April 1986), pp.80-83, ISSN 0040-5752

McHale, NA. (1983). Environmental induction of high frequency 2n pollen formation in diploid *Solanum*. *Canadian Journal of Genetics and Cytology*, Vol.25, No.6, (December 1983), pp.609-615, ISSN 0008-4093

Meredith, MR.; Michaelson-Yeates, TPT.; Ougham, H. & Thomas, H. (1995). *Trifolium ambiguum* as a source of variation in the breeding of white clover. *Euphytica*, Vol.82, No.2, (January 1995), pp.185-191, ISSN 0014-2336.

Mishra, MK. (1997). Stomatal characteristics at different ploidy levels in *Coffea* L. *Annals of Botany*, Vol.80, No.5, (November 1997), pp.689-692, ISSN 0305-7364

Mittelsten Scheid, O.; Jakovleva, L.; Afsar, K.; Maluszynska, J. & Paszkowski, J. (1996). A change in ploidy can modify epigenetic silencing. PNAS, Vol.93, No.9, (July 1996), pp.7114-7119, ISSN 0027-8424

Mok, DW. & Peloquin, SJ. (1975a). Breeding value of 2n pollen (diploandroids) in tetraploid x diploid crosses in potatoes. Theoretical and Applied Genetics, Vol.46, No.6, (January 1975), pp.307-314, ISSN 0040-5752

Mok, DW. & Peloquin, SJ. (1975b). The inheritance of three mechanisms of diplandroid (2n pollen) formation in diploid potatoes. Heredity, Vol.35, No.3, (December 1975), pp.157-164, ISSN 0018-067X

Mok, DW.; Peloquin, SJ. & Tarn, TR. (1975). Cytology of potato triploids producing 2n pollen. American Journal of Potato research, Vol.52, No.6, (June 1975), pp.171-174, ISSN 1099-209X

Morgan, WG., Thomas, H. & Humphreys, MW. (1995). Unreduced gametes in interspecific hybrids in the Festuca/Lolium complex. Plant Breeding, Vol.114, No.3, (June 1995), pp.267-268, ISSN 0179-9541

Negri, V. & Lemmi, G. (1998). Effect on selection and temperature stress on the production of 2n gametes in Lotus tenuis. Plant Breeding, Vol.117, No.4, (September 1998), pp.345-349, ISSN 0179-9541

Negri, V. & Veronesi, F. (1989). Evidence for the excistence of 2n gametes in Lotus tenuis Wald. Et Kit (2n = 2x = 12); their relevance in evolution of breeding of Lotus corniculatus L. (2n = 4x = 24). Theoretical and Applied Genetics, Vol.78, No.3, (September 1989), pp.400-404, ISSN 0040-5752

Nelson, MN.; Mason., A.; Castello, M-C.; Thomson, L.; Yan, G. & Cowling, WA. (2009). Microspore culture preferentially selects unreduced (2n) gametes from an interspecific hybrid of Brassica napus L. × Brassica carinata Braun. Theoretical and Applied Genetics, Vol.119, No.3, (August 2009), pp.497-505, ISSN 0040-5752

Nimura, M.; Kato, J.; Mii, M. & Ohishi, K. (2008). Cross-compatibility and the polyploidy of progenies reciprocal backcrosses between diploid carnation (Dianthus caryophyllus L.) and its amphidiploid with Dianthus japonicus Thunb. Scientia Horticulturae, vol. 115, No.2 (January 2008), pp. 183-189, ISSN 0304-4238

Okazaki, K.; Kurimoto, K.; Miyajima, I.; Enami, A.; Mizuochi, H.; Matsumoto, Y. & Ohya, H. (2005). Induction of 2n pollen by arresting the meiotic process with nitrous oxide gas. Euphytica, Vol. 143, No.1-2, (January 2005), pp.101-114, ISSN 0014-2336

Ortiz, R. (2002). Analytical breeding. Acta Horticulturae, Vol.622, (August 2003), pp.235-247 ISSN 0567-7572

Ortiz, R. & Peloquin, SJ. (1991). Breeding for 2n egg production in haploid x species 2x potato hybrids. American Journal of Potato research, Vol.68, No.10, (October 1991), pp.691-703, ISSN 1099-209X

Ortiz, R. & Peloquin, SJ. (1992). Recurrent selection for 2n gamete production in 2x potatoes. Journal of Genetics and Breeding, Vol.46, pp.383-390, ISSN 0394-9257

Ortiz, R. & Vuylsteke, D. (1995). Factors influencing seed set in triploid Musa spp. L. and production of euploid hybrids. Annals of Botany, Vol.75, No.2, (February 1995), pp.151-155, ISSN 0305-7364

Ortiz, R. (1997). Occurrence and inheritance of 2n pollen in Musa. Annals of Botany, Vol.79, No.4, (April 1997), pp.449-453, ISSN 0305-7364

Ortiz, R., Ulburghs, F. & Okoro, JU. (1998). Seasonal variation of apparent male fertility and 2n pollen production in plantain and banana. *Hortscience*, Vol.33, No.1, pp.146-148, ISSN 0018-5345

Osborn, T.; Pires, J.; Birchler, J.; Auger, D.; Chen, Z.; Lee, H.; Comai, L.; Madlung, A.; Doerge, R.; Colot, V. & Martienssen, R (2003). Understanding mechanisms of novel gene expression in polyploids. *Trends in Genetics*, Vol.19, No.3, (March 2003), pp.141-147, ISSN 0168-9525

Östergren, G. (1954). Polyploids and aneuploids of *Crepis capilaris* by treatment with nitrous oxide. *Genetica*, Vol.27, No.1, (December 1955), pp.54-64, ISSN 0016-6707

Otto, SP. & Whitton, J. (2000). Polyploid incidence and evolution. *Annual Review of Genetics*, Vol.34, (December 2000), pp.401-437, ISSN 0066-4197

Pagliarini, MS.; Takayama, SY.; de Freitas, PM.; Carraro, LR.; Adamowski, EV.; Silva, N. & Batista, LAR. (1999). Failure of cytokinesis and 2n gamete formation in Brazialian accessions *Paspalum. Euphytica*, Vol.108, No.2, (August 1999), pp.129-135, ISSN 0014-2336

Pan, G.; Zhou, Y.; Fowke, LC. & Wang, H. (2004). An efficient method for flow cytometric analysis of pollen and detection of 2n nuclei in *Brassica napus* pollen. *Plant Cell Reports*, Vol.23, No.4, (October 2004), pp.196-202, ISSN 0721-7714

Parrot, WA. & Smith, RR. (1984). Production of 2n pollen in red clover. *Crop Science*, Vol.24, No.3, (May 1984), pp.469-472, ISSN 0011-183X

Parrot, WA. & Smith RR. (1986). Recurrent selection for 2n pollen formation in red clover. *Crop Science*, Vol.26, No.6, (November 1986), pp.1132-1135, ISSN 0011-183X.

Pécrix, Y.; Rallo, G.; Folzer, H.; Cigna, M.; Gudin, S. & Le Bris, M. (2011). Polyploidization mechanisms: temperature environment can induce diploid gamete formation in *Rosa* sp. *Journal of Experimental Botany*, Vol.62, No.10, (June 2011), pp.3587-3597, ISSN 0022-0957

Peloquin, SJ.; Boiteux, LS.; Simon, PW. & Jansky, SH. (2008). A chromosome-specific estimate of transmission of heterozygosity by 2n gametes in potato. *Journal of Heredity*, Vol.99, No.2, (March-April 2008), pp.177-181, ISSN 0022-1503

Peloquin, SJ. & Ortiz, R. (1992). Techniques for introgressing unadapted germplasm to breeding populations. In: *Plant breeding in the 1990s*, Stalker, TP. & Murphy, JP. (eds.), 485-507, CAB International, ISBN 0851987176, Wallingford, UK

Peloquin, SJ.; Boiteux, LS. & Carputo, D. (1999). Meiotic mutants in potato: valuable variants. *Genetics*, Vol.153, No.4, (December 1999), pp.1493-1499, ISSN 0016-6731.

Pichot, C. & El Maâtaoui, M. (2000). Unreduced diploid nuclei in *Cupressus dupreziana* A. Camus pollen. *Theoretical and Applied Genetics*, Vol.101, No.4, (September 2000), pp.574-579, ISSN 0040-5752

Pikaard, C. (1999). Nucleolar dominance and silencing of transcription. *Trends in Plant Science*, Vol.4, No.12, (December 1999), pp.478-483, ISSN 1360-1385.

Qu, L. & Hancock, JF. (1995) Nature of 2n gamete formation and mode of inheritance in interspecific hybrids of diploid *Vaccinium darrowi* and tetraploid *V. corymbosum. Theoretical and Applied Genetics*, Vol.91, No.8, (December 1995), pp.1309-1315, ISSN 0040-5752

Ramanna, MS. (1983). First division restitution gametes through fertile desynaptic mutants of potato. *Euphytica*, Vol.32, No.2, (June 1983), pp.337-350, ISSN 0014-2336

Ramanna, MS. & Jacobsen, E. (2003). Relevance of sexual polyploidization for crop improvement-a review. *Euphytica*, Vol.133, No.1, (July 2003), pp.3-18, ISSN 0014-2336

Ramanna, MS.; Kuipers, AGJ. & Jacobsen, E. (2003). Occurrence of numerically unreduced (2n) gametes in *Alstroemeria* interspecific hybrids and their significance for sexual polyploidization. *Euphytica*, Vol.133, No.1, (July 2003), pp.95-106, ISSN 0014-2336

Ramsey, J. (2007). Unreduced gametes and neopolyploids in natural populations of *Achillea borealis* (*Asteraceae*). *Heredity*, Vol.98, No.3, (March 2007), pp.143-150, ISSN 0018-067X

Ramsey, J. & Schemske, DW. (1998). Pathways, mechanisms and rates of polyploidy formation in the flowering plants. *Annual Review of Ecology and Systematics*, Vol.29, pp.267-501, ISSN 0066-4162

Rhoades, MM. & Dempsey, E. (1966). Induction of chromosome doubling by the elongate gene in maize. *Genetics*, Vol.54, No.2, (August 1966), pp.505-522, ISSN 0016-6731

Roberts, AV. (2007). The use of bead beating to prepare suspensions of nuclei for flow cytometry from fresh leaves, hebarium leaves, petals and pollen. *Cytometry*, Vol.71A, No.12, (December 2007), pp.1039-1044, ISSN 1552-4930

Roberts, MA.; Reader, SM.; Dalgliesh, C.; Miller, TE.; Foote, TN.; Fish, LJ.; Snape, JW. & Moore, G. (1999). Induction and characterization of *Ph1* mutants. *Genetics*, Vol.153, No.4, (December 1999), pp.1909-1918, ISSN 0016-6731

Russel, SD. (1991). Isolation and characterization of sperm cells in flowering plants. *Annual Review of Plant Physiology and Plant Molecular Biology*, Vol.42, pp.189-204, (June 1991), ISSN 1040-2519

Sanford, JC. (1983). Ploidy manipulations. In: *Methods in fruit breeding*, Janick, J. & Moore, J.N. (eds.), 100-123, Perdue University Press, ISSN 9780911198638, West Lafayette

Sato, T.; Miyoshi, K. & Okazaki, K. (2010). Induction of 2n gametes and 4n embryo in *Lilium* (*Lilium* x *formolongi* Hort.) by nitrous oxide gas treatment. *Acta Horticulturae*, Vol.855, (February 2010), pp.243-248, ISSN 0567-7572

Shu-Ye, J.& Srinivasan, R. (2010) Natural and artificial mutants as valuable resources for functional genomics and molecular breeding *International journal of biologicam sciences*. Vol. 6, No. 3 pp. 228-251, ISSN 1449-2288

Singhal, VK. & Kumar, P. (2008). Impact of cytomixis on meiosis, pollen viability and pollen size in wild populations of Himalayan poppy (*Meconopsis aculeate* Royle). *Journal of Biosciences*, Vol.33, No.3, (September 2008), pp.371-380, ISSN 0250-5991.

Sugiura, A.; Ohkuma, T.; Choi, YA. & Tao R. (2000). Production of nonaploid (2n = 9x) japanese persimmons (*Diospyros kaki*) by pollination with unreduced (2n = 6x) pollen and embryo rescue culture. *Journal of the American Society for Horticultural Science*, Vol.125, No.5, (September 2000), pp.609-614, ISSN 0003-1062

Simioni, C.; Schifino-Wittmann, MT. & Dall'Agnol, M. (2006). Sexual polyploidization in red clover. *Scientia Agricola*, Vol.63, No.1, (January 2006), pp.26-31, ISSN 0103-9016

Simon, CJ. & Sanford, JC. (1990). Separation of 2n potato pollen from a heterogeneous pollen mixture by velocity sedimentation. *Hortscience*, Vol.25, No.3, (March 1990), pp.342-344, ISSN 0018-5345

Skiebe, K. (1958). Die bedeutung von unreduzierten gameten für die polyploïdiezüchtüng bei der Fliederprimeln (*Primula malacoides* Franchet). *Theoretical and Applied Genetics*, Vol.28, No.8, (August 1958), pp.353-359, ISSN 0040-5752

Soltis, PS. & Soltis, DE. (2000). The role of genetic and genomic attributes in the success of polyploids. *PNAS*, Vol.97, No.13, (June 2000), pp.7051-7057, ISSN 0027-8424

Ssebuliba, RN.; Tenkouano, A. & Pillay, M. (2008). Male fertility and occurrence of 2n gametes in East African Highland bananas (*Musa spp.*). *Euphytica*, Vol.164, No.1, (November 2008), ISSN 0014-2336

Stebbins, G. (1971). Chromosomal evolution of higher plants. Edward Arnold Ltd, ISBN 0713122870, London, UK

Stephens, LC. (1998). Formation of unreduced pollen by an *Impatiens hawkeri* x *Platypetala* interspecific hybrid. *Hereditas*, Vol.128, No.3, (August 1998), pp.251-255, ISSN 0018-0661

Takamura, T. & Miyajima, I. (2002). Origin of tetraploid progenies in 4x x 2x crosses of Cyclamen (*Cyclamen persicum* Mill.). *Journal of the Japanese Society for Horticultural Science*, Vol.71, No.5, pp.632-637, ISSN 0013-7626

Takayashi, C.; Leitch, IJ.; Ryan, A., Bennett, MD. & Brandham, PE. (1997). The use of genomic in situ hybridization (GISH) of recombinant chromosomes by a partially fertile bigeneric hybrid, *Gasteria lutzii* x *Aloe aristata* (Aloaceae), to its progeny. *Chromosoma*, Vol.104, No.6, (April 1997), pp.342-348, ISSN 0009-5915

Taschetto, OM. & Pagliarini, MS. (2003). Occurrence of 2n and jumbo pollen in the Brazilian ginseng (*Pfaffia glomerata* and *P. tuberosa*). *Euphytica*, Vol.133, No.1, (July 2003), pp.139-145, ISSN 0014-2336

Tavoletti, S., Mariani, A. & Veronesi, F. (1991a). Cytological analysis of macro- and microsporogenesis of a diploid alfalfa clone producing male and female 2n gametes. *Crop Science*, Vol. 31, No.5, (September 1991), pp.1258-1263, ISSN 0011-183X

Tavoletti, S.; Mariani, A. & Veronesi, F. (1991b). Phenotypic recurrent selection for 2n pollen and 2n egg production in diploid alfalfa. *Euphytica*, Vol.57, No.2, (September 1991), pp.97-102. ISSN 0014-2336

Tel-Zur, N.; Abbo, S.; Bar-Zvi, D.; Mizrahi, Y. (2003). Chromosome Doubling in Vine Cacti Hybrids. *Journal of Heredity*, Vol.94, No.4, (July 2003), pp.329–333, ISSN 0022-1503

Thomas, H. (1993). Chromosome manipulation and polyploidy. In: *Plant Breeding Principles and Prospects*, Hayward, MD.; Bosemark, NO.; Rosmagosa, I. (eds.), 79-92, Chapman and Hall, ISBN 0412433907, New York.

Thompson, JN.; Nuismer, SL. & Merg, K. (2004). Plant polyploidy and the evolutionary ecology of plant/animal interactions. *Biological Journal of the Linnean Society*, Vol.82, No.4, (August 2004), pp.511-519, ISSN 0024-4066

Trojak-Goluch, A. & Berbeć, A. (2003). Cytological investigations of the interspecific hybrids of *Nicotiana tabacum* x *N. glauca* Grah. *Journal of Applied Genetics*, Vol.44, No.1, pp.45-54, ISSN 1234-1983

Van Huylenbroeck, J.; Leus, L. & Van Bockstaele, E. (2005). Interploidy crosses in roses: use of triploids. *Acta Horticulturae*, Vol.690, (September 2005), pp.109-112, ISSN 0567-7572

Van Huylenbroeck, JM.; De Riek, J. & De Loose, M. (2000). Genetic relationships among *Hibiscus syriacus*, *Hibiscus sinosyriacus* and *Hibiscus paramutabilis* revealed by AFLP, morphology and ploidy analysis. *Genetic Resources and Crop Evolution*, Vol. 47, No.3, (June 2000), pp.335-343, ISSN 0925-9864

Van Laere, K.; Dewitte, A.; Van Huylenbroeck, J. & Van Bockstaele, E. (2009). Evidence for the occurrence of unreduced gametes in interspecific hybrids of *Hibiscus*. *The Journal of Horticultural Science and Biotechnology*, Vol.84:, No.2, (March 2009), pp.240-247, ISSN 1462-0316

Van Laere, K.; França, SC.; Vansteenkiste, H.; Van Huylenbroeck, J.; Steppe, K. & Van Labeke, M.-C. (2011). Influence of ploidy level on morphology, growth and drought susceptibility in *Spathyphyllum wallisii*. *Acta Physiologiae Plantarum*, Vol.33, No.4, pp.1149-1156, ISSN 0137-5881

Van Tuyl, JM.; de Vries, JN.; Bino, RJ. & Kwakkenbos, AM. (1989). Identification of 2n pollen producing interspecific hybrids of *Lilium* using Flow Cytometry. *Cytologia*, Vol.54, No.4, pp.737-745, ISSN 0011-4545

Vandenhout, H.; Ortiz, R.; Vuylsteke, D.; Swennen, R.; Bai, KV. (1995). Effect of ploidy on stomatal and other quantitative traits in plantain and banana hybrids. *Euphytica*, Vol.83, No.2, (June 1995), pp.117-122, ISSN 0014-2336

Veilleux, R. (1985). Diploid and polyploid gametes in crop plants: mechanisms of formation and utilization in plant breeding. In: *Plant Breeding reviews (Vol.3)*, Janick, J. (ed.), 253-288, AVI Publishing Company, ISBN 0870554875, Westport, USA

Veilleux, RE. & Lauer, FI. (1981). Variation for 2n pollen production in clones of *Solanum phureja* Juz. and Buk. *Theoretical and Applied Genetics*, Vol.59, No.2, (March 1981), pp.95-100, ISSN 0040-5752

Veilleux, RE.; McHale, NA. & Lauer, FI. (1982). Unreduced gametes in diploid *Solanum*. Frequencies and types of spindle abnormalities. *Canadian Journal of Genetics and Cytology*, Vol.24, No.3, (June 1982), pp.301-314, ISSN 0008-4093

Veronesi, F.; Mariani, A. & Bingham, ET. (1986). Unreduced gametes in diploid *Medicago* and their importance in alfalfa breeding. *Theoretical and Applied Genetics*, Vol.72, No.1, (April 1986), pp.37-41, ISSN 0040-5752

Vorsa, N. & Rowland, LJ. (1997). Estimation of 2n gametophyte heterozygosity in a diploid blueberry (*Vaccinium darrowi* camp) clone using RAPDs. *Journal of Heredity*, Vol.88, No.5, (September 1997), pp.423-426, ISSN 0022-1503

Wang, W.; Guo, Q.; Xiang, S.; Xiaolin, L. & Guolu, L. (2003). Study on the effect of hot shock treatment on the occurrence frequency of 2n pollen of loquat trees. *Journal of Fruit Science*, Vol.4, pp.284-286, ISSN 1009-9980

Wang, J.; Kang, X. & Zhu, Q. (2010). Variation in pollen formation and its cytological mechanism in an allotriploid white poplar. *Tree Genetics and Genomes*, Vol.6, No.2, (February 2010), pp.281-290, ISSN 1614-2942

Watanabe, K. & Peloquin, SJ. (1989). Occurrence of 2n pollen and *ps* gene frequencies in cultivated groups and their related wild species in tuber bearing Solanums. *Theoretical and Applied Genetics*, Vol.78, No.3, (September 1989), pp.329-336, ISSN 0040-5752

Werner, JE. & Peloquin, SJ. (1987). Frequency and mechanisms of 2n egg formation in haploid tuberosum –wild species F1 hybrids. *American Journal of Potato Research*, Vol.64, No.12, pp.641-654, ISSN 1099-209X

Werner, JE. & Peloquin, SJ. (1991a). Yield and tuber characteristics of 4x progeny from 2x x 2x crosses. *Potato Research*, Vol.34, No.3, (September 1991), pp.261-267, ISSN 0014-3065

Werner, JE. & Peloquin, SJ. (1991b). Significance of allelic diversity and 2n gametes for approaching maximum heterozygosity in 4x potatoes. *Euphytica*, Vol.58, No.1, (October 1991), pp.21-29, ISSN 0014-2336

Wu, H.; Zheng, S.; He, Y.; Yan, G.; Bi, Y. & Zhu, Y. (2007). Diploid female gametes induced by colchicines in oriental lilies. *Scientia Horticulturae*, Vol.114, No.1, (September 2007), pp.50-53, ISSN 0304-4238.

Xiao, Y-Q.; Zheng, S-X.; Long, C-L.; Zheng, L.; Guan, W-L. & Zhao, Y. (2007). Initial study on 2n-gametes induction of *Strelitzia reginae*. *Journal of Yunnan Agricultural University*, Vol.22, No.4, pp.475-479, ISSN 1004-390X

Xu, H.; Weterings, K.; Vriezen, W.; Feron, R.; Xue, Y.; Derksen, J. & Mariani, C. (2002). Isolation and characterization of male-germ-cell transcripts in *Nicotiana tabacum*. *Sexual Plant Reproduction*, Vol.14, No.6, pp.339-346, ISSN 0934-0882

Xu, SJ. & Joppa, LR. (1995). Mechanism and inheritance of first division restitution in hybrids of wheat, rye and *Aegilops squarrosa*. *Genome*, Vol.38, No.3, (June 1995), pp.607-615, ISSN 0831-2796

Xu, SJ. & Joppa, LR. (2000). First-division restitution in hybrids of Langdon durum disomic substitution lines with rye and *Aegilops squarrosa*. *Plant Breeding*, Vol.119, No.3, (June 2000), pp.233-241, ISSN 0179-9541

Xu, L-Q.; Zhang, Q-L. & Luo, Z-R. (2008). Occurrence and cytological mechanism of 2n pollen formation in Chinese *Diospyros spp.* (Ebenaceae) staminate germplasm. *The Journal of Horticultural Science and Biotechnology*, Vol.83, No.5, (September 2008), pp.668-672, ISSN 1462-0316

Xue, Z.; Liu, P. & Liu, M. (2011). Cytological mechanism of 2n pollen formation in Chinese jujube (*Ziziphus jujube* Mill. 'Linglingzao'). *Euphytica*, doi: 10.1007/s10681-011-0461-7, , ISSN 0014-2336

Yan, G.; Ross Ferguson, A.; McNeilage, MA. & Murray, BG. (1997). Numerically unreduced (2n) gametes and sexual polyploidization in *Actinidia*. *Euphytica*, Vol.96, No.2, (July 1997), pp.267-272, ISSN 0014-2336

Yan, L.; Zhang, X. & Liu, G. (2000). Occurrence of unreduced gametes and ploidy restoration in haploid *Capsicum annuum* L. *Journal of Horticultural Science and Biotechnology*, Vol.75, No.2, (March 2000), pp.195-197, ISSN 1462-0316

Yang, M.; Hu, Y.; Lohdi, M.; McCombie, WR. & Ma, H. (1999). The *Arabidopsis* SKP1-LIKE1 gene is essential for male meiosis and can control homologue separation. PNAS, Vol.96, No.20, (September 1999), pp.11416-11421, ISSN 0027-8424

Ye, Z-w.; Du, J-h; Su, M-s.; Li, L-l. & Zhang, S-l. (2010). Effects of high temperature on the microsporogenesis and pollen development of peach. *Acta Horticulturae Sinica*, Vol.3, pp.355-362, ISSN: 0513-353X

Zhang, G.; Campenot, MK.; McGann, LE. & Cass, DD. (1992). Flow cytometric characteristics of sperm cells isolated from pollen of *Zea Mays* L. *Plant Physiology*, Vol.99, No.1, (May 1992), pp.54-59, ISSN 0032-0889

Zhang, X. & Liu, G. (1998). Induction of 2n pollen in *Prunus* by heat shock. *Acta Horticulturae Sinica*, Vol.3, pp.392-293, ISSN 0513-353X

Zhang, Z. & Kang, X. (2010). Cytological characteristics of numerically unreduced pollen production in *Populus tomentosa* Carr. *Euphytica*, Vol.173, No.2, (May 2010), pp. 151-159, ISSN 0014-2336

Zhang, L.-Q.; Liu, D-C.; Zheng, Y-L.; Yan, Z-H.; Dai, S-F.; Li, Y-F.; Jiang, Q.; Ye, Y-Q. & Yen,Y. (2010). Frequent occurrence of unreduced gametes in *Triticum turgidum– Aegilops tauschii* hybrids. *Euphytica*, Vol.172, No.2, (March 2010), pp.285–294, ISSN 0014-2336

Zhang, L.-Q.; Yen,Y.; Zheng, Y-L. & Liu, D-C. (2007a). Meiotic restriction in emmer wheat is controlled by one or more nuclear genes that continue to function in derived lines. *Sexual Plant Reproduction*, Vol.20, No.3, (September 2007), pp.159-166, ISSN 0934-0882

Zhang, Z.; Kang, X., Zhang, P.; Li,Y. & Wang, J. (2007b). Incidence and molecular markers of 2n pollen in *Populus tometosa* Carr. *Euphytica*, Vol. 154, pp.145-152, ISSN 0014-2336

Zhong, C.; Zhang, S-n.; Yu, X-h.; Li, Y.; Hou, X-l. & Li, S-j. (2010). Studies on the Induction of 2n Gamete in Chinese Cabbage and the Production of Tetraploid by Sexual Polyploidization. *Acta Horticulturae Sinica*, Vol.37, No.11, pp.1789-1795, ISSN 0513-353X

Zhou, S.; Ramanna, MS.; Visser, RGF. & Van Tuyl, JM. (2008). Analysis of the meiosis in the F1 hybrids of Longiflorum x Asiatic (LA) of lilies (*Lilium*) using genomic in situ hybridisation. *Journal of Genetics and Genomics*, Vol.35, No.11, (November 2008), pp.687-695, ISSN 1673-8527

Zlesak, DC. (2009). Pollen diameter and guard cell length as predictor of ploidy in diverse rose cultivars, species, and breeding lines. In: *Floriculture and Ornamental Biotechnology (Vol.3, special issue 1 'Roses')*, Zlesak, DC. (ed.), 53-70, Globalsciencebooks, ISBN 9784903313283, Middlesex, UK

# Part 3

# Molecular Markers and Breeding

# 7

# Genomics-Assisted Plant Breeding in the 21st Century: Technological Advances and Progress

Siva P. Kumpatla[1], Ramesh Buyyarapu[1],
Ibrokhim Y. Abdurakhmonov[2] and Jafar A. Mammadov[1]
*[1]Department of Trait Genetics and Technologies, Dow AgroSciences LLC,*
*[2]Center of Genomic Technologies, Institute of Genetics and Plant*
*Experimental Biology, Academy of Sciences of Uzbekistan,*
*[1]USA*
*[2]Uzbekistan*

## 1. Introduction

One of the key global challenges of the 21st century is the production of enough food for the increasing world population. As per some recent reports the global population will continue to grow with some 9 billion people by the middle of the current century and the world will need 70 to 100% more food by that time (Godfray et al., 2010 and references therein). Agricultural productivity needs to be increased while addressing the issues of scarcity of arable land and water, impact of changing climate and preservation of natural resources. Improvement of crop yields on available agricultural land requires concerted efforts using modern scientific and technological advances in multiple disciplines (Hubert et al., 2010). Two such disciplines that have revolutionized crop improvement in the recent decades are molecular breeding and plant genomics. While the availability and application of molecular markers have accelerated the pace and precision of plant genetics and breeding, the introduction of a multitude of "omics" tools has provided unprecedented ability to dissect the molecular and genetic basis of traits as well as the characterization of whole genomes.

Molecular markers have occupied center stage in plant genetics since late 1980s. The advent of markers based on simple sequence repeats (SSRs) and single nucleotide polymorphisms (SNPs) and the availability of high-throughput (HTP) genotyping platforms have further accelerated the generation of dense genetic linkage maps and the routine use of the markers for marker-assisted breeding in several crops (Collard and Mackill, 2008). However, despite the routine use of markers for genome-wide profiling and trait-specific marker-assisted selection (MAS), breeding of crops with many traits of interest such as yield, improved nutritive value and resistance to several biotic and abiotic stresses is still a challenge due to complex inheritance of these traits. Therefore, there is a dire need for the molecular dissection of these traits in the context of the whole genome. This is where plant genomics plays a key role by providing the knowledge base required for the understanding and improvement of these traits. Genome is defined as a haploid (single set) content of all of the hereditary information of an organism, and genomics is the scientific discipline that studies the genome at the structural and functional levels towards understanding the genetic basis of inheritance, molecular basis of important intragenomic biological phenomena and the evolutionary history of genes. Plant genomics has enormous potential to revolutionize crop improvement by providing extensive knowledge from the analysis of genomes which in turn can be used for rapid and efficient plant breeding towards crop improvement. In the following sections we have reviewed prominent genomics tools and how technological advances in these as well as associated tools are contributing to the progress towards genomics-assisted plant breeding of 21st century.

## 2. Genomics tools and technologies and their applications

### 2.1 Structural genomics: random, targeted and whole genome approaches

Structural genomics is an approach in molecular genetics that enables researchers to detect segments of DNA with allelic variations, correlate those polymorphisms with phenotypic data and determine causative mutations underlying important traits. The scope of "structural genomics" discussed here needs to be distinguished from that coined by protein community where similarly-named approach has been used to investigate the comprehensive repertoire of protein folds to infer molecular functions of the proteins (Burley et al., 1999). Although the main goal of structural genomics is similar in both cases, i.e. from structure to function, researchers use different paths to achieve the final goal.

### 2.1.1 Molecular markers: development and applications

Structural allelic alterations, or polymorphisms, of a genome can be grouped into three major categories that include differences in the number of tandem repeats at a particular locus [microsatellites, or simple sequence repeats (SSRs)] (Weber and May, 1989), segmental insertions/deletions (InDels) (Ophir and Graur, 1997) and single nucleotide substitutions [single nucleotide polymorphisms (SNPs)] (Wang et al., 1998). In order to detect and track allelic variations in progeny, the scientific community has been developing genetic tools, called molecular markers, since the late 1980s (Botstein et al., 1980). Although SSRs, InDels and SNPs are the three major allelic variations discovered so far, a plethora of molecular markers have been developed to detect the above-mentioned polymorphisms (Bernardo, 2008; Gupta et al., 1999). The main drivers for the evolution of molecular markers have been throughput, level of reproducibility and cost reduction (Bernardo, 2008). Depending on the detection method and throughput, all molecular markers can be divided into three major groups: (1) low-throughput, hybridization-based markers such as restriction fragment length polymorphism [RFLP (Botstein et al., 1980)], (2) medium-throughput PCR-based markers, that include random amplification of polymorphic DNA (RAPD) (Welsh and McClelland, 1990; Williams et al., 1990), amplified fragment length polymorphism (AFLP) (Vos et al., 1995) and SSRs (Wang et al., 1998), and (3) HTP sequence-based markers: SNPs (Wang et al., 1998). In late eighties, RFLPs were the most popular molecular markers and were widely used in plant molecular genetics, because they were reproducible and co-dominant. However, the detection of RFLPs was very expensive, labor- and time-consuming process, which made these markers eventually obsolete. Additionally, RFLP markers were not amenable for automation. Invention of PCR technology and application of this method for the rapid detection of polymorphisms overthrew low-throughput RFLP markers, and new generation of PCR-based markers emerged in the beginning of 1990s. RAPD, AFLP and SSR markers are the major PCR-based markers that research community has been using in various plant systems. RAPDs were able to simultaneously detect polymorphic loci in various regions of a genome. However, they were anonymous and the level of their reproducibility was very low due to the non-specific binding of short, random primers. Although AFLPs were anonymous too, the level of their reproducibility and sensitivity was very high owing to the longer +1 and +3 selective primers and the presence of discriminatory nucleotides at 3' end of each primer. That is why AFLP markers are still popular in molecular genetics research in crops with little to zero reference genome sequence available (Zhang et al., 2011). However, AFLP markers did not find widespread application in molecular genetics and molecular breeding applications, because the detection method was too long, laborious and not amenable to automation. Therefore, it was not surprising that in the beginning of 21ˢᵗ century SSR markers were declared as "markers of choice" (Powell et al., 1996). SSRs were no longer anonymous; they were highly-reproducible, highly-polymorphic, and amenable to automation. Despite the cost of detection remaining high, SSR markers pervaded all areas of plant molecular genetics and breeding. However, during the last five years, the hegemony of medium-throughput SSRs was eventually broken by SNP markers. First developed for human genome, SNPs have proven universal and are the most abundant forms of genetic variation among individuals within a species (Rafalski, 2002). Although SNPs are less polymorphic than SSR markers because of their bi-allelic nature, they easily compensate this drawback by being abundant, ubiquitous and amenable to high and ultra-high-throughput automation. Since SNPs are currently the most widely used markers in plant molecular genetics and breeding, they are discussed in great detail in the following sections.

### 2.1.2 SNP markers

Development of SNP markers usually consists of two parts: SNP discovery and SNP validation. SNP discovery in crops is not an easy task because of genome complexity and often the lack of reference genome sequences. Even in crops such as maize (*Zea mays*), where a reference genome sequence is available, large scale SNP discovery efforts are still impeded by the highly repetitive (Meyers et al., 2001) and duplicated (Gaut and Doebley, 1997) nature of the genome. In order to avoid repetitive sequences, maize researchers have focused on the discovery of SNPs within coding sequences by re-sequencing amplicons derived from unigenes (Wright et al., 2005) or by *in silico* mining of SNPs within ESTs (Batley et al., 2003). The advantage of these approaches is the detection of gene-based SNPs. However, both approaches have some drawbacks: they are low-throughput and are unable to detect SNPs located in low-copy non-coding regions and intergenic spaces. Additionally, amplicon re-sequencing is an expensive and labor intensive procedure (Ganal et al., 2009). The recent emergence of next generation sequencing (NGS) technologies such as 454 Life Sciences (Roche Applied Science, Indianapolis, IN), Hiseq (Illumina, San Diego, CA), SOLiD and Ion Torrent (Life Technologies Corporation, Carlsbad, CA) have elevated expectations towards the rapid genome-wide identification of a large number of SNPs at a much lower price tag (Mardis, 2008a). However, efficient application of these technologies for SNP discovery in a given crop depends on the availability of the reference genome sequence (Ganal et al., 2009) as well as the level of genome complexity. For instance, in maize, the availability of a reference sequence does not guarantee a painless SNP discovery using NGS technologies. The complexity and existence of re-arrangements in the maize genome complicate the assembly of short-read NGS sequences and their alignment to the reference genome (Morozova and Marra, 2008). Thus, the reduction of genome complexity becomes an important prerequisite for the genome-wide discovery of true SNPs in crops with and without reference genome using sequencing by synthesis (SBS) technologies. Several genome complexity reduction techniques have been developed, including High $C_0t$ (DNA renaturation kinetics $C_0t$) selection (Yuan et al., 2003), methylation filtering (Emberton et al., 2005; Palmer et al., 2003), and microarray-based genomic selection (Okou et al., 2007). However, a majority of these techniques mainly reduce the number of repetitive sequences and are ineffective in the recognition and elimination of paralogues and homoeologues, which cause the detection of false-positive SNPs. Recently, computational SNP calling methods were developed that can drastically reduce the number of false SNPs resulting from the alignment of duplicated sequences and re-sequencing errors (Baird et al., 2008; Barbazuk et al., 2007; Gore et al., 2009; Van Orsouw et al., 2007). Hence, the availability of reference sequences, the application of genome complexity reduction techniques and NGS technologies coupled with post-re-sequencing computational treatment become important prerequisites for genome-wide detection of SNPs in complex genomes.

### 2.1.3 SNP validation and modern genotyping platforms and chemistries

The availability of reference sequence and sophisticated software do not always guarantee that the discovered SNP can be converted into a valid marker. In order to insure that the discovered SNP is a Mendelian locus, it has to be validated. The validation of a marker is a process of designing an assay based on the discovered polymorphism and genotyping a panel of diverse germplasm or segregating population. Segregating population is more informative as a validation panel than a collection of unrelated lines, because it not only allows inspection of the discriminatory ability of a marker but also its segregation patterns and ratios which helps researcher to understand whether it is a Mendelian locus or a duplicated/repetitive sequence that escaped software-filter (Mammadov et al., 2010).

In plants, SNPs can be validated using flexible and HTP assays, chemistries and genotyping platforms, including Illumina's BeadArray technology-based GoldenGate (GG) (Fan et al., 2003) and Infinium assays (Steemers and Gunderson, 2007), Life Technologies' TaqMan (Livak et al., 1995) assay coupled with OpenArray platform (TaqMan OpenArray Genotyping system, Product bulletin, 2010) and KBiosciences' Competitive Allele Specific PCR (KASPar) complemented with the SNP Line platform (SNP Line XL; http://www.kbioscience.co.uk). These modern genotyping assays and platforms differ from each other in chemistry, cost and throughput of samples to genotype and number of SNPs to validate. The choice of chemistry and genotyping platform depends on many factors that include the length of SNP fragment sequence, overall number of SNPs to genotype and finally the funds available to the research unit because most of these chemistries remain cost-intensive. Below is the summary of four SNP genotyping assays and platforms, which have been widely used in academia and industry.

## 2.1.3.1 Illumina's BeadArray platform

The Illumina's BeadArray platform (Fan et al., 2003) is capable of validation of a large number of SNPs in parallel by combining several technologies. The core of the technology is a collection of 3-micron silica beads that get self assembled in the wells, which are etched on the surface of a miniaturized matrix (either fiber optic bundles or planar silica slides) and evenly spaced at ~5.7 micron distance. Each bead is covered with hundreds of thousands of copies of a specific oligonucleotide that act as the capture sequences in one of Illumina's assays such as GG and Infinium. A high-resolution confocal scanner (iScan) is engineered to read arrays and generate intensity data, which is converted into genotypic data by reliable genotype-calling software, GenomeStudio. GG and Infinium are highly multiplexed chemistries and can genotype a maximum of 3,072 SNPs and ~1.1 million attempted bead types, respectively, in a single reaction without adverse effect on allele discrimination. All previous multiplexing efforts by various companies and academic labs had limited success mostly because of the interactions of primers and discrimination of alleles during amplification. In GG and Infinium, primers do not interact with each other. In GG assay all 3,072 loci are amplified with the same trio of universal primers (namely, P1, P2 and P3). In addition, allele discrimination occurs prior to PCR. Last but not least, small, newly synthesized DNA fragments, not entire genomic DNA, serve as templates for PCR amplification, which dramatically reduces the complexity of PCR reactions. Although the whole SNP genotyping process using the GG assay takes three days, the GG assay is a combination of simple molecular biology techniques, which is easy to follow and implement. In contrast to GG where all 3,072 assays [oligo pool assay (OPA)] are manufactured as a suspension in a single tube, in case of Infinium all assays are immobilized via beads on the surface of a chip. Depending on SNP type, two types of assay can be designed: Infinium I and Infinium II. Infinium I is designed for [A/G] and [T/C] SNPs and requires one bead type per allele (two bead types per SNP), while Infinium II is designed for all other SNPs and requires only one bead type for both alleles. That is why, the calculation of the price of Infinium assay is based on number of attempted bead types but not SNPs. The entire set of attempted bead types is called iSelect, which is an equivalent of OPA in GG assay. In both assays, the number of samples processed per day is restricted to three 96-well plates because of limited capacity of a liquid handler TECAN, which is a part of automation in BeadArray technology. Thus, BeadArray technology coupled with GG and Infinium assays is a robust and high-throughput platform designed to validate a large number of SNPs with relatively small number of samples: minimum 24 (one beadchip) and maximum 288 samples (12 beadchips).

### 2.1.3.2 The OpenArray technology

The chemistry of OpenArray technology (Brenan and Morrison, 2005) is based on Life Technologies' end-point TaqMan assay. The arrays require assays and samples on a small OpenArray plate. The physical size of an OpenArray plate is 1/8 of the size of 384-well plate. However, unique structural design of OpenArray plate allows accommodation of 3,072 assays within one plate, which is equal to the capacity of eight 384-well plates. The plate has 48 subarrays and each subarray has 64 holes, where nano-volumes of DNA get loaded. Another great feature of this platform is that it is very flexible and allows the user to array different SNP vs. sample combinations, including 64 x 48, 128 x 24, 256 x 12, 192 x 16, 32 x 96 and 16 x 144 formats. Another important feature of the OpenArray plate is that the hydrophobic and hydrophilic coatings of the surface and holes, respectively, enable reagents to stay in the bottomless through-holes via capillary action. OpenArray plates are preloaded with assay reagents by a vendor, and sent to the end-user to load DNA samples. The throughput of SNP genotyping using OpenArray technology can be greatly increased by attaching slide towers on top of the DNA engine, i.e. thermocycler. Each slide tower can harbor 32 slides. Throughput can be increased by using several two to three thermocyclers with slide-towers. For example, if 128 x 24 format is used, 128 SNPs can be validated with 2,048 (32 x 24 x 2) samples per day using two thermocyclers. In contrast to BeadArray platform, OpenArray technology can validate relatively small subset of SNPs with larger number of samples, which makes it very attractive for marker-assisted breeding projects, where gene or QTL region must be tracked by a few markers within a large number of samples.

### 2.1.3.3 KBiosciences' competitive allele specific PCR (KASPar)

In addition to above platforms, the Competitive Allele Specific PCR (KASPar) from KBios-ciences (Hoddesdon, Hertfordshire, UK) (http://www.kbioscience.co.uk ) is widely used in SNP validation, although it does not have any multiplexing capabilities. However, this chemistry is becoming widely used in SNP validation. KASPar assay uses a technique based on allele specific oligo extension and fluorescence resonance energy transfer (FRET) for signal generation. The fluorescent reporting system comprises of four single-labeled oligonucleotides that hybridize to one another in free solution to form a fluorescent quenched pair which upon introduction of complementary sequences generates a measurable signal. The kit requires two components, the assay mix (a mixture of three unlabelled primers: two allele specific oligos and one common reverse locus specific oligo) and the reaction mix (the other components required for PCR, including the universal fluorescent reporting system and Taq polymerase). KASPar is a very flexible assay, because SNP validation can be carried out in a variety of formats and the chemistry has been shown to function well in 96, 384 and 1536-well plates. One of the most attractive features of KASPar is the cost effectiveness and the duration of the synthesis of the assay. One KASPar assay will cost ~$15 and results can be delivered next day. Compared to KASPar assay, cost per one TaqMan and GG assays as well as Infinium bead type will be around $400, $42 and ~$9, respectively. Duration of synthesis of Taqman assay, GG OPA and Infinium iSelect is two, six and nine weeks, respectively. Also, depending on the size of the validation panel, GG and Infinium assays might not be suitable to validate SNPs, because Illumina imposes minimum sample order limitation per OPA and iSelect, which are 480 and 1152 samples, respectively. Finally, the choice of chemistry and genotyping platform for validation will also depend on the length of the context sequence based on which one can develop an assay. The length of context sequence is a crucial factor because most of the modern genotyping chemistries have a strict requirement for the length of the template strand. For example, 70 nucleotide (nt) short reads generated by Illumina's HiSeq NGS instrument can be suitable for validation using GG, Infinium and KASPar assays, which require minimum 50 nt template sequence from both sides of a SNP. At the same time, HiSeq output might not be suitable for TaqMan assay design which requires a longer input sequence (100 nt), because it needs enough space for a probe and two oligos flanking the SNP. Thus, there is no ideal genotyping platform and assay that a researcher can leverage for SNP validation. The choice of assay and genotyping system will depend on the number of SNPs, length of the template sequence, sample size, time-sensitivity of a project and the funds available to the researcher.

## 2.1.4 SNP Application

When SNP markers have passed the validation step, they are considered as viable markers and ready for use in various areas of molecular genetics and plant breeding, including gene/QTL mapping, linkage-disequilibrium-based association mapping, map-based gene/QTL cloning, germplasm characterization, genetic diagnostics, event characterization, marker-assisted trait introgression, and finally marker-assisted selection (MAS). In order to conduct most of the above-mentioned SNP applications, researcher must know the order of the markers on chromosomes, which can be obtained by constructing recombination-based genetic linkage maps. Genetic mapping is carried out using segregating populations, including $F_2$, backcross, recombinant inbred lines (RILs) or doubled haploids (DHs). Currently, most of the major crops possess genetic maps densely saturated with molecular markers. Publicly available genetic linkage maps that are constructed solely based on SNPs currently exist for rice (http://www.gramene.org) and maize (http://www.maizegdb.org) only. Remaining crops have genetic maps, which have been constructed by means of SNPs in combination with other markers such as SSRs and RFLPs that include barley, wheat, sorghum (http://wheat.pw.usda.gov/ggpages/map_shortlist.html) and soybean (http://soybase.org/).

## 2.2 Other key "Omics" tools needed for structural genomics work

Genotypic and the corresponding phenotypic data are the two major components required for understanding the genetic basis of traits through genetic linkage analysis. While advances in molecular marker fingerprinting and next generation sequencing are enabling economical and HTP genotyping of samples (Peleman and van der Voort, 2003), phenotyping of a large number of samples under field conditions is still a bottleneck. It is a laborious and expensive task and is a serious drawback for the dissection of complex and dynamic traits such as abiotic stress tolerance, yield and nutrient use efficiency where data needs to be collected form really large populations for efficient genetic analysis. Because of this drawback, many research efforts to date have treated dynamic traits as static traits and have relied on only one measurement for analysis, that too on small populations. Gathering multiple data points is even difficult for traits for which root measurements are needed. In order to address this situation, several HTP phenotyping techniques have been conceived and implemented. The discipline focused on developing such HTP phenotyping tools and platforms is termed as 'phenomics'. An example of HTP phenotyping is a near-infrared spectroscopy equipment, mounted on agricultural harvesters that can be used to collect spectral information about the plants during the harvesting of field trials (Montes et al., 2006). Spectral information thus collected can be condensed into a single near-infrared spectrum and analyzed using calibration models for the determination of information on several traits. Spectral reflectance of plant canopy using light curtains and spectral reflectance sensors mounted on a tractor is another phenotyping technique which non-invasively monitors several dynamic and complex traits such as biomass accumulation (Montes et al., 2011). The domain of phenomics concerned with the measurement of phenome (measurement of physical and biochemical attributes or phenotypes of traits of interest) has seen commendable efforts in the recent years in the automation of plant phenotyping. Several automated platforms and approaches such as Phenopsis (automated growth chambers for growing 504 pots of *Arabidopsis thaliana* at a time), Phenodyn (for simulating drought conditions and measuring transpiration and growth), Growscreen (digital imaging and processing for growth rate determination), Traitmill (fully automated growth facility) and LemnaTec (automated greenhouse) now exist for the HTP collection of plant phenotypic data on several traits of interest [reviewed in Kolukisaoglu and Thurow (2010)]. Availability of such automated phenotyping methods holds a great promise for the molecular and genetic dissection of complex traits by integrating this information with that of multiple datasets resulting from HTP genotyping as well as diverse 'omics' efforts.

## 2.3 Next generation sequencing

Improvements in crop productivity require adoption of new breeding technologies. Integration of genomic and transcriptomic data provides an opportunity to generate newer molecular resources for improved breeding technologies and crop improvement. Availability of DNA/RNA sequence information is highly critical to develop such resources. Until recently, sequencing efforts were dominated by Sanger sequencing method. Initial draft of human genome sequence was generated using BAC-by-BAC approach using Sanger's sequencing method by investing approximately three billion dollars into Human Genome Project (Venter et al., 2001). The availability of human genome reference sequence paved the way to a multitude of applications including detection of structural and copy number variations to understand the underlying genetic and epigenetic mechanisms. Though Sanger sequencing method dominated the industry for almost two decades and still considered the gold standard for sequencing, its limitations, especially with respect to throughput and cost, necessitated high demand for new and improved technologies for sequencing large and complex genomes. With advances made in the fields of microfluidics, microscale imaging, detection and computational tools, alternative sequencing technologies with increased throughput and lower sequencing cost are continuously emerging. These alternative technologies to Sanger's sequencing can be collectively termed as Next generation sequencing (NGS) technologies (Varshney et al., 2009). Since NGS technologies are impacting several 'omics' efforts, they are discussed in extended detail below.

### 2.3.1 NGS technologies

The advent of NGS technologies has changed the dynamics and the pace of genomic research in humans, plants, animals and microorganisms because of their rapid, inexpensive and highly accurate sequencing capabilities. Unlike Sanger sequencing method which depends upon capillary electrophoresis, these NGS technologies are highly dependent on massive parallel sequencing, high resolution imaging, and complex algorithms to deconvolute the signal data to generate sequence data. NGS technologies offer a wide variety of applications such as whole genome *de novo* and re-sequencing, transcriptome sequencing (RNA-seq), microRNA sequencing, amplicon sequencing, targeted sequencing, chromatin immunoprecipitated DNA sequencing (ChIP-seq), methylome sequencing and many others. Before dwelling into the use of this wide variety of NGS applications for crop improvement, various NGS technologies and their capabilities are briefly reviewed first.

Current NGS technologies can be broadly grouped into long and short read length technologies based on the number of bases they can sequence in a single sequencing reaction. Long read length technologies are preferred for applications involving *de novo* sequencing while short read length technologies are relatively inexpensive and mostly used for re-sequencing applications. Most of the NGS technologies monitor millions of sequencing reactions in parallel and thus result in a massive amount of sequencing data. The output capacities of these instruments outpaced the development of computational tools and hardware for data processing needs. Sophisticated computer programs are created to handle and process large amounts of sequencing data before final data analysis. Several bioinformatics tools were designed for diverse purposes such as *de novo* sequence assembly, mapping sequences to an existing reference genome sequence, mutation detection and annotation. Long read technologies include Roche/454 GS FLX and Pacific Biosciences RS systems while short read technologies include Illumina Genome Analyzer IIx, HiSeq 2000, MiSeq, Life Technologies' SOLiD™ system, Helicos Genetic Analysis system and Life technologies/Ion Torrent Personal Genome Machine (PGM). Mardis (2008b) and Metzker (2009) provided detailed reviews of these NGS technologies. NGS technologies that are widely used at present are briefly reviewed below and sequencing capabilities of instruments are summarized in Table 1.

### 2.3.1.1 Roche/454 GS FLX – pyrosequencing

This is the first NGS technology commercially introduced and is based on pyrosequencing method (Margulies et al., 2005). This technology is relatively rapid and inexpensive as it omits the expensive *in vivo* sub-cloning of sheared fragments for template amplification. Instead of cloning, sheared fragments are attached to microbeads and amplified in an emulsion-based PCR. These microbeads are further distributed to a fiber optic slide (PicoTiterPlate™), where the four dNTPs are added in turns. In pyrosequencing, the DNA sequence is determined by analyzing the fluorescence emitted by the activity of luciferase during the process of template extension by a single nucleotide addition. The fluorescence emitted is captured by a high resolution CCD camera for each type of nucleotide passed in a flow cycle. The intensity of the fluorescence is proportional to the number of nucleotides integrated in each step. The first commercial 454 instrument was able to generate >25 milion bases in short reads of 100 bp or more per 4 hr run. With the improvements in sequencing chemistry, PicoTiterPlate (PTP), reagent volumes and the number of nucleotide flow cycles in the instrument, the current GS FLX plus instrument was able to achieve an average read length of ~750 bp across 1 – 1.5 million sequences in ~20 hr runtime. Long read length capabilities of this instrument enable *de novo* sequencing of genomes and transcriptomes with ease compared to short read technologies. However, this technology is prone to sequencing errors in the homopolymer regions. Since the advent of 454 sequencing technology, there are ~1331 peer reviewed publications as of July, 2011 (http://454.com/publications/all-publications. asp) across a wide range of topics.

### 2.3.1.2 Illumina Genome Analyzer/HiSeq/MiSeq – sequencing-by-synthesis

Illumina sequencing method utilizes clonal array formation and proprietary reversible terminator reaction chemistry for rapid, accurate and large scale sequencing. DNA template fragments were immobilized in an 8-channel microfabricated flow cell where they were amplified up to 1000 copies in close proximity by bridge amplification method. Sequencing-by-Synthesis uses all four fluorescently labeled nucleotides to sequence millions of clusters on the flow cell surface. The fluorescent label in each nucleotide blocks the 3'–OH group and thus acts as a terminator for polymerase extension. At the incorporation of each nucleotide,

| Platform | Template preparation | Method | No. of reads | Read length (bases) | Run time | Throughput (Gb) | Advantages | Disadvantages |
|---|---|---|---|---|---|---|---|---|
| Roche/454's GS FLX + | emPCR | Pyrosequencing | 1 - 1.5 million | 750 | 18-20 hrs | 0.75 | Longer readlengths, faster run time, highly useful for de novo sequencing applications | Higher reagent and per base sequencing cost |
| Illumina HiSeq 2000 | bridge amplification | Sequencing-by-synthesis using reversible terminator nucleotides | $3^1$ or $6^2$ billion | 100 | 6 or 11 days | $150^1$ or $300^2$ per flow cell | Very high throughput, lower per base sequecing cost and useful for resequencing applications | Lower multiplexing capabilities |
| Life technologies / ABI's SOLiD 5500 | emPCR | Sequencing-by-Ligation using di-base probes | 1.4 billion | 35-75 | 7 or 14 days | 90 | HTP, lower per base cost and useful for resequencing applications | Long run times |
| Polonator G.007 | emPCR | Sequencing-by-Ligation using non-cleavable probes | 7-12 million | 26 | 5 days | 12 | Open source and less expensive NGS platform | Shortest read lengths |
| Helicos Biosciences Heliscope | single molecule | True single molecule sequencing | 0.6 - 1 billion | 35 | 8 days | 37 | Non-biased representation of templates and high throughput | High error rates |
| Pacific Bioscience's RS | single molecule | Single molecule real time sequencing | 30,000-50,000 | 1000 - 10000 | 30 min | 0.03 - 0.05 per SMRT cell | Extra longer read lengths, short run times, highly useful for de novo sequencing | High error rates |
| Ion Torrent PGM | bead based | Semiconductor sequencing | 1 - 12 million$^\Phi$ | 200 | 2 hrs | 0.01 - 1 | Shorter run times, relatively cheap NGS technology, Useful for amplicon sequencing | NA |

* Based on the information provided by Metzker (2009) and other company web resources

[1] Single end read chemistry, [2] paired end read chemistry

$\Phi$ Sequence capacity change with the type of chip used for sequencing.

Table 1. Comparison of NGS technologies and capabilities *

fluorescent dye is imaged to identify the dye and then the label is enzymatically cleaved to allow the incorporation of next base (Bentley et al., 2008; Ju et al., 2006). As each nucleotide base incorporation is a unique event, the error rate in homopolymer regions is minimal compared to 454 pyrosequencing method (http://www.illumina.com/technology/ sequencing_technology.ilmn). Illumina has a range of sequencing instruments that can generate from ~1 Gibabase (Gb) from ~3-6 million sequences (MiSeq) and up to 600 Gb from 6 billion paired end reads per two flow cells (HiSeq 2000) in a single sequencing run. Though the output capabilities of Illumina sequencing instruments are large, they also take longer sequencing time from 3 – 11 days depending on the machine, single end or paired end protocol and number of flow cycles. This technology has revolutionized the pace of re-sequencing efforts in human and other genomes besides bringing down per base cost to a bare minimum. As of July, 2011, there are ~1746 peer reviewed publications that have used this technology.

### 2.3.1.3 Life technologies SOLiD™ – Sequencing-by-Ligation

Life technologies, previously Applied Biosystems, developed another short read sequencing technology which utilizes sequencing-by-ligation method. Template DNA fragments are clonally amplified in an emulsion PCR reaction similar to that of 454 sequencing and the clonal bead populations are covalently bound to a slide by 3′ modification of the beads. During the sequencing reaction, a fluorescently labeled di-base probe hybridizes to the complementary sequence adjacent to primed template and DNA ligase enzyme joins the dye-labeled probe to the primer. After the non-ligated probes are washed off, fluorescence is imaged to identify the nucleotides incorporated at first and second base (http:// www. appliedbiosystems.com/absite/us/en/home/applications-technologies/solid-nextgenerati on-sequencing/next-generation-systems/solid-sequencing-chemistry.html). The cycle can be repeated either by using cleavable probes to remove the fluorescent dye and regenerate a 5′ -$PO_4$ group for subsequent ligation cycles or by removing and hybridizing a new primer to the template (Metzker, 2009; Valouev et al., 2008). SOLiD 5500, the recent version of this technology can generate up to 90 Gb of sequence data from ~1.4 billion reads of 35-75 bases in length over ~7 days of time. Due to its massive outputs and short read length capabilities, this system is heavily used for re-sequencing and RNA-Seq applications.

### 2.3.1.4 Pacific Biosciences *RS* – SMRT™ (single molecule real time) sequencing

Pacific Biosciences developed SMRT technology which implements detection of fluorescently labeled nucleotides as they are incorporated over a single DNA molecule in real time. A single Φ-29 DNA polymerase enzyme molecule, a highly processive and strand displacing enzyme, is immobilized in a small hole called zero-mode wave guide (ZMW) to process the extension of a single molecule of primed DNA template (Eid et al., 2009). Four color phospholinked dye labeled nucleotides are used in this process and their fluorescence is quenched until they are incorporated during the sequencing reaction (Korlach et al., 2008). In the ZMW, as each nucleotide is incorporated by the anchored DNA polymerase, the phospholinked dye label is cleaved and its fluorescence light pulses are captured by four single photon sensitive cameras in the sequencing instrument (Lundquist et al., 2008). The real time light pulse information coming from 75000 ZMWs in a SMRT cell is converted to A, C, G, or T based on quality metrics to provide the sequencing information. The biggest advantage of this technology is longer read lengths of ~1000 – 10000 bases which facilitates easy sequence assemblies especially for *de novo* sequencing applications. As the sequencing reaction in a SMRT cell is monitored in real time, each typical sequencing run requires as little as 30 minutes compared to other technologies which can take up to 11 days. Strobe sequencing was used to achieve higher read lengths with higher accuracy (Lo et al., 2011). Though the cost of sequencing is relatively cheap, observed sequencing error rates are higher compared to other NGS technologies.

### 2.3.1.5 Life technologies/Ion Torrent PGM – Semi conductor sequencing

Ion Torrent PGM machine uses semi conductor technology with simple, non-fluorescent sequencing chemistry to generate the sequencing information. It is based on the detection of $H^+$ ions released (pH change) during a natural polymerase reaction using an ion sensor underneath the micro machined wells in a semiconductor chip, each containing a different DNA template. As each nucleotide flows in one at a time during the sequencing reaction, pH change is observed in all wells where the complementary nucleotide is incorporated (Pennisi, 2010). Change in pH is relative to the number of bases added to the template strand, and thus can sequence the homopolymer regions. As there is no involvement of fluorescent labeled nucleotides or imaging, incorporation of each nucleotide is recorded in seconds and the cost of sequencing is relatively cheap compared to other NGS technologies (http://www.iontorrent.com/technology-scalability-simplicity-speed/). Current read lengths are ~200 bp and each run takes about 2 hrs.

Existing and emerging NGS technologies are helping to bring down the sequencing costs towards making personalized genome services, personalized medicine and other applications possible in the near future. Third generation sequencing technologies such as Oxford's nanopore sequencing and VisiGen's nano sequencing technologies are currently being developed and would help the genome research more affordable than any time before.

### 2.3.2 NGS Applications for crop research

Widely available and cost-effective NGS technologies enabled many exciting opportunities for crop research in plants with or without a reference genome. Availability of reference genome/transcriptome sequence greatly enhances our ability to decipher the underlying molecular mechanisms of a trait, understand the gene regulatory mechanisms, determine gene expression differences and variations in expressed gene sequences, and other structural variations such as copy number variations (CNV) and presence-absence variations (PAV). NGS technologies can be applied to answer a wide variety of biological questions such as sequencing of complete genomes and transcriptomes and genome wide analysis of DNA-protein interactions (Bräutigam and Gowik, 2010). To facilitate crop improvement, NGS and other accessory technologies can be used for whole genome sequencing, transcriptome sequencing, genome wide and candidate gene marker development, targeted enrichment and sequencing and other applications. These NGS technologies even hold promise for a methodological leap towards genotyping–by–sequencing (GBS) and genetic mapping applications. Analysis of NGS data from genome wide association studies, transcriptomics and epigenomics in combination with data from proteomics, metabolomics and other 'omics' can provide an integrative systems biology approach to understand the regulation of complex traits.

## 2.3.2.1 Whole genome de novo / re-sequencing

Recently, whole genome sequencing efforts of many plant species including model and non-model crop species gained momentum due to lower sequencing costs and turnaround time. Genomes of model plant species such as Arabidopsis, rice and maize have been sequenced using Sanger sequencing method. *De novo* and re-sequencing of the genomes of several crop species with and without reference genome sequence are currently being accomplished by various NGS technologies. NGS technologies typically employ either multiplexed BAC pool sequencing or shotgun sequencing approaches to derive the whole genome sequence. Despite the availability of sophisticated assembly software, *de novo* assembly of large, complex and highly repetitive genomes poses enormous challenges to generate genomic reference sequence. To overcome the challenges of *de novo* sequence assembly of genomes, it is ideal to build a genome scaffold using long read length technologies and then use the short read length technologies to support the consensus sequence and thus minimize sequencing errors. Gaps generated during sequence assembly can be mitigated by using paired end and mate pair library sequencing in both long and short read length technologies. Information from the recent plant genome sequencing efforts using NGS technologies is summarized in Table 2. Genome re-sequencing efforts are currently in progress for crop species with reference genome such as corn and rice to understand important agronomic traits. These efforts are using structural, CNVs and PAVs and comparative genomics approaches for understanding variations, especially in closely related cultivars. Short read length technologies are routinely used for re-sequencing applications and re-sequence data is analyzed by mapping the reads back to genome scaffold to identify different kinds of variations in genic and non-genic regions.

| Plant Species | Ploidy | Genome size (Mb) | Sequencing Technology | Reference |
|---|---|---|---|---|
| *Theobroma cacao* (Cocoa) | diploid | 430 | Roche 454 | Argout et al. (2011) |
| *Malus × domestica* (domesticated apple) | diploid | 743 | Sanger paired end / Roche 454 | Velasco et al. (2010) |
| *Fragaria vesca* (Woodland Strawberry) | diploid | 240 | Rohe 454 / Illumina GA / ABI SOLiD | Shulaev et al. (2011) |
| *Vitis vinifera ssp. sativa* (Grapevine) | diploid | 504 | Sanger paired end / Illumina GA | Velasco et al. (2007) |
| *Jatropha curcas* | diploid | 410 | Sanger WGS / Roche 454 / Illumina GA | Zieler et al. (2010) |
| *Elaeis guineensis* (Oil Palm) | diploid | 1700 | Roche 454 | Zieler et al. (2010) |
| *Gossypium raimondii* (cotton) | diploid | 880 | Roche 454 / Illumina GA | http://www.jgi.doe.gov/sequencing/why/ gossypium.html |
| *Triticum aestivum* ('Chinese Spring' wheat) | Hexaploid | 16000 | Roche 454 | http://www.wheatgenome.org |
| *Musa* spp (Banana) | diploid | 550-650 | Sanger / Roche 454 / Illumina GA | Hribova et al. (2009) |

Table 2. Examples of whole genome sequencing in crop species.

## 2.3.2.2 Transcriptome / siRNA / miRNA sequencing

Transcriptome sequencing provides information about functional genes in an organism and helps in gene discovery. Collection of mRNA from different tissues and different stages of plant growth provides a comprehensive set of expressed genes in the template libraries for transcriptional profiling of even non-model organisms. Such libraries can be *de novo* or re-sequenced using NGS technologies more efficiently compared to earlier gene cloning and sequencing methodologies. Although Sanger sequencing method provides longer EST contigs, detection of allelic variations in the gene sequences is an expensive effort. The assembly becomes increasingly difficult when the read length gets shorter and shorter, which is the most compelling reason for choosing a long read technology for *de novo* sequencing. Many protocols are currently available to prepare the template libraries in normalized and non-normalized fashion either from cDNA or directly from mRNA/total RNA for diverse applications. Due to its longer read length capabilities and improved protocols to overcome the inherent problems of homopolymer region sequencing, 454 sequencing is well suited for *de novo* sequencing of EST libraries. Gene regulatory networks and pathways could be easily developed using transcriptome profiling experiments. Transcriptome data is highly useful not only to know the gene content and transcriptional status in various tissues but also helps in identifying SSRs and SNPs in the genic regions, which can be converted to gene-based markers (Narina et al., 2011).

Comparative transcriptomic approaches using NGS technologies can be applied to find functional gene homologs and orthologs in non-model organisms to support the gene discovery efforts. In several model plant species, EST libraries have been deeply sequenced and annotated to provide information on reference transcriptome and led to the creation of transcriptome databases (Morozova and Marra, 2008). For crop species with such annotated reference transcriptome dataset, RNA-seq and serial analysis of gene expression (SAGE) experiments using short read length NGS technologies could provide the expression levels of various genes very cost effectively. Illumina and Life technologies can be used for efficient sequencing of novel variants. They can also facilitate detection of homolog and paralogs of functional genes due to their high data throughput. Apart from expressed genes, other RNA molecules such as microRNA (miRNA), short interfering RNA (siRNA) are also present in the cell and are involved in the regulation of gene expression. Novel miRNA molecules can be easily sequenced from different tissues using NGS technologies to understand the mechanisms of gene regulation. For example, miRNA molecules thus detected during a biotic or abiotic stress condition were utilized to develop improved cultivars through transgenic approaches (Sindhu et al., 2009).

## 2.3.2.3 Molecular marker development

Genetic variation is the key for implementing molecular breeding approaches in any crop improvement project. Genetic variation is usually detected by identifying the polymorphisms exhibited at restriction site, as fragment lengths, or at single nucleotide levels either in genic or intergenic regions of the genome. Traditionally, the development of markers such as microsatellites, RFLPs and AFLPs was a costly iterative process that involved time-consuming cloning and primer design steps that could not easily be parallelized. In recent years, SNPs have been the markers of choice for the researchers due to their high abundance and amenability for automation and HTP genotyping capabilities.

However, prior availability of sequencing information is absolutely necessary to identify and design assays using SNPs. Genomic and transcriptomic resources can be easily generated by NGS technologies to rapidly and cost-effectively develop molecular markers such as SNPs and SSRs. NGS technologies have been recently used for whole genome and re-sequencing projects where the genomes of several specimens were sequenced to discover large numbers of SNPs for exploring within-species diversity, constructing haplotype maps and performing genome-wide association studies (GWAS) (Elshire et al., 2011).

Genome wide marker development is often achieved by comparing sequences from either whole genome re-sequencing efforts or genome complexity reduction approaches. SNP marker development in plant species with reference genome is relatively easier where the NGS data from different genotypes is mapped against the reference genome to identify SNPs. However, overall size and structure of plant genomes constitute a major hurdle for non-model crop species. Sequencing the whole genome of every individual in a population is costly and often unnecessary, as many biological questions can be answered using polymorphisms that are measured in a subset of genomic regions. Sequencing of libraries generated from reduced representation or target enrichment techniques as well as the DNA fragments resulting from the application of restriction-site associated DNA (RAD) or Complexity Reduction of Polymorphic Sequences (CRoPS) approaches are some of the methods for sampling and sequencing a small set of genome-wide regions without sequencing the entire genome and all these processes are often coupled with NGS technologies. Genome complexity reduction approaches greatly help marker development and have been used for SNP marker discovery in multiple crop and animal species (Davey et al., 2011). Complexity reduction methods include restriction digestion of genomic DNA using methylation sensitive and other restriction enzymes to exclude the repetitive regions and retroposon/transposon sequences during sequencing; and target enrichment for selective sequencing of regions of interest (Deschamps and Campbell, 2010).

Transcriptome data could also be used as a resource for detecting genetic variants in many crop species (Hamilton et al., 2011; Oliver et al., 2011). NGS technologies have been routinely used to generate huge EST datasets by RNA-seq experiments, which can be used for identifying SNPs in functional genes. Comparison of transcriptome datasets from parental genotypes could derive polymorphic SNPs. SNPs derived from gene sequences have higher significance compared to those of non-genic regions, as they can be directly associated with the gene function. Libraries enriched with PCR amplicons from target gene regions were sequenced by Roche 454 technology to use the NGS data as a resource to detect SNP markers in sugarcane, a complex polyploid crop species (Bundock et al., 2009). Bioinformatics tools such as AutoSNP (Wang and Liu, 2011), HaploSNPer (Tang et al., 2008) have been designed to detect the variations in NGS data by mathematical calculations of minor allele frequency or haplotype information. Minor allele frequency can be used as a measure to identify candidate SNPs in simple diploid species while calculation of haplotype information improves the SNP confidence in polyploid crop species such as potato and cotton.

### 2.3.2.4 Genotyping-by-sequencing and genetic mapping

The development of PCR based markers has revolutionized marker development and genotyping procedures to identify QTL regions associated with important traits. However, these markers, although still widely used, have shown growing limitations in chromosomal coverage, time, and cost effectiveness. The development of genomics concepts and tools and genome-based HTP strategies has provided an alternative approach to marker based mapping approaches. The NGS technologies coupled with the growing number of genome sequences opens the opportunity to redesign genotyping strategies for more effective genetic mapping and genome analysis. Although array-based genotyping methods such as Illumina Infinium iSelect assays provide HTP genotyping ability, it is laborious, time-consuming, and expensive to design, produce, and process arrays suited for specific mapping populations (Huang et al., 2009). Current SNP genotyping technologies often interrogate two alleles at a polymorphic site and this could limit genotyping of other alleles, especially in diverse natural populations, for association mapping.

Advances in next generation technologies have driven the costs of DNA sequencing down to the point that genotyping-by-sequencing (GBS) is now feasible for high diversity, large genome species. The new sequencing techniques not only increase sequencing throughput by several orders of magnitude but also allow simultaneous sequencing of a large number of samples using a multiplexed sequencing. These recent technical advances have paved the way for the development of a sequencing-based HTP genotyping method that combines the advantages of time and cost effectiveness, dense marker coverage, high mapping accuracy and resolution, and more comparable genome and genetic maps among mapping populations and organisms (Elshire et al., 2011).

GBS strategies often depend upon the resources available and type of mapping population used in the study. Availability of reference genome is always encouraged, but not an absolute requirement to implement GBS approaches. GBS experiments are often complemented with sequencing of samples in a mapping population in a multiplexed format and HTP manner to derive the genotyping information. This type of SNP data differ from that of traditional genetic markers primarily in two aspects. First, it is often not the case that all members of a recombinant population can be scored at a given SNP site. Second, an individual SNP site is no longer a reliable marker or locus for genotyping due to several potential sources of sequence errors. To overcome these difficulties, bioinformatics and statistical tools are used to validate the SNPs in a genotype at a specific locus and such tools included sliding window approach where the SNP information is confirmed not only at a single polymorphic site, but also in the flanking regions by verifying the haplotype information.

In rice, Huang et al. (2009) first demonstrated a HTP GBS method by whole genome re-sequencing in a 150 recombinant inbred line (RIL) population, where 16 samples were multiplexed per lane, and a total of 112 samples per flow cell were sequenced using Illumina GA. In this case, the mapping population was derived from only a set of two parents vs. *indica* and *japonica* cultivars. Sequence alignment of the population data and validation of the SNPs using sliding window approach provided the genotype calls for the population. Another parent independent GBS approach was also implemented in rice by Xie et al. (2010), where an ultra-high density linkage map was constructed by low coverage sequencing of mapping population. In this study, genotype calls in the population were derived by maximum parsimonious inference of recombination assisted by Hidden Markov Model (HMM) and were validated with Bayesian inference. This approach can be implemented in crops with no reference genome.

Whole genome re-sequencing is not always an option to implement GBS especially in crop species with large, complex, and repetitive genomes. Although GBS is fairly straightforward for small genomes, target enrichment or reduction of genome complexity must be employed to ensure sufficient overlap in sequence coverage for species with large genomes. Reducing genome complexity with restriction enzymes is relatively easy and reproducible compared to other target enrichment methods such as use of long range PCR, molecular inversion and capture probes etc. Elshire et al. (2011) applied complexity reduction approaches using *Ape*KI, a type II restriction endonuclease, to generate reduced representation libraries and then generated sequence data across these libraries from RIL mapping populations in both maize and barley. They analyzed the data to demonstrate GBS as a proof of concept for routine mapping and QTL identification studies. These studies illustrate and promise eventual application of GBS in introgression programs for traits of interest.

In human research, multiple cancer and other disease traits were investigated by GWAS using NGS technologies and that was possible due to the existence of narrow genetic variation in humans. Though GBS is an attractive option for populations derived from a set of parents, its application is very challenging for association mapping studies in plant species due to huge variations existing in natural populations. In GBS, variations are typically detected by aligning to the reference genome, but in natural populations, the variations are not limited to SNPs but also have PAVs. Detection of PAVs becomes exceptionally difficult unless comparative genome hybridization (CGH) approaches are applied along with NGS. Complex computational tools and deep sequencing data would help to overcome these problems.

### 2.3.2.5 Targeted sequencing, Methylation profiling and DNA-protein interactions

NGS technologies are often paired up with multiple accessory molecular biology methods to achieve the project-specific goals more efficiently and cost-effectively. Target enrichment is one of those accessory molecular techniques often used in NGS applications. Target enrichment methods mainly help to derive the NGS data from targeted regions such as candidate genes/exome regions and QTL regions and reduce the noise from unwanted regions in an experiment. Several commercial technologies are available for target enrichment and they include Roche/Nimblegen sequence capture arrays, Agilent SureSelect™ platform, RainDance technologies' RainStorm™ microdroplet-based PCR technology and Fluidigm Access Array technologies. For a review of these technologies and applications in NGS refer to Mamanova et al. (2010). The NGS data derived from targeted regions could be used for variant detection, identification of gene analogs and paralogs, SNP discovery in QTL/exome regions (Nijman et al., 2010) and also for fine mapping efforts. Gene sequences are usually conserved and exome enrichment methods help to enrich the gene regions in libraries from genomic DNA by not only capturing exon regions but also the intron and other flanking regions next to the gene sequence. This helps to detect the mutations in the genes and intron-exon junctions to identify the splice variants (Ng et al., 2009).

Genome-wide sequence data should greatly facilitate our understanding of complex phenomena, such as heterosis and epigenetics, which have implications for crop genetics and breeding (Varshney et al., 2009). Expression of the genome is influenced by chromatin structure, which is governed by processes often associated with epigenetic regulation, namely histone variants, histone post-translational modifications, and DNA methylation. Developmental and environmental signals can induce epigenetic modifications in the genome, and thus, the single genome in a plant cell gives rise to multiple epigenomes in response to developmental and environmental cues. N-terminal regions of nucleosome core complex histones undergo various post-translational modifications, namely acetylation, phosphorylation and ubiquitination that enhance transcription, while biotinylation and sumoylation repress genes (Chinnusamy and Zhu, 2009). Such modifications can be easily detected by combining the chromatin immunoprecipitation (ChiP) procedures with NGS technologies to analyze genome-wide histone modifications.

Methylation of cytosine bases in DNA provides a layer of epigenetic control in many eukaryotes that has important implications for normal biology and disease. Therefore, profiling DNA methylation across the genome is vital to understand the influence of epigenetics (Laird, 2010). Determination of DNA methylation patterns usually requires the use of methylation-sensitive restriction endonucleases or affinity chromatography with methyl-binding proteins, or anti-mC antibodies. Reinders et al. (2008) have used bisulfite conversion for genome-wide DNA methylation profiling. In Arabidopsis, DNA methylation patterns and effects of methylation mutants were studied using Illumina GA sequencing technology by Cokus et al. (2008).

NGS technologies have been leading genome sequencing initiatives across many non-model and orphan crops in recent years to answer the complex biological questions. Newer NGS technologies are being developed and implemented to meet the ever increasing needs of research community and to solve the complex puzzles of nature. NGS methods are being extended to study population genetics, evolutionary biology, molecular ecology, host-pathogen interactions, organellar development, genotype-phenotype interactions and many others. NGS can also accelerate the development of better transformation technologies to modify genes and transform plants easily. In a nut shell, NGS technologies have already demonstrated significant impact on crop breeding and would certainly help to transform the practices and pace of molecular breeding of crops.

## 2.4 Functional genomics

Functional genomics is the field of molecular biology that utilizes the vast wealth of data produced by genome sequencing projects to understand the gene functions, and their interactions. It is often referred to the study of the genes, their functions, interactions, and regulation to provide a biological function in an organism. Functional genomics mainly focuses on dynamic aspects such as gene transcription, translation, their interaction with other genes and proteins to define the gene function and their regulation. Functional genomics helps to understand the mechanism of a biological function and usually involves combination of both transcriptomics and proteomics.

Genome-wide expression analysis is rapidly becoming an essential tool for identifying and analyzing genes involved in, or controlling, various biological processes ranging from development to responses to environmental cues (Breyne and Zabeau, 2001). Transcriptional profiling studies routinely generate huge EST datasets using sequencing technologies to understand the biological significance of the genes. Availability of gene function information enables various applications of functional genomics. Assignment of gene function (annotation) in most instances is facilitated by comparing them with the genes of known function. Gene prediction modeling tools such as FGENESH, GENESCAN,

GLIMMER, SNAP are used to predict the coding regions (exons) in the genome and compare the translated protein to existing protein database for finding the gene function (Korf, 2004). Gene annotation tools such as BLAST2GO use the BLAST algorithm to find the similarity with the existing gene information to derive the gene annotation information (Conesa et al., 2005). Analysis of transcriptomic data using these bioinformatics tools helps gene discovery and associated pathways.

While structural genomics uses genetic variations to understand the phenotypic changes, functional genomics often uses gene expression differences to understand the same. Gene expression differences are usually measured by estimating mRNA expression either by relative or absolute quantification methods. These methods frequently involve PCR-based, hybridization-based or sequencing-based approaches. Differential gene expression of known genes are usually characterized using quantitative PCR (qPCR), microarray technologies, serial analysis of gene expression (SAGE) and RNA-seq methods. Variations in gene expression data could be used to generate expression QTL (eQTL) information, similar to that of genetic markers. Different techniques and methods that are routinely used for gene expression studies are briefly reviewed below.

## 2.4.1 Quantitative PCR

Measurement of gene expression (RNA) has been used extensively in monitoring biological responses to various environmental conditions. Quantitative gene analysis has been used for detecting the CNVs of a particular gene in the genome or in the transcriptome. PCR method has revolutionized many aspects of molecular biology including gene quantification. PCR protocols are modified and optimized by using either fluorescent probes or dyes in the reaction mixture to obtain accurate quantification of genes in the input DNA or RNA, and these procedures can be collectively called as quantitative PCR (qPCR). The qPCR approaches frequently use either fluorescent probes such as TaqMan® probes (Applied Biosystems) which detect a specific PCR product as it accumulates during PCR cycles, or fluorescent dyes such as SYBR Green which detects all double stranded DNA in a PCR reaction.

In TaqMan® assays, probes are designed specific to a target gene along with a pair of primer sequences in flanking regions of a probe. TaqMan® probes while hybridized to the target gene during PCR do not emit the fluorescence, but as the DNA extension is continued by the DNA polymerase using flanking primers in the PCR reaction, these hybridized probes are displaced and emit the fluorescence thereby facilitating the quantification of the target gene. TaqMan® probes have been routinely used for quantification of genes in the genome for allelic discrimination and zygosity studies and also in reverse transcribed mRNA/cDNA to study the gene expression. Using different reporter dyes, one can quantify two or more genes in the same PCR reaction. However, poor probe design could result in false positive signals. SYBR Green assays exploit the double stranded binding ability of the dye molecules during a PCR reaction. As the target gene product is accumulated in the PCR reaction, the fluorescence emitted by the dye increases and thus can quantify the gene. Though this procedure is simple to set up, there is no specificity to the target gene as it quantifies the entire double stranded DNA in the PCR reaction. There are several publications in the literature that employed these techniques.

## 2.4.2 Microarray technology

Parallel quantification of large numbers of mRNA transcripts for studying the regulation of gene expression was made possible by microarray technologies. The use of microarrays to analyze gene expression on a global level has recently received a great deal of attention. This should allow new understanding of gene signaling and regulatory networks that operate in various cell processes. The principle of a microarray experiment, as opposed to the classical northern-blotting analysis, is that mRNA from a given cell line or tissue is used to generate a labeled sample, sometimes termed the 'target', which is hybridized in parallel to a large number of DNA sequences, immobilized on a solid surface in an ordered array. Tens of thousands of transcript species can be detected and quantified simultaneously (Schulze and Downward, 2001).

The probes used in this technology could be cDNA fragments generated by PCR or synthetic oligonucleotides and these probes vary in their length based upon the instrumentation and technology available for synthesizing these arrays. Also, probes can be designed to represent the most unique part of a given transcript, making the detection of closely related genes or splice variants possible. To understand complex traits in crops, microarrays can be designed from existing sources of EST/functional gene information within the same crop or could leverage others with better resources. For example, Yang et al. (2007) utilized the microarrays designed from the model plant *Arabidopsis thaliana* to conduct transcriptional profiling experiments in canola for a disease condition.

Generally, in these experiments, mRNA from cells or tissue is extracted, converted to cDNA and labeled, hybridized to the DNA elements on the surface of the array, and detected by phospho-imaging or fluorescence scanning. The use of different fluorescent dyes (such as Cy3 and Cy5) allows mRNAs from two different cell populations or tissues to be labeled in different colors, mixed and hybridized to the same array, which results in competitive binding of the target to the arrayed sequences. After hybridization and washing, the slide is scanned at two different wavelengths corresponding to the dyes used, and the intensity of the same spot in both channels is compared. This results in the measurement of the ratio of transcript levels for each gene represented on the array. Microarrays can be used to investigate the changes in expression at single gene or across the whole genes to infer the changes in the phenotype. Analysis of expression differences at multigene level would facilitate understanding of gene regulatory pathways and gene-to-gene interactions besides providing the information of up or down regulation to a particular experimental condition. Several statistical and data analysis tools are available to interpret the microarray data (Schulze and Downward, 2001).

In traditional QTL analyses, linkage mapping leads to the detection of genomic regions which are associated with phenotypic variations within a population. Genetical genomics employs this same approach, except that the phenotypes are levels in gene expression resulting in the detection of expression QTL (eQTL). The eQTL do not necessarily result from sequence polymorphisms proximal to the gene being measured (cis-acting) but could result from differences in genes unlinked to the target. In these cases, the eQTL function in a trans-acting manner (Holloway et al., 2011). When the gene expression data is collected from a specialized tissue to understand the phenotype of the trait, we can use that data as a marker system to derive the eQTL information. In a recent genome wide eQTL study,

Holloway et al. (2011) utilized expression data from 50,000 maize genes to identify both cis-acting and trans-acting genetic elements that cooperate to regulate gene expression in maize crown roots and described the pitfalls of detecting false cis- or trans-acting eQTL in the absence of perfect genomic sequences from both parents. Multiple examples of eQTL analysis were reported in many plants and crop species (Druka et al., 2010). Despite many advantages, microarray technologies have their limitations, because expression profiling is conducted for only a limited set of genes that are currently known and we cannot detect the influence of unknown genes for the phenotype.

### 2.4.3 Serial analysis of gene expression (SAGE)

The SAGE technique is based on counting sequence tags of 14–15 bases from cDNA libraries. These tags are generally derived from either 5′ or 3′ end of the expressed genes by restriction enzyme and can be up to 22 bases. Earlier, these tags used to be ligated to each other to create longer stretches of tags (Super-SAGE) and then sequenced by Sanger sequencing. But now, with the use of NGS short read technologies, these tags can be individually sequenced to generate the expression information. Each unique tag can represent a copy of the gene in the cDNA and thus by counting these tags, the gene expression can be quantified. The principal advantage of SAGE is that it gives an absolute measure of gene expression instead of measuring relative expression levels. Indeed, by counting the number of tags from each cDNA, one obtains an accurate measure of the number of transcripts present in the mRNA sample. This technology has been widely used to monitor gene expression in human cell cultures and tissue samples, but was used sporadically in plants. The principal limitation of SAGE is the need to sequence large numbers of tags in order to monitor scarcely expressed genes. Another drawback of SAGE is that the obtained tags are very short and hence not always unambiguous. Gene identification on the basis of short sequence tags relies on the availability of large databases of well-characterized ESTs (Breyne and Zabeau, 2001).

### 2.4.4 RNA-Seq/digital gene expression

Recent developments in NGS technologies have transformed the way through which quantitative transcriptomics can be done. Due to their massively parallel sequencing abilities, these NGS technologies have been driving down the sequencing costs and time required to generate large amounts of sequencing data. Using these NGS technologies, RNA content of the cells can be directly sequenced without requiring any of the traditional cloning associated with EST sequencing. This approach, called "RNA-seq", can generate quantitative expression scores that are comparable to microarrays, with the added benefit that the entire transcriptome is surveyed without the requirement of *a priori* knowledge of transcribed regions. One key advantage of this technique is that not only quantitative expression measures can be made, but transcript structures including alternatively spliced transcript isoforms, can also be identified (Wilhelm and Landry, 2009).

RNA-seq procedures usually involve generation of multiplexed sequencing libraries from the mRNA/total RNA followed by high throughput sequencing. Short read sequencing instruments are more cost-effective for RNA-seq studies. Using the efficient bioinformatics tools, the sequence data is mapped to the reference genome to provide the multiple sequence alignment. Removal of repetitive sequences from the dataset would improve the mapping process. Sequence data can be converted into expression data in different ways: i) by simply adding the number of reads which fall within the co-ordinates of each element (either exon or gene), and then normalizing the data for the length of the element; ii) by calculating a sequence score for each nucleotide in the genome based on the number of reads which cover each base position, and again normalizing for element lengths (Wilhelm and Landry, 2009); iii) by calculating RPKM (reads mapping to the genome per kilobase of transcript per million reads sequenced) values (Mortazavi et al., 2008) using a mathematical formula and use them as a measure of gene expression. These processes are often referred to as 'Digital Gene Expression'.

Compared to microarrays, the limits of the dynamic range measured in RNA-seq experiments are only determined by the amount of sequencing obtained. This means that through the continued sequencing of a given library, it should be possible to eventually measure the expression of every transcript present and so the "dynamic range" only represents the actual biological diversity of the transcriptomes. To implement RNA-seq experiments effectively, availability of accurate annotation of the reference genome is necessary. This is particularly challenging for higher eukaryotes especially plant species with large genomes. However, RNA-Seq promises the gene quantification for studying complex phenotypic traits in cost-effective way in the near future for many crop species.

## 2.5 Comparative genomics

Comparative genomics relates the structure and function of genomes of evolutionarily close species. It is a tool that helps researchers to study complex genomes of plants by leveraging sequence information of related species with smaller and less complicated genomes. However, in 1990s and at the beginning of 21st century when no reference sequences existed, comparative genomics was limited to comparative mapping. Both in dicot and monocot plants, collinearity at micro level, i.e. the same order of the molecular markers in genetic maps, were observed. In dicots, comparative sequencing approach revealed collinearity in gene order within several chromosomal segments of Arabidopsis, Capsella and Brassica genomes (Rossberg et al., 2001). Later Schranz et al. (2006) integrated all comparative genomics data in Brassicaceae and constructed a set of 24 genomics blocks, which represented the conserved segments of ancestral karyotype, *Arabidopsis thaliana*, and *Brassica rapa*. In monocots, comparative genetic mapping in oats, maize, rice, barley, wheat, sorghum, sugarcane and fox millet resulted in the construction of the "Crop Circle", which placed the small genome of rice in the center of the circle and aligned with maps of crops with larger genomes (Gale and Devos, 1998). Comparative mapping revealed 30 blocks of rice genome that could be found within genomes of other crops (Devos, 2005).

Sequencing of several model plants, including Arabidopsis, rice, *Medicago truncatula, Lotus japonicum* and Brachypodium, as well as crops such as maize, rice (both a model and a crop plant), sorghum and soybean confirmed, in general, the existence of synteny between genomes of related species at DNA level, which was, however, reported to be less obvious because of the unique patterns of distribution of repetitive sequences, duplications, insertions and deletion of genes (Dubcovsky et al., 2001). Nevertheless, comparative genomics has been a valuable tool for the development of molecular resources for crops and the identification of key genes for crop improvement.

Owing to the collinearity between rice and other cereals and *Arabidopsis thaliana* and Brassicas, genomic resources of these model plants were leveraged to boost map-based cloning of genes from the genomes of other cereals with larger and complex genomes (Salse et al., 2008). Although the model plant, *A. thaliana,* and the field crop, rice, are the only plants with completely sequenced genomes, availability of high quality draft genome sequences of Brachypodium, sorghum and maize are believed to provide even more opportunities in gene and QTL discovery in orphan crops with zero to little genomic resources (Mayer et al., 2011).

Molecular markers and EST collection of the above-mentioned crops, have been widely used to saturate genomic regions of other crops to narrow down the location of economically important QTL and genes. Using rice and Brachypodium ESTs, a powdery mildew resistance gene Ml3D232 was isolated from bread wheat (*Triticum aestivum*) (Zhang et al., 2010). Wheat tiller inhibition gene, tin3, was mapped using molecular markers developed based on ESTs from syntenic region of rice. Comparative analysis revealed collinear regions between perennial ryegrass (*Lolium perenne L.*), *B. distachyon* and sorghum, which facilitated cloning of the self-incompatibility genes in the former   (Shinozuka et al., 2010). Using synteny between rice chromosome 9 and Italian ryegrass LG5 and rice EST-based molecular markers, the location of LMPi1 gene conferring resistance to grey leaf spot was delimited to short chromosomal segment of the latter (Takahashi et al., 2010). Using homology between *B. rapa* and *A. thaliana,* the TRANSPARENT TESTA GLABRA 1 (TTG1) gene controlling both hairiness and seed coat color traits in Brassica species was isolated (Zhang et al., 2009). Sequences from Arabidopsis chromosome1 were used as RFLP probes to genetically map fertility restorer gene, *Rfp,* in *B. napus* genome.

Comparative genomics has been an indispensable tool to study evolution of genomes and gene families. *Brachypodium distachyon* has been widely used to study the evolution of important traits in barley and wheat, including flowering time pathways and (1,3;1,4)-b-D-glucans in plant cell walls (Higgins et al., 2010). Recently, the genomes of Brachypodium, rice and sorghum were used to assign 32,000 barley genes to the corresponding individual chromosomes (Mayer et al., 2011). Genome of Arabidopsis expanded its value to study xylem genomics of conifers (Xinguo et al., 2010). Two legume species, *Medicago truncatula* and *Lotus japonica,* are being sequenced to study genetic background of legume-specific phenomena such as symbiotic nitrogen fixation (Sato et al., 2008). Genome of *M. truncatula* should also facilitate the assembly of next generation sequence data in closely related taxa such as alfalfa (Young and Udvardi, 2009). Tremendous progress in sequencing technologies is opening new avenues for comparative genomics and enabling researchers to do genome-wide comparisons. Current trends indicate that NGS technologies might change the focus of comparative studies by shifting them more towards evolutionary genomics rather than synteny-based gene-cloning or marker development. No matter what direction comparative genomics will take in the future, it is certain that it has the potential to broaden our current understanding of complex biological processes. An understanding of complex traits and processes such as adaptation of plants to biotic and abiotic stresses, yield, gene regulation, polyploidy and the influence of natural selection on gene and protein function can be translated into development of new strategies for crop improvement.

## 2.6 Genetical genomics

The concept and strategy of genetical genomics, outlined by Jansen and Nap (2001), aims to rapidly identify key gene targets by super-imposing gene expression data on that of genetic mapping. Although molecular markers linked to quantitative trait loci (QTL), the genomic regions genetically determined to be associated with the observed phenotypic variation, can be used in marker-assisted breeding programs, in order to identify the gene(s) underlying the QTLs, comparison of gene expression differences between contrasting lines is very important. Genetical genomics approach involves expression profiling and molecular marker based genotyping of all individuals in a segregating population. It is followed by a comprehensive analysis using all statistical tools that are normally used in the analysis of quantitative trait loci. These analyses result in the identification of expression QTL (eQTL). Based on the nature of gene expression variation, there are two types of eQTL, cis-eQTL and trans-eQTL. If the variation or polymorphism is located near the gene, it is classified as a cis-eQTL, whereas if the source of variation is located at a distant location in the genome then it is a trans-eQTL. Genetical genomics has been applied to a broad range of organisms and all studies have demonstrated the power of combining gene expression data with that of genetic analysis to fine tune pathways involved in complex phenotypes thereby enabling the identification of key genes (Joosen et al., 2009). Genetical genomics studies benefit greatly with the availability of reference genome sequence as well as with the use of large populations. In a genetical genomics study of the model plant Arabidopsis where a 162-line RIL population was used and data analyzed in conjunction with the genome sequence, researchers successfully predicted key regulators of flowering time and circadian rhythms through the construction of genetic regulatory network by combining eQTL mapping and regulator candidate gene selection (Keurentjes et al., 2007). Examples also exist where synteny of the target genome with other species can be used in genetical genomics. For example, in a wheat study, synteny with the sequenced rice genome was used to map the eQTL for seed quality parameters (Jordan et al., 2007). The use of genetical genomics is on the rise. Several technological advances such as expression profiling using next generation sequencing and the availability of several types of 'omics' data sets are enabling rapid construction of biological networks for not only identifying key genes for phenotypic variations but also for understanding the pathways and systems.

## 2.7 Systems biology

Systems biology is the study of interactions among biological components using models and/or networks to integrate genes, metabolites, proteins, regulatory elements and other biological components (Yuan et al., 2008). Although integration of data from multiple areas of research is not a new concept in biological research, the availability of huge amounts of diverse sets of data obtained from modern, HTP 'omics' technologies has renewed interest in systems biology. In particular, next generation genome sequencing as well as HTP profiling of transcripts, proteins, metabolites, phenotypes etc. are providing the necessary raw material that can enable us to construct interaction networks of genes, their products and many associated players. It must be emphasized that the goal of systems biology exceeds that of the omics components in that it looks holistically at the biological systems with respect to the structure and dynamics.

Considering the complexity and diversity of datasets analyzed and integrated in systems biology, strategy for these projects can be broadly divided into 4 steps. Step 1 is the development of a sound experimental strategy and collection of reliable and reproducible data. Step 2 is the annotation of the data components, for example, genes and proteins investigated in the study. While gene ontology system and other molecular function information resources are used for the functional classification of genes/proteins, pathway databases (Bauer-Mehren et al., 2009) such as KEGG (Kyoto Encyclopedia of Genes and Genomes; http://www.genome.jp/kegg/), Reactome (http://www.reactome.org), PANTHER (Protein ANalysis THrough Evolutionary Relationships; http://www. pantherdb.org/) and plant metabolic pathway databases ( http://plantcyc. org/ ) can be used for obtaining pathway information of the candidates. In step 3, the annotation information from step 2 is used for the generation of mathematical models and networks based on the associations and interactions observed in the datasets in comparison to known pathway information from the databases. Step 4 is the model development and validation where generated models are validated, and then variations introduced and models revised and validated again. Several software tools are used for this iterative cycle of model development – validation – perturbation – validation (Endler et al., 2009). In addition to the tools needed in these steps, availability of efficient platforms and infrastructure that can archive and support analysis of data from diverse sources and labs is a critical requirement for any systems biology project.

There can be multiple types of output expected from systems biology depending on the datasets used for integration. The output can either be a snap shot of a system with respect to a key gene or protein such as gene regulatory or biochemical networks or it can be based on the quantitative or qualitative dynamics of the system based on multiple perturbations. The most popular outcomes to date have been the construction of biological networks that include gene and transcriptional regulatory networks and interactome networks. A major promise of systems biology for crop improvement lies in the area of understanding quantitative traits and abiotic and biotic stresses and examples currently exist where systems biology approach has been explored in these areas. Cooper et al. (2003) developed a network of genes associated with developmental and stress responses in rice by measuring the gene expression changes with respect to environmental, biological and stress treatments related to interaction domains for 200 proteins from stressed and developing tissues. Data obtained from this study resulted in the identification of several stress response genes and were also found to be useful for the prediction of gene function in monocots and dicots. Use of systems biology in unraveling plant defense response has to deal with several dynamics related layers before key resistance gene candidates can be identified. In a study where mutant analysis and whole genome gene expression profiling was used for determining the regulatory roles of WRKY genes in systemic acquired resistance (SAR), a 'regulatory node' was identified in the transcriptional regulatory network controlling SAR (Wang et al., 2006). Although reports that utilized input from several omics for a detailed analysis of quantitative traits are yet to appear, approaches that utilized genome wide eQTL or metabolite QTL (mQTL) do exist that led to the construction of networks and identification of key metabolic QTLs that could be leveraged for crop improvement (Schauer et al., 2006). Current efforts in this area also include extending the impact of systems biology in understanding plant communities using a holistic approach by merging systems biology and systems ecology in order to improve agricultural productivity(Keurentjes et al., 2011).

## 2.8 Bioinformatics

The field of genomics has grown leaps and bounds during the recent decades. While the development and implementation of HTP genome sequencing and more than a dozen omics technologies revolutionized our ability to study whole genomes and biological systems, one important capability that has also played a major role in this unprecedented growth is bioinformatics. The explosive growth of information from the biological and genomics research has accentuated the need for computational requirements. The development of computational biology tools enabled rapid and HTP analysis of large amounts of data generated by research community while the creation of comprehensive databases made the information available globally, further enhancing the pace of research. A recent review by Mochida and Shinozaki (2010) on various genomics and bioinformatics resources available and how they are helping in biological research provides a greater appreciation for the impactful contributions of bioinformatics over the years. Systems biology is one area where we need additional technological advances as well as improvements in our computational tools in order to efficiently analyze, integrate and interpret huge data sets resulting from dynamic biological systems and currently efforts are underway to this end.

## 3. Molecular breeding approaches

### 3.1 Mapping and map-based cloning of QTL

The availability of dense genetic maps with informative markers makes it possible to map genes and QTL. Numerous studies have been dedicated to mapping QTL governing major agronomically important traits. As of July 06, 2011, there were about 24,000 research articles in this direction in Google Scholar. However, majority of those studies ended at the mapping step and did not pursue the ultimate goal of cloning the gene(s) underlying the QTL due to many reasons including lack of funding, insufficient genomic resources, inadequate phenotypic data collection methodology, availability of experimental designs with limited detection power and finally complexity of the trait and the genome. Due to these factors, QTL cloning in crops remains a very challenging process. Researchers have to undergo multiple labor-, time- and cost-intensive steps prior to answering the question "to clone or not to clone plant QTL" (Salvi and Tuberosa, 2005, 2007). What are those steps and challenges?

The first step in every map-based cloning (MBC) project is genetic mapping of QTL using a small mapping population (200-300 individuals) and identifying flanking markers. Provided that phenotypic data collection methodology and experimental design are at adequate level, the main constraint that researchers face at this initial step is very large confidence interval (CI) of QTL, which can span 10-30 cM (Kearsey and Farquhar, 1998). However, even in cases where the CI is limited to a few centimorgans, the interpretation of this genetic distance is not a straight forward process. The reason is that the genetic distance largely depends on the rate of recombination frequency in the region and the size of the mapping population used to construct the linkage map. In many cases, markers that flank QTL are physically far from the target, and the interval between markers contains a large number of genes. Depending on the length of DNA segment spanning QTL, strategies to reduce the distance between markers and QTL have to be designed. One of the strategies to narrow down the CI is the right choice of molecular markers for QTL mapping. Majority of QTL mapping studies have been using a pool of publicly available markers, which are not always informative for a particular bi-parental cross under investigation. Another most important factor that can condition the success of QTL mapping and subsequent cloning is knowledge of all possible allelic variations existing between parents. For instance, most of the public SNP markers in maize are developed from B73 and Mo17 cultivars. Taking into account massive intraspecific variations among maize inbred lines (Eichten et al., 2011; Fu and Dooner, 2002; Springer et al., 2009), SNPs developed from a few lines will capture only a small portion of all allelic variations happening between parents of a cross designed for a QTL study. Consequently, there is a big chance that majority of allelic variations, including the causative mutation between parents of this cross will be missing. In order to avoid this situation, re-sequencing of genomes of both parents and discovery of allelic variations in low and single copy regions could be implemented using NGS technologies coupled with genome complexity reduction techniques. Discovered cross-specific polymorphisms can later be converted into any modern SNP genotyping assay (Mammadov et al., 2010; Trebbi et al., 2011). For instance, technologies such as complexity reduction of polymorphic sequences (CRoPS™) (Van Orsouw et al., 2007) or restriction site associated DNA (RAD) (Baird et al., 2008) can be successfully applied to generate cross-specific SNPs (Mammadov et al., 2010). Depending on the organism, this approach may result in the validation of about 1000 robust cross-specific markers, which can be later combined with public SNPs and used for mapping. NimbleGen Sequence Capture technology (Roche Applied Science) has also been used for the detection of cross-specific polymorphisms within low and single copy sequences (Springer et al., 2009). However, this technology can be applied only to crops for which reference genome sequences are available because of the necessity to design capture probes. The approach of development of cross-specific markers increases the precision of QTL mapping and consequently narrows down CI and can lead to the detection of causative mutation(s).

If the heritability and the effect of QTL governing the trait are high and CI is narrow enough, then the development of large mapping populations and the creation of high-resolution or fine mapping can be sufficient (Jiao et al., 2010). However, in many cases there is a risk that the detected major QTL is in reality represented by several co-segregating loci with minor effects. In order to eliminate the effects of other co-segregating alleles affecting the phenotype, target locus is recommended to be Mendelized through the painfully long and expensive process of developing near isogenic lines (NILs) (Kearsey and Farquhar, 1998). In QTL-NIL the target QTL will behave as a single Mendelian gene. Development of NILs begins with the same population where all QTL were mapped, and can be carried out by marker-assisted backcross introgression. QTL Mendelization is believed to give an answer on the feasibility of cloning the target locus. In some cases, after Mendelization, the effect of target QTL may become so negligible that further cloning activities will not make any sense. However, if the high-resolution reveals that target QTL still explains the large portion of phenotypic variation, the physical mapping of the locus must be implemented (Saito et al., 2004).

In order to physically anchor the QTL, bacterial artificial chromosome (BAC) library has to be developed, which is a very important prerequisite for map-based cloning. Researchers use markers flanking the gene of interest as probes to implement chromosome walking (Tanksley et al., 1995). When the price of sequencing was high, chromosome walking was a very tedious process, which included several rounds of marker-library hybridization, identification of new BAC clones spanning the region between the flanking markers, BAC fingerprinting to construct contigs, identification of minimum tiling path (MTP) and development of new markers from the extreme left and right BAC clones representing the

MTP. Chromosome walking had not been a straight forward process either. It was especially complicated in crops with complex genomes. In many cases, the isolation of BAC ends resulted in the dissection of repetitive sequences which were of no use in designing a new probe for the subsequent library hybridization. Currently, instead of doing BAC fingerprinting, NGS technology allows direct sequencing of all identified BACs and construction of BAC contigs based on sequence similarity (Schneeberger and Weigel, 2011). Availability of a reference sequence simplifies the assembly of BAC clone sequences and contig construction. Lack of reference sequence will force researchers to construct contigs *de novo*, which is not an easy task taking into account the complexity of a plant genome. From this point of view, MBC in crops with available reference genome sequence is supposed to be easier than in crops with no genome sequence available. However, this is not always the case because success of MBC depends not only on available structural genomics resources but also the complexity of a trait and availability of adequate phenotypic data collection methodology. Because QTL mapping results from the comparison of marker and phenotypic data, the accuracy of the latter is of great importance. Although accurate phenotypic data is equally important both for QTL and single gene cloning, the former is more sensitive to the robustness of phenotypic data due to their dependence on environmental conditions. Nowadays, precision phenotypic data collection remains the major bottleneck for the successful QTL mapping and cloning.

If the position of a QTL is delimited to one BAC clone, this large insert clone must be sequenced to identify the candidate genes. Ideally, homologous BACs from both parents should be sequenced, because this will facilitate the identification of gene candidates. Sequencing of a BAC clone using NGS technology is a very straightforward process if the target crop has reference genome sequence. For example, sequencing can be done using fairly cost-effective HiSeq instrument (Illumina, San Diego, CA), which generates small 70-100 bp fragments which can later be mapped back to reference genome to facilitate the assembly. In orphan crops sequencing efforts are impeded due to the absence of reference sequence. Sequencing using traditional Sanger could be the only choice. However, it is very expensive and a long process. Another alternative is sequencing the same BAC clone with two NGS instruments in parallel such as GS20 Sequencer (454/Roche) and Illumina's HiSeq, which generate long (1 kb) and short (100 bp) sequences, respectively. A combination of short and long sequences may improve the assembly of a BAC clone from a crop with no reference sequence.

When a BAC clone is sequenced and assembled, next step is sequence analysis of the large insert clone. Sequence analysis of the BAC clone is necessary to reveal gene candidates. Normally, the entire BAC clone first gets scanned for the repetitive sequences using publicly available Repeat_Masker software (http://www.repeatmasker.org/cgi-bin/WEBRepeat Masker). The Repeat_Masker output can be used as an input template for any protein prediction software, including FGENESH (http://linux1.softberry.com/ berry.phtml?topic =fgenesh&group=programs&subgroup=gfind) and GenScan (http:// genes.mit.edu/ GENSCAN.html) and GenMark (http://exon.gatech.edu/). Finally, all predicted open reading frames (ORFs) must be BLASTed against protein database using 'blastp' algorithm to reveal their biological functions. Depending on the genome size and repetitive sequence content one BAC clone may have dozen or so genes. The number of candidate genes can be reduced by aligning homologous BAC clones from both parents. Further, any allelic variations between two parents within target BAC could be converted into KASPar assay and bulk segregant analysis (BSA) can be performed. If BSA does not reduce the number of candidate genes to one, then causative mutation can be identified through functional prediction. For example, if the target gene confers resistance to a disease and among the remaining candidate genes there is an NBS-LRR gene, which belongs to one of the classes of defense-related genes, then it might be considered as a gene of interest. However, if the researcher wants to confirm the gene candidate using classical methods, then either further increase in the resolution within a locus or functional characterization of each predicted protein is recommended. Both methods are very expensive and labor- and time-intensive. Functional characterization of gene candidates can be done using several methods including genetic complementation through transformation technique, down-regulation of a gene via RNAi, complementation of a known mutant or by marker-assisted trait introgression (Borevitz and Chory 2004).

## 3.2 QTL cloning through association mapping

One of the major limitations of QTL cloning using bi-parental mapping approach is insufficient amount of meiotic recombination in mapping populations such as F2, DH and RIL, which lead to a strong statistical association of QTL with the block of markers that physically span large chromosomal segments. QTL-mapping approach requires that specific mapping populations, usually consisting of several hundred F2 or RIL progeny, be developed from each germplasm accession to be examined for important genes effecting traits of interest. Each population must be genotyped using hundreds, perhaps thousands of molecular markers. This population development and marker screening is extremely time-consuming, high-risk and expensive work - prohibitively expensive if dozens, let alone hundreds or thousands, of germplasm accessions are to be examined (Abdurakhmonov et al., 2008; Abdurakhmonov and Abdukarimov, 2008). However, geneticists mapping complex traits in the human genome have circumvented the need for large F2 or RI mapping populations (which are not available in humans) by making use of information contained within the genetic recombinations that have occurred in typical human populations during the course of recent evolution. Genetic loci linked to a specific disease will show historically reduced level of recombination (non equilibrium) with specific alleles at different loci controlling particular genetic variation in a population (Abdurakhmonov et al., 2008). This "linkage disequilibrium (LD)" can be detected statistically, and has been used to map and eventually clone a number of genes underlying complex genetic traits in humans (Schulze and McMahon, 2002; Weiss and Clark, 2002). LD-Mapping, referred also as Association mapping (AM), is an alternative approach to a now classical QTL-mapping in bi-parental mapping populations because it overcomes the problem related to the lack of recombination events. AM population is composed of genetically un-related individuals with unknown pedigrees and accumulates larger number of historical recombination events that occurred in the past (Nordborg and Tavaré, 2002). Multiple increases in number of recombination events will break the block of markers that is associated with QTL and increase the resolution of QTL region. Association Mapping is linkage disequilibrium (LD)-based association while bi-parental approach is a genetic recombination-based mapping. AM attempts to reveal a significant association between a trait and a gene, or a molecular marker, or block of molecular markers, which are at LD. Marker and trait are in disequilibrium if they are truly linked to each other and historically have been passing from generation to generation together. The theory of AM is based on the idea that LD tends to be preserved over many generations between loci, which are linked to one another and form haplotypes. The higher the LD, the tighter is the linkage between markers. However, in bi-parental mapping, because of lack of recombination event, several haplotypes could be grouped into one linkage and be statistically associated with QTL, which decreases the resolution of the map. To summarize, AM theoretically has several advantages over classical bi-parental approach: (1) increased mapping resolution; (2) availability of more genetic variability for marker-trait correlations and detection of multiple-alleles simultaneously; (3) elimination of the necessity to develop large populations for fine mapping which saves expenses and time; (4) one AM population can be leveraged for dissection of many traits, while bi-parental crosses are dead-end and in many cases can be used to study one trait only and (5) feasibility of employing historical phenotypic data collected over many years (Zhu et al., 2008).

However, not everything is that smooth and painless in the implementation of AM. This approach is often criticized for (1) detecting a large number of false-positive QTL due to population confounding effects and 2) influence of allele frequency distributions (rare and minor allele frequencies) of functional polymorphisms to the power of the detected associations (Abdurakhmonov and Abdukarimov, 2008; Aranzana et al., 2005; Stich and Melchinger, 2010). In order to avoid false positives, several factors have to be taken into consideration. Structure of the population must be carefully analyzed using various computational methods. Too structured population with too many sub-groups will detect pseudo LDs between loci that in reality are not linked, and cause false-positive association between a marker and a trait. In order to avoid this, prior to AM implementation, the population must be analyzed for the presence of hidden sub-structures. One of the popular software programs that researchers use to resolve this issue is publicly available software called STRUCTURE. Removal of rare alleles is a choice in AM to reduce false-positives (Abdurakhmonov and Abdukarimov, 2008), but studies showed that most phenotypic variations are due to rare alleles (Stich and Melchinger, 2010), suggesting importance of these rare alleles in tagging biologically meaningful associations. Both structured population and rare allele frequency issues can be greatly minimized by creating segregating populations and performing genetic crosses between several reference populations with known allele frequencies for functional polymorphisms. Such approach is referred to as nested association mapping (NAM) and NAM populations greatly enhance the power of association mapping in plants (Stich and Melchinger, 2010).

Because AM is LD-based, another consideration is the rate of LD decay in a crop under the study. The pattern of LD throughout the genome will determine the appropriate marker density for whole genome scanning (Yu et al., 2008). The longer the haplotype, the lesser is the marker density needed because all markers within a haplotype will behave similarly. In contrary, if the segment of a chromosome is characterized by the presence of short haplotypes then the density of markers has to be increased correspondingly. Additionally, the rate of LD decay with physical and genetic distances is important to determine the maximum resolution that can be achieved for association mapping. Length of haplotypes depends on a genome, the number of loci investigated, and the reproductive history of a population. It was previously reported that in maize, LD decay distance was on an average less that 2000bp (Remington et al., 2001). Later studies suggested that in commercial inbred lines, LD decay may span more than 100-500 Kb (Jung et al., 2004; Tian et al., 2011). Recently developed first-generation haplotype map of maize presented the evidence for much longer haplotypes spanning several million bases (Gore et al., 2009), which was later supported by Mammadov et al. (2010). In cultivated barley, LD has been reported to span from 1 cM to 10 cM (Rostoks et al., 2006). Use of different molecular markers, can also significantly change the length of LD. In rice, it was indicated that LD decays within 1 cM or less using SNP markers (Agrama and Eizenga, 2008), whereas others reported 20-30 cM length of LD while using simple sequence repeat markers (SSRs) (Jin et al., 2010). Differences in the rate of LD decay within a crop could be explained by the nature of unit it was represented. Centimorgan is a unit of recombination, and the rate of recombination has proven to be not uniform across the genome. Consequently, 30 cM LD span in one region of rice genome physically may carry the same value as 1 cM LD span in another region of the genome. Physical distance seems to be more realistic way to designate LD decay. However, absence of reference sequence in some crops limits researcher to centimorgans only. Now-a-days researchers successfully use AM to study genetics of complex traits in all major crops, including rice, maize (Poland et al., 2011), barley (Massman et al., 2011), soybean (Wang et al., 2008) and canola (Honsdorf et al., 2010) and other crops (Abdurakhmonov and Abdukarimov, 2008; Stich and Melchinger, 2010). The common feature of these studies is the detection of large number of QTL with small effects. Whether or not the information on these QTL can be translated into real use in crop improvement is unclear and the answer is yet to come.

To date there are only a few successful QTL cloning studies. Most of them have been reported in rice, which is not surprising because rice genome has been sequenced since 2004 (Table 3). Additionally rice genome is smaller and less complex. QTL cloning is in progress in maize and soybean, two crops that have reference genome sequence available. Undoubtedly, the progress in sequencing technologies, availability of reference genome in major crops, rapid evolution in high-throughput polymorphism detection platforms and bioinformatics tools and last, but not least, the development of accurate phenotypic data collection methodology will increase the precision of QTL mapping in major and orphan crops bringing us closer to the discovery of the Holy Grail of molecular geneticists, causative mutations, that will be subsequently translated into robust diagnostic tools to implement marker-assisted selection.

### 3.3 Marker-assisted selection

Marker-assisted selection (MAS) as a process refers to the selection of superior genotypes using molecular markers. MAS is thought to have substantial advantages over conventional phenotypic selection because the latter could be (1) unreliable when the expression of the trait is environmentally dependent, (2) biologically deadline-sensitive, (3) expensive and difficult to screen and (4) subject to the mercy of weather. In contrast to phenotypic selection, MAS (1) does not rely on environmental conditions because it detects the structural polymorphisms at molecular level, (2) requires leaf tissue collected at seedling stage, which is very useful for traits that are expressed at later stages of development and which also helps to avoid adverse weather conditions that could kill the plant at adult stage, (3) could be cheaper and less labor intensive, (4) allows selection in off-season nurseries and has a potential to accelerate breeding process.

| Crop | Gene | Trait | Function | Reference |
|------|------|-------|----------|-----------|
| Rice | *Ctb1* | Cold tolerance | F-box protein | Saito et al 2010 |
| | OsSPL14 | Panicle number, grain productivity | Transcription factor, a protein similar to Squamosa promoter binding protein | Jiao et al., 2010; Miura et al., 2010 |
| | *ERECT PANICLE 3* | Erect Panicle architecture | F-box protein | Piao et al 2009 |
| | DEP1 | Panicle number, grain number | Gain-of-function mutation causing truncation of a phosphatidylethanolamine-binding protein-like domain protein | Huang et al., 2009 |
| | *SK1/SK2* | Deepwater tolerance | Ethilene responsive factors | Hattori et al., 2009 |
| | Mt1 | Rice tillering | carotenoid cleavage dioxygenase 8 (CCD8) | Zhoua et al 2009 |
| | qSW5(GW5) | Grain width and weight | novel nuclear protein likely acts in the ubiquitin-proteasome pathway to regulate cell division during seed development | Shomura et al., 2008; Weng, et al., 2008 |
| | Ghd7 | Grain number, plant height, heading date | CCT (CO, CO-LIKE and TIMING OF CAB1) domain protein | Xue et al., 2008 |
| | GW2 | Grain width and weight | RING-type protein with E3 ubiquitin ligase activity | Song et al., 2007 |
| | *TAC1* | Tiller angle | Unknown | Yu et al., 2007 |
| | GS3 | Grain weight and length | VWFC membrane protein | Fan et al., 2006 |
| | *sh4* | Shattering | Transcription factor | Li et al., 2006a |
| | *qSH1* | Shattering | BEL1-homeobox | Konishi et al., 2006 |
| | *Sub1A* | Submergence tolerance | Transcription factor | Xu et al., 2006 |
| | Gn1a | Grain productivity | Cytokinin oxidase | Ashikari et al., 2005 |
| | *PSR1* | Regenerability | Nitrite reductase | Nishimura et al. 2005 |
| | *qUVR-10* | UV resistance | CDP photlyase | Ueda et al. 2005 |
| | *SKC1* | Salt tolerance | HKT transporter | Ren et al., 2005 |
| | *Ehd1* | Heading date | B-type response regulator | Doi et al., 2004 |

|  |  |  |  |  |
|---|---|---|---|---|
|  | Hd3a | Heading date | Unknown protein | Kojima et al., 2002 |
|  | Hd6 | Heading date | Protein kinase | Takahashi et al., 2001 |
|  | Hd1 | Heading date | Transcription factor | Yano et al., 2000 |
| Maize | Vgt1 | Flowering time | Non-coding sequence | Salvi et al. 2007 |
|  | Tga1 | Glume architecture | Transcription factor | Wang et al 2005 |
|  | Tb1 | Plant architecture | Transcription factor | Doebley et al 1995, 1997 |
|  | DGAT | High oil content | acyl-CoA:diacylglycerol acyltransferase | Zheng et al 2008 |
| Soybean | E3 locus | Flowering time | Phytochrome A | Watanabe et al 2009 |

Table 3. QTL cloning studies in the literature

In a review article, Xu and Crouch (2008) demonstrated an interesting chronology on the evolution of MAS as a technology. According to them, the term was coined in the mid – eighties by Beckmann and Soller (1986) as a technology that might have a potential use in plant breeding. Ten years later, MAS was already considered as a possible technology to tag genes (Concibido et al. 1996). Although as of June 27, 2011, according to Google Scholar, there were about 32,300 articles containing the keyword "marker-assisted selection", most of them still have been referring to potential application of MAS in plant breeding. A vast majority of those publications were from academia. Although private sector does not normally release the details of their breeding methodologies to public domain, several articles on successful application of MAS in the development of varieties of maize (Ragot et al., 2007) and soybean (Cahill and Schmidt, 2004; Crosbie et al., 2003) came mainly from industry. Fairly low impact of academic research in developing varieties using MAS can be explained by the lack of funding to complete the entire marker development pipeline (MDP), which can be long-term and cost-intensive task. MDP includes several steps such as (1) population development, (2) initial QTL mapping, (3) QTL validation (testing in several locations and years and implementing fine mapping) and (4) marker validation (development of inexpensive but high-throughput assays that are amenable to automation) (Collard and Mackill, 2008). Every step of the development of markers linked to QTL are associated with numerous constraints which may take several years and substantial funding to resolve. In the 1990s it was believed that molecular markers identified at step 2 were enough for successful MAS. In the 1990s it was observed that markers that were previously declared as tightly linked were failing to confirm the phenotype at advanced stages of MAS. One of the main reasons for the failure of a marker in MAS, which was identified at pre-fine mapping step, was the inconsistency in QTL mapping. Detection of QTL within one year and in one location was proved to be not enough to claim the robust QTL location because the expression of latter has been environmentally dependent. Thus, QTL validation and confirmation was required, which foresaw QTL mapping based on data collected within several years and multiple locations. Molecular markers that were tightly linked to QTL and were consistent across several years and locations did have a potential in MAS. However, even after QTL validation, so called "tightly linked marker" hardly met the expectations because the confidence interval (CI) of QTL peak is so large that it is very difficult to predict the real distance between marker and QTL. Moreover, there are several hundreds of candidate genes within CI, and it is impossible to predict which gene explains the phenotypic variation. Genetic proximity of a marker to QTL depends on two factors such as the size of the population and the region of a genome where the marker and QTL are residing. If the size of the mapping population is small (100-200 individuals), then the claims of having a marker closely-linked to a gene are barely valid, because fine mapping will identify many crossing-over events happening within marker-gene complex. Occurrence of the recombination events between marker and QTL makes the marker unable to track a target. This type of situation is especially true for the regions of the genome with high rate of recombination frequency. However, certain regions of genome exhibit very low rate of recombination frequency. If QTL was mapped to low recombination frequency region, then it will be very difficult to prove that a marker tagging QTL is indeed physically close to the locus. In some cases, even fine mapping will not help to break the linkage between marker and QTL. Availability of reference genome sequence is helpful to define the physical proximity of the marker to a gene. However, even physical proximity may not insure successful MAS. There are examples showing that out of several mutations, which occurred within a target gene, only one of them, the causative, can be converted into viable assay to leverage in MAS (Zheng et al., 2008). In human molecular genetics, mostly causative mutations have been used to develop diagnostic tools to detect diseases. However, in plants development of gene-based diagnostic tools for MAS has been limited to major crops with available reference genome sequence, e.g. rice and maize (Chen et al., 2010; Zheng et al., 2008). With respect to orphan crops, including wheat and barley, it may require several years before gene-based molecular markers derived from QTL cloning can be used in MAS.

Tracking QTL using molecular markers is just one of the MAS applications in plant breeding. This application uses mostly one or a couple of markers ideally developed based on causative mutations. Applications of MAS in plant breeding were grouped into five broad categories (Collard and Mackill, 2008): (1) marker-assisted germplasm evaluation including pedigree verification, purity assessment, evaluation of genetic diversity, identification of heterotic patterns and event characterization; (2) marker-assisted trait introgression, (3) marker-assisted pyramiding of genes and (4) genomic selection (GS). The nature of the MAS-based molecular breeding projects determines the marker and sample throughput and consequently requires specific marker genotyping technologies. Most of the contemporary marker genotyping technologies are oriented towards SNP detection, because SNPs are amenable for high-throughput automation, and are preferred type of polymorphism in molecular genetics research projects because of their abundance and resolution (Chagné et al., 2007). Majority of SNP genotyping technologies that have been described in this chapter were originally developed for SNP detection in human genetics research. However, the rapid growth and expansion of agribusiness, challenges to 'increase the slope' and 'stay competitive' forced major seed companies to adapt those technologies to plant genome and use in high-throughput SNP genotyping for MAS projects. Although MAS projects are diverse, in terms of sample and marker throughput, they all can be divided into two major groups with opposite tasks: (1) projects that deal with large sample volume (>10,000 plants) to be genotyped with a few markers (1-96 SNPs) and (2) projects that require genotyping a fewer samples (1-300) with large number of SNPs (384 to several millions of SNPs). The projects that fall into the first group are related to categories such as marker-assisted germplasm evaluation, marker-assisted trait introgression and marker-assisted gene pyramiding. The categories of MAS-based projects such as genome wide selection (GWS) for complex traits that fall into the opposite category will require genotyping of several millions of SNPs in fairly small subset of samples. SNP genotyping platforms that would match the requirements of the project from group 1 could include OpenArray platform coupled with TaqMan chemistry and KBiosciences' Competitive Allele Specific PCR (KASPar) complemented with the SNP Line platform (SNP Line XL, Kbiosciences, Hoddesdon, England). The latter has proven to be more cost effective and flexible compared to TaqMan assay (Chen et al., 2010). The second group of projects can be implemented using Illumina's BeadArray technology coupled with GG and Infinium assays. Current throughput of Infinium assay is ~1.1 MM SNPs per iSelect. However, GWS might require several millions of SNPs depending on the complexity of genome and its LD decay rate. If this is the case, then genotyping-by-sequencing (Elshire et al., 2011) could be another alternative.

### 3.4 Genomic selection approach towards breeding complex traits

Current MAS strategies fit the breeding programs for traits with high heritability and are governed by a single gene or one major QTL that explains large portion of the phenotypic variability. However, the application of MAS for breeding traits with complex genetics based on the interaction of multiple QTL with minor effects has been inefficient. Examples of complex traits are yield, drought tolerance, and nitrogen and water use efficiency. In classical MAS projects researchers use molecular markers that show statistically significant association with a phenotype and are linked to major QTL. Because minor QTL have small effects on phenotype, they have not been applicable in MAS. Meuwissen et al. (2001) described a new methodology in plant breeding, called genomic selection (GS) that was believed to solve problems related to MAS of complex traits. This methodology also applies to molecular markers but in different fashion. Unlike MAS, in GS markers are not used for tracking a trait. In GS high density marker coverage is needed to potentially have all QTL in LD with at least one marker. Then the comprehensive information on all possible loci, haplotypes and marker effects across the entire genome is used to calculate genomic estimated breeding value (GEBV) of a particular line in the breeding population.

Genomic selection of superior lines can be carried out within any breeding population. In order to enable successful GS, the experimental population must be identified. The population should not be necessarily derived from bi-parental cross but must be representative of selection candidates in the breeding program to which GS will be applied (Heffner et al., 2009). Experimental population must be genotyped with large number of markers. Taking into account the low cost of sequencing, the best choice is the implementation of genotyping-by-sequencing which will yield maximum number of polymorphisms. The sequence of the two events, i.e. phenotypic and genotypic data collection, is arbitrary and can be done in parallel. When both phenotypic and genotypic data are ready, one can start "training" molecular markers (Zhong S. 2009). In order to train

GS model, the effect of each marker is calculated computationally. The effect of a marker is represented by a number with a positive or negative sign that indicates the positive or negative effect, respectively, of a particular locus to phenotype. When the effects of all markers are known, they are considered "trained" and ready to assess any breeding population different from the experimental one for the same trait. Availability of trained GS model does not require the collection of phenotypic data from new breeding populations. The same set of "trained" markers will be used to genotype a new breeding population. Based on genotypic data, the known effects of each marker will be summed and GEBV of each line will be calculated. The higher the GEBV value of an individual line, the more likely that this line will be selected and advanced in the breeding cycle. Thus, GS using high-density marker coverage enables to capture QTL with major and minor effects and eliminates the need to collect phenotypic data in all breeding cycles. Also, the application of GS was demonstrated to reduce the number of breeding cycles and increases the annual gain (Heffner et al., 2009). One of the problems of GS is the level of GEBV accuracy. Simulation studies based on simulated and empirical data demonstrated that GEBV accuracy could be within 0.62-0.85. Heffner et al. (2009) used previously reported GEBV accuracy of 0.53 and reported three- and two-fold annual gain in maize and winter barley, respectively.

The obvious advantages of GS over traditional MAS have been successfully proven in animal breeding (Hayes and Goddard, 2010). Rapid evolution of sequencing technologies and high-throughput SNP genotyping systems are enabling generation and validation of millions of markers, giving a "cautious optimism" for successful application of GS in plant breeding of complex traits. Thus, considering current application level and success in various crops, MAS technology still remains in its development stages but attractive for 21st century breeding. Successful efforts, as a wake-up call, further require incorporation of abovementioned advances in large-scale modern genotyping, precise phenotyping, statistically improved genetic mapping and data analysis as well as genome characterization of the crop species.

## 4. Genomics efforts in understanding molecular basis of plant growth, development and traits of interest towards crop improvement

### 4.1 Genomics tools for understanding natural variation

The availability of Next generation sequencing (NGS) technologies has paved the way for discovery of genetic variation at whole genome level of multiple genotypes. Ossowski et al. (2008) have demonstrated that even the short reads derived from Illumina Genome Analyzer could reveal most of the sequence variations in A. thaliana strains/accessions. The 1001 Arabidopsis Genomes project that is currently ongoing aims to discover the whole genome sequence variation in 1001 distinct accessions of Arabidopsis. The wealth of information from this project enables large scale genetic and functional analyses to address key biological phenomena and leverage that information for improvement in cultivated crop species. NGS technologies have also enabled other whole genome exploratory research on variation detection such as genome-wide DNA methylation detection (Lister et al., 2008), mutation mapping (Ossowski et al., 2008) and DNA-protein interactions (Bernatavichute et al., 2008). Section 2.3.2.3 provides additional details on how NGS technologies are being leveraged for variation detection and HTP molecular marker development. Together these efforts are instrumental in developing molecular markers and other diagnostic tools thereby enabling or accelerating molecular breeding.

## 4.2 Genomics efforts in understanding root growth, development and architecture

Water and nutrient uptake by roots plays a significant role in the growth of plants. In addition to providing anchorage in the soil, roots can adapt developmentally and physiologically to environmental changes. Efforts in the past to understand the molecular basis of root development have focused on single mutant analysis. While these approaches shed light on the cell type patterning in root, they have revealed the complex interactions underlying root growth and development accentuating the need for the use of exhaustive global "omics" analyses. Recent availability of a root expression map (Brady et al., 2007), root proteome (Baerenfaller et al., 2008) and environment-specific expression data are revealing complex transcriptional and pot-transcriptional pathways in root development (Iyer-Pascuzzi et al., 2009). These efforts and the initiatives to integrate the data from multiple "omics" studies are paving the way for the understanding of root biology across plant species.

In a recent review Hochholdinger and Tuberosa (2009) summarized the latest results on the genetic and genomic dissection of maize root development and on the cloning of underlying genes using root architecture mutants. Maize root system has complex architecture and is controlled by many genes. Characterization of *rtcs* mutant (rootless concerning crown and seminal roots) and the map-based cloning of underlying *RTCS* gene revealed that this codes for a transcription factor involved in early events responsible for root initiation. Similarly analysis of *rth1* and *rth3* mutants (roothairless 1 and 3) demonstrated their involvement in root hair elongation and the corresponding genes were found to be parts of machinery responsible for tethering exocytotic vesicles (rth1) (Wen et al., 2005) and cell expansion and cell wall biosynthesis related processes. While QTL mapping has identified some regions that influence root features and thereby yield, the use of 'omics' technologies provided unprecedented capability in obtaining significant insights into maize root development. For example, comparative laser capture microdissection (LCM) gene expression profiles of primary root meristem and root cap cells identified gene clusters linked to transport, environmental interactions and hormonal and carbohydrate signaling (Jiang et al., 2006). Similarly, comparison of LCM microarray profiles between pericle cells of wild-type and mutant rum1 (rootless with undetectable meristems 1) seedlings revealed a set of genes related to signal transduction, cell cycle, transcription and translation that are probably linked to lateral root initiation (Woll et al., 2005). In another transcriptome study that analyzed maize root responses towards environmental stimuli, highly differentially expressed transcripts were found to be those arising from reactive oxygen species (ROS) and carbon metabolism in root tips and elongation zone (Spollen et al., 2008). Comparative proteome analysis of maize roots from mutants and wild type as well as before and after a given treatment led to the identification of proteins that are likely to be associated with influence of lateral roots on the proteome composition of the primary root, phosphorus depletion and water deficit (reviewed in Hochholdinger & Tuberosa, 2009).

## 4.3 Genomics-based dissection of molecular basis of biomass production and Cell wall composition

Yield is the most important yet one of the most intriguing traits in agriculture. Despite its economic importance very little is known about the mechanisms underlying yield. One approach to identify the candidate genes responsible for yield is the use of information from model plants such as Arabidopsis and leverage this information in cultivated crop plants. Using genetics and genomics approaches candidate genes were identified in Arabidopsis, many of which had a significant effect on the biomass production through an increase in the size of leaves or roots (Gonzalez et al., 2009). Genes thus identified belonged to different functional classes that include transcriptional factors, translational regulators (protein synthesis and modification), signaling pathways, hormonal regulation, cell division and expansion. Examples now exist where some of the candidate genes belonging to these categories have been demonstrated to positively influence yield based on transgenic studies (Wu et al., 2008). In order to get a better handle on yield, it is proposed to employ a systems biology approach for obtaining additional 'omics' data and generating an integrated network of pathways in which the candidate genes are key players.

One of the key distinguishing features of grasses is the presence of $(1,3;1,4)$-β-D-glucans in their cell walls. These $(1,3;1,4)$-β-D-glucans are almost exclusively distributed within Poaceae where they are present in both primary and secondary walls. Considering the undesirable characteristics of barley $(1,3;1,4)$-β-D-glucans in malting and brewing industries as well as in animal feeds, many researchers set out to investigate the molecular mechanism underlying the accumulation of this class of polysaccharides in barley by developing molecular markers and mapping QTL. Mapping efforts led to the identification of a region on barley chromosome 2 that controls the production of $(1,3;1,4)$-β-D-glucans (Han et al., 1995). Although many biochemical efforts during 1980s and 1990s focused on the isolation of $(1,3;1,4)$-β-D-glucan synthase, the genes coding for these synthases could not be identified due to issues in the purification of these enzymes. In the recent years, the beneficial effects of $(1,3;1,4)$-β-D-glucans in human health (Wood, 2007) as well as their ability to positively influence biofuel industry through increased biomass production led to the implementation of molecular and genomics approaches for rapid identification of genes underlying these glucans (Fincher, 2009). In order to identify the genes and proteins responsible for the biosynthesis of β-glucans, Burton et al. (2006) have used the mapping information of Han et al. (1995) along with the information on conserved genome structure, gene collinearity or synteny between barley and rice, for which complete genome sequence is available. Using this approach they have successfully demonstrated the presence of a rice locus corresponding to mapped barley region and showed that the rice genome contains six cellulose synthase like (CslF) genes, thus identifying strong candidate genes for β-glucan biosynthesis.

## 4.4 Dissecting leaf architecture using genomics

Photosynthesis, the process through which plants harvest light energy and convert it into the building blocks of life is an extremely important biological phenomenon. While details on the process of photosynthesis have been worked out over the years, we are at the beginning in understanding the genetic and molecular basis of this complex process. Many efforts are currently underway in dissecting different components of photosynthesis. Using a high-throughput Illumina sequencing approach, Li et al. (2010) have analyzed maize leaf transcriptome and identified differential mRNA processing events for most maize genes. Their data revealed maize transcriptome to be a dynamic one with transcripts for primary cell wall and basic cellular metabolism at the leaf base transitioning to transcripts for secondary cell wall biosynthesis and $C_4$ photosynthetic development toward the tip. They found that as the leaf develops, large numbers of genes are turned on and off. Such information could not be obtained prior to the availability of massively parallel techniques such as the next generation sequencing. The comprehensive information from this and other studies will serve as the foundation for a systems biology approach for the understanding of photosynthetic development of maize.

## 4.5 Genomics for improving abiotic stress tolerance of crops

Abiotic stresses have become major concerns for global crop production and conventional approaches for developing tolerant cultivars have been difficult due to the complex inheritance of stress tolerance traits. For example, drought is the most recalcitrant abiotic stress trait for crop improvement due to its quantitative genetic inheritance and the involvement of multiple physiological effects on the ultimate yield (Passioura, 2002). Recent years have seen a renewed interested in understanding the molecular basis of drought tolerance with many studies reporting the mapping of QTL underlying this trait and the availability of high-throughput sequencing and associated computational tools are providing new avenues for the characterization of this complex trait (Tuberosa and Salvi, 2006). Genomics efforts to date in drought tolerance research could be broadly divided into two approaches. In one approach, QTL maps were combined with maps containing genic information or annotated genome sequence (Varshney et al., 2005) for the rapid identification of candidate genes. Use of this approach for the analysis of root trait QTL along with EST and cDNA screening has identified OsEXP2 and EGase genes in rice that were found to be involved in cell expansion (Zheng et al., 2003). The second approach is the employment of high-throughput transcriptomic profiling to investigate the changes in gene expression in response to drought. A recurring theme based on the comparison of multiple transcriptomics studies is the central role of transcription factors (TFs) in drought as well as the complex hierarchy of regulatory networks that modulate the tissue-specific differential expression of candidate genes (Yamaguchi-Shinozaki and Shinozaki, 2005). Proteomic profiling has also revealed several lead candidates for drought resistance in rice and maize. The actin depolymerizing factor (ADF) in rice has displayed most significant drought-induced fluctuations with its concentration increasing in leaves (especially in leaf blades and sheath) and roots after exposure to dehydration in drought-tolerant cultivars (Ali and Komatsu, 2006). In another approach, proteomic profiling was carried out on a mapping population to identify protein quantity loci (PQLs) i.e., QTLs influencing quantity of protein. Such an analysis in maize led to the identification of a putative transcription factor gene (Asr1) that co-localized with a PQL for the ASR1 protein and a QTL for ASI and leaf senescence (Jeanneau et al., 2002).

Salt tolerance is another complex trait threatening crop production in many countries worldwide and genomics efforts are underway to dissect this trait. Sanchez et al. (2011) have used comparative functional genomics techniques such as ionomics, transcriptomics and metabolomics to distinguish genotype-specific transcriptional and metabolic changes from those of true salinity responses leading to the identification of conserved and tolerance-specific responses towards achieving salinity tolerance across species.

## 5. Conclusion

Plant breeding has a major role to play in increasing global food production while tackling the issues of limited land and water resources and changing climate. While the molecular era has laid the foundation for molecular breeding during the last quarter of twentieth century, the advent of genomics tools and technologies has been providing unprecedented capabilities for understanding the molecular basis of plant growth, development and key traits towards improving crop productivity in the 21st century. A multitude of omics and associated HTP technologies are enabling systematic dissection and understanding of plants that was not possible previously. The knowledge derived from such efforts will certainly be useful in developing "designer plants" that can yield better through improved growth and ability to withstand biotic and abiotic stresses. In addition to the insights, derived from individual or a combination of omics technologies applied to specific traits of interest, the renewal of 'holistic' systems biology concept and genome-wide measurements of components of interest certainly has the potential in dissecting the molecular, biochemical, physiological and evolutionary basis of traits and biological phenomena. This holds a great promise for crop improvement. Continued development of 'omics' technologies and computational tools for accurate analysis, and integration and interpretation of massive amounts of data are key challenges that need to be addressed to reap the full potential of genomics and systems biology approaches. The progress made so far through marker-assisted breeding and genomics and the promising technological breakthroughs are certainly paving the way for "Genomics-assisted breeding" in the 21st century!

## 6. References

Abdurakhmonov, I., Kohel, R., Yu, J., Pepper, A., Abdullaev, A., Kushanov, F., Salakhutdinov, I., Buriev, Z., Saha, S., and Scheffler, B. (2008). Molecular diversity and association mapping of fiber quality traits in exotic G. hirsutum L. germplasm. *Genomics* Vol. 92, No. 6, pp. 478-487, ISSN 0888-7543.

Abdurakhmonov, I. Y., and Abdukarimov, A. (2008). Application of association mapping to understanding the genetic diversity of plant germplasm resources. *Int J Plant Genomics* Vol. 574927.

Agrama, H., and Eizenga, G. (2008). Molecular diversity and genome-wide linkage disequilibrium patterns in a worldwide collection of Oryza sativa and its wild relatives. *Euphytica* Vol. 160, No. 3, pp. 339-355, ISSN 0014-2336.

Ali, G. M., and Komatsu, S. (2006). Proteomic analysis of rice leaf sheath during drought stress. *Journal of proteome research* Vol. 5, No. 2, pp. 396-403, ISSN 1535-3893.

Aranzana, M. J., Kim, S., Zhao, K., Bakker, E., Horton, M., Jakob, K., Lister, C., Molitor, J., Shindo, C., and Tang, C. (2005). Genome-wide association mapping in Arabidopsis identifies previously known flowering time and pathogen resistance genes. *PLoS genetics* Vol. 1, No. 5, pp. e60, ISSN 1553-7404.

Argout, X., Salse, J., Aury, J.-M., Guiltinan, M. J., Droc, G., Gouzy, J., Allegre, M., Chaparro, C., Legavre, T., Maximova, S. N., Abrouk, M., Murat, F., Fouet, O., Poulain, J., Ruiz, M., Roguet, Y., Rodier-Goud, M., Barbosa-Neto, J. F., Sabot, F., Kudrna, D.,

Ammiraju, J. S. S., Schuster, S. C., Carlson, J. E., Sallet, E., Schiex, T., Dievart, A., Kramer, M., Gelley, L., Shi, Z., Berard, A., Viot, C., Boccara, M., Risterucci, A. M., Guignon, V., Sabau, X., Axtell, M. J., Ma, Z., Zhang, Y., Brown, S., Bourge, M., Golser, W., Song, X., Clement, D., Rivallan, R., Tahi, M., Akaza, J. M., Pitollat, B., Gramacho, K., D'Hont, A., Brunel, D., Infante, D., Kebe, I., Costet, P., Wing, R., McCombie, W. R., Guiderdoni, E., Quetier, F., Panaud, O., Wincker, P., Bocs, S., and Lanaud, C. (2011). The genome of Theobroma cacao. *Nat Genet* Vol. 43, No. 2, pp. 101-108, ISSN 1061-4036.

Baerenfaller, K., Grossmann, J., Grobei, M. A., Hull, R., Hirsch-Hoffmann, M., Yalovsky, S., Zimmermann, P., Grossniklaus, U., Gruissem, W., and Baginsky, S. (2008). Genome-scale proteomics reveals Arabidopsis thaliana gene models and proteome dynamics. *Science* Vol. 320, No. 5878, pp. 938, ISSN 0036-8075.

Baird, N. A., Etter, P. D., Atwood, T. S., Currey, M. C., Shiver, A. L., Lewis, Z. A., Selker, E. U., Cresko, W. A., and Johnson, E. A. (2008). Rapid SNP discovery and genetic mapping using sequenced RAD markers. *PLoS ONE* Vol. 3, No. 10, pp. e3376, ISSN 1932-6203.

Barbazuk, W. B., Emrich, S. J., Chen, H. D., Li, L., and Schnable, P. S. (2007). SNP discovery via 454 transcriptome sequencing. *The Plant Journal* Vol. 51, No. 5, pp. 910-918, ISSN 1365-313X.

Batley, J., Barker, G., O'Sullivan, H., Edwards, K. J., and Edwards, D. (2003). Mining for single nucleotide polymorphisms and insertions/deletions in maize expressed sequence tag data. *Plant physiology* Vol. 132, No. 1, pp. 84, ISSN 0032-0889.

Bauer-Mehren, A., Furlong, L. I., and Sanz, F. (2009). Pathway databases and tools for their exploitation: benefits, current limitations and challenges. *Molecular systems biology* Vol. 5, No. 1.

Beckmann, J., and Soller, M. (1986). Restriction fragment length polymorphisms in plant genetic improvement. *Oxford surveys of plant molecular and cell biology.* Vol. 3, No., pp. 196-250.

Bentley, D. R., Balasubramanian, S., Swerdlow, H. P., Smith, G. P., Milton, J., Brown, C. G., Hall, K. P., Evers, D. J., Barnes, C. L., and Bignell, H. R. (2008). Accurate whole human genome sequencing using reversible terminator chemistry. *Nature* Vol. 456, No. 7218, pp. 53-59, ISSN 0028-0836.

Bernardo, R. (2008). Molecular markers and selection for complex traits in plants: learning from the last 20 years. *Crop Science* Vol. 48, No. 5, pp. 1649-1664.

Bernatavichute, Y. V., Zhang, X., Cokus, S., Pellegrini, M., and Jacobsen, S. E. (2008). Genome-wide association of histone H3 lysine nine methylation with CHG DNA methylation in Arabidopsis thaliana. *PLoS ONE* Vol. 3, No. 9, pp. e3156, ISSN 1932-6203.

Botstein, D., White, R. L., Skolnick, M., and Davis, R. W. (1980). Construction of a genetic linkage map in man using restriction fragment length polymorphisms. *American Journal of Human Genetics* Vol. 32, No. 3, pp. 314.

Brady, S. M., Orlando, D. A., Lee, J. Y., Wang, J. Y., Koch, J., Dinneny, J. R., Mace, D., Ohler, U., and Benfey, P. N. (2007). A high-resolution root spatiotemporal map reveals dominant expression patterns. *Science* Vol. 318, No. 5851, pp. 801, ISSN 0036-8075.

Bräutigam, A., and Gowik, U. (2010). What can next generation sequencing do for you? Next generation sequencing as a valuable tool in plant research. *Plant Biology* Vol. 12, No. 6, pp. 831-841, ISSN 1438-8677.

Brenan, C., and Morrison, T. (2005). High throughput, nanoliter quantitative PCR. *Drug Discovery Today: Technologies* Vol. 2, No. 3, pp. 247-253, ISSN 1740-6749.

Breyne, P., and Zabeau, M. (2001). Genome-wide expression analysis of plant cell cycle modulated genes. *Current opinion in plant biology* Vol. 4, No. 2, pp. 136-142, ISSN 1369-5266.

Bundock, P. C., Eliott, F. G., Ablett, G., Benson, A. D., Casu, R. E., Aitken, K. S., and Henry, R. J. (2009). Targeted single nucleotide polymorphism (SNP) discovery in a highly polyploid plant species using 454 sequencing. *Plant biotechnology journal* Vol. 7, No. 4, pp. 347-354, ISSN 1467-7652.

Burley, S. K., Almo, S. C., Bonanno, J. B., Capel, M., Chance, M. R., Gaasterland, T., Lin, D., Sali, A., Studier, F. W., and Swaminathan, S. (1999). Structural genomics: beyond the human genome project. *Nat Genet* Vol. 23, No. 2, pp. 151-7,(Oct), ISSN 1061-4036.

Burton, R. A., Wilson, S. M., Hrmova, M., Harvey, A. J., Shirley, N. J., Medhurst, A., Stone, B. A., Newbigin, E. J., Bacic, A., and Fincher, G. B. (2006). Cellulose synthase-like CslF genes mediate the synthesis of cell wall (1, 3; 1, 4)-ß-D-glucans. *Science* Vol. 311, No. 5769, pp. 1940, ISSN 0036-8075.

Cahill, D. J., and Schmidt, D. H. (2004). Use of marker assisted selection in a product development breeding program. *Proc. 4th Int. Crop. Sci. Cong* Vol. 26.

Chagné, D., Batley, J., Edwards, D., and Forster, J. W. (2007). Single Nucleotide Polymorphism Genotyping in Plants. *In* "Association Mapping in Plants" (N. C. Oraguzie, E. H. A. Rikkerink, S. E. Gardiner and H. N. Silva, eds.), pp. 77-94. Springer New York.

Chen, W., Mingus, J., Mammadov, J., Backlund, J. E., Greene, T., Thompson , S., and Kumpatla, S. (2010). *In* "International Plant & Animal Genomes XVIII Conference", San Diego.

Chinnusamy, V., and Zhu, J. K. (2009). Epigenetic regulation of stress responses in plants. *Current opinion in plant biology* Vol. 12, No. 2, pp. 133-139, ISSN 1369-5266.

Cokus, S. J., Feng, S., Zhang, X., Chen, Z., Merriman, B., Haudenschild, C. D., Pradhan, S., Nelson, S. F., Pellegrini, M., and Jacobsen, S. E. (2008). Shotgun bisulfite sequencing of the Arabidopsis genome reveals DNA methylation patterning. *Nature* Vol. 452, No. 7184, pp. 215.

Collard, B. C., and Mackill, D. J. (2008). Marker-assisted selection: an approach for precision plant breeding in the twenty-first century. *Philos Trans R Soc Lond B Biol Sci* Vol. 363, No. 1491, pp. 557-72,(Feb 12), ISSN 0962-8436.

Concibido, V., Denny, R., Lange, D., Orf, J., and Young, N. (1996). RFLP mapping and marker-assisted selection of soybean cyst nematode resistance in PI 209332. *Crop Science* Vol. 36, No. 6, pp. 1643-1650, ISSN 0011-183X.

Conesa, A., Götz, S., García-Gómez, J. M., Terol, J., Talón, M., and Robles, M. (2005). Blast2GO: a universal tool for annotation, visualization and analysis in functional genomics research. *Bioinformatics* Vol. 21, No. 18, pp. 3674, ISSN 1367-4803.

Cooper, B., Clarke, J. D., Budworth, P., Kreps, J., Hutchison, D., Park, S., Guimil, S., Dunn, M., Luginbühl, P., and Ellero, C. (2003). A network of rice genes associated with stress response and seed development. *Proceedings of the National Academy of Sciences* Vol. 100, No. 8, pp. 4945, ISSN 0027-8424.

Crosbie, T. M., Eathington, S. R., Johnson Sr, G. R., Edwards, M., Reiter, R., Stark, S., Mohanty, R. G., Oyervides, M., Buehler, R. E., and Walker, A. K. (2003). Plant breeding: Past, present, and future. pp. 3-50. Wiley Online Library.

Davey, J. W., Hohenlohe, P. A., Etter, P. D., Boone, J. Q., Catchen, J. M., and Blaxter, M. L. (2011). Genome-wide genetic marker discovery and genotyping using next-generation sequencing. *Nat Rev Genet* Vol. 12, No. 7, pp. 499-510, ISSN 1471-0056.

Deschamps, S., and Campbell, M. (2010). Utilization of next-generation sequencing platforms in plant genomics and genetic variant discovery. *Molecular Breeding* Vol. 25, No. 4, pp. 553-570, ISSN 1380-3743.

Devos, K. M. (2005). Updating the 'Crop Circle'. *Current opinion in plant biology* Vol. 8, No. 2, pp. 155-162, ISSN 1369-5266.

Druka, A., Potokina, E., Luo, Z., Jiang, N., Chen, X., Kearsey, M., and Waugh, R. (2010). Expression quantitative trait loci analysis in plants. *Plant biotechnology journal* Vol. 8, No. 1, pp. 10-27, ISSN 1467-7652.

Dubcovsky, J., Ramakrishna, W., SanMiguel, P. J., Busso, C. S., Yan, L., Shiloff, B. A., and Bennetzen, J. L. (2001). Comparative sequence analysis of colinear barley and rice bacterial artificial chromosomes. *Plant physiology* Vol. 125, No. 3, pp. 1342, ISSN 0032-0889.

Eichten, S. R., Foerster, J., de Leon, N., Ying, K., Yeh, C. T., Liu, S., Jeddeloh, J., Schnable, P., Kaeppler, S. M., and Springer, N. M. (2011). B73-Mo17 near isogenic lines (NILs) demonstrate dispersed structural variation in maize. *Plant physiology* Vol., No., ISSN 0032-0889.

Eid, J., Fehr, A., Gray, J., Luong, K., Lyle, J., Otto, G., Peluso, P., Rank, D., Baybayan, P., and Bettman, B. (2009). Real-time DNA sequencing from single polymerase molecules. *Science* Vol. 323, No. 5910, pp. 133, ISSN 0036-8075.

Elshire, R. J., Glaubitz, J. C., Sun, Q., Poland, J. A., Kawamoto, K., Buckler, E. S., and Mitchell, S. E. (2011). A Robust, Simple Genotyping-by-Sequencing (GBS) Approach for High Diversity Species. *PLoS ONE* Vol. 6, No. 5, pp. e19379.

Emberton, J., Ma, J., Yuan, Y., SanMiguel, P., and Bennetzen, J. L. (2005). Gene enrichment in maize with hypomethylated partial restriction (HMPR) libraries. *Genome Res* Vol. 15, No. 10, pp. 1441, ISSN 1088-9051.

Endler, L., Rodriguez, N., Juty, N., Chelliah, V., Laibe, C., Li, C., and Le Novère, N. (2009). Designing and encoding models for synthetic biology. *Journal of The Royal Society Interface* Vol. 6, No. Suppl 4, pp. S405, ISSN 1742-5689.

Fan, J. B., Oliphant, A., Shen, R., Kermani, B., Garcia, F., Gunderson, K., Hansen, M., Steemers, F., Butler, S., and Deloukas, P. (2003). Highly parallel SNP genotyping. Vol. 68, pp. 69. Cold Spring Harbor Laboratory Press.

Fincher, G. B. (2009). Exploring the evolution of (1, 3; 1, 4)-[beta]-d-glucans in plant cell walls: comparative genomics can help! *Current opinion in plant biology* Vol. 12, No. 2, pp. 140-147, ISSN 1369-5266.

Fu, H., and Dooner, H. K. (2002). Intraspecific violation of genetic colinearity and its implications in maize. *Proceedings of the National Academy of Sciences* Vol. 99, No. 14, pp. 9573, ISSN 0027-8424.

Gale, M. D., and Devos, K. M. (1998). Comparative genetics in the grasses. *Proceedings of the National Academy of Sciences* Vol. 95, No. 5, pp. 1971, ISSN 0027-8424.

Ganal, M. W., Altmann, T., and Röder, M. S. (2009). SNP identification in crop plants. *Current opinion in plant biology* Vol. 12, No. 2, pp. 211-217, ISSN 1369-5266.

Gaut, B. S., and Doebley, J. F. (1997). DNA sequence evidence for the segmental allotetraploid origin of maize. *Proceedings of the National Academy of Sciences* Vol. 94, No. 13, pp. 6809, ISSN 0027-8424.

Godfray, H. C., Beddington, J. R., Crute, I. R., Haddad, L., Lawrence, D., Muir, J. F., Pretty, J., Robinson, S., Thomas, S. M., and Toulmin, C. (2010). Food security: the challenge of feeding 9 billion people. *Science* Vol. 327, No. 5967, pp. 812-8,(Feb 12), ISSN 1095-9203.

Gonzalez, N., Beemster, G. T. S., and Inzé, D. (2009). David and Goliath: what can the tiny weed Arabidopsis teach us to improve biomass production in crops? *Current opinion in plant biology* Vol. 12, No. 2, pp. 157-164, ISSN 1369-5266.

Gore, M. A. W., Ersoz, M. H., Bouffard, E. S., Szekeres, P., Jarvie, E. S., Hurwitz, T. P., Narechania, B. L., Harkins, A., Grills, T. T., and Ware, G. S. (2009). Large-scale discovery of gene-enriched SNPs. *The Plant Genome* Vol. 2, No. 2, pp. 121, ISSN 1940-3372.

Gupta, P., Varshney, R., Sharma, P., and Ramesh, B. (1999). Molecular markers and their applications in wheat breeding. *Plant breeding* Vol. 118, No. 5, pp. 369-390, ISSN 1439-0523.

Hamilton, J., Hansey, C., Whitty, B., Stoffel, K., Massa, A., Van Deynze, A., De Jong, W., Douches, D., and Buell, C. R. (2011). Single nucleotide polymorphism discovery in elite north american potato germplasm. *BMC Genomics* Vol. 12, No. 1, pp. 302, ISSN 1471-2164.

Han, F., Ullrich, S., Chirat, S., Menteur, S., Jestin, L., Sarrafi, A., Hayes, P., Jones, B., Blake, T., and Wesenberg, D. (1995). Mapping of b-glucan content and b-glucanase activity loci in barley grain and malt. *Theoretical and Applied Genetics* Vol. 91, No. 6, pp. 921-927, ISSN 0040-5752.

Hayes, B., and Goddard, M. (2010). Genome-wide association and genomic selection in animal breeding. *Genome* Vol. 53, No. 11, pp. 876-883, ISSN 0831-2796.

Heffner, E. L., Sorrells, M. E., and Jannink, J. L. (2009). Genomic selection for crop improvement. *Crop Science*, Vol. 49, No. 1, pp 1-12.

Higgins, J. A., Bailey, P. C., and Laurie, D. A. (2010). Comparative genomics of flowering time pathways using Brachypodium distachyon as a model for the temperate grasses. *PLoS ONE* Vol. 5, No. 4, pp. e10065, ISSN 1932-6203.

Hochholdinger, F., and Tuberosa, R. (2009). Genetic and genomic dissection of maize root development and architecture. *Current opinion in plant biology* Vol. 12, No. 2, pp. 172-177, ISSN 1369-5266.

Holloway, B., Luck, S., Beatty, M., Rafalski, J. A., and Li, B. (2011). Genome-wide expression quantitative trait loci (eQTL) analysis in maize. *BMC Genomics* Vol. 12, No. 1, pp. 336, ISSN 1471-2164.

Honsdorf, N., Becker, H. C., and Ecke, W. (2010). Association mapping for phenological, morphological, and quality traits in canola quality winter rapeseed (Brassica napus L.). *Genome* Vol. 53, No. 11, pp. 899-907, ISSN 0831-2796.

Hribova, E., Neumann, P., Macas, J., and Dolezel, J. (2009). Analysis of genome structure and organization in banana (Musa acuminata) using 454 sequencing. *Plant and Animal Genomes XVII. San Diego, CA.*

Huang, X., Feng, Q., Qian, Q., Zhao, Q., Wang, L., Wang, A., Guan, J., Fan, D., Weng, Q., and Huang, T. (2009). High-throughput genotyping by whole-genome resequencing. *Genome Res* Vol. 19, No. 6, pp. 1068, ISSN 1088-9051.

Hubert, B., Rosegrant, M., van Boekel, M., and Ortiz, R. (2010). The future of food: scenarios for 2050. *Crop Science* Vol. 50.

Iyer-Pascuzzi, A., Simpson, J., Herrera-Estrella, L., and Benfey, P. N. (2009). Functional genomics of root growth and development in Arabidopsis. *Current opinion in plant biology* Vol. 12, No. 2, pp. 165-171, ISSN 1369-5266.

Jansen, R. C., and Nap, J. P. (2001). Genetical genomics: the added value from segregation. *Trends in Genetics* Vol. 17, No. 7, pp. 388-391, ISSN 0168-9525.

Jeanneau, M., Gerentes, D., Foueillassar, X., Zivy, M., Vidal, J., Toppan, A., and Perez, P. (2002). Improvement of drought tolerance in maize: towards the functional validation of the Zm-Asr1 gene and increase of water use efficiency by over-expressing C4-PEPC. *Biochimie* Vol. 84, No. 11, pp. 1127-1135, ISSN 0300-9084.

Jiang, K., Zhang, S., Lee, S., Tsai, G., Kim, K., Huang, H., Chilcott, C., Zhu, T., and Feldman, L. J. (2006). Transcription profile analyses identify genes and pathways central to root cap functions in maize. *Plant molecular biology* Vol. 60, No. 3, pp. 343-363, ISSN 0167-4412.

Jiao, Y., Wang, Y., Xue, D., Wang, J., Yan, M., Liu, G., Dong, G., Zeng, D., Lu, Z., and Zhu, X. (2010). Regulation of OsSPL14 by OsmiR156 defines ideal plant architecture in rice. *Nature genetics* Vol. 42, No. 6, pp. 541-544, ISSN 1061-4036.

Jin, L., Lu, Y., Xiao, P., Sun, M., Corke, H., and Bao, J. (2010). Genetic diversity and population structure of a diverse set of rice germplasm for association mapping. *TAG Theoretical and Applied Genetics* Vol. 121, No. 3, pp. 475-487, ISSN 0040-5752.

Joosen, R., Ligterink, W., Hilhorst, H., and Keurentjes, J. (2009). Advances in genetical genomics of plants. *Current Genomics* Vol. 10, No. 8, pp. 540.

Jordan, M. C., Somers, D. J., and Banks, T. W. (2007). Identifying regions of the wheat genome controlling seed development by mapping expression quantitative trait loci†. *Plant biotechnology journal* Vol. 5, No. 3, pp. 442-453, ISSN 1467-7652.

Ju, J., Kim, D. H., Bi, L., Meng, Q., Bai, X., Li, Z., Li, X., Marma, M. S., Shi, S., and Wu, J. (2006). Four-color DNA sequencing by synthesis using cleavable fluorescent nucleotide reversible terminators. *Proceedings of the National Academy of Sciences* Vol. 103, No. 52, pp. 19635, ISSN 0027-8424.

Jung, M., Ching, A., Bhattramakki, D., Dolan, M., Tingey, S., Morgante, M., and Rafalski, A. (2004). Linkage disequilibrium and sequence diversity in a 500-kbp region around the adh1 locus in elite maize germplasm. *TAG Theoretical and Applied Genetics* Vol. 109, No. 4, pp. 681-689, ISSN 0040-5752.

Kearsey, M., and Farquhar, A. (1998). QTL analysis in plants; where are we now? *Heredity* Vol. 80, No. 2, pp. 137-142, ISSN 0018-067X.

Keurentjes, J. J. B., Angenent, G. C., Dicke, M., Santos, V. A. P., Molenaar, J., and van der Putten, W. H. (2011). Redefining plant systems biology: from cell to ecosystem. *Trends in Plant Science* Vol., No., ISSN 1360-1385.

Keurentjes, J. J. B., Fu, J., Terpstra, I. R., Garcia, J. M., Van Den Ackerveken, G., Snoek, L. B., Peeters, A. J. M., Vreugdenhil, D., Koornneef, M., and Jansen, R. C. (2007). Regulatory network construction in Arabidopsis by using genome-wide gene expression quantitative trait loci. *Proceedings of the National Academy of Sciences* Vol. 104, No. 5, pp. 1708, ISSN 0027-8424.

Kolukisaoglu, Ü., and Thurow, K. (2010). Future and frontiers of automated screening in plant sciences. *Plant Science* Vol. 178, No. 6, pp. 476-484, ISSN 0168-9452.

Korf, I. (2004). Gene finding in novel genomes. *BMC Bioinformatics* Vol. 5, No. 1, pp. 59, ISSN 1471-2105.

Korlach, J., Bibillo, A., Wegener, J., Peluso, P., Pham, T. T., Park, I., Clark, S., Otto, G. A., and Turner, S. W. (2008). Long, processive enzymatic DNA synthesis using 100% dye-labeled terminal phosphate-linked nucleotides. *Nucleosides, Nucleotides and Nucleic Acids* Vol. 27, No. 9, pp. 1072-1082, ISSN 1525-7770.

Laird, P. W. (2010). Principles and challenges of genome-wide DNA methylation analysis. *Nature Reviews Genetics* Vol. 11, No. 3, pp. 191-203, ISSN 1471-0056.

Li, P., Ponnala, L., Gandotra, N., Wang, L., Si, Y., Tausta, S. L., Kebrom, T. H., Provart, N., Patel, R., and Myers, C. R. (2010). The developmental dynamics of the maize leaf transcriptome. *Nature genetics* Vol., No., ISSN 1061-4036.

Lister, R., O'Malley, R. C., Tonti-Filippini, J., Gregory, B. D., Berry, C. C., Millar, A. H., and Ecker, J. R. (2008). Highly integrated single-base resolution maps of the epigenome in Arabidopsis. *Cell* Vol. 133, No. 3, pp. 523-536, ISSN 0092-8674.

Livak, K. J., Flood, S., Marmaro, J., Giusti, W., and Deetz, K. (1995). Oligonucleotides with fluorescent dyes at opposite ends provide a quenched probe system useful for detecting PCR product and nucleic acid hybridization. *Genome Res* Vol. 4, No. 6, pp. 357, ISSN 1088-9051.

Lo, C., Bashir, A., Bansal, V., and Bafna, V. (2011). Strobe sequence design for Haplotype assembly. *BMC Bioinformatics* Vol. 12, No. Suppl 1, pp. S24, ISSN 1471-2105.

Lundquist, P. M., Zhong, C. F., Zhao, P., Tomaney, A. B., Peluso, P. S., Dixon, J., Bettman, B., Lacroix, Y., Kwo, D. P., and McCullough, E. (2008). Parallel confocal detection of single molecules in real time. *Optics letters* Vol. 33, No. 9, pp. 1026-1028, ISSN 1539-4794.

Mamanova, L., Coffey, A. J., Scott, C. E., Kozarewa, I., Turner, E. H., Kumar, A., Howard, E., Shendure, J., and Turner, D. J. (2010). Target-enrichment strategies for next-generation sequencing. *Nature methods* Vol. 7, No. 2, pp. 111-118, ISSN 1548-7091.

Mammadov, J. A., Chen, W., Ren, R., Pai, R., Marchione, W., Yalçin, F., Witsenboer, H., Greene, T. W., Thompson, S. A., and Kumpatla, S. P. (2010). Development of highly polymorphic SNP markers from the complexity reduced portion of maize [Zea mays L.] genome for use in marker-assisted breeding. *TAG Theoretical and Applied Genetics* Vol. 121, No. 3, pp. 577-588, ISSN 0040-5752.

Mardis, E. R. (2008a). The impact of next-generation sequencing technology on genetics. *Trends in Genetics* Vol. 24, No. 3, pp. 133-141, ISSN 0168-9525.

Mardis, E. R. (2008b). Next-generation DNA sequencing methods. *Annu. Rev. Genomics Hum. Genet.* Vol. 9, No., pp. 387-402, ISSN 1527-8204.

Margulies, M., Egholm, M., Altman, W. E., Attiya, S., Bader, J. S., Bemben, L. A., Berka, J., Braverman, M. S., Chen, Y. J., Chen, Z., Dewell, S. B., Du, L., Fierro, J. M., Gomes, X. V., Godwin, B. C., He, W., Helgesen, S., Ho, C. H., Irzyk, G. P., Jando, S. C.,

Alenquer, M. L., Jarvie, T. P., Jirage, K. B., Kim, J. B., Knight, J. R., Lanza, J. R., Leamon, J. H., Lefkowitz, S. M., Lei, M., Li, J., Lohman, K. L., Lu, H., Makhijani, V. B., McDade, K. E., McKenna, M. P., Myers, E. W., Nickerson, E., Nobile, J. R., Plant, R., Puc, B. P., Ronan, M. T., Roth, G. T., Sarkis, G. J., Simons, J. F., Simpson, J. W., Srinivasan, M., Tartaro, K. R., Tomasz, A., Vogt, K. A., Volkmer, G. A., Wang, S. H., Wang, Y., Weiner, M. P., Yu, P., Begley, R. F., and Rothberg, J. M. (2005). Genome sequencing in microfabricated high-density picolitre reactors. *Nature* Vol. 437, No. 7057, pp. 376-80,(Sep 15), ISSN 1476-4687.

Massman, J., Cooper, B., Horsley, R., Neate, S., Dill-Macky, R., Chao, S., Dong, Y., Schwarz, P., Muehlbauer, G., and Smith, K. (2011). Genome-wide association mapping of Fusarium head blight resistance in contemporary barley breeding germplasm. *Molecular Breeding* Vol. 27, No. 4, pp. 439-454, ISSN 1380-3743.

Mayer, K. F. X., Martis, M., Hedley, P. E., Šimková, H., Liu, H., Morris, J. A., Steuernagel, B., Taudien, S., Roessner, S., and Gundlach, H. (2011). Unlocking the barley genome by chromosomal and comparative genomics. *The Plant Cell Online* Vol. 23, No. 4, pp. 1249, ISSN 1040-4651.

Metzker, M. L. (2009). Sequencing technologies—the next generation. *Nature Reviews Genetics* Vol. 11, No. 1, pp. 31-46, ISSN 1471-0056.

Meuwissen, T., Hayes, B., and Goddard, M. (2001). Prediction of total genetic value using genome-wide dense marker maps. *Genetics* Vol. 157, No. 4, pp. 1819.

Meyers, B. C., Tingey, S. V., and Morgante, M. (2001). Abundance, distribution, and transcriptional activity of repetitive elements in the maize genome. *Genome Res* Vol. 11, No. 10, pp. 1660, ISSN 1088-9051.

Mochida, K., and Shinozaki, K. (2010). Genomics and bioinformatics resources for crop improvement. *Plant and Cell Physiology* Vol. 51, No. 4, pp. 497, ISSN 0032-0781.

Montes, J., Paul, C., Kusterer, B., and Melchinger, A. (2006). Near infrared spectroscopy to measure maize grain composition on plot combine harvesters: evaluation of calibration techniques, mathematical transformations and scatter corrections. *Journal of near infrared spectroscopy* Vol. 14, No. 6, pp. 387-394, ISSN 0967-0335.

Montes, J., Technow, F., Dhillon, B., Mauch, F., and Melchinger, A. (2011). High-throughput non-destructive biomass determination during early plant development in maize under field conditions. *Field Crops Research* Vol., No., ISSN 0378-4290.

Morozova, O., and Marra, M. A. (2008). Applications of next-generation sequencing technologies in functional genomics. *Genomics* Vol. 92, No. 5, pp. 255-264, ISSN 0888-7543.

Mortazavi, A., Williams, B. A., McCue, K., Schaeffer, L., and Wold, B. (2008). Mapping and quantifying mammalian transcriptomes by RNA-Seq. *Nature methods* Vol. 5, No. 7, pp. 621-628.

Narina, S., Buyyarapu, R., Kottapalli, K., Sartie, A., Ali, M., Robert, A., Hodeba, M., Sayre, B., and Scheffler, B. (2011). Generation and analysis of Expressed Sequence Tags (ESTs) for marker development in yam (Dioscorea alata L.). *BMC Genomics* Vol. 12, No. 1, pp. 100, ISSN 1471-2164.

Ng, S. B., Buckingham, K. J., Lee, C., Bigham, A. W., Tabor, H. K., Dent, K. M., Huff, C. D., Shannon, P. T., Jabs, E. W., and Nickerson, D. A. (2009). Exome sequencing identifies the cause of a mendelian disorder. *Nature genetics* Vol. 42, No. 1, pp. 30-35, ISSN 1061-4036.

Nijman, I. J., Mokry, M., van Boxtel, R., Toonen, P., de Bruijn, E., and Cuppen, E. (2010). Mutation discovery by targeted genomic enrichment of multiplexed barcoded samples. *Nature methods* Vol. 7, No. 11, pp. 913-915, ISSN 1548-7091.

Nordborg, M., and Tavaré, S. (2002). Linkage disequilibrium: what history has to tell us. *Trends in Genetics* Vol. 18, No. 2, pp. 83-90, ISSN 0168-9525.

Okou, D. T., Steinberg, K. M., Middle, C., Cutler, D. J., Albert, T. J., and Zwick, M. E. (2007). Microarray-based genomic selection for high-throughput resequencing. *Nature methods* Vol. 4, No. 11, pp. 907-909, ISSN 1548-7091.

Oliver, R., Lazo, G., Lutz, J., Rubenfield, M., Tinker, N., Anderson, J., Wisniewski Morehead, N., Adhikary, D., Jellen, E., Maughan, P. J., Brown Guedira, G., Chao, S., Beattie, A., Carson, M., Rines, H., Obert, D., Bonman, J. M., and Jackson, E. (2011). Model SNP development for complex genomes based on hexaploid oat using high-throughput 454 sequencing technology. *BMC Genomics* Vol. 12, No. 1, pp. 77, ISSN 1471-2164.

Ophir, R., and Graur, D. (1997). Patterns and rates of indel evolution in processed pseudogenes from humans and murids. *Gene* Vol. 205, No. 1-2, pp. 191-202, ISSN 0378-1119.

Ossowski, S., Schneeberger, K., Clark, R. M., Lanz, C., Warthmann, N., and Weigel, D. (2008). Sequencing of natural strains of Arabidopsis thaliana with short reads. *Genome Res* Vol. 18, No. 12, pp. 2024, ISSN 1088-9051.

Palmer, L. E., Rabinowicz, P. D., O'Shaughnessy, A. L., Balija, V. S., Nascimento, L. U., Dike, S., de la Bastide, M., Martienssen, R. A., and McCombie, W. R. (2003). Maize genome sequencing by methylation filtration. *Science* Vol. 302, No. 5653, pp. 2115, ISSN 0036-8075.

Passioura, J. B. (2002). Review: Environmental biology and crop improvement. *Functional Plant Biology* Vol. 29, No. 5, pp. 537-546, ISSN 1445-4416.

Peleman, J. D., and van der Voort, J. R. (2003). Breeding by design. *Trends in Plant Science* Vol. 8, No. 7, pp. 330-334, ISSN 1360-1385.

Pennisi, E. (2010). Semiconductors inspire new sequencing technologies. *Science* Vol. 327, No. 5970, pp. 1190, ISSN 0036-8075.

Poland, J. A., Bradbury, P. J., Buckler, E. S., and Nelson, R. J. (2011). Genome-wide nested association mapping of quantitative resistance to northern leaf blight in maize. *Proceedings of the National Academy of Sciences* Vol. 108, No. 17, pp. 6893, ISSN 0027-8424.

Powell, W., Machray, G. C., and Provan, J. (1996). Polymorphism revealed by simple sequence repeats. *Trends in Plant Science* Vol. 1, No. 7, pp. 215-222, ISSN 1360-1385.

Rafalski, J. A. (2002). Novel genetic mapping tools in plants: SNPs and LD-based approaches. *Plant Science* Vol. 162, No. 3, pp. 329-333, ISSN 0168-9452.

Ragot, M., Lee, M., Guimarães, E., Ruane, J., Scherf, B., Sonnino, A., and Dargie, J. (2007). Marker-assisted selection in maize: current status, potential, limitations and perspertives from the private and public sectors. *Marker-Assisted Selection, Current Status and Future Perspectives in Crops, Livestock, Forestry and Fish* Vol., No., pp. 117-150.

Reinders, J., Delucinge Vivier, C., Theiler, G., Chollet, D., Descombes, P., and Paszkowski, J. (2008). Genome-wide, high-resolution DNA methylation profiling using bisulfite-mediated cytosine conversion. *Genome Res* Vol. 18, No. 3, pp. 469, ISSN 1088-9051.

Remington, D. L., Thornsberry, J. M., Matsuoka, Y., Wilson, L. M., Whitt, S. R., Doebley, J., Kresovich, S., Goodman, M. M., and Buckler, E. S. (2001). Structure of linkage disequilibrium and phenotypic associations in the maize genome. *Proceedings of the National Academy of Sciences* Vol. 98, No. 20, pp. 11479, ISSN 0027-8424.

Rossberg, M., Theres, K., Acarkan, A., Herrero, R., Schmitt, T., Schumacher, K., Schmitz, G., and Schmidt, R. (2001). Comparative sequence analysis reveals extensive microcolinearity in the lateral suppressor regions of the tomato, Arabidopsis, and Capsella genomes. *The Plant Cell Online* Vol. 13, No. 4, pp. 979, ISSN 1040-4651.

Rostoks, N., Ramsay, L., MacKenzie, K., Cardle, L., Bhat, P. R., Roose, M. L., Svensson, J. T., Stein, N., Varshney, R. K., and Marshall, D. F. (2006). Recent history of artificial outcrossing facilitates whole-genome association mapping in elite inbred crop varieties. *Proceedings of the National Academy of Sciences* Vol. 103, No. 49, pp. 18656, ISSN 0027-8424.

Saito, K., Hayano-Saito, Y., Maruyama-Funatsuki, W., Sato, Y., and Kato, A. (2004). Physical mapping and putative candidate gene identification of a quantitative trait locus Ctb1 for cold tolerance at the booting stage of rice. *TAG Theoretical and Applied Genetics* Vol. 109, No. 3, pp. 515-522, ISSN 0040-5752.

Salse, J., Bolot, S., Throude, M., Jouffe, V., Piegu, B., Quraishi, U. M., Calcagno, T., Cooke, R., Delseny, M., and Feuillet, C. (2008). Identification and characterization of shared duplications between rice and wheat provide new insight into grass genome evolution. *The Plant Cell Online* Vol. 20, No. 1, pp. 11, ISSN 1040-4651.

Salvi, S., and Tuberosa, R. (2005). To clone or not to clone plant QTLs: present and future challenges. *Trends in Plant Science* Vol. 10, No. 6, pp. 297-304, ISSN 1360-1385.

Salvi, S., and Tuberosa, R. (2007). Cloning QTLs in plants. *Genomics-assisted crop improvement* Vol., No., pp. 207-225.

Sanchez, D. H., Pieckenstain, F. L., Szymanski, J., Erban, A., Bromke, M., Hannah, M. A., Kraemer, U., Kopka, J., and Udvardi, M. K. (2011). Comparative Functional Genomics of Salt Stress in Related Model and Cultivated Plants Identifies and Overcomes Limitations to Translational Genomics. *PLoS ONE* Vol. 6, No. 2, pp. e17094, ISSN 1932-6203.

Sato, S., Nakamura, Y., Kaneko, T., Asamizu, E., Kato, T., Nakao, M., Sasamoto, S., Watanabe, A., Ono, A., and Kawashima, K. (2008). Genome structure of the legume, Lotus japonicus. *DNA research* Vol. 15, No. 4, pp. 227, ISSN 1340-2838.

Schauer, N., Semel, Y., Roessner, U., Gur, A., Balbo, I., Carrari, F., Pleban, T., Perez-Melis, A., Bruedigam, C., and Kopka, J. (2006). Comprehensive metabolic profiling and phenotyping of interspecific introgression lines for tomato improvement. *Nature biotechnology* Vol. 24, No. 4, pp. 447-454, ISSN 1087-0156.

Schneeberger, K., and Weigel, D. (2011). Fast-forward genetics enabled by new sequencing technologies. *Trends in Plant Science* Vol., No., ISSN 1360-1385.

Schranz, M. E., Lysak, M. A., and Mitchell-Olds, T. (2006). The ABC's of comparative genomics in the Brassicaceae: building blocks of crucifer genomes. *Trends in Plant Science* Vol. 11, No. 11, pp. 535-542, ISSN 1360-1385.

Schulze, A., and Downward, J. (2001). Navigating gene expression using microarrays-a technology review. *Nature Cell Biology* Vol. 3, No. 8, pp. 190-195, ISSN 1465-7392.

Schulze, T. G., and McMahon, F. J. (2002). Genetic association mapping at the crossroads: which test and why? Overview and practical guidelines. *American journal of medical genetics* Vol. 114, No. 1, pp. 1-11, ISSN 1096-8628.

Shinozuka, H., Cogan, N. O. I., Smith, K. F., Spangenberg, G. C., and Forster, J. W. (2010). Fine-scale comparative genetic and physical mapping supports map-based cloning strategies for the self-incompatibility loci of perennial ryegrass (Lolium perenne L.). *Plant molecular biology* Vol. 72, No. 3, pp. 343-355, ISSN 0167-4412.

Shulaev, V., Sargent, D. J., Crowhurst, R. N., Mockler, T. C., Folkerts, O., Delcher, A. L., Jaiswal, P., Mockaitis, K., Liston, A., Mane, S. P., Burns, P., Davis, T. M., Slovin, J. P., Bassil, N., Hellens, R. P., Evans, C., Harkins, T., Kodira, C., Desany, B., Crasta, O. R., Jensen, R. V., Allan, A. C., Michael, T. P., Setubal, J. C., Celton, J.-M., Rees, D. J. G., Williams, K. P., Holt, S. H., Rojas, J. J. R., Chatterjee, M., Liu, B., Silva, H., Meisel, L., Adato, A., Filichkin, S. A., Troggio, M., Viola, R., Ashman, T.-L., Wang, H., Dharmawardhana, P., Elser, J., Raja, R., Priest, H. D., Bryant, D. W., Fox, S. E., Givan, S. A., Wilhelm, L. J., Naithani, S., Christoffels, A., Salama, D. Y., Carter, J., Girona, E. L., Zdepski, A., Wang, W., Kerstetter, R. A., Schwab, W., Korban, S. S., Davik, J., Monfort, A., Denoyes-Rothan, B., Arus, P., Mittler, R., Flinn, B., Aharoni, A., Bennetzen, J. L., Salzberg, S. L., Dickerman, A. W., Velasco, R., Borodovsky, M., Veilleux, R. E., and Folta, K. M. (2011). The genome of woodland strawberry (Fragaria vesca). *Nat Genet* Vol. 43, No. 2, pp. 109-116, ISSN 1061-4036.

Sindhu, A. S., Maier, T. R., Mitchum, M. G., Hussey, R. S., Davis, E. L., and Baum, T. J. (2009). Effective and specific in planta RNAi in cyst nematodes: expression interference of four parasitism genes reduces parasitic success. *Journal of experimental botany* Vol. 60, No. 1, pp. 315, ISSN 0022-0957.

Spollen, W., Tao, W., Valliyodan, B., Chen, K., Hejlek, L., Kim, J. J., LeNoble, M., Zhu, J., Bohnert, H., and Henderson, D. (2008). Spatial distribution of transcript changes in the maize primary root elongation zone at low water potential. *BMC plant biology* Vol. 8, No. 1, pp. 32, ISSN 1471-2229.

Springer, N. M., Ying, K., Fu, Y., Ji, T., Yeh, C. T., Jia, Y., Wu, W., Richmond, T., Kitzman, J., and Rosenbaum, H. (2009). Maize inbreds exhibit high levels of copy number variation (CNV) and presence/absence variation (PAV) in genome content. *PLoS genetics* Vol. 5, No. 11, pp. e1000734, ISSN 1553-7404.

Steemers, F. J., and Gunderson, K. L. (2007). Whole genome genotyping technologies on the BeadArray™ platform. *Biotechnology journal* Vol. 2, No. 1, pp. 41-49, ISSN 1860-7314.

Stich, B., and Melchinger, A. (2010). An introduction to association mapping in plants. *CAB Reviews* Vol..

Takahashi, W., Miura, Y., Sasaki, T., and Takamizo, T. (2010). Targeted mapping of rice ESTs to the LmPi1 locus for grey leaf spot resistance in Italian ryegrass. *European journal of plant pathology* Vol. 126, No. 3, pp. 333-342, ISSN 0929-1873.

Tang, J., Leunissen, J. A. M., Voorrips, R. E., Van Der Linden, C. G., and Vosman, B. (2008). HaploSNPer: a web-based allele and SNP detection tool. *BMC genetics* Vol. 9, No. 1, pp. 23, ISSN 1471-2156.

Tanksley, S. D., Ganal, M. W., and Martin, G. B. (1995). Chromosome landing: a paradigm for map-based gene cloning in plants with large genomes. *Trends in Genetics* Vol. 11, No. 2, pp. 63-68, ISSN 0168-9525.

Tian, F., Bradbury, P. J., Brown, P. J., Hung, H., Sun, Q., Flint-Garcia, S., Rocheford, T. R., McMullen, M. D., Holland, J. B., and Buckler, E. S. (2011). Genome-wide association study of leaf architecture in the maize nested association mapping population. *Nature genetics* Vol., No., ISSN 1061-4036.

Trebbi, D., Maccaferri, M., de Heer, P., Sørensen, A., Giuliani, S., Salvi, S., Sanguineti, M. C., Massi, A., van der Vossen, E. A. G., and Tuberosa, R. (2011). High-throughput SNP discovery and genotyping in durum wheat (Triticum durum Desf.). *TAG Theoretical and Applied Genetics* Vol., No., pp. 1-15, ISSN 0040-5752.

Tuberosa, R., and Salvi, S. (2006). Genomics-based approaches to improve drought tolerance of crops. *Trends in Plant Science* Vol. 11, No. 8, pp. 405-412, ISSN 1360-1385.

Valouev, A., Ichikawa, J., Tonthat, T., Stuart, J., Ranade, S., Peckham, H., Zeng, K., Malek, J. A., Costa, G., and McKernan, K. (2008). A high-resolution, nucleosome position map of C. elegans reveals a lack of universal sequence-dictated positioning. *Genome Res* Vol. 18, No. 7, pp. 1051, ISSN 1088-9051.

Van Orsouw, N. J., Hogers, R. C. J., Janssen, A., Yalcin, F., Snoeijers, S., Verstege, E., Schneiders, H., Van Der Poel, H., Van Oeveren, J., and Verstegen, H. (2007). Complexity reduction of polymorphic sequences (CRoPS™): a novel approach for large-scale polymorphism discovery in complex genomes. *PLoS ONE* Vol. 2, No. 11, pp. e1172, ISSN 1932-6203.

Varshney, R. K., Graner, A., and Sorrells, M. E. (2005). Genomics-assisted breeding for crop improvement. *Trends in Plant Science* Vol. 10, No. 12, pp. 621-630, ISSN 1360-1385.

Varshney, R. K., Nayak, S. N., May, G. D., and Jackson, S. A. (2009). Next-generation sequencing technologies and their implications for crop genetics and breeding. *Trends in Biotechnology* Vol. 27, No. 9, pp. 522-530, ISSN 0167-7799.

Velasco, R., Zharkikh, A., Affourtit, J., Dhingra, A., Cestaro, A., Kalyanaraman, A., Fontana, P., Bhatnagar, S. K., Troggio, M., and Pruss, D. (2010). The genome of the domesticated apple (Malus [times] domestica Borkh.). *Nature genetics* Vol. 42, No. 10, pp. 833-839, ISSN 1061-4036.

Velasco, R., Zharkikh, A., Troggio, M., Cartwright, D. A., Cestaro, A., Pruss, D., Pindo, M., FitzGerald, L. M., Vezzulli, S., and Reid, J. (2007). A high quality draft consensus sequence of the genome of a heterozygous grapevine variety. *PLoS ONE* Vol. 2, No. 12, pp. e1326, ISSN 1932-6203.

Venter, J. C., Adams, M. D., Myers, E. W., Li, P. W., Mural, R. J., Sutton, G. G., Smith, H. O., Yandell, M., Evans, C. A., Holt, R. A., Gocayne, J. D., Amanatides, P., Ballew, R. M., Huson, D. H., Wortman, J. R., Zhang, Q., Kodira, C. D., Zheng, X. H., Chen, L., Skupski, M., Subramanian, G., Thomas, P. D., Zhang, J., Gabor Miklos, G. L., Nelson, C., Broder, S., Clark, A. G., Nadeau, J., McKusick, V. A., Zinder, N., Levine, A. J., Roberts, R. J., Simon, M., Slayman, C., Hunkapiller, M., Bolanos, R., Delcher, A., Dew, I., Fasulo, D., Flanigan, M., Florea, L., Halpern, A., Hannenhalli, S., Kravitz, S., Levy, S., Mobarry, C., Reinert, K., Remington, K., Abu-Threideh, J., Beasley, E., Biddick, K., Bonazzi, V., Brandon, R., Cargill, M., Chandramouliswaran, I., Charlab, R., Chaturvedi, K., Deng, Z., Francesco, V. D., Dunn, P., Eilbeck, K., Evangelista, C., Gabrielian, A. E., Gan, W., Ge, W., Gong, F., Gu, Z., Guan, P., Heiman, T. J., Higgins, M. E., Ji, R.-R., Ke, Z., Ketchum, K. A., Lai, Z., Lei, Y., Li, Z., Li, J., Liang, Y., Lin, X., Lu, F., Merkulov, G. V., Milshina, N., Moore, H. M., Naik, A. K., Narayan, V. A., Neelam, B., Nusskern, D., Rusch, D. B.,

Salzberg, S., Shao, W., Shue, B., Sun, J., Wang, Z. Y., Wang, A., Wang, X., Wang, J., Wei, M.-H., Wides, R., Xiao, C., Yan, C., et al. (2001). The Sequence of the Human Genome. *Science* Vol. 291, No. 5507, pp. 1304-1351.

Vos, P., Hogers, R., Bleeker, M., Reijans, M., Lee, T., Hornes, M., Friters, A., Pot, J., Paleman, J., and Kuiper, M. (1995). AFLP: a new technique for DNA fingerprinting. *Nucleic Acids Res* Vol. 23, No. 21, pp. 4407, ISSN 0305-1048.

Wang, D., Amornsiripanitch, N., and Dong, X. (2006). A genomic approach to identify regulatory nodes in the transcriptional network of systemic acquired resistance in plants. *PLoS Pathogens* Vol. 2, No. 11, pp. e123, ISSN 1553-7374.

Wang, D. G., Fan, J. B., Siao, C. J., Berno, A., Young, P., Sapolsky, R., Ghandour, G., Perkins, N., Winchester, E., and Spencer, J. (1998). Large-scale identification, mapping, and genotyping of single-nucleotide polymorphisms in the human genome. *Science* Vol. 280, No. 5366, pp. 1077, ISSN 0036-8075.

Wang, J., McClean, P. E., Lee, R., Goos, R. J., and Helms, T. (2008). Association mapping of iron deficiency chlorosis loci in soybean (Glycine max L. Merr.) advanced breeding lines. *TAG Theoretical and Applied Genetics* Vol. 116, No. 6, pp. 777-787, ISSN 0040-5752.

Wang, S., and Liu, Z. (2011). SNP Discovery through EST Data Mining. *In* "Next Generation Sequencing and Whole Genome Selection in Aquaculture", pp. 91-108. Wiley-Blackwell.

Weber, J. L., and May, P. E. (1989). Abundant class of human DNA polymorphisms which can be typed using the polymerase chain reaction. *American Journal of Human Genetics* Vol. 44, No. 3, pp. 388.

Weiss, K. M., and Clark, A. G. (2002). Linkage disequilibrium and the mapping of complex human traits. *Trends in Genetics* Vol. 18, No. 1, pp. 19-24, ISSN 0168-9525.

Welsh, J., and McClelland, M. (1990). Fingerprinting genomes using PCR with arbitrary primers. *Nucleic Acids Res* Vol. 18, No. 24, pp. 7213, ISSN 0305-1048.

Wen, T. J., Hochholdinger, F., Sauer, M., Bruce, W., and Schnable, P. S. (2005). The roothairless1 gene of maize encodes a homolog of sec3, which is involved in polar exocytosis. *Plant physiology* Vol. 138, No. 3, pp. 1637, ISSN 0032-0889.

Wilhelm, B. T., and Landry, J. R. (2009). RNA-Seq--quantitative measurement of expression through massively parallel RNA-sequencing. *Methods* Vol. 48, No. 3, pp. 249-257, ISSN 1046-2023.

Williams, J. G. K., Kubelik, A. R., Livak, K. J., Rafalski, J. A., and Tingey, S. V. (1990). DNA polymorphisms amplified by arbitrary primers are useful as genetic markers. *Nucleic Acids Res* Vol. 18, No. 22, pp. 6531, ISSN 0305-1048.

Woll, K., Borsuk, L. A., Stransky, H., Nettleton, D., Schnable, P. S., and Hochholdinger, F. (2005). Isolation, characterization, and pericycle-specific transcriptome analyses of the novel maize lateral and seminal root initiation mutant rum1. *Plant physiology* Vol. 139, No. 3, pp. 1255, ISSN 0032-0889.

Wood, P. J. (2007). Cereal [beta]-glucans in diet and health. *Journal of Cereal Science* Vol. 46, No. 3, pp. 230-238, ISSN 0733-5210.

Wright, S. I., Bi, I. V., Schroeder, S. G., Yamasaki, M., Doebley, J. F., McMullen, M. D., and Gaut, B. S. (2005). The effects of artificial selection on the maize genome. *Science* Vol. 308, No. 5726, pp. 1310, ISSN 0036-8075.

Wu, C., Trieu, A., Radhakrishnan, P., Kwok, S. F., Harris, S., Zhang, K., Wang, J., Wan, J., Zhai, H., and Takatsuto, S. (2008). Brassinosteroids regulate grain filling in rice. *The Plant Cell Online* Vol. 20, No. 8, pp. 2130, ISSN 1040-4651.

Xie, W., Feng, Q., Yu, H., Huang, X., Zhao, Q., Xing, Y., Yu, S., Han, B., and Zhang, Q. (2010). Parent-independent genotyping for constructing an ultrahigh-density linkage map based on population sequencing. *Proceedings of the National Academy of Sciences* Vol. 107, No. 23, pp. 10578.

Xinguo, L., Harry, W., and Simon, S. (2010). Comparative genomics reveals conservative evolution of the xylem transcriptome in vascular plants. *BMC Evolutionary Biology* Vol. 10, No., ISSN 1471-2148.

Xu, Y., and Crouch, J. H. (2008). Marker-assisted selection in plant breeding: from publications to practice. *Crop Science* Vol. 48, No. 2, pp. 391–407.

Yamaguchi-Shinozaki, K., and Shinozaki, K. (2005). Organization of cis-acting regulatory elements in osmotic-and cold-stress-responsive promoters. *Trends in Plant Science* Vol. 10, No. 2, pp. 88-94, ISSN 1360-1385.

Yang, B., Srivastava, S., Deyholos, M. K., and Kav, N. N. V. (2007). Transcriptional profiling of canola (Brassica napus L.) responses to the fungal pathogen Sclerotinia sclerotiorum. *Plant Science* Vol. 173, No. 2, pp. 156-171, ISSN 0168-9452.

Young, N. D., and Udvardi, M. (2009). Translating Medicago truncatula genomics to crop legumes. *Current opinion in plant biology* Vol. 12, No. 2, pp. 193-201, ISSN 1369-5266.

Yu, J., Holland, J. B., McMullen, M. D., and Buckler, E. S. (2008). Genetic design and statistical power of nested association mapping in maize. *Genetics* Vol. 178, No. 1, pp. 539, ISSN 0016-6731.

Yuan, J. S., Galbraith, D. W., Dai, S. Y., Griffin, P., and Stewart Jr, C. N. (2008). Plant systems biology comes of age. *Trends in Plant Science* Vol. 13, No. 4, pp. 165-171, ISSN 1360-1385.

Yuan, Y., SanMiguel, P. J., and Bennetzen, J. L. (2003). High Cot sequence analysis of the maize genome. *The Plant Journal* Vol. 34, No. 2, pp. 249-255, ISSN 1365-313X.

Zhang, H., Guan, H., Li, J., Zhu, J., Xie, C., Zhou, Y., Duan, X., Yang, T., Sun, Q., and Liu, Z. (2010). Genetic and comparative genomics mapping reveals that a powdery mildew resistance gene Ml3D232 originating from wild emmer co-segregates with an NBS-LRR analog in common wheat (Triticum aestivum L.). *TAG Theoretical and Applied Genetics* Vol., No., pp. 1-9, ISSN 0040-5752.

Zhang, J., Lu, Y., Yuan, Y., Zhang, X., Geng, J., Chen, Y., Cloutier, S., McVetty, P. B. E., and Li, G. (2009). Map-based cloning and characterization of a gene controlling hairiness and seed coat color traits in Brassica rapa. *Plant molecular biology* Vol. 69, No. 5, pp. 553-563, ISSN 0167-4412.

Zhang, Z., Guo, X., Liu, B., Tang, L., and Chen, F. (2011). Genetic diversity and genetic relationship of Jatropha curcas between China and Southeast Asian revealed by amplified fragment length polymorphisms. *African Journal of Biotechnology* Vol. 10, No. 15, pp. 2825-2832, ISSN 1684-5315.

Zheng, B., Yang, L., Zhang, W., Mao, C., Wu, Y., Yi, K., Liu, F., and Wu, P. (2003). Mapping QTLs and candidate genes for rice root traits under different water-supply conditions and comparative analysis across three populations. *TAG Theoretical and Applied Genetics* Vol. 107, No. 8, pp. 1505-1515, ISSN 0040-5752.

Zheng, P., Allen, W. B., Roesler, K., Williams, M. E., Zhang, S., Li, J., Glassman, K., Ranch, J., Nubel, D., and Solawetz, W. (2008). A phenylalanine in DGAT is a key determinant of oil content and composition in maize. *Nature genetics* Vol. 40, No. 3, pp. 367-372, ISSN 1061-4036.

Zhu, C. G., Buckler, M., and Yu, E. S. (2008). Status and prospects of association mapping in plants. *The Plant Genome* Vol. 1, No. 1, pp. 5, ISSN 1940-3372.

Zieler, H., Richardson, T., Schwartz, A., Herrgard, M., Lomelin, D., Mathur, E., Cheah, S., Tee, T., Lee, W., and Chua, K. (2010). Whole-Genome shotgun sequencing of the oil palm and Jatropha genomes. pp. 9-13.

# 8

# A Multiplex Fluorescent PCR Assay in Molecular Breeding of Oilseed Rape

Katarzyna Mikolajczyk[1], Iwona Bartkowiak-Broda[1], Wieslawa Poplawska[1],
Stanislaw Spasibionek[1], Agnieszka Dobrzycka[1] and Miroslawa Dabert[2]
*[1]Plant Breeding and Acclimatization Institute – National Research Institute,*
*Research Division in Poznan,*
*[2]Molecular Biology Techniques Laboratory, Faculty of Biology,*
*Adam Mickiewicz University in Poznan*
*Poland*

## 1. Introduction

### 1.1 Oilseed rape as an important oil crop

Oilseed rape (*Brassica napus* L. var *oleifera*) is the second-most important oil crop in the world and it is a predominant one in Europe, with respect to seed oil production. The current seed yield of almost 60 metric tones (MT) makes above 13% of the world oilseeds production. Due to oil crop market demands, rapeseed oil production permanently increases, not only for nutritional purposes but also for biodiesel production, according to the promoting the development of renewable energy European Commission Directives. The EU-27 countries are the most important producers of oilseed rape, with the leading contributors, such as Germany, France, Poland, Great Britain and the Czech Republic. The other important oilseed rape producers are: China, Canada, India, and Ukraine. *B. napus* is an allotetraploid (amphidiploid) species with an AACC genome (2n = 38), which is derived from ancestral genomes of turnip, *B. rapa* syn. *campestris* (AA, 2n = 20) and cabbage, *B. oleracea* (CC, 2n = 18), according to the „Triangle of U" (U, 1935). The *B. napus* haploid genome (AC) consists of 19 chromosomes deriving from *B. rapa* (fom A1 to A10) and from *B. oleracea* (C1 to C9) (http//www.brassica.info.resource/maps/lg-assignments.php).

Seeds of oilseed rape are a valuable source of oil (45% of seed mass) and protein (20%). The discovery of the zero erucic acid (C22:1) lines in spring fodder variety Liho (Steffansson et al., 1961; Stefansson & Hougen, 1964) and low glucosinolates content in Polish spring variety Bronowski (Downey & Roebbelen, 1989; Krzymanski, 1968, 1970) were crucial milestones in oilseed rape breeding for seed yield quality. As a result of over fifty years of intensive breeding, superior cultivars with no erucic acid (C22:1) content in seed oil and with a very low glucosinolates content in seed meal have been developed and introduced into production. Those cultivars were named as double-low, double-zero (00), or canola (canola-type) ones. Oil of double-low cultivars is characterized by low content of saturated fatty acids and relatively high amount of C18 unsaturated fatty acids with 2:1 linoleic (C18:2) to linolenic (C18:3) acid ratio (Table 1). In addition, the presence of natural antioxidants (tocopherols) makes this oil an optimal and universal component of human diet used as salad oil, for salad dressing, short deep frying and margarine production (Snowdon et al., 2007). For nonfood purposes, canola oil may be used as a raw material for methyl ester (biodiesel) production, industrial lubricants, surface active agents for detergent and soap production, as well as for biodegradable plastics (Snowdon et al., 2007).

| Type of oilseed rape | Fatty acid content [%] | | | | | | |
|---|---|---|---|---|---|---|---|
| | Saturated | Oleic | Linoleic | Linolenic | Eicosenic | Erucic | other |
| | (C16:0 + C18:0) | (C18:1) | (C18:2) | (C18:3) | (C20:1) | (C22:1) | |
| High erucic/ traditional | 4 | 11 | 12 | 9 | 8 | 52 | 4 |
| 00/ canola | 6 | 60 | 21 | 10 | 1 | 1 | 1 |
| Low linolenic | 6 | 61 | 28 | 3 | 1 | - | 1 |
| HOLL | 5 | 84 | 5 | 3 | 1 | - | 2 |

Table 1. *B. napus* seed oil fatty acid composition (according to Wittkop et al., 2009)

## 1.2 The main breeding goals for oilseed rape

The C18:1 oleic acid is thermostable and appropriate for deep frying. The C18:2 linoleic acid with two double bonds provides nutritional benefits, whereas the C18:3 linolenic acid with three double bonds leads to instability and rapid oxidation. This reduces the shelf life of products (Barker et al., 2007, and references therein). Therefore, reduced level of polyunsaturated fatty acids, especially C18:3 linolenic acid, and increased content of monounsaturated C18:1 oleic acid provide higher oil stability. According to the demands of oil crop market, the development of high oleic (HO) and low linolenic (LL) cultivars is one of the major breeding goals. LL mutant of spring oilseed rape, M11 was obtained by ethyl methanesulfonate (EMS) treatment of the Canadian cultivar Oro (Rakow, 1973; Roebbelen & Nitsch, 1975). Subsequently, low linolenic cultivars Stellar (Scarth et al., 1988) and Apollo (Scarth et al., 1995) were developed as a result of recombinant breeding of the M11 mutant line. Canola mutant inbred lines with high oleic (≥75%) at the expense of polyunsaturated fatty acids (≤6%) were developed by Auld et al. (1992). Another *B. napus* breeding line with modified fatty acid composition is the Dow AgroScience (DAS) proprietary HOLL (high oleic and low linolenic) mutant line DMS100 derived from the line AG019 (Hu et al., 2006, and references therein). New winter canola oilseed rape mutant lines were selected by Spasibionek (2003) and used for development of stable inbred lines with high oleic (≥75%) and low linolenic (≤3%) acid content (Spasibionek 2006; 2008). High oleic canola lines (75%-85%) were described by Falentin et al. (2007).

## 1.3 Hybrid breeding methods and molecular markers for oilseed rape hybrid breeding programs

In major rapeseed growing areas, hybrids represent an increasing proportion of the registered and cultivated varieties (Wittkop et al., 2009). In Europe, the oilseed rape hybrid breeding is based mainly on two male sterility systems: the ogura-INRA CMS (cytoplasmic male sterility) and the MSL-NPZ Lembke genic male sterility, whereas the *ogura* system is characterized by stable expression of male sterility in different genetic backgrounds and under different environmental conditions.

In order to improve the poor agronomic value of new breeding materials with changed fatty acid composition developed by Spasibionek (2006) and to increase their seed and oil yield, they were introduced into new genetic background by crosses with high yielding cultivars and lines. Moreover, they were implemented into hybrid breeding, in which pollination controlling cytoplasmic male sterility (CMS) systems including male sterile cytoplasm and an appropriate restoring male fertility gene are used to produce F1 hybrid seeds. The new mutant lines were crossed with F1 hybrid components, *i. e.* the male-sterile and the restorer lines, in order to develop high-yielding single-cross hybrids with the desired traits.

An effective CMS system used for oilseed rape F1 hybrid seed production on commercial scale is the alloplasmic *ogura* radish CMS which completely ensures cross-pollination (Bartkowiak-Broda et al., 1979). It was originally found in radish (*Raphanus sativus* L.) by Ogura (1968) and transferred to *B.oleracea* and *B.napus* by interspecific crosses (Bannerot et al., 1974). Male-sterile *B.napus* cybrids were then produced throughout protoplast fusion (Pelletier et al., 1983) to generate male sterile lines with minor defects (Pelletier et al., 1987). *Ogura* CMS oilseed rape plants have phenotypically distinctive flowers with underdeveloped anthers. On molecular level, it is a result of the expression of mitochondrial locus *orf138* that is present in male sterile and absent in male fertile normal plant revealed by physical mapping studies (Bonhomme et al., 1992; Krishnasamy & Makaroff, 1993). Primers specific for 5′ and 3′ ends of the *orf138* nucleotide sequences (Krishnasamy & Makaroff, 1993) were used for PCR-based identification of the *ogura* CMS cytoplasm during the fusion experiments of leaf protoplasts from fertile cabbage and cold-tolerant *ogura* CMS broccoli lines (Sigareva & Earle, 1997). The *orf138*-specific primer pair was applied by our group for monitoring of the *ogura* CMS cytoplasm in *B. napus* breeding programs (Fig. 1, panel „CMS″) (Mikolajczyk et al., 1998).

To obtain hybrid seeds, nuclear fertility restorer genes are required, which are present in native CMS-restorer systems. In turn, for identifying the *Rfo* restorer gene, the 1 kb SCAR (sequence characterized amplified region) marker, which we named as "C02" (Fig. 1, panel „Rfo″) (Mikolajczyk et al, 2008) was developed by conversion of the OPC02$_{1150}$ RAPD (random amplified polymorphic DNA) marker tightly linked to the *Rfo* gene (Delourme et

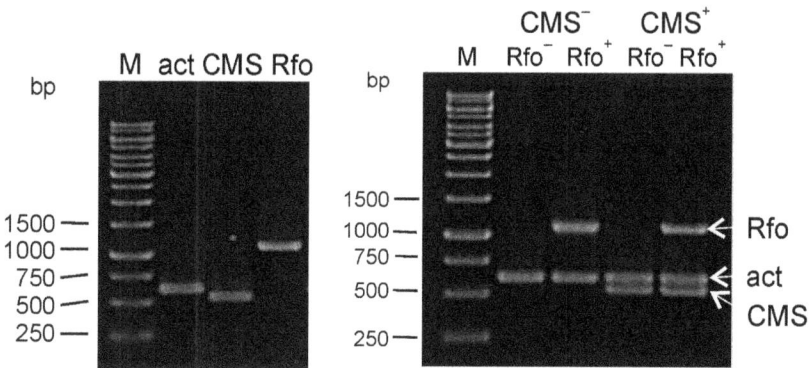

Fig. 1. Amplification of actin internal standard (act) and the *Rfo* and *ogura* CMS SCAR markers by separate (*on the left*) and multiplex (*right*) PCRs (Mikolajczyk et al., 2010a).

al., 1994). Both SCAR markers were applied for identification of the *ogura* male-sterile cytoplasm and the *Rfo* gene in *B. napus* F hybrid components, F hybrids, as well as among the *ogura* CMS and *Rfo* restorer recombinants obtained as a result of crosses with high yielding and stress-resistant cultivars. The use of those markers proved to be very useful, due to phenotypic identity of F1 hybrids, *Rfo* lines and *Rfo* recombinants, as well as the possibility of genotyping plants at the early stages of plant development.

To improve the effectiveness of the method and to reduce the costs, the multiplex PCR method was applied (Fig. 1, on the right) based on simultaneous amplification of both SCAR markers with an internal standard, a 600 bp conservative region of an actin 7 gene fragment (Figure 1, panel „act") (Mikolajczyk et al., 2010a).

The low linolenic mutant genotypes were monitored with the use of the developed SNaPshot assay (Mikolajczyk et al., 2010b), detecting wild-type and mutant alleles of the *FAD3* desaturase genes in the AC allotetraploid genome of *B. napus*. The *FAD3* genes encode for encoplasmic delta-15 linoleate desaturase responsible for desaturation of linoleic acid (C18:2) into linolenic acid. As a result of cloning and sequencing of *FAD3* genes from wild-type and LL mutant *B. napus* plants, we reported two point mutations (*BnaA.FAD3* and *BnaC.FAD3*) responsible for disruption of the *FAD3* genes expression and function (Mikolajczyk et al., 2010b) in the new LL mutant rapeseed line (Spasibionek, 2006). One point mutation comprised a C to T transition in the mutant *bnaA.fad3* gene leading to a possible Arg to Cys substitution. Another is a G to A transition in the 5′ donor splice site of the mutant *bnaC.fad3* gene disrupting intron 6th splicing. We developed genetic markers for monitoring *FAD3* alleles in breeding programs. The detection of wild-type and mutant *FAD3* alleles comprises two steps: independent PCR amplification of short SNP fragments and a detection of the SNPs based on microsequencing method (SNaPshot) with the use of allele-specific primers (Mikolajczyk et al., 2010b). The SNaPshot assay enabled precise and unambiguous detection of this allelic variability.

The developed multiplex PCR detecting the *ogura* CMS and the *Rfo* restorer gene along with the SNaPshot analysis for monitoring wild-type and mutant *FAD3* alleles have been very useful for the precise determining of almost 700 of individual plants. This helped to select desired genotypes for further breeding of new high-yielding lines with changed fatty acid composition. With the use of molecular markers the selection process is more time- and cost-effective.

Despite their usefulness, using both assays separately may generate errors. The analysis of a large number of individuals in independent assays increases the costs as well. To make the genotyping analysis more effective, we developed a new fluorescent multiplex PCR combined with SNaPshot detection for identification of the *Rfo* restorer gene, the *ogura* CMS, and the wild-type and mutant low linolenic genotypes in one assay. This new method is easy to adapt to high-throughput genotyping.

## 2. Material and methods

### 2.1 Plant material

The plant material used in this study were *B. napus* cultivars, recombinant and mutant lines, as well as the *ogura* CMS system F1 hybrids and F1 hybrid components, developed at the Plant Breeding and Acclimatization Institute – National Research Institute (NRI) in Poznan,

| Cat. | Line | Parent(s) | Generation | Number of plants | Owned by/ *ref. |
|---|---|---|---|---|---|
| 1. | Recombinant inbred lines | LL M681 (PN1712) and *Rfo* line PN 5-4 | F3 (*Rfo* x LL M681) | 10 | PBAI-NRI |
| 2. | Recombinant inbred lines | HO M10464 (PN1704), LL M681 (PN1712) and *Rfo* line PN 5-4 | F3 (LL M681 x HO M 10464) x *Rfo* | 34 | PBAI-NRI |
| 3. | Recombinant lines | *ogura* CMS DH line 66-64-68/05 and DH LL M681 (DH219) | LL *ogura* CMS F1 hybrid component | 6 | PBAI-NRI |
| 4. | Recombinant lines | DH *Rfo* line PN544 and DH LL M681 (DH219) | LL *Rfo* F1 hybrid component | 7 | PBAI-NRI |
| 5. | *ogura* CMS line | MS120 | multiplication | 1 | PBAI-NRI |
| 6. | *ogura* CMS line | CMS PN66 | multiplication | 4 | PBAI-NRI |
| 7. | *Rfo* DH line PN5/4 | BO 20-48 | DH | 1 | PBAI-NRI |
| 8. | *Rfo* DH line PN492 | *Rfo* DH line PN17-5 | DH | 1 | PBAI-NRI |
| 9. | DH *Rfo* lines: 337DHR2 and 345DHR2 | no description | DH | 2 | PBAI-NRI |
| 10. | *Rfo* line PN17/8 | *Rfo* line PN17-5 | multiplication | 1 | PBAI-NRI |
| 11. | *Rfo* line R44/3i/07 | no description | multiplication | 1 | PBAI-NRI |
| 12. | HO *Rfo* recombinant line PN1280 | *Rfo* DH line PN544 and HO line PN2185 | *Rfo* PN544 x HO PN2185 | 1 | PBAI-NRI |
| 13. | HOLL-type new mutant DH line 321-2 | canola-type line PN5282 | mutagenesis | 1 | *Spasibionek 2008 |
| 14. | LL mutant DH lines: 1044/2 and 1050/6 | LL M681 (PN1712) | DH | 2 | PBAI-NRI |
| 15. | HO mutant DH lines: 1704/5 and 1704/60 | HO M10464 (PN 1704) | DH | 2 | PBAI-NRI |
| 16. | Recombinant line A2/17 | LL cultivar. Apollo and canola-type line PN1775 | Apollo x PN1775 | 1 | PBAI-NRI |
| 17. | F1 hybrid cultivar. Poznaniak | no description | F1 hybrid | 1 | PBC Strzelce-Borowo Ltd. |
| 18. | F1 hybrid line PN600 | CMS PN66 and *Rfo* DH line PN5-4 | F1 hybrid | 1 | PBAI-NRI |
| 19. | F1 hybrid line PN594 | CMS PN64 and *Rfo* DH line PN17-5 | F1 hybrid | 1 | PBAI-NRI |
| test-1 | *Rfo* and *ogura* CMS lines | no description | no description | 113 | PBC Smolice Ltd. |
| test-2 | *Rfo* and *ogura* CMS lines | no description | no description | 507 | PBC Strzelce Ltd. |

Table 2. Plant material used in this study. "Cat." – plant category; "PBAI-NRI" – Plant Breeding and Acclimatization Institute - National Research Institute, Research Division in Poznan, Poland; "PBC" – Plant Breeding Company.

Poland and also at the Plant Breeding Company Ltd. Strzelce – Division at Borowo and at the Plant Breeding Company Ltd. Smolice – Division at Bakow, as it is presented in the Table 2. In total, 698 individual plants of different genetic background were analyzed.

## 2.2 Molecular methods

### 2.2.1 DNA extraction

Total genomic DNA was prepared from plant leaves, according to the CTAB extraction method (Doyle & Doyle, 1990). For one sample, approximately 50–100 mg of young leaf tissue was put into a 1.5 ml tube and ground thoroughly with a teflon pestle in 0.75 ml of 7.5 pH washing buffer containing: sorbitol 0.5 M, Tris 0.1 M, $Na_2EDTA$ 0.07 M, and $Na_2S_2O_3$ 0.02 M. Then, following centrifugation of the suspension (at 12 000 x g for 2 min.), the supernatant was removed and the washed pellet was resuspended in 0.75 ml of the CTAB buffer (8.0 pH Tris HCl 0.1 M, NaCl 1.4 M, CTAB 2%, $Na_2EDTA$ 0.02 M, PVP 40 000 1%) for 0.5 h extraction at 65 °C. Subsequently, the equal volume of chlorophorm/octanol (24:1) solution was added and the suspension was shaken gently for 10 min. The aqueous and organic phases were separated by centrifugation at 12 000 x g for 10 min., then the aqueous phase was put into a fresh tube and nucleic acids were precipitated with 2/3 volume of isoporpanol. After centrifugation, the supernatant was removed and the pellet was air-dried. Then, 210 µl of RNase A solution (40 µg/ ml) was added and after 1 h of incubation at 37 °C, DNA was precipitated with 2/3 volume of isopropanol in the presence of 0.9 M NaCl. After centrifugation and removing of the supernatant, the pellet was washed with 70% etanol for 15 min. Then, the ethanol was removed and the DNA sample was air-dried and resuspended in approximately 100 µl of TE (10; 0.1) buffer containing 8.0 pH Tris 0.01 M and 8.0 pH $Na_2EDTA$ 0.0001 M. The obtained DNA sample contained about 10 ng of DNA in 1 µl of solution.

### 2.2.2 Fluorescent multiplex-PCR for *Rfo* and *ogura* CMS markers

Primers for amplification of shortened fragments of the *Rfo* restorer, the *ogura* CMS, and *B. napus actin 7* genes were developed based on nucleotide sequences of PCR products generated with primers designed previously for the multiplex PCR assay (Mikołajczyk et al., 2010a). In each primer pair, the forward primer was labeled at its 5′ end with the fluorescent dye 6-Carboxyl-X-Rhodamine (Rox) (see Table 3 for primer details). The lengths of the shortened amplicons varied from 97 bp for *ogura* CMS to 115 for *actin 7* (Table 3). PCR mixtures were prepared as described above, but the amplification was carried out by using the following parameters: 5 min at 95°C; 29 cycles of 30 s at 95°C, 90 s at 50°C, and 30 s at 72°C; and a final extension of 30 min at 65°C. After the amplification, PCR products were cleaned with FAST alkaline phosphatase and exonuclease I (*exoI*) as described in the paragraph 2.2.3. The samples were diluted with 50 µl of sterile deionized water (MQ; Millipore, USA) before capillary electrophoresis.

### 2.2.3 PCR amplification of *BnaA.FAD3* and *BnaC.FAD3* gene fragments

Target DNA fragments comprising polymorphic sites of *BnaA.FAD3* and *BnaC.FAD3* wild-type and mutant alleles were amplified in two independent reactions with the use of locus-specific PCR primer pairs (FAD3Af/FAD3Ar and FAD3Cf/FAD3Cr, respectively) developed

| Primer | Mod. (5′) | Sequence (5′-3′) | Locus | Product (bp) | Use | Reference |
|---|---|---|---|---|---|---|
| FAD3Af | | CATCATCATGGTCACGATGATAAGT | BnaA.FAD3 | 189 | template for SNaPshot analysis | Mikolajczyk et al. 2010 |
| FAD3Ar | | GAAGATCCCGTAATCTCTATCAAT | | | | |
| shFAD3Cf | | CATCATCATGGTCACGATGATAAGC | BnaC.FAD3 | 187 | template for SNaPshot analysis | Mikolajczyk et al. 2010 |
| shFAD3Cr | | GAAGATCCCGTAATCTCTATCAAC | | | | |
| Act-rox | ROX | CTCGACTCTGGTGATGGTGTG | actin 7 | 115 | internal PCR control | this study |
| ActR5 | | TTCATTAGAGAATCCGTGAGA | | | | |
| CMS-rox | ROX | TTCGAAAAAGGTAATCATTG | orf 138 (ogura CMS) | 97 | ogura CMS marker | this study |
| CMSp2 | | GTCGTTATCGACCTCGCAAGG | | | | |
| Res-rox | ROX | TGTAACATAAGAAACGCTTGGT | Rfo | 107 | restorer gene marker | this study |
| C02p3 | | TTGGCGCATCCTAAATTCAATC | | | | |
| mutA-1f | (A)6 | TGTACAATAATAGGAATGGAGTTATTTA | BnaA.FAD3 | 35 | SNaPshot analysis | Mikolajczyk et al. 2010 |
| mutC-45F | (A)24 | TGCCTTGGTACAGAGGCAAG | BnaC.FAD3 | 45 | SNaPshot analysis | Mikolajczyk et al. 2010 |

Table 3. Oligonucleotides used in this study. "Mod." – 5′-terminal modification of the oligonucleotide; "Rox" – 6-Carboxyl-X-Rhodamine; "(A)6" – poly(A)-tail of 6 nucleotides; "(A)24" - poly(A)-tail of 24 nucleotides.

previously by Mikolajczyk et al. (2010b) for the analysis of splicing variant (see Table 3 for primer details). The PCR was carried out in a 96-well plate (Brandt, Wertheim, Germany) sealed with silicone compression mat (Axygen, Union City, CA, USA) in a reaction volume of 6 µl containing 2.5 µl of Type-it Microsatellite PCR Kit (Qiagen, Hilden, Germany), 0.2 mM of each primer, and 1 µl of DNA template (50-100 ng). Amplification was performed on Applied Biosystems thermal cyclers (Verity 96-Well, GeneAmp 9700, and 2720 TC) using the following PCR program: 1 cycle of 5 min at 95°C, followed by 35 cycles of 30 s at 95°C and 90 s at 65°C, and a final extension of 10 min at 65°C. After the amplification, PCR products were cleaned with exonuclease I and alkaline phosphatase to remove free nucleotides and primers: 5 µl containing 1 U of FAST alkaline phosphatase and 2 U of *exoI* (Fermentas, Vilnius, Lithuania) were combined with 6 µl of the PCR product and incubated for 1 h at 37°C, followed by denaturation step of 15 min at 80°C.

### 2.2.4 Detection of *BnaA.FAD3* and *BnaC.FAD3* alleles

Both polymorphic sites were analysed independently by single-base primer extension reaction (microsequencing) with primers varying in length as described previously by Mikolajczyk et al. (2010b). The first oligonucleotide, mutA-1f (35 nt), was used for detection of alleles in the locus *BnaA.FAD3*, the second, mutC-45F (45 nt), was used for the locus *BnaC.FAD3* (see Table 3 for primer details). Primer extension reaction was performed separately for each locus using 3 µl *exoI*/FAST treated PCR product as template in a total volume of 10 µl containing 2 µl of the SNaPshot Ready Reaction Mix (Applied Biosystems, Foster City, CA, USA) and 0.2 mM primer. The following microsequencing protocol was applied: 35 cycles of 10 s at 96°C, 5 s at 50°C, and 30 s at 60°C. After the reaction, 5 µl containing 1 U of FAST alkaline phosphatase was added to the each sample and incubated at 37°C for 15 min.

| | Reagent volumes (µl) for | | | PCR program | | |
|---|---|---|---|---|---|---|
| | 1 sample | 16 samples | 96 plate | Temp. (°C) | Time | Cycles |
| **1. Fluorescent multiplex-PCR for *Rfo* and *ogura* CMS markers.** | | | | | | |
| Type-it PCR Kit (2X) | 2.5 | 40 | 250 | 95 | 5 min | |
| Primer Act-rox (10 µM) | 0.1 | 1.6 | 10 | 95 | 30 s | |
| Primer ActR5 (10 µM) | 0.1 | 1.6 | 10 | 50 | 90 s | 29 |
| Primer CMS-rox (10 µM) | 0.1 | 1.6 | 10 | 72 | 30 s | |
| Primer CMSp2 (10 µM) | 0.1 | 1.6 | 10 | 65 | 30 min | |
| Primer Res-rox (10 µM) | 0.1 | 1.6 | 10 | 4 | hold | |
| Primer C02p3 (10 µM) | 0.1 | 1.6 | 10 | | | |
| MQ water | 1.9 | 32 | 190 | | | |
| Total volume: | 5 | 81.6 | 500 | | | |

Dispense 5 µl of reaction mix into each well and add 1 µl of genomic DNA. After the reaction, dilute with 50 µl of MQ water and store at 4°C until use.

| | | | | | | |
|---|---|---|---|---|---|---|
| **2. PCR amplification of *BnaA.FAD3* and *BnaC.FA3* gene fragments for SNaPshot analysis.** | | | | | | |
| Type-it PCR Kit (2X) | 2.5 | 40 | 250 | 95 | 5 min | |
| FAD3Af or FAD3Cf (10 µM) | 0.1 | 1.6 | 10 | 95 | 30 s | 35 |
| FAD3Ar or FAD3Cr (10 µM) | 0.1 | 1.6 | 10 | 65 | 90 s | |
| MQ water | 2.3 | 38 | 230 | 65 | 10 min | |
| Total volume: | 5 | 81.2 | 500 | 4 | hold | |

Dispense 5 µl of reaction mix into each well and add 1 µl of genomic DNA.

| | | | | | | |
|---|---|---|---|---|---|---|
| **3. Exonuclease I and alkaline phosphatase cleaning.** | | | | | | |
| FAST (1U/µl) | 1 | 16 | 100 | 37 | 60 min | |
| *exo*I (20U/µl) | 0.1 | 1.7 | 10 | 80 | 15 min | |
| exonuclease buffer (10X) | 0.5 | 8 | 50 | 4 | hold | |
| MQ water | 3.4 | 55 | 340 | | | |
| Total volume: | 5 | 80.7 | 500 | | | |

Dispense 5 µl of reaction mix into each well.

| | | | | | | |
|---|---|---|---|---|---|---|
| **4. Detection of *BnaA.FAD3* and *BnaC.FAD* alleles by the use of SNaPshot analysis.** | | | | | | |
| SNaPshot-mix (5X) | 1 | 16 | 100 | 96 | 10 s | |
| Primer mutA-1f or mutC-45F (10 µM) | 0.2 | 3.2 | 20 | 96 | 10 s | |
| Sequencing Buffer (5X)* | 1 | 16 | 100 | 50 | 5 s | 35 |
| MQ water | 4.8 | 80 | 480 | 60 | 30 s | |
| Total volume: | 7 | 115.2 | 700 | 4 | hold | |

Dispense 7 µl of reaction mix into each well and add 3µl of PCR reaction from step 2 after *exo*I and FAST cleaning.

| | | | | | | |
|---|---|---|---|---|---|---|
| **5. Alkaline phosphatase cleaning.** | | | | | | |
| FAST (1U/µl) | 0.5 | 8 | 50 | 37 | 15 min | |
| MQ water | 4.5 | 74 | 450 | 80 | 15 min | |
| Total volume: | 5 | 82 | 500 | 4 | hold | |

Dispense 5 µl of reaction mix into each well.

| | | | | | | |
|---|---|---|---|---|---|---|
| **6. Capillary electrophoresis.** | | | | | | |
| HiDi formamide | 9 | 145 | 1000 | 95 | 5 min | |
| GeneScan-120 LIZ | 0.2 | 3.2 | 25 | 4 | hold | |

Dispense 9 µl of reaction mix into each well and add 0.5 µl alkaline phosphatase cleaned SNaPshot reaction for *BnaA.FAD3*, 0.5 µl alkaline phosphatase cleaned SNaPshot reaction for *BnaC.FAD3*, and 0.5 µl fluorescent multiplex-PCR for *Rfo* and *ogura* CMS markers diluted with MQ water.

Table 4. Reaction components, volumes, and conditions for PCR amplifications, incubations, and capillary electrophoresis. *Sequencing Buffer (5X): 400 mM Tris-HCl (pH 9.0) and 10 mM MgCl₂.

## 2.2.5 Capillary electrophoresis and genotype scoring

The samples for electrophoresis containing 0.5 µl of each microsequencing reaction, 0,5 µl of water-diluted fluorescent multiplex-PCR, 0.2 µl of GeneScan-120 LIZ size standard (Applied Biosystems), and 9 µl of HiDi formamide (Applied Biosystems) were denatured for 5 min at 95°C and separated by capillary electrophoresis on an ABI Prism 3130XL Genetic Analyser (Applied Biosystems). Injection was performed at 1.2 kV for 23 s. Separation was carried out at 15 kV, 60°C using 36-cm capillaries containing POP7 polymer. Detection was performed using the dye set E5 in order to process the data from the 5 fluorescent dyes (dR110, dR6G, dTAMRA, dROX, and LIZ). The *Rfo* (Rfo), *ogura* CMS (CMS), and *actin 7* (Act) gene fragments as well as the alleles of *BnaA.FAD3* (A-wild, A-mut) and *BnaC.FAD3* (C-wild, C-mut) were automatically visualized and scored using the GeneMapper 3.7 software (Applied Biosystems). The components of the reactions and the conditions concerning PCR amplifications, incubations, and capillary electrophoresis are presented in Table 4.

# 3. Results

We designed a multiplex fluorescent PCR test for the detection and identification of the *Rfo* restorer gene, the *ogura* male sterile cytoplasm internally controlled by amplification of the *actin 7* gene fragment of similar, but longer, length (Fig. 2, „Rfo", „CMS", „Act" in the upper panel). The fluorescently labeled PCR products and the specific oligonucleotide probe primers generated during SNaPshot analysis (Fig. 2, „A-wild", „A-mut", „C-wild", „C-mut") were detected simultaneously in the same capillary during electrophoresis in the ABI Prism genetic analyzer. Using this method it is possible to detect all possible genotypes at genotyped loci. The following are presented as examples in Fig. 3: heterozygous at both *FAD3* loci with CMS and *Rfo* traits (sample D015), homozygous for the low linolenic mutant alleles at both *FAD3* loci with CMS and *Rfo* traits (D011), homozygous for the mutant allele at *BnaA.FAD3* locus and heterozygous at *BnaC.FAD3* with CMS but without *Rfo* trait (Rob-10), heterozygous at *BnaA.FAD3* locus and homozygous for the mutant allele at *BnaC.FAD3* locus with *Rfo* but without CMS trait (D035), and the wild-type genotype, which is homozygous for the wild-type alleles at both *FAD3* loci and has no CMS and *Rfo* traits (G001).

First, 190 plants belonging to 19 categories of recombinant lines (Table 2), previously phenotyped and/or genotyped, were used to test the accuracy and reproducibility of the new multiplex fluorescent assay. The results were compared with the previously analyzed genotypes (scored genotypes for selected plants as examples are presented in Table 5).

Using the new SNaPshot analysis in combination with the multiplex fluorescent assay, SNPs were detected in 95 plants analyzed previously for allelic variation in *FAD3* genes. Among 190 SNP sites, 187 were scored accurately and in accordance with seed oil fatty acid composition determined by gas liquid chromatography (data not shown). Similarly, new fluorescent multiplex PCR was effective for detection of CMS and *Rfo* traits. Furthermore, the fluorescent assay was applied to the analysis of *Rfo* and *ogura* CMS lines included in breeding programs in plant breeding companies (test-1 and test-2, Table 2). The results obtained from the analysis of 620 plants were consistent with the previous genotyping results obtained by conventional multiplex PCR (Fig. 3) and in accordance with breeders' predictions, revealing the fluorescent multiplex PCR assay as a sensitive tool for detection of CMS and *Rfo* traits in oilseed rape.

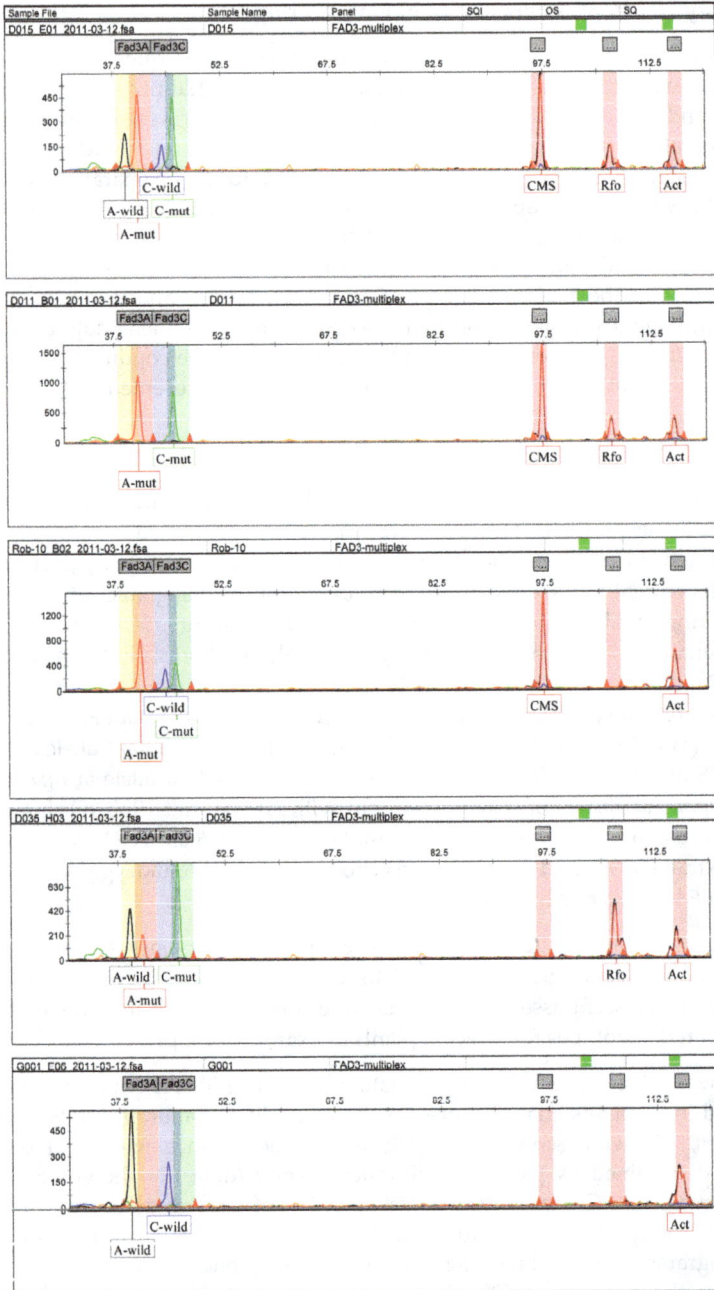

Fig. 2. The use of fluorescent labeled specific PCR products (red peaks *on the right*) together with specific oligonucleotide probe primers (black, red, green, and blue peaks *on the left*) generated during SNaPshot analysis. See text for details.

| | | Previously analyzed genotypes | | | | New multiplex fluorescent test | | | | | | |
| | | FAD3 genotype | | SCAR | | FAD3A | | FAD3C | | CMS | Rfo | Act |
| Cat. | Plant | loc.A | loc.C | CMS | C02 | Allele 1 | Allele 2 | Allele 1 | Allele 2 | Allele 1 | Allele 1 | Allele 1 |
| 5 | G006 | AA | CC | CMS | absent | A-wild | A-wild | C-wild | C-wild | CMS | absent | Act |
| 6 | G030 | AA | CC | CMS | absent | A-wild | A-wild | C-wild | C-wild | CMS | absent | Act |
| 6 | G039 | AA | CC | CMS | absent | A-wild | A-wild | C-wild | C-wild | CMS | absent | Act |
| 7 | G031 | AA | CC | CMS | Rfo | A-wild | A-wild | C-wild | C-wild | CMS | Rfo | Act |
| 8 | G048 | AA | CC | CMS | Rfo | A-wild | A-wild | C-wild | C-wild | CMS | Rfo | Act |
| 9 | G007 | AA | CC | CMS | Rfo | A-wild | A-wild | C-wild | C-wild | CMS | Rfo | Act |
| 12 | G055 | AA | CC | CMS | Rfo | A-wild | A-wild | C-wild | C-wild | CMS | Rfo | Act |
| 13 | G001 | AA | Cc | absent | absent | A-wild | A-wild | C-wild | C-wild | absent | absent | Act |
| 14 | G002 | aa | cc | absent | absent | A-mut | A-mut | C-mut | C-mut | absent | absent | Act |
| 14 | G003 | aa | cc | absent | absent | A-mut | A-mut | C-mut | C-mut | absent | absent | Act |
| 15 | G004 | AA | CC | absent | absent | A-wild | A-wild | C-wild | C-wild | absent | absent | Act |
| 16 | G009 | aa | cc | absent | absent | A-mut | A-mut | C-mut | C-mut | absent | absent | Act |
| 17 | G045 | AA | CC | CMS | Rfo | A-wild | A-wild | C-wild | C-wild | CMS | Rfo | Act |
| 18 | G032 | AA | CC | CMS | Rfo | A-wild | A-wild | C-wild | C-wild | CMS | Rfo | Act |
| 19 | G043 | AA | CC | CMS | Rfo | A-wild | A-wild | C-wild | C-wild | CMS | Rfo | Act |
| test-2 | W001 | AA | CC | CMS | Rfo | A-wild | A-wild | C-wild | C-wild | CMS | Rfo | Act |
| test-2 | W084 | AA | CC | absent | absent | A-wild | A-wild | C-wild | C-wild | absent | absent | Act |
| test-2 | Y008 | AA | CC | CMS | Rfo | A-wild | A-wild | C-wild | C-wild | CMS | Rfo | Act |
| test-2 | Y065 | AA | CC | absent | Rfo | A-wild | A-wild | C-wild | C-wild | absent | Rfo | Act |
| test-2 | Y066 | AA | CC | absent | absent | A-wild | A-wild | C-wild | C-wild | absent | absent | Act |
| test-2 | Y073 | AA | CC | absent | Rfo | A-wild | A-wild | C-wild | C-wild | absent | Rfo | Act |
| test-2 | Y075 | AA | CC | absent | absent | A-wild | A-wild | C-wild | C-wild | absent | absent | Act |
| test-2 | Y076 | AA | CC | absent | Rfo | A-wild | A-wild | C-wild | C-wild | absent | Rfo | Act |
| test-2 | Y077 | AA | CC | absent | Rfo | A-wild | A-wild | C-wild | C-wild | absent | Rfo | Act |
| test-2 | Y090 | AA | CC | absent | Rfo | A-wild | A-wild | C-wild | C-wild | absent | Rfo | Act |
| test-2 | Y116 | AA | CC | CMS | absent | A-wild | A-wild | C-wild | C-wild | CMS | absent | Act |

Table 5. Comparison of plant genotyping results using separate SNaPshot analysis and conventional multiplex PCR (on the left) with the new multiplex fluorescent test (on the right). Only selected plants are presented as examples. "Cat." – plant category, "loc.A" – alleles at locus *BnaA.FAD3*, "loc.C" – alleles at locus *BnaC.FAD3*, "A, C" – wild-type alleles, "a, c" – mutant alleles. See Table 2 for plant category details.

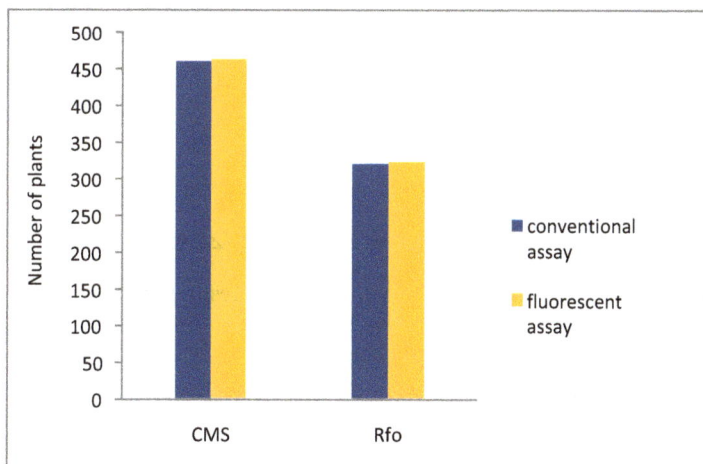

Fig. 3. Comparison of the number of CMS and *Rfo* traits detected in analyzed plants using conventional (*blue*) and fluorescent (*yellow*) assays.

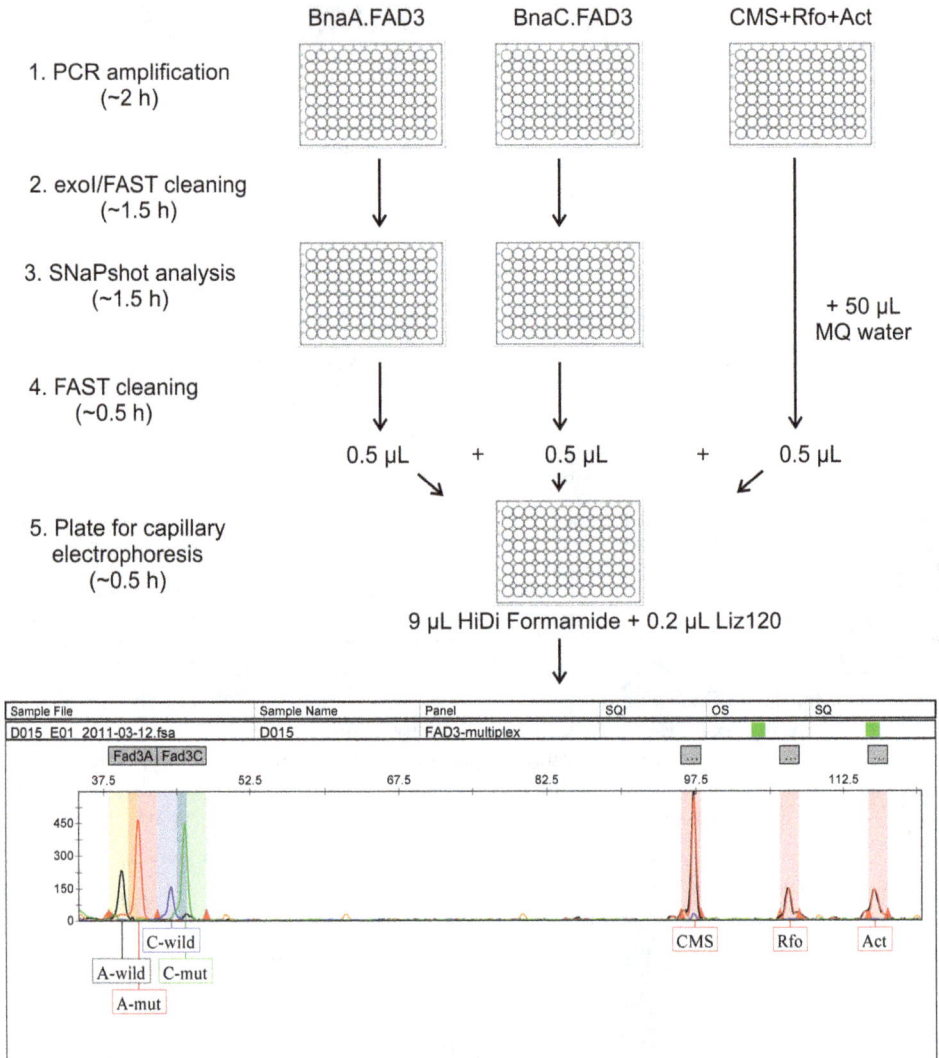

6. Capillary electrophoresis and automated allele scoring (~2 h for 96 samples, ABI 3130XL)

Fig. 4. Multiplex fluorescent assay for detection of the *Rfo* restorer gene, the *ogura* male sterile cytoplasm and the low linolenic mutant genotypes in oilseed rape hybrid breeding. The assay can be performed within one working day.

## 4. Discussion

In the method described previously (Mikolajczyk et al., 2010b), much longer PCR products were used as templates for microsequencing in the search for allele-specific SNPs for the low linolenic mutant genotype of winter oilseed rape. The templates used in the former assay were generated using one locus-specific PCR primer (*forward*, from the 5′-end) while the second primer (*reverse*, from the 3′-end) was the same for both loci. The long length of the PCR products (1.1–1.34 kb) and non-specific reverse primer both could affect the efficiency of amplification. In fact, while assaying SNP polymorphism in plants from breeding experiments we found relatively lower amplification rate in case of the *BnaA.FAD3* locus which could be correlated with the size of PCR product (1.34 kb). In the new assay, two short PCR (ca. 190 bp) amplicons are generated for each target by using locus-specific primer pairs. The level of fluorescence resulting from the amount of primer-extended products generated in the SNaPshot reaction shows that both loci are amplified at the same rate. The new amplification method is more effective, faster and very efficient.

A similar G to A substitution in the 5′ donor splice site associated with LL phenotype was detected in the mutant *BnaC.fad3* of the canola mutant line DMS100 gene by Hu et al. (2006). The authors invented a method of this SNP detection based on hybridization-involving assay and real-time PCR technology. However, our new test combines cytoplasmic male sterility and low linolenic markers for the first time.

In plant studies, a multiplex fluorescent PCR method is applied for high throughput genetic mapping and measurement of the extent of diversity within and between cultivars using SSR (multiplex simple sequence repeat) markers. Up to now, identification of cytoplasm type and fertility restorer of rapeseed accessions for hybrid breeding has been performed using multiplex PCR method, followed by conventional gel electrophoresis. A simple multiplex PCR was applied by Zhao et al. (2010) to distinguish the existing common cytoplasm resources, Pol, Nap, Cam, and Ogu in rapeseed. In their test, four pairs of specific primers were used for the appropriate mitochondrial DNAs identification in addition to an internal control for the presence of nuclear DNA. According to our knowledge, the method presented in this chapter is the first assay combining the multiplex fluorescent PCR with SNaPshot analysis to be applied for plant molecular breeding.

## 5. Conclusion

The detection of the restorer gene, the *ogura* male sterile cytoplasm, and low linolenic mutant genotypes by multiplex fluorescence PCR combined with SNaPshot method is a practical alternative to classic methods of phenotype prediction. Starting with DNA, this method is fast with a turnaround time of 8 hours with mean reagent cost around $2 per marker detected. Moreover, the assay could be extended by increasing or changing SNP and SCAR markers included in the test.

## 6. Acknowledgments

This work was partially supported by the Research Grant No. 4201401 (2007-2013) "Progress in plant production" sponsored by the Polish Ministry of Agriculture and Rural Development.

## 7. References

Auld, D. L., Heikken, M. K., Erickson, D. A., Sernyk, J. L., & Romero, J. E. (1992). Rapeseed mutants with reduced levels of polyunsaturated fatty acids and increased levels of oleic acid. *Crop Sci.*, Vol. 32, (3), pp. 657-662

Bannerot H., Boulidard L., Cauderon Y., & Tempe J. (1974). Cytoplasmic male sterility transfer from *Raphanus* to *Brassica*. *Proceedings of EUCARPIA Meet. Crop. Sect. Crucirefae* 25, pp. 52-54

Barker, G. C., Larson, T. R., Graham, I. A., Lynn, J. R., & Graham, J. K. (2007). Novel Insights into Seed Fatty Acid Synthesis and Modification Pathways from Genetic Diversity and Quantitative Trait Loci Analysis of the *Brassica* C Genome[1][OA]. *Plant Physiology,* Vol. 144, (4), pp. 1827-1842

Bartkowiak-Broda, I., Rouselle P., & Renard M. (1979). Investigations of two kinds of cytoplasmic male sterility in rapeseed (*Brassica napus* L.). *Genetica Polonica*, Vol. 20, (4), pp. 487-497

Bonhomme, S., Budar F., Lancelin, D., Small, I., Defrance, M.-C., & Pelletier G. (1992). Sequence and transcript analysis of the *Nco2.5* Ogura-specific fragment correlated with cytoplasmic male sterility in *Brassica* hybrids. *Mol. Gen. Genet.,* Vol. 235, (2/3), pp. 340-348

Delourme, R., Bouchereau, A., Hubert, N., Renard, M., & Landry, B.S. (1994). Identification of RAPD markers linked to a fertility restorer gene for the *Ogura* radish cytoplasmic male sterility of rapeseed (*Brassica napus* L.). *Theor Appl Genet,* Vol. 88, (6/7), pp. 741-748

Doyle, J.J., & Doyle, J.L. (1990). Isolation of plant DNA from fresh tissue. *Focus,* Vol. 12, pp. 13-15

Downey, R.K., & Roebbelen G. (1989). Brassica species, In: *Oil Crops of the World*, G. Roebbelen, R. K. Downey & A. Ashri, (Eds.), pp. 339-382, McGraw-Hill Publishing Co., New York

Falentin, C., Brégeon, M., Lucas, M. O., Deschamps, M., Leprince, F., Fournier, M. T., Delourme, R., & Renard M. (2007). Identification of *fad2* mutations and development of allele-specific markers for high oleic acid content in rapeseed (*Brassica napus* L.), *Proceedings of the 12th International Rapeseed Congress, vol. II Biotechnology*, pp. 117-119, Wuhan, China, March 2007

Hu, X., Sullivan, M.L., Gupta, M., & Thompson S.A. (2006). Mapping of the loci controlling oleic and linolenic acid contents and development of *fad2* and *fad3* allele-specific markers in canola (*Brassica napus* L.) *Theor Appl Genet,* Vol. 113, (3), pp. 497-507

Krishnasamy, S., & Makaroff, C.A. (1993). Characterization of the radish mitochondrial *orf*B locus: possible relationship with male sterility in Ogura radish. *Curr. Genet.,* Vol. 24, (1/2), pp. 156-163

Krzymanski, J. (1968). Variation in thioglucosides in rapeseed meal (*Brassica napus*). Meeting of the Associate Commitees of National Research Council in Plant Breeding. Winnipeg, Manitoba, Canada 20.02.1968

Krzymanski, J. (1970). Genetyczne mozliwości ulepszania skladu chemicznego nasion rzepaku ozimego./ *Changes of genetical improvement of chemical composition of winter oilseed rape (Brassica napus) seeds* (In Polish). *Hodowla Roslin Aklimatyzacja i Nasiennictwo*, Vol. 14, (2), pp. 95-133

Mikolajczyk, K., Matuszczak, M., Pietka, T., Bartkowiak-Broda, I., & Krzymanski J. (1998). *The use of DNA markers for hybrid component analysis* (in Polish, abstract in English). *Rośliny Oleiste – Oilseed Crops*, Vol. XIX (2), pp. 463-471

Mikolajczyk, K., Dabert, M., Nowakowska, J., Podkowinski, J., Poplawska, W., & Bartkowiak-Broda I. (2008). Conversion of the RAPD OPC02[1150] marker of the *Rfo* restorer gene into a SCAR marker for rapid selection of oilseed rape. *Plant Breeding*, Vol. 127, (6), pp. 647-649

Mikolajczyk, K., Dobrzycka, A., Podkowinski, J., Poplawska W., Spasibionek S., & Bartkowiak-Broda, I. (2010a). A multiplex PCR assai for identification of the *ogura* male sterile cytoplazm and the *Rfo* restorer gene among oilseed rape breeding forms. *Rosliny Oleiste – Oilseed Crops*, Vol. XXXI, (2), pp. 201-210

Mikolajczyk, K., Dabert, M., Karlowski, W. M., Spasibionek, S., Nowakowska, J., Cegielska-Taras, T., & Bartkowiak-Broda I. (2010b). Allele-specific SNP markers for the new low linolenic mutant genotype of winter oilseed rape. *Plant Breeding*, Vol. 129, (5), pp. 502-507

Ogura, H. (1968). Studies on the new male sterility in Japanese radish, with special references to the utilisation of this sterility towards the practical raising of hybrid seeds. *Mem. Fac. Agr. Kogoshima Univ.*, Vol. 6, pp. 39-78

Pelletier, G., Primard, C., Vedel, F., Chetrit, P., Remy, R., Rousselle, P., & Renard M. (1983). Intergeneric cytoplasm hybridization in *Cruciferae* by protoplast fusion. *Mol. Gen. Genet.*, Vol. 191, pp. 244-250

Pelletier, G., Primard, C., Vede,l F., Chetrit, P., Renard, M., Pellan-Delourme, R., & Mesquida J. (1987). Molecular phenotype and genetic characterization of mitochondrial recombinants in rapeseed. *Proceedings of the 7th International Rapeseed Congress*, Poznan, Poland, pp. 113-118

Rakow, G. (1973). Selektion auf Linol- und Linolensaeuregehalt in Rapssamen nach mutagener Behandlung. *Z. Pflanzenzuechtg*, Vol. 69, pp. 62-82

Roebbelen, G., & Nitsch, A. (1975). Genetical and physiological investigations on mutants for polynoic fatty acids in rapeseed, *Brassica napus* L. *Z. Pflanzenzuechtg*, Vol. 75, pp. 93-105

Scarth, R., McVetty, P.B.E., Rimme,r S.R., & Stefansson, B.R. (1988). 'Stellar' low linolenic-high linoleic acid summer rape. *Can. J. Plant Sci.*, Vol. 68 (April 1988), pp. 509-511

Scarth, R., Rimmer, S.R., & McVetty, P.B.E. (1995). Apollo low linolenic acid summer rape. Can. J. Plant Sci. 75, (1): 203-204

Sigareva M.A., Earle E.D. (1997) Direct transfer of a cold-tolerant Ogura male-sterile cytoplasm into cabbage (*Brassica oleracea* ssp. *capitata*) via protoplast fusion. *Theor. Appl. Genet.*, Vol. 94, (2), pp. 213-220

Snowdon, R., Luehs, W., & Friedt, W. (2007). Oilseed rape, In: *Genome Mapping and Molecular Breeding in Plants*, K. Chittaranjan, (Ed.), 55-114, Springer-Verlag Berlin Heidelberg, ISBN-13 978-3-540-34387-5, Leipzig, Germany

Spasibionek, S. (2006). New mutants of winter rapeseed (Brassica napus L.) with changed fatty acid composition. *Plant Breding*, Vol. 125, (3), pp. 259-267

Spasibionek, S., Krzymański, J., & Bartkowiak-Broda, I. (2003). Mutants of Brassica napus with changed fatty acids composition. *Proceedings of the 11th International Rapeseed Congress*, pp. 221-224, Copenhagen, Denmark, July 2003.

Spasibionek, S. (2008). Variability of fatty acid composition in seed oil of winter rapeseed (Brassica napus L.) developed using mutagenesis. *Rośliny Oleiste - Oilseed Crops*, Vol. XXIX, (2), pp. 161-168

Stefansson, B.R. Hougen, F.W., & Downey, R.K. (1961). Note on the isolation of rape plants with seed oil free from erucic acid. *Can. J. Plant Sci.*, Vol. 41, pp. 218-219

Stefansson, B.R., & Hougen, F.W. (1964). Selection of rape plants (*Brassica napus*) with seed oil practically free of erucic acid. *Can. J. Plant Sci.*, Vol. 44, (4), pp. 359-364

U, N. (1935). Genome analysis in *Brassica* with special reference to the experimental formation of *B. napus* peculiar mode of fertilization. *Jpn. J. Bot.*, Vol. 7, pp. 389-452

Wittkop, B., Snowdon, R., Friedt, W. (2009). Status and perspectives of breeding for enhanced yield and quality of oilseed crops for Europe. *Euphytica*, Vol. 170, (1/2), pp. 131-140

Zhao, H. X., Li, Z. J., Hu, S. W., Sun, G. L., Chang, J. .J, & Zhang, Z. H. (2010). Identification of cytoplasm types in rapeseed (Brassica napus L.) accessions by a multiplex PCR assay. *Theor Appl Genet*, Vol. 121, (4), pp. 643-50

# 9

# Marker Assisted Characterization in *Tigridia pavonia* (L.f) DC

José Luis Piña-Escutia, Luis Miguel Vázquez-García
and Amaury Martín Arzate-Fernández
*Centro de Investigación y Estudios Avanzados en Fitomejoramiento, Facultad de Ciencias
Agrícolas, Universidad Autónoma del Estado de México,
México*

## 1. Introduction

Conventional plant breeding programs largely depend on phenotypic selection, breeders' experience, and knowledge of plant genetics for traits of agronomic importance. A large amount of biological data is available from genetic studies of a crop of interest related to important target traits in crop breeding, which may in turn directly assist genotypic selection (Xu & Crouch, 2008). However, conventional breeding is time consuming and often testing procedures may be complex, unreliable or expensive due to the nature of the target traits (e.g. abiotic stresses) or the target environment (Choudhary et al., 2008).

Marker assisted characterization offers such a possibility by adopting a wide range of novel approaches to improving the selection strategies in horticultural plant breeding (Ibitoye & Akin-Idowu, 2010). The genetic markers involved in marker assisted characterization include morphological, biochemical and DNA fingerprints which can be employed in selection and identification of closely related genotypes.

Many variants of an enzyme, referred to as isozymes, can be resolved by electrophoresis and are very useful genetic markers. The alleles of most isozyme markers segregate in a codominant manner and rarely show epistatic interactions, which allows accumulation of many polymorphic isozyme loci in a single $F_2$ population and increases the efficiency of gene mapping. Once the map location of isozyme genes are known, they can be used efficiently as biochemical markers to tag other genes for morphological, physiological and phytopathological traits (Tanksley & Rick, 1980). However, only a relatively small number of protein variants may exist between the two parents and this limits the total number of protein loci that can actually be scored in a given mapping population (Young, 1999).

DNA markers have been used in the identification of varieties and characterization of genotypes because they offer a fast screening and a more precise discriminatory power (Vicente & Fulton, 2003).

With the discovery of restriction enzymes and the polymerase chain reaction (PCR), different markers systems such as Restriction Fragment Length Polymorphisms (RFLP), Random Amplified Polymorphic DNAs (RAPD), Amplified Fragment Length Polymorphisms (AFLP),

Simple Sequence Repeats or microsatellites (SSR), Single Nucleotide Polymorphism (SNP), Inter-Simple Sequence Repeats (ISSR) and others have been developed to visualize the composition of organisms at the DNA level and obtain their genetic fingerprints. Among these markers, RFLPs, AFLPs, SSRs and SNPs have been reported as highly polymorphic and reproducible. However RFLPs require large amount of DNA, are laborious intensive, time consuming and mostly require radioactively labeled probes, whereas AFLPs and SSRs are quite costly also require high resolution electrophoresis or automated sequences. On the other hand, although SNPs are less time consuming against rest of the markers and highly amenable to automation, initial cost involved is quite high (Jehan and Lakhanpaul, 2006). So, although marker informativeness is an important element when comparing different assay systems, other factors such as cost per assay, level of skills required, and reliability of assays should also be considered. PCR molecular markers like ISSR and RAPD would be an option because of the lower level of skill required, low costs per assays and the ready availability of primers can scan the entire genome and make more efficient the genotype characterization. Thus, due to their characteristics and efficiency for detecting polymorphism, the ISSR and RAPD markers have been successfully used to calculate the intra or inter-specific genetic diversity in different domesticated and wild species (Arzate-Fernández et al., 2005b; Escandón et al., 2005a; Tapia et al., 2005; Kumar et al., 2006; Luna-Paez et al., 2007; Marotti et al., 2007; Muthusamy et al., 2008; Ye et al., 2008).

The genus *Tigridia* comprises plants with highly colored flowers that exhibit great morphological variation, making many species potentially valuable as cultivated plants. *Tigridia pavonia* (L.f.) DC, also known as tiger flower or oceloxóchitl, is a species native to México which was used by the Aztecs for ornamental, nutritional and medicinal purposes (Hernández, 1959). Due to the great variability of the coloring of its flower and its ornamental potential, it is one of the principal phytogenetic resources of this country, considered the center of greatest genetic diversity of this species. In spite of fact that in Mexico, the potential for sustainable *T. pavonia* utilization is enormous, its remains largely unexploited due to the little information on the resource abundance and distribution, whereas in different countries of Europe, Asia and Australia this species is widely distributed and commercialized as a garden plant (Vázquez-García et al., 2001). Thus, genetic diversity studies are needed to provide information for its efficient conservation as well as for possible future use in breeding programs.

The widely distributed species generally present morphological and physiological variation, as well as in the genetic structure of their populations (Wen & Hsiao, 2001). Likewise, the dissemination in different regions could cause that the same variety or cultivar may be known under different names in different countries (and even within a given country) and different varieties sometimes may appear under the same name. Therefore, the study of phenotypic and genetic diversity to identify groups with similar genotypes is important for conserving, evaluating and exploiting genetic resources. Also, these studies are useful for determining the uniqueness and distinctness of the phenotypic and genetic constitution of genotypes with the purpose of protecting the breeder's intellectual property rights (Franco et al., 2001).

The focus of this chapter is in a more applied direction, encompassing the potential of the morphological, isozymes and DNA markers for characterization of nine *T. pavonia* varieties to their management into selection and breeding programme.

## 2. Materials

Nine botanical varieties of *Tigridia pavonia* (L.f.) DC, collected in three localities of the State of Mexico and registered in the National Seed Inspection and Certification Service (SNICS) were used in this study (Figure 1). The Sandra variety (TGD-008-030408) was collected in the municipality of Tenancingo, Carolina (TGD-002-030408), Trinidad (TGD-009-030408), Penelope (TGD-006-030408), Angeles (TGD-001-030408), Dulce (TGD-003-030408) and Mariana (TGD-005-030408), in the municipality of Temascaltepec and Samaria (TGD-007-030408) and Gloria (TGD-004-030408) in the municipality of Temoaya. All of the varieties were cultivated in a rustic greenhouse, at the Faculty of Agricultural Sciences (FCAgr) of the Autonomous University of State of Mexico (UAEM).

Fig.1. Nine botanical varieties of *Tigridia pavonia* (L. f.) DC used in the morphological and molecular characterization assay

## 3. Methodology

### 3.1 Morphological markers

To make the matrix of morphological data, 20 quantitative and one qualitative characters (Table 1) were evaluated among the nine *T. pavonia* varieties. Nine characters were previously published by Vázquez-García et al. (2001). Each character was evaluated in 14 individuals of each variety.

| Character | Mean | Max | Min | Standar Deviation | CV (%) | P |
|---|---|---|---|---|---|---|
| 1. Number of flowers per shoot | 9.32 | 17.00 | 3.00 | 2.22 | 23.23 | 0.0753 |
| 2. Length of external tepal | 8.65 | 11.70 | 7.00 | 0.88 | 7.87 | <0.0001 |
| 3. Length of internal tepal | 4.42 | 5.20 | 3.70 | 0.31 | 6.60 | <0.0004 |
| 4. Type of internal tepal (oval formed, sharp pointed, lanceolated, rounded) | 1.92 | 4.00 | 1.00 | 1.23 | 22.72 | <0.0001 |
| 5. Length of staminal column | 4.49 | 5.40 | 3.80 | 0.29 | 4.48 | <0.0001 |
| 6. Distance from the base of the anther to the stigma | 1.05 | 2.50 | 0.30 | 0.54 | 20.46 | <0.0001 |
| 7. Length of anther | 1.59 | 2.00 | 1.00 | 0.21 | 7.40 | <0.0001 |
| 8. Length of the reproductive part | 6.31 | 7.30 | 5.10 | 0.49 | 4.42 | <0.0001 |
| 9. Number of fruits per shoot | 4.84 | 13.00 | 1.00 | 2.32 | 34.44 | <0.0001 |
| 10. Number of shoot per bulb | 4.00 | 6.00 | 3.00 | 0.98 | 18.64 | <0.0001 |
| 11. Length of shoot | 60.03 | 96.50 | 38.60 | 18.68 | 7.03 | <0.0001 |
| 12. Nodes per shoot | 3.84 | 5.00 | 2.00 | 0.44 | 11.14 | 0.0240 |
| 13. Length of the internode | 8.18 | 22.00 | 0 | 6.15 | 39.24 | <0.0001 |
| 14. Number of branches per shoot | 1.98 | 3.00 | 0.90 | 0.72 | 25.03 | <0.0001 |
| 15. Length of the leaf | 34.13 | 46.00 | 16.00 | 6.78 | 8.23 | <0.0001 |
| 16. Width of the leaf | 3.53 | 5.60 | 2.50 | 0.88 | 7.46 | <0.0001 |
| 17. Length of the pseudopetiole | 7.55 | 11.00 | 6.00 | 1.32 | 6.16 | <0.0001 |
| 18. Floral scapus length | 11.21 | 28.00 | 3.00 | 7.29 | 10.89 | <0.0001 |
| 19. Floral scapus thickness | 0.55 | 1.00 | 0.30 | 0.22 | 6.33 | <0.0001 |
| 20. Length of the bract | 11.29 | 24.50 | 8.00 | 3.88 | 16.60 | <0.0001 |
| 21. Bract width | 2.85 | 3.50 | 2.20 | 0.38 | 4.03 | <0.0001 |

Table 1. Morphological characters and basic statistical data of nine varieties of *T. pavonia* (L. f.) DC.

### 3.2 Isozyme markers

For isozyme markers, two individuals of each *T. pavonia* variety were used. Leaf sections were taken from each individual eight to 12 weeks after budding initiated. All of the samples were processed in the laboratory of Plant Molecular Biology at the FCAgr of the UAEM.

Approximately 50 mg of fresh leaf tissue from each sample was homogenized using a plastic chopstick in 50 μL extraction buffer Tris-HCl 0.1M, 2% EDTA-2Na, 50.4% glycerol, 3.2% Tween-80, and 0.8% DTT, pH 7.5 (Arzate-Fernández et al., 2005a).

The samples were preserved at −20 °C until analysis. Two buffer systems of histidine-citric acid, pH 5.7 and pH 6.5 (H-AC) and Tris-citrate/Tris-histidine, pH 8.5 (T-C/T-H) were used following the procedures of Glaszmann et al. (1988) and Stuber et al. (1988).

The 18 isozyme systems evaluated were aspartate amino transferase (AAT; EC 2.6.1.1), acid phosphatase (ACP; EC 13.1.3.2), dehydrogenase alcohol (ADH; EC 1.1.1.1), aminopeptidase (AMP; EC 3.4.11.1), catalase (CAT; EC 1.11.1.6), endopeptidase (ENP; EC 3.4.23.6), esterase (EST; EC 3.1.1.1), formate dehydrogenase (FDH; EC 1.2.1.2), glutamate dehydrogenase (GDH; EC 1.4.1.2), glucose-6-phosphate dehydrogenase (G-6-PDH; EC 1.1.1.49), malic enzyme (MAL; EC 1.1.1.40), malate dehydrogenase (MDH; EC 1.1.1.37), phosphoglucose dehydrogenase (PGD; EC 1.1.1.44), phosphoglucose isomerase (PGI; EC 5.3.1 9), phosphoglucomutase (PGM; EC 2.7.5.1), phosphohexose isomerase (PHI; EC 5.3.1.8), peroxidase (POX; EC 1.11.1.7), and shikimate dehydrogenase (SKD; EC 1.1.1.25). The samples were loaded into a 12% hydrolyzed potato starch gel and electrophoresis was conducted at 4 °C for a mean duration of 3 h, 40-50 mA and 100-150 v. Each sample was run at least three times to verify reproducibility. Enzyme staining was performed according to the procedures described by Torres et al. (1978), Vallejos (1983), Glaszmann et al. (1988), Stuber et al. (1988) Wendel & Weeden (1990) and Ishikawa (1994).

### 3.3 DNA markers

Young leaves were collected of two individuals of each variety (in the same development stage), and stored at -20 °C prior to DNA extraction. The genomic DNA was extracted of approximately 100 mg of leaf tissue of each variety of T. pavonia. The extraction procedure was the CTAB as reported by Zhou et al. (1999). The DNA samples were stored to -20 °C prior to analysis. The polymerase chain reaction (PCR) was made in a final volume of 10 μL with: 1 μL of DNA (10 ng), 1 μL of 10X PCR buffer with ammonium: (15 mM), 0.5 μL of MgCl$_2$ (15 mM), 1 μL of dNTPs (10 mM), 1 μL of the primer (20 mM) and 0.1 units of the enzyme Taq DNA polymerase.

### 3.3.1 ISSR markers

For ISSR markers, five primers of the anchored microsatellites type were used (3′-ASSR) (Yamagishi et al., 2002). In each primer, the anchor consisted of a triplicate of distinct sequence (Table 4). The amplification conditions for the primers 3′-ASSR02 and 3′ -ASSR15 were those described by Arzate-Fernández et al. (2005b), and consisted of an initial cycle of 9 min at 94 °C, 1 min at 46 °C and 1 min at 72 °C and a final cycle of 9 min at 94 °C, 1 min at 46 °C and 10 min at 72 °C. For primers 3′-ASSR20, 3′-ASSR29 and 3′-ASSR35, the amplification cycles were those used by Yamagishi et al. (2002) and consisted of an initial cycle of 9 min at 94 °C, 1 min at 46 °C and 1 min at 72 °C and a final cycle of 10 min at 72 °C.

### 3.3.2 RAPD markers

For RAPD markers, five 10 base primers (Yamagishi, 1995), five 15 base primers (Yamagishi et al., 2002), and five 20 base primers (Debener & Mattiesch, 1998) were used (Table 5). The amplification conditions for RAPD primers of 10 base were the reported by Yamagishi (1995). The program for 15 base primers used in this study was followed according to Yamagishi et al. (2002) with minor modifications (40 cycles of 94 °C for 1 min, 53 °C for 3 min and 72 °C for 2 min). The PCR cycle conditions for 20 base primers were performed according to Debener & Mattiesch (1998).

The amplification of DNA fragments for ISSR and RAPD markers was made in a thermocycler (Mastercycler gradient, Eppendorf, Germany). The separation of the fragments was made in horizontal electrophoresis. A molecular marker of 100 to 3000 pb molecular weight was used. The running conditions for each sample were 100 V and 120 mA for 80 min, and the observation of the fragments was made in a transilluminator UVP.

### 3.4 Statistical analysis

For morphological markers, the 21 morphological characters were evaluated through an analysis of variance and a completely randomized design with 14 replicates, with the statistical program SAS version 8.0. With the data generated of the morphological characters, a binary matrix was made following the criteria described by Vicente & Fulton (2003); that is, the value of 1 was assigned if the character was present and 0 if it was absent.

For isozyme markers, a record of banding patterns was constructed and each band was assigned to a binary matrix as absent, 0, or present, 1.

For ISSR and RAPD markers, each band generated by each primer was considered as an independent locus calculated manually; that is, the value of 1 was assigned for the presence of a band and 0 for its absence. The total number fragments (FT), polymorphic fragments (FP), percentage of polymorphism (%P) and the Nei's genetic distance ($G_D$) were calculated using the program POPGENE version 1.32 (Yeh & Boyle, 1999). Also, capacity of each primer to differentiate the nine varieties under study was evaluated through the resolution power ($Rp$), according to Prevost & Wilkinson (1999).

In RAPD markers, a simple correlation analysis was performed to investigate the correlation between the length primer and the polymorphism generated by each primer group.

In all cases (morphological, isozyme and DNA markers), to determine genetic relationships and variety grouping, a dendrogram was constructed with the data using POPGENE with the UPGMA method (modified from the NEIGHBOR procedure of PHYLIP, version 3.5) (Felsenstein, 1990) based on the Nei (1972) matrix genetic distances. UPGMA is one of the simplest and most commonly used hierarchical clustering algorithms. It receives as input a set of elements and a dissimilarity matrix which contains pairwise distances between all elements, and returns a hierarchy of clusters on this set.

## 4. Results and discussion

### 4.1 Morphological markers

The values of the mean, maximum, minimum, standard deviation (SD) and coefficient of variation (CV) were calculated for each one of the 21 morphological characters evaluated in the nine varieties (Table 1). According to the values of the CV, the characters number of flowers per shoot, type of internal tepal, distance from the base of the anther to the stigma, number of fruits per shoot, length of the internode and number of branches per shoot showed a slightly high variation level. In contrast, the characters that presented low variation levels among the varieties were the length of the staminal column, length of the reproductive part and bract width.

The statistical analysis showed that the number of flowers per shoot and nodes per shoot were not significant. However, for the other 19 characters there were highly significant differences (p ≤ 0.0001) (Table 1). These results show the high phenotypic diversity among the varieties analyzed, and can be used for the selection of parental lines within a plant breeding program.

The varieties formed two groups based on their morphological characteristics (Figure 2). Group I with average of $G_D$ = 0.42 consisted of the varieties Carolina, Trinidad, Mariana, Angeles, Sandra, Penelope and Dulce, while group II included the varieties Gloria and Samaria, with an average of $G_D$= 0.32.

The morphological characters (18 of 21 equal) showed the close genetic relationship between the varieties Trinidad and Mariana ($G_D$ = 0.28). In contrast, the least related varieties were Carolina and Samaria, as well as Penélope and Samaria ($G_D$ = 0.80), with 4 and 5 of 21 similar morphological characters, respectively.

According to Vázquez-García et al. (2001), the varieties Gloria and Samaria can be easily distinguished by shoot length. In the present study, the results were similar and both varieties showed the highest values for length of shoot, length of internode, floral scapus length and length of the bract, as well as bract width.

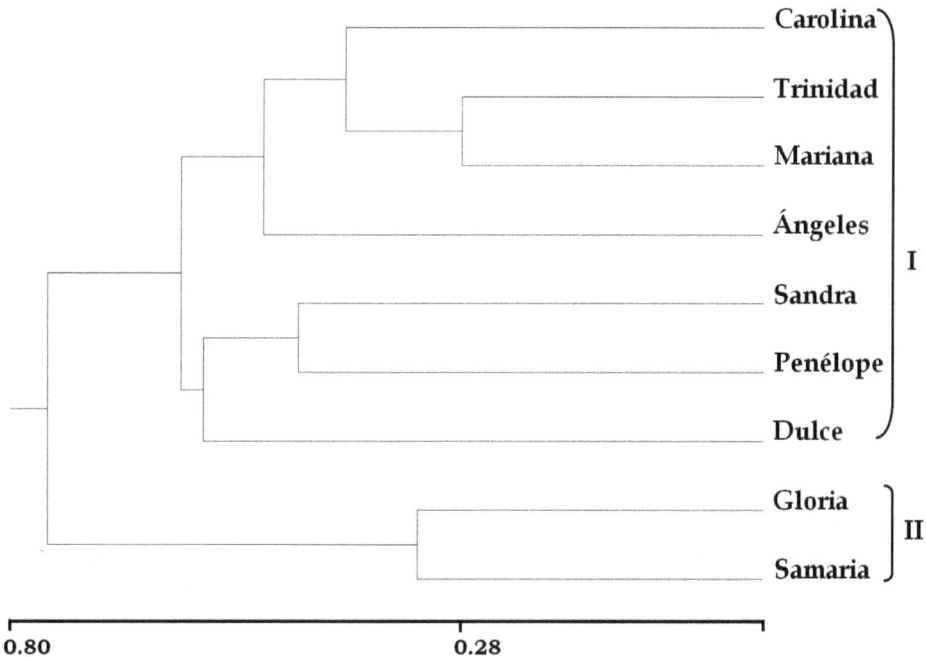

Fig. 2. Dendrogram of nine varieties of T. pavonia based on the genetic distance of Nei (1972), with the method UPGMA. The bar indicates the genetic distance among the varieties and the groups, calculated from morphological data.

## 4.2 Isozyme markers

The nine *T. pavonia* varieties evaluated were wild and free-pollinating. Therefore, they were considered to be heterozygotic; this agrees with the diverse isozyme banding patterns (IBPs) observed. Of the 18 isozymes tested, 12 could be stained, but only nine had sufficient resolution and clarity to be recorded and studied. Some bands were omitted because of their poor resolution. The isozymes ACP, MDH, PGI, PHI and PGM had better resolution when examined with the H-AC system, pH 5.7, and the isozymes AAT and PGD in the same system, but at pH 6.5. In contrast, the isozymes CAT and POX were better in the T-C/T-H system. Of the selected isozymes, nine zimograms in which 32 bands arranged in 20 IBP were identified: two for AAT, two for ACP, two for CAT, four for MDH, one for PGD, one for POX, three for PGI, three for PHI, and two for PGM (Figure 3).

In this study most of the varieties could be identified by the IBP obtained in each isozyme, as we describe in the following sections.

### 4.2.1 Aspartate amino transferase

AAT showed five active bands (Figure 3A) and varieties with four bands arranged in two banding patterns were detected. Pattern 1 was observed in the Angeles, Carolina, Dulce, Mariana, Penelope, and Sandra varieties, while the Gloria, Samaria, and Trinidad varieties exhibited pattern 2 (Table 2).

### 4.2.2 Acid phosphatase

ACP revealed four active bands (Figure 3B) and varieties with one or four bands arranged in two banding patterns were detected. The Angeles, Carolina, Dulce, Gloria, Mariana, Penelope, Sandra, and Trinidad varieties exhibited pattern 1 with four bands. Pattern 2 had a single band and was seen only in the Samaria variety (Table 2).

### 4.2.3 Catalase

CAT produced two active bands (Figure 3C) and varieties with one or two bands arranged in two banding patterns were detected. Pattern 1 with two bands was observed in Carolina, Dulce, Mariana, Penelope, and Samaria. Pattern 2 was observed in Angeles, Gloria, Sandra, and Trinidad (Table 2).

### 4.2.4 Malate dehydrogenase

MDH showed six active bands (Figure 3D) and varieties with three or four bands in four different arrangements were detected to produce four banding patterns. Patterns 1, 2, and 3 were composed of four bands. Pattern 1 was observed in Angeles, Carolina, Dulce, Mariana, Sandra, and Trinidad, while Penelope and Gloria exhibited patterns 2 and 3. Pattern 4 had three active bands and was observed in Samaria (Table 2).

### 4.2.5 Phosphoglucose dehydrogenase

PGD produced two active bands (Figure 3E) arranged in a single pattern, so that it was not possible to differentiate any of the varieties (Table 2).

### 4.2.6 Peroxidase

POX produced two active bands (Figure 3F) arranged in a single pattern, and thus it was not possible to differentiate varieties (Table 2).

### 4.2.7 Phosphoglucose isomerase

PGI produced four active bands (Figure 3G) and varieties with two, three or four bands arranged to produce three banding patterns were detected. Pattern 1 had four bands, and was observed in Angeles, Carolina, Dulce, Penelope, Samaria, Sandra, and Trinidad. Pattern 2 had three bands and was observed in Mariana. Pattern 3 had two bands and was observed in Gloria (Table 2).

### 4.2.8 Phosphohexose isomerase

PHI produced three active bands (Figure 3H) and varieties with two or three active bands in three arrangement producing three banding patterns were detected. Pattern 1 had three bands and was observed in ANG, CAR, DUL, PEN, SAN, and TRI. Patterns 2 and 3 had two bands each. Varieties GLO and SAM exhibited pattern 2 and variety MAR pattern 3 (Table 2).

### 4.2.9 Phosphoglucomutase

PGM produced four active bands (Figure 3I) and varieties with three or four bands arranged in two banding patterns were observed. Pattern 1 had four bands and was observed only in Gloria. Pattern 2 had three bands and was observed in Angeles, Carolina, Dulce, Mariana, Penelope, Samaria, Sandra, and Trinidad (Table 2).

As can be seen in Table 2, the nine isozymes produced sufficient polymorphic bands to distinguish most of the *T. pavonia* varieties evaluated. Of all the isozymes tested, MDH produced the highest number (4) of IBP and, therefore, the highest percentage of polymorphism (18.75%). Based on the observed IBP, this isozyme allowed to differentiate a larger number of varieties. In contrast, the isozymes PGD and POX produced only one IBP, and therefore neither isozyme was useful for differentiating the varieties used.

Twenty-nine bands derived from seven isozymes were only considered: five for AAT, four for ACP, two for CAT, six for MDH, four for PGI, three for PHI, and four for PGM (Figure 3). A dendrogram (Figure 4) was generated from the genetic relationship among these bands, clustering the varieties in two main groups. Group 1 comprises seven varieties: Angeles, Sandra, Carolina, Dulce, Mariana, Penelope, and Trinidad, with an average $G_D$ of 0.074. Group II only includes the Gloria and Samaria varieties, with a $G_D$ of 0.374.

Genetic analysis yielded an average genetic distance $G_D$ of 0.194 among the nine *T. pavonia* varieties (Table 3). As observed in Table 3, the $G_D$ between Angeles and Sandra varieties and between Carolina and Dulce was 0.0, revealing genetic similarity and, in spite of the number of isozymes tested, it was not possible to distinguish one variety from the other. In contrast, Vázquez-García et al. (2001) characterized these four varieties morphologically, differentiating them by flower color: white (Angeles), red (Sandra), pink (Carolina), and yellow (Dulce). The varieties with the highest $G_D$ value (0.421) were Gloria-Mariana, Gloria-Penelope, and Mariana-Samaria.

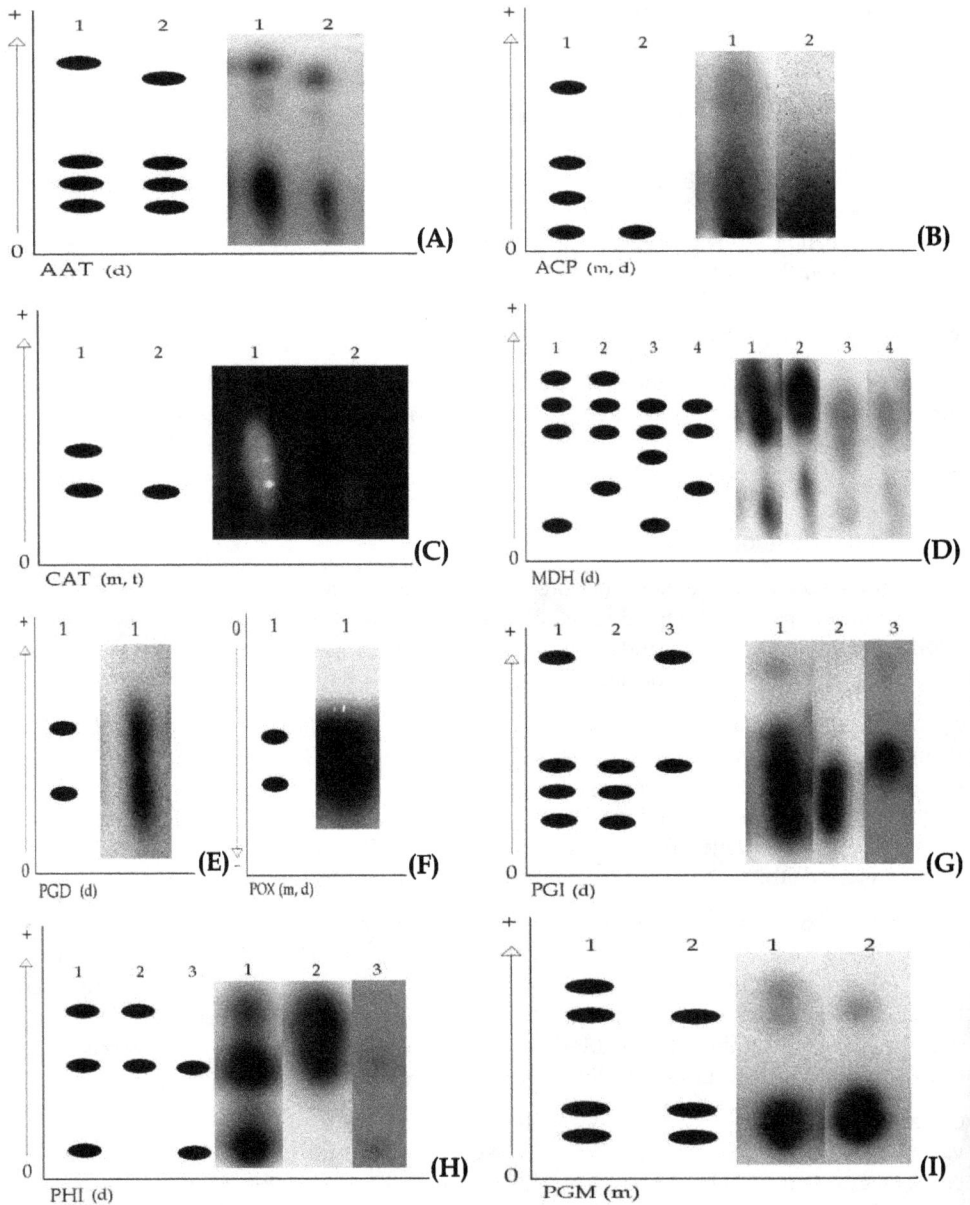

Fig. 3. Schematic representation of isozyme banding patterns (IBP) detected in nine *T. pavonia* (L.f.) DC. varieties, produced by A) AAT; B) ACP; C) CAT; D) MDH; E) PGD; F) POX; G) PGI; H) PHI; I) PGM. (m=monomer, d=dimer, t=trimer. "o" represents origin of electrophoresis.

| Variety | Isozyme and pattern | | | | | | | | |
|---|---|---|---|---|---|---|---|---|---|
|  | AAT | ACP | CAT | MDH | PGD | POX | PGI | PHI | PGM |
| Ángeles[1] | 1 | 1 | 2 | 1 | 1 | 1 | 1 | 1 | 2 |
| Carolina[2] | 1 | 1 | 1 | 1 | 1 | 1 | 1 | 1 | 2 |
| Dulce[2] | 1 | 1 | 1 | 1 | 1 | 1 | 1 | 1 | 2 |
| Gloria | 2 | 1 | 2 | 3 | 1 | 1 | 3 | 2 | 1 |
| Mariana | 1 | 1 | 1 | 1 | 1 | 1 | 2 | 3 | 2 |
| Penélope | 1 | 1 | 1 | 2 | 1 | 1 | 1 | 1 | 2 |
| Samaria | 2 | 2 | 1 | 4 | 1 | § | 1 | 2 | 2 |
| Sandra | 1 | 1 | 2 | 1 | 1 | 1 | 1 | 1 | 2 |
| Trinidad | 2 | 1 | 2 | 1 | 1 | 1 | 1 | 1 | 2 |

Table 2. Isozyme characterization of nine *T. pavonia* (L.f.) DC. varieties using nine isozyme systems. The number indicates the isozyme banding pattern (IBP) of each variety. AAT, aspartate amino transferase; ACP, acid phosphatase; CAT, catalase; MDH, malate dehydrogenase; PGD, phosphoglucose dehydrogenase; PGI, phosphoglucose isomerase; PGM, phosphoglucomutase; PHI, phoshohexose isomerase; POX, peroxidase. [1,2] varieties with the same number can not be distinguished with the evaluated isozymes. § It was not possible determinate the IBP.

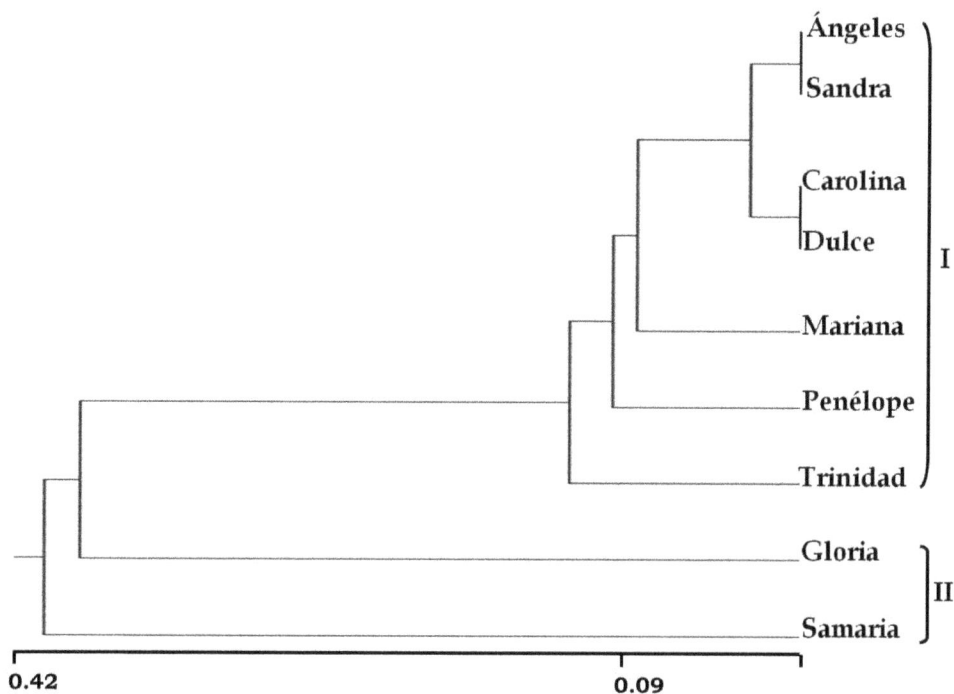

Fig. 4. Dendrogram of nine varieties of *Tigridia pavonia* based on Nei (1972) genetic distance, UPGMA method. The bar indicates the genetic distance among the varieties and the groups obtained with isozyme markers.

| Variety | ANG | CAR | DUL | GLO | MAR | PEN | SAM | SAN |
|---------|-----|-----|-----|-----|-----|-----|-----|-----|
| ANG |  |  |  |  |  |  |  |  |
| CAR | 0.031 |  |  |  |  |  |  |  |
| DUL | 0.031 | 0.000 |  |  |  |  |  |  |
| GLO | 0.287 | 0.330 | 0.330 |  |  |  |  |  |
| MAR | 0.098 | 0.064 | 0.064 | 0.421 |  |  |  |  |
| PEN | 0.098 | 0.064 | 0.064 | 0.421 | 0.133 |  |  |  |
| SAM | 0.374 | 0.330 | 0.330 | 0.374 | 0.421 | 0.246 |  |  |
| SAN | 0.000 | 0.031 | 0.031 | 0.287 | 0.098 | 0.098 | 0.374 |  |
| TRI | 0.064 | 0.098 | 0.098 | 0.207 | 0.169 | 0.169 | 0.287 | 0.064 |

Table 3. Estimation of genetic distance ($G_D$), according to the formula of Nei (1972), for the nine varieties of *T. pavonia*.

## 4.3 DNA markers

### 4.3.1 ISSR markers

The total number of fragments reproducible with the five ASSR primers was 40, with an interval of 350 to 1900 pb in the size of the amplified fragments. Of the 40 fragments, 35 were polymorphic, with an average of seven per primer. The percentage of polymorphic fragments varied from 66.6 to 100%, with an average of 87.5% (Table 4). These results are similar with those reported by Yamagishi et al. (2002) and Arzate-Fernández et al. (2005b), who used these ASSR primers and proved their efficiency for detecting high polymorphism percentage in different species of the genus *Lilium*.

One aspect that favors the effectiveness of the ASSR primers as well as the type of the motive replicate in its sequence, is the sequence of its anchor. The CT motive sequences produce higher polymorphism with respect to the AT replicates (Pradeep et al., 2002; Hu et al., 2003), which despite being the most abundant in the plant genomes, have the disadvantage that the amplification of DNA fragments is low; this may be due to the semi-complementarity of the primer in the alignment stage of the PCR (Fang & Roose, 1997). In this study, the level of distinction among the variables depended on the sequence of the anchor of the primer. Thus, it was possible to obtain 100% polymorphism among the varieties of *T. pavonia* with primers ASSR02 and ASSR35 (Table 4).

| Primer | Sequence (5'-3') | AF | PF | %P | *Rp* | FS (pb) |
|--------|------------------|----|----|-----|------|---------|
| 3´-ASSR02 | CTCTCTCTCTCTCT ATC | 9 | 9 | 100 | 3.12 | 500-1300 |
| 3´-ASSR15 | CTCTCTCTCTCTCT ATG | 8 | 6 | 85.7 | 2.2 | 350-1200 |
| 3´-ASSR20 | CTCTCTCTCTCTCT GCA | 7 | 6 | 85.7 | 2.2 | 600-1800 |
| 3´-ASSR29 | CTCTCTCTCTCTCT GTA | 6 | 4 | 66.6 | 1.8 | 550-1400 |
| 3´-ASSR35 | CTCTCTCTCTCTCT TGA | 10 | 10 | 100 | 5.54 | 350-1900 |
| Total or mean |  | 40 | 35 | 87.5 | 1.97 | 450 |

Table 4. Characteristics of the ISSR fragments amplified in nine varieties of *T. pavonia* (L.f.) DC.: total amplified fragments (AF), polymorphic fragments (PF), percentage of polymorphism (%P), resolving power (*Rp*) and size of the amplified fragments (FS) for each ASSR primer.

Prevost & Wilkinson (1999) described the resolution power (*Rp*) as a useful tool for evaluating the capacity of a primer in the distinction of various genotypes. In our results, the highest values of *Rp* were for the primers ASSR02 and ASSR35 (Table 4), with which the nine varieties were distinguished from each other. Escandón et al. (2005a) obtained similar results with just two ASSR primers and generated specific profiles for 18 of 21 collections of *Jacaranda mimosifolia*. The efficiency of the ASSR primers for discriminating genotypes in the inter-varietal level has also been reported in *Solanum tuberosum* (Prevost & Wilkinson, 1999), *Jacaranda mimosifolia* (Pérez de la Torre et al., 2003), *Nierembergia linaeriefolia* (Escandón et al., 2005b) and *Ficus carica* (Guasmi et al., 2006), among others.

The dendrogram generated with the ISSR data (Figure 5) formed two groups of varieties: group I included Carolina, Ángeles, Trinidad, Sandra, Dulce, Mariana and Penélope with an average $G_D = 0.19$ whereas the group II included Gloria and Samaria with a $G_D = 0.39$. The highest genetic association ($G_D = 0.07$) was found between the varieties Carolina and Ángeles, while the least related were Gloria and Dulce ($G_D = 0.91$).

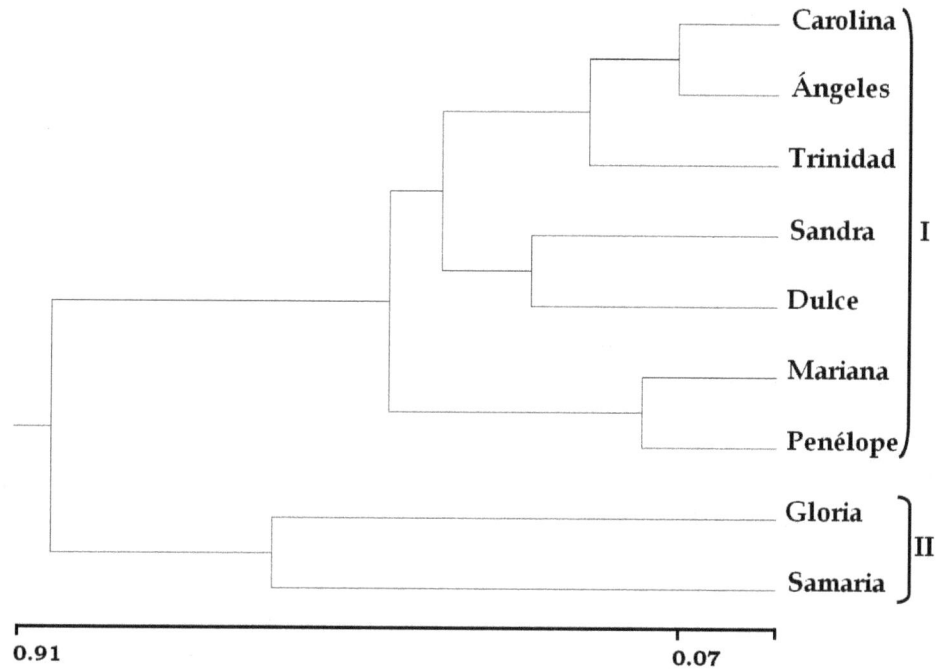

Fig. 5. Dendrogram of nine varieties of *T. pavonia* based on the genetic distance of Nei (1972), with the method UPGMA. The bar indicates the genetic distance among the varieties and the groups, calculated from ISSR data.

### 4.3.2 RAPD markers

In the present study, PCR amplification with 10, 15 and 20 length base RAPD primers led to reproducible fragment patterns for all varieties of *T. pavonia* evaluated. The majority of those RAPD fragments ranged from 250 to 2800 bp (Figure 6). For 10 base primers, the average of total and the polymorphic fragments generated per primer were 8.8 and 8.2, respectively, whereas for 15 base primers, those values were 11 and 10.4, respectively, and for 20 base primers, 9.4 and 9, respectively (Table 5).

Generally, in RAPD analysis 10 base primers are preferred, nevertheless, because of the high annealing temperature applied in PCR reaction than their original, some 10 base primers could not hybridize with template DNA, generating only a few minor bands or none (Yamagishi et al., 2002), while the advantage of long primers is the smaller number of fragments containing repetitive DNA, thereby increasing the specificity and reproducibility of DNA fragments (Debener & Mattiesch, 1998). In the present study, the positive correlation (r = 0.99; p ≤ 0.05) between primer length and the percentage of polymorphism generated per each primer was observed, indicating that the efficiency of primers to generate polymorphic fragments it increased by primer length. Thus, 15 or 20 base primers generated more DNA fragments (55), and a greater number of polymorphic fragments (52), in comparison with those obtained by the 10 base primers (44, and 41, respectively) (Table 5). The high efficiency of long primers to generate a large number of RAPD markers has been also reported in other species as *Lilium* spp. (Yamagishi et al., 2002), *Vitis vinifera* (Solouki et al., 2007) and *Matricaria chamomilla* (Solouki et al., 2008).

| Primer | %GC | AF | PF | %P | *Rp* |
|--------|-----|----|----|----|------|
| Y24* | 70 | 8 | 6 | 75 | 1.6 |
| Y29* | 70 | 10 | 10 | 100 | 5.1 |
| Y37* | 70 | 3 | 2 | 66.6 | 0.9 |
| Y38* | 70 | 15 | 15 | 100 | 7.5 |
| Y41* | 80 | 8 | 8 | 100 | 3.3 |
| Average | 72 | 8.8 | 8.2 | 93.1 | 3.6 |
| P619** | 66 | 13 | 13 | 100 | 5.8 |
| P625** | 66 | 8 | 6 | 75 | 2.6 |
| P628** | 60 | 4 | 3 | 75 | 1.3 |
| P635** | 73 | 17 | 17 | 100 | 8.9 |
| P647** | 60 | 13 | 13 | 100 | 6.4 |
| Average | 65 | 11 | 10.4 | 94.5 | 5 |
| P495*** | 60 | 11 | 10 | 90.9 | 3.9 |
| P496*** | 60 | 6 | 5 | 83.3 | 2.6 |
| P497*** | 60 | 8 | 8 | 100 | 3.8 |
| P498*** | 60 | 12 | 12 | 100 | 4.9 |
| P500*** | 60 | 10 | 10 | 100 | 3.8 |
| Average | 60 | 9.4 | 9 | 95.7 | 3.8 |

Table 5. Percentage of GC (%GC), total amplified fragments (AF), polymorphic fragments (PF), percentage of polymorphism (%P) and resolving power (*Rp*) for each RAPD primer used. *10 base (Yamagishi, 1995). **15 base (Yamagishi et al., 2002). ***20 base (Debener & Mattiesch 1998).

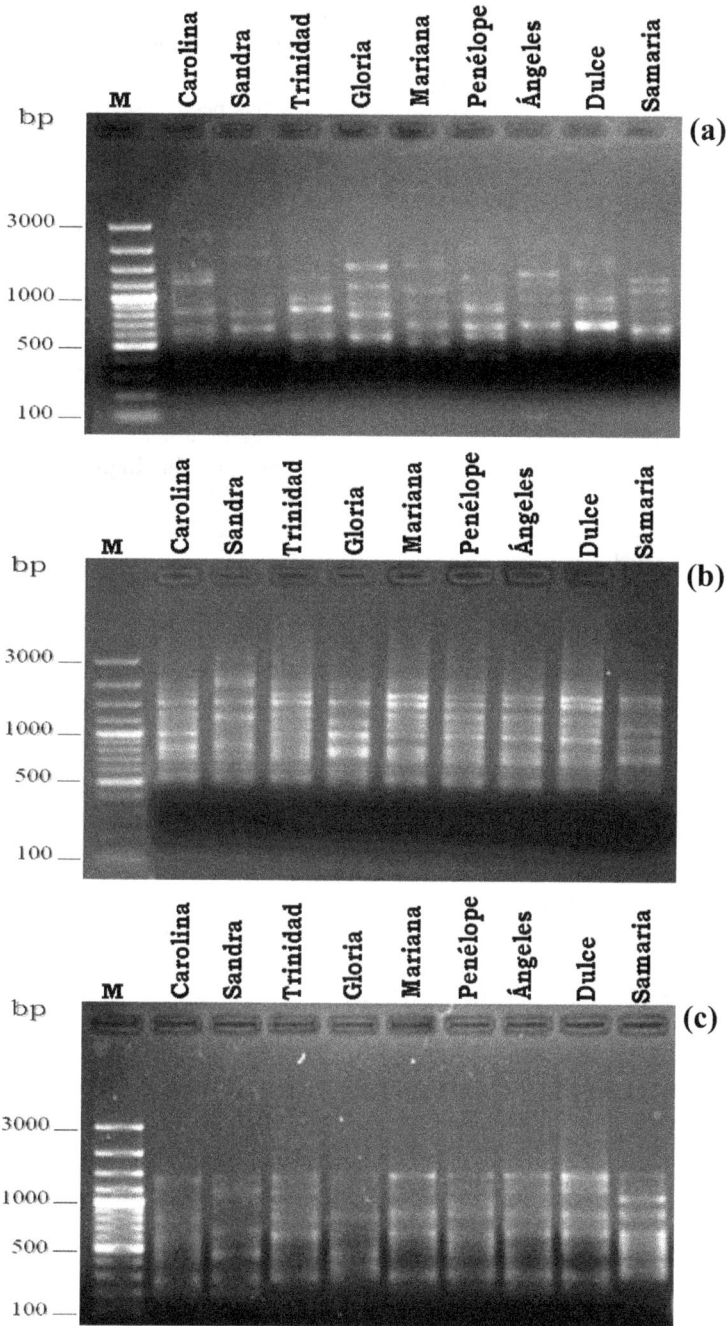

Fig. 6. RAPD profiles of nine varieties of *T. pavonia* generated by the 10 base primer Y38 (a), 15 base primer P647 (b) and 20 base primer P497 (c).

Although it is no clear exactly why the long primers produced more polymorphic bands, it has been reported that the GC content may be a factor that determine the efficiency of a primer (Solouki et al., 2007), since GC content is associated with annealing temperature and related to generation of more DNA fragments. Thus, it has been observed that with long primers with a lower GC content, the efficiency for amplifying polymorphic bands is higher, in comparison with short primers with a higher GC content (Ye et al., 1996; Solouki et al., 2007). According to this, it is possible that the GC content has favored the major efficiency of the 15 (65 % GC) and 20 (60 % GC) base primers used in the present study, because more DNA fragments were amplified and also a greater percentage of polymorphism, in comparison with those amplified with 10 base primers with 72 % GC content (Table 5). Our results are closely similar with those reported by Solouki et al., (2008) where it were obtained more DNA fragments and 100 % of polymorphism, using a long primer with low GC content.

The high efficiency of long primers in the genetic differentiation of T. pavonia varieties was also confirmed with the measurement of resolving power. The highest values of Rp belonged to the 15 and 20 base primers (5 and 3.8, respectively) (Table 5), indicating a better distinction of the varieties. So, these results also confirm the utility of the Rp as measure of capacity of a primer to discriminate among closely related individuals as was pointed out by Prevost &Wilkinson (1999) and Escandón et al. (2007).

The dendrograms based on UPGMA analysis of the 10, 15, 20 and the pooled (10, 15 and 20) RAPD data showed the genetic differentiation of the nine varieties of T. pavonia (Figure 7). In the dendrogram generated with decamers, the $G_D$ among the varieties ranged from 0.20 to 0.69, with an average of 0.24. For the 15 base dendrogram, the $G_D$ ranged from 0.13 to 0.78, with an average of 0.32, while the dendrogram of the 20 base showed a range of $G_D$ of 0.11 to 0.67, with an average of 0.28, and finally the dendrogram with pooled data showed a range of $G_D$ of 0.20 to 0.68, with an average of 0.24.

All four dendrograms clearly grouped the varieties in two major clusters. In the dendrograms of 10 and 15 bases, and in the pooled, the cluster I consisted of the varieties Carolina, Dulce, Trinidad, Penelope, Angeles, Mariana, and Sandra, whereas the cluster II grouped the varieties Gloria and Samaria. The close relationship observed in these dendrograms, among the varieties collected in Tenancingo and Temascaltepec municipalities, might be the result of having some common morphological characters such as: number of shoots by bulb, number of nodes by shoot, length of leaf, and number flowers by shoot. The separation of Gloria and Samaria collected in Temoaya, could be due to the highest values for length of shoot, length of internode, and length of floral escape, besides that the flowering of these varieties is delayed in comparison with the rest of them, as it was reported in our previous study (Piña-Escutia et al., 2010).

On the other hand, the dendrogram developed with 20 base primers also grouped the cluster I similar to the dendrograms generated with other primers sets, except for the variety Sandra, which was grouped with the varieties Gloria and Samaria in the cluster II. Thus, this dendrogram showed a best differentiation of the varieties evaluated, confirming the utility of long primers in generation high polymorphism and genetic discrimination of plant species, which was also reported by Solouki et al. (2007), and Solouki et al. (2008).

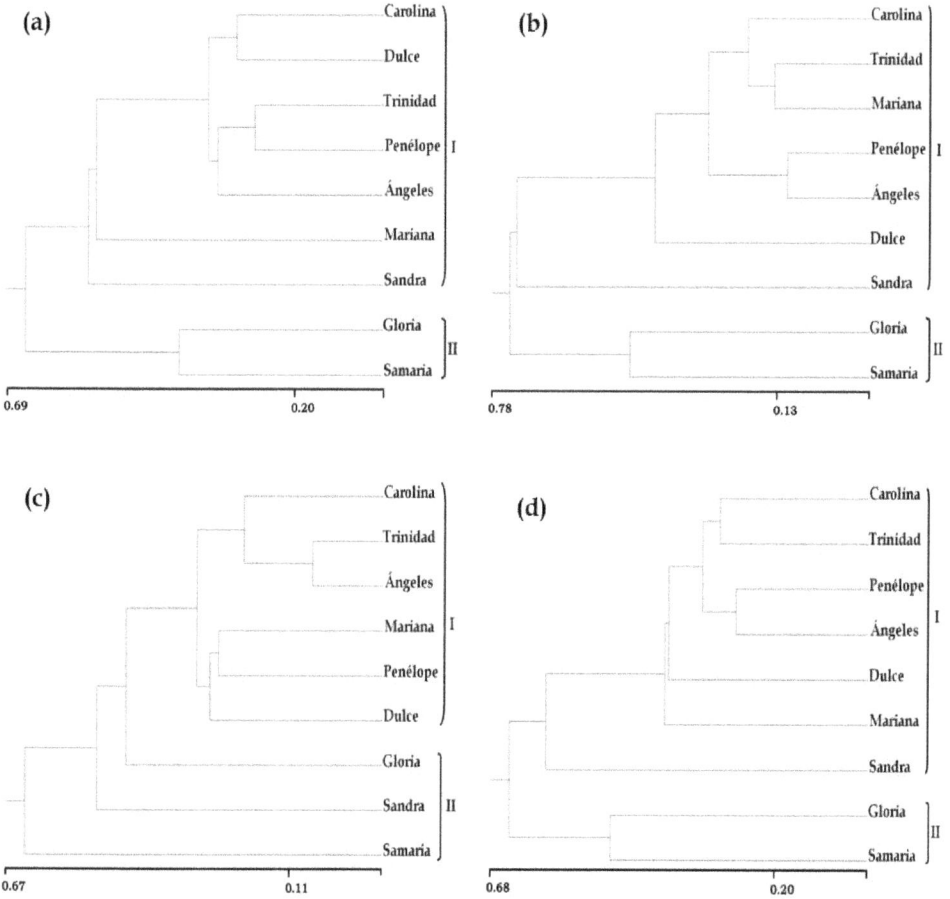

Fig. 7. Dendogram generated using UPGMA analysis, showing relationships between nine varieties of *T. pavonia*, based on RAPD data of 10 (a), 15 (b), 20 (c) and the pooled (10, 15 and 20) base of length primer (d). The bar indicates the genetic distance among the varieties and the groups.

## 5. General conclusions and perspectives

The assessment of genetic variability of Mexican native species is fundamental for the conservation, and plant variety protection. In this review we have conducting the morphological analysis in combination with the molecular assays to distinguish *T. pavonia* varieties. The analyses of the morphological and DNA markers (ISSR and RAPD) seem to be more efficient than the isozymes, because whereas the former showed polymorphism among the nine varieties evaluated and permit the distinction among them, the biochemical characterization revealed that the varieties Angeles and Sandra as well as Carolina and Dulce were the same, indicating the low discrimination power of isozymes which can be due to inadequate genome coverage. So, our results indicate that the marker assisted characterization can be applied to complement existing conventional breeding programmes and, despite its application in genetic improvement of horticultural crops is limited, their use as balance between molecular and non-molecular methods can offers opportunities of improved phenotypic selection in the future.

## 6. Acknowledgment

We are grateful to RED TIGRIDIA (SINAREFI, México) for covering the publishing payment.

## 7. References

Arzate-Fernández, A.M.; Mejía G, C.O.; Nakazaki, T.; Okumoto, Y. & Tanisaka, T. (2005a). Isozyme electrophoretic characterization of 29 related cultivars of lily (*Lilium* spp.). *Plant Breeding* Vol.124, No. 1, (February 2005), pp. 71-78. ISSN 1437-0523

Arzate-Fernández, A.M., Miwa, M.; Shimada, T.; Tonekura, T. & Ogawa, K. (2005b). Genetic diversity of miyamasukashi-yuri (*Lilium maculatum* Thumb. Var. Bukosanense), an endemic populations endangered species at Mount Buko, Saitama, Japan. *Plant Species Biology* Vol.20, No.1, (March 2005), pp. 57-65. ISSN 1442-1984

Choudhary, K.; Choudhary O.P. & Shekhawat, N.S. (2008). Marker assisted selection: a novel approach for crop improvement. *American-Eurasian Journal of Agronomy* Vol.1, No.2, (August 2008) pp. 26-30. ISSN 1995-896X

Debener, T. & Mattiesch, I. (1998). Effective pairwise combination of long primers for RAPD analysis in roses. *Plant Breeding* Vol.117, No.2, (May 1998), pp. 147-151. ISSN 1437-0523

Escandón, A.; Pérez de la Torre, M.; Acevedo, A.; Marucci-Poltri, S. & Mijayima, I. (2005a). Anchored ISSR as molecular marker to characterize accesions of *Jacaranda mimosifolia* L. Don. *Acta Horticulturae* Vol.683, No.1, pp. 121-127. ISSN 0567-7572

Escandón, A.; Pérez de la Torre, M.; Soto, M.S. & Zelener, N. (2005b). Identificación de clones selectos de *Nierembergia linariaefolia* mediante microsatélites anclados. *Revista de Investigaciones Agropecuarias* Vol.34, No.1, (Abril 2005) pp. 5-17. ISSN 1669-2614, Available from http://www.inta.gov.ar/ediciones/ria/34_1/01.pdf

Fang, D.Q. & Roose, M.L. (1997). Identification of closely related citrus cultivars with inter-simple sequence repeat markers. *Theoretical and Applied Genetics* Vol.95, No.3, (August 1997), pp. 408-417. ISSN 1432-2242

Felsenstein, J. (1990). Phylip Manual, version 3.3. University Herbarium, University of California, Berkley, CA, USA.

Franco, J.; Crossa, J.; Ribaut, J.M.; Betran, J; Warburton, M.L. & Khairallah, M. (2001). A method for combining molecular markers and phenotypic attributes for classifying plant genotypes. *Theoretical and Applied Genetics* Vol.103, No.6-7, (November 2001), pp. 944-952. ISSN 1432-2242

Glaszmann, J.C.; De los Reyes B.G. & Khush, G.S. (1988). Electrophoretic variation or isozymes in plumules of rice (*Oriza sativa* L.) a key to the identification of 76 alleles at 24 loci. *IRRI Research Paper Series* Vol.134, No.11, (November 1988), pp. 1-14. ISSN 0115-3862

Guasmi, F.; Ferchichi, A.; Farés, K. & Touil, L. (2006). Identification and differentiation of *Ficus carica* L. cultivars using Inter simple sequence repeat markers. *African Journal of Biotechnology* Vol.5, No.15, (August 2006), pp. 1370-1374. ISSN 1684-5315 Available                                                            from http://www.academicjournals.org/AJB/PDF/pdf2006/3Aug/Guasmi%20et%20al .pdf

Hernández, F. (1959). Historia natural de la Nueva España. Universidad Nacional Autónoma de México. Tomo II. Vol. I. 476.

Hu, J.; Nakatani, M.; García, L.A.; Kuranouchi, T. & Fujimura, T. (2003). Genetic analysis of sweetpotato and wild relatives using inter-simple sequence repeats (ISSRs). *Breeding Science* Vol.53, No.4, (December 2003), pp. 297-304. ISSN 1437-3735 Available from http://www.jstage.jst.go.jp/article/jsbbs/53/4/297/_pdf

Ibitoye, D.O. & Akin-Idowu P. E. (2010). Marker-assisted-selection (MAS): A fast track to increase genetic gain in horticultural crop breeding. *African Journal of Biotechnology* Vol.9, No.52, (December 2010), pp. 8889-8895. ISSN 1684-5315 Available from http://www.academicjournals.org/AJB/PDF/pdf2010/21DecConf/Ibitoye%20an d%20Akin-Idowu.pdf

Ishikawa, R. (1994). Genetical studies on isozyme genes in rice (in japanese with english summary). *Bulletin Faculty of Agriculture Hirosaki University* Vol.57, pp. 105-180. ISSN 0073-229X

Jehan, T. & Lakhanpaul, S. (2006). Single nucleotide polymorphism (SNP)-Methods and applications in plant genetics: A review. *Indian Journal of Biotechnology* Vol.5, No.4, (October 2006), pp. 435-439. ISSN 0972-5849 Available from http://nopr.niscair.res.in/bitstream/123456789/5608/1/IJBT%205%284%29%2043 5-459.pdf

Kumar, S.; Prasad, K.V. & Choudhary, M.L. (2006). Detection of genetic variability among *Chrysanthemum* radiomutants using RAPD markers. *Current Science* Vol.90, No.8 (April 2006), pp. 1108-1113. ISSN 001-3891 Available from http://www.ias.ac.in/currsci/apr252006/1108.pdf

Luna-Paez, A.; Valadez-Moctezuma, E.; Barrientos-Priego, A.F. & Gallegos-Vázquez, C. (2007). Caracterización de *Opuntia* spp. mediante semilla con marcadores RAPD e

ISSR y su posible uso para diferenciación. *Journal of the Professional Association for Cactus Development* Vol.9, (June 2007), pp. 43-59. ISSN 1938-6648 Available from http://www.jpacd.org/?modulo=JS&ID=10

Marotti, I.; Bonetti, A.; Minelli, M.; Catizone, P. & Dinelli, G. (2007) Characterization of some Italian common bean (*Phaseolus vulgaris* L.) landraces by RAPD, semi-random and ISSR molecular markers. *Genetic Resources and Crop Evolution* Vol.54, No.1, (February 2007), pp. 175-188. ISSN 1573-5109

Muthusamy, S.; Kanagarajan, S. & Ponnusamy, S. (2008). Efficiency of RAPD and ISSR markers system in accessing genetic variation of rice bean (*Vigna umbellate*) landraces. *Electronic Journal of Biotechnology* Vol.11, No.3, (July 2008), pp. 1-10. ISSN 0717-3458                    Available                    from http://www.ejbiotechnology.cl/content/vol11/issue3/full/8/index.html

Nei M. 1972 Genetic distance between populations. *American Naturalist* 106: 282-283.

Pérez de la Torre, M.; Acevedo, A.; Serpa, J.C.; Mijayima, I. & Escandón, A.S. (2003). Puesta a punto de la técnica de microsatélites anclados para la caracterización de individuos selectos de jacaranda. En: Mascarini, L., F. Vilella., and E. WRIGHT, (eds). *Floricultura en la Argentina*. Investigación y Tecnología de Producción. 3-12.

Piña-Escutia, J.L.; Vences-Contreras, C.; Gutiérrez-Martínez, M.G.; Vázquez-García, L.M. & Arzate-Fernández, A.M. (2010). Morphological and molecular characterization of nine botanical varieties of *Tigridia pavonia* (L.f.) DC. *Agrociencia* Vol.44, No.2, (Febrero-Marzo 2010), pp. 147-158. ISSN 1405-3195 Available from http://www.colpos.mx/agrocien/Bimestral/2010/feb-mar/art-3.pdf

Pradeep, R.M.; Sarla, N. & Siddiq, E.A. (2002). Inter simple sequence repeat (ISSR) polymorphism and its application in plant breeding. *Euphytica* Vol.128, No.1, (November 2002), pp. 9-17. ISSN 1573-5060

Prevost, A. & Wilkinson, M.J. (1999). A new system of comparing PCR primers applied to ISSR fingerprint of potato cultivars. *Theoretical and Applied Genetics* Vol.98, No.1 (January 1999), pp. 107-112. ISSN 1432-2242

Solouki, M.; Rigi Nazhad, N.; Vignani, R.; Ali Siahsar, B.; Kamaladini, H. & Emamjomeh, A. (2007). Polymorphism of some native sistan grapes assesed by long and short primers for RAPD markers. *Pakistan Journal of Biological Sciences* Vol. 10, No.12, (August    2007),    pp.    1996-2001.    ISSN    1028-8880    Available    from http://scialert.net/pdfs/pjbs/2007/1996-2001.pdf..

Solouki, M.; Mehdikhani, H.; Zeinali H. & Emamjomeh, A. A. (2008). Study of genetic diversity in chamomile (*Matricaria chamomilla*) based on morphological traits and molecular markers. *Scientia Horticulturae* Vol.117, No. 3, (July 2008), pp. 281–287. ISSN 0304-4238

Stuber, C.W.; Wendel, J.; Goodman, M. & Smith, J. (1988). Techniques and scoring procedures for starch gel electrophoresis of enzymes from maize (*Zea mays* L.) Tech. Bull. 286. North Carolina State University. Raleigh, North Carolina. 217.

Tanksley, S.D. & Rick, C.M. (1980). Isozyme gene linkage map of the tomato: applications in genetics and breeding. *Theoretical and Applied Genetics* Vol.58, no.2 (March 1980), pp. 161-170. ISSN 1432-2242

Tapia, C.E.; A. Guillén, H. & Gutiérrez, E.M.A. (2005). Caracterización genética de materiales de piña (*Ananas* spp.) mediante RAPD e ISSR. *Revista Fitotecnia Mexicana* Vol.28, No.3, (Julio-Septiembre 2005), pp. 187-194. ISSN 0187-7380 Available from http://www.revistafitotecniamexicana.org/documentos/28-3/2a.pdf

Torres, A.M., Soost, R.K., & Diedenhofen, U. (1978). Leaf isozymes as genetic markers in citrus. *American Journal of Botany* Vol.65, No.8, (September 1978), pp. 869-881. ISSN 1537-2197

Vallejos, E. (1983). Enzyme activity staining. In: Tanksley, S. D., and T. J. Orton (eds). *Isozymes in Plant Genetics and Breeding*, Part A. Elsevier, Amsterdam, Netherlands. 459-416.

Vázquez-García, L.M.; Przybyla, A.A.; De la Cruz, T.E.; Torres, N.H. & Rodríguez, G. (2001). Morphological description of nine botanical varieties of *Tigridia pavonia* (L. F.) Ker. Gawl. *Journal of Applied Botany* Vol.75, No.1-2, pp. 14-19. ISSN 1439-040X

Vicente, M.C. & Fulton, T. (2003). Tecnologías de Marcadores Moleculares para Estudios de Diversidad Genética de Plantas: Módulo de Aprendizaje. Illus. Nelly Giraldo. Instituto Internacional de Recursos Fitogenéticos (IPGRI), Roma, Italia.1: 1-52.

Wen, C.S. & Hsiao, J.Y. (2001). Altitudinal genetic differentiation and diversity of Taiwan Lily (*Lilium longiflorum* var. *formosanum*; liliaceae) using RAPD markers and morphological characters. *International Journal of Plant Science* Vol.162, No.2, (March 2001), pp. 287-295. ISSN 1058-5893

Wendel, J.F. & Weeden, N.F. (1990). Visualization and interpretation of plant isozymes. In: Soltis D. E., and P. S. Soltis. (eds). *Isozymes in plant biology*. Dioscorides Press. Portland, Oregon. pp: 5-45.

Xu, Y. & Crouch J.H. (2008). Marker-assisted selection in plant breeding: from publications to practice. *Crop Science* Vol.48, No.2, (March-April 2008), pp. 391-407. ISSN 1435-0653 Available from doi: 10.2135/cropsci2007.04.0191

Yamagishi, M.; Abe, H.; Nakano, M. & Nakatsuka, A. (2002). PCR-based molecular markers in Asiatic hybrid lily. *Scientia Horticulturae* Vol.96, No.1-4, (December 2002), pp. 225–234. ISSN 0304-4238

Yamagishi, M. (1995). Detection of section-specific random amplified polymorphic DNA (RAPD) markers in *Lilium*. *Theoretical and Applied Genetics* Vol.91, No.6-7, (November 1995), pp. 830-835. ISSN 1432-2242

Ye, G.N.; Hemmat, M.; Lodhi, M.A.; Weeden, N.F. & Reisch, B.I. (1996). Long primers for RAPD mapping and fingerprinting of grape and pear. *Biotechniques* Vol.20, No.3, (March 1996), pp. 368–371. ISSN 0736-6205 Available from http://www.biotechniques.com/multimedia/archive/00053/19962003368_53006a.pdf

Ye, Y.M.; Zhang, J.W.; Ning, G.G. & Bao, M.Z. (2008). A comparative analysis of the genetic diversity between inbred lines of *Zinnia elegans* using morphological traits and RAPD and ISSR markers. *Scientia Horticulturae* Vol.118, No.1, (September 2008), pp. 1-7. ISSN 0304-4238

Yeh, F.C. & Boyle, T.J.B. (1999). Population genetics analysis of codominant and dominant markers and quantitative traits. *Belgium Journal of Botany* Vol.129, pp. 157.

Young, N.D. (1999). A cautiously optimistic vision for marker assisted breeding. *Molecular Breeding* Vol.5, No.6, (December 1999), pp. 505-510. ISSN 1572-9788

Zhou, Z.; Miwa, M. & Hogetsu, T. (1999). Analysis of genetic structure of a *Suillus grevillei* population in a Laris kaemferi stand by polymorphism of inter-simple sequence repeat (ISSR). *New Phytologist* Vol.144, No.1, (October 1999), pp. 55-63. ISSN 1469-8137 Available from http://onlinelibrary.wiley.com/doi/10.1046/j.1469-8137.1999.00504.x/pdf

# 10

# Molecular Markers to Access Genetic Diversity of Castor Bean: Current Status and Prospects for Breeding Purposes

Santelmo Vasconcelos[1], Alberto V. C. Onofre[1], Máira Milani[2],
Ana Maria Benko-Iseppon[1] and Ana Christina Brasileiro-Vidal[1]
*[1]Laboratory of Plant Genetics and Biotechnology, Federal University of Pernambuco,*
*[2]Embrapa Algodão, Campina Grande*
*Brazil*

## 1. Introduction

The spurge family (Euphorbiaceae) is one of the most diverse and numerous clades of the angiosperms, including several species of great economic importance as rubber tree (*Hevea brasiliensis*), cassava (*Manihot esculenta*), and some oil seed crops, as candlenut (*Aleurites moluccana*), physic nut (*Jatropha curcas*) and castor bean (*Ricinus communis*). Castor bean, the single member of the African genus *Ricinus* (subfamily Acalyphoideae), presents a wide variation regarding vegetative traits such as leaf and stem colors, number and size of leaf lobes and presence of wax covering the stem (Popova & Moshkin, 1986; Savy-Filho, 2005; Webster, 1994; see Fig. 1). Depending on the environmental conditions, even the vegetative habit may vary, although it is more likely in a shrubby form (Webster, 1994). However, the most conspicuous variability is related to reproductive characters, as color shape and size of seeds, number of flowers per raceme, peduncle length and fruit dehiscence (Figs. 1 and 2) as described by Popova & Moshkin (1986).

Castor oil, which has a long history of use for medicinal purposes (see Gaginella et al., 1998), has been considered a promising raw material for the production of renewable energy in tropical countries. Besides, castor bean has been traditionally cultivated for the production of lubricants and paints (see Berman et al., 2011; Ogunniyi, 2006; Scholz & Silva, 2008). Mainly in the semi-arid regions, a xerophytic-like as castor bean can be grown in areas with higher farming limitations, not intended for other crops (Ogunniyi, 2006). Furthermore, the biodiesel derived from castor oil has several advantages over other vegetable oils due to the presence of 5% more oxygen, low levels of residual phosphorus and carbon, high cetan number, solubility in alcohol and absence of aromatic hydrocarbons (Ogunniyi, 2006; Scholz & Silva, 2008). The high viscosity of the castor oil is due to the high percentage of ricinoleic acid (a hydroxycarboxylic acid), which is a limiting factor for the use of pure castor bean diesel in the engines (Pinzi et al., 2009). However, the employment of this biodiesel blended with petrodiesel can be exploited in regions with severe winter. This is a highly recommended procedure because of its low freezing point and the lubricant power afforded by castor oil, as well as all other advantages associated to the utilization of renewable energy resources (see Berman et al., 2011; Demirbas, 2007; Ogunniyi, 2006; Pinzi et al., 2009; Singh, 2011).

Fig. 1. Different raceme types observed in castor bean accessions held by Embrapa Algodão (Brazil). Inflorescences of a cultivar ('BRS Nordestina') and a dwarf lineage ('CSRD-2') are shown in (a) and (b), respectively. Observe in (b) a pistillate flower with red stigmas in the left-superior corner and a multi-staminate flower in the right-inferior corner. A large raceme, typical of the cultivar 'BRS Energia', is shown in (c). Racemes with long peduncles of the accessions 'CNPAM 93-168' and 'BRS Nordestina' are shown in (d) and (e), respectively. Compact raceme characteristic of a castor bean subespontaneous population from northeastern Brazil, (Buíque – PE) (f), and of the cultivar 'BRS Paraguaçu' (g). Spineless fruits of the lineage 'BRA 10740' in (h).

The development of new cultivars with traits of interest and adapted to specific microclimates is only possible when there is available knowledge about the extant genetic diversity of the species (Gepts, 2004). Despite the recent publication of the castor bean genome (Chan et al., 2010), little is known about the actual genetic diversity of this species. Genetic diversity analyses of castor bean germplasm collections worldwide have showed low levels of variability and lack of geographically structured genetic populations, regardless of a marker system used (e.g. Allan et al., 2008; Foster et al., 2010; Qiu et al., 2010). Thus, the remarkable phenotypic variation observed in castor bean do not seem to reflect a high genetic diversity, similarly to the reported for physic nut, in which variations in epigenetic mechanisms may have a more important role in the diversity of the species than genetic variability per se (Yi et al., 2010).

Fig. 2. Representation of the variability in color and size observed among seeds of castor bean.

In this work, we provide a review on the current status of genetic diversity analysis in castor bean. Moreover, we present the results of our data mining efforts on screening for genomic simple sequence repeat (SSR) primers in addition to the previously reported expressed sequence tag-SSR (EST-SSR) sequences. We performed genotyping among castor bean accessions with inter simple sequence repeat (ISSR) primers from the University of British Columbia (UBC) set and primer combinations from amplified fragment length polymorphism (AFLP) Starter Primer Kit (Invitrogen, Carlsbad, USA) for amplicon generation. Furthermore, we have tested the characterization of distribution of large microsatellite clusters along the castor bean chromosomes by means of fluorescent *in situ* hybridization (FISH). Our results in addition to compiled data from literature will be highly useful for breeding programs, providing information about genetic diversity and tools for genetic mapping in this important crop.

## 2. Diversity analyses with molecular markers in castor bean

Several molecular markers are available for germplasm characterization and identification of cultivated plant varieties. The profile analysis of multilocus DNA markers, also called DNA fingerprinting, is a potential source of informative marker bands, which allows a reliable differentiation among cultivars (Tanya et al., 2011), wild populations (Andrade et al., 2009), species and even related genera (Simon et al., 2007). Additionally, molecular markers are very stable, in contrast to morphological characters, which may be influenced by environmental factors and having continuous variation and high plasticity (Weising et al., 2005).

Unlike other important oilseed crops, as oil palm (*Elaeis guineensis*), soybean (*Glycine max*), sunflower (*Helianthus annuus*), and some Euphorbiaceae species, as cassava and rubber tree, castor bean diversity is still poorly characterized by means of molecular marker systems (see Billotte et al., 2010; Feng et al., 2009; Sayama et al., 2011; Sraphet et al., 2011; Talia et al., 2010). In fact, the species had been overlooked until the late 2000s, when analyses regarding genetic diversity of germplasm collections were first published (see Allan et al., 2008). However, castor bean was the first member of the Euphorbiaceae family with the whole genome published (Chan et al., 2010), a fact that will be of great importance for characterizing the genetic base of the species.

## 2.1 Genetic diversity characterization with dominant markers

AFLP, ISSR and random amplified polymorphic DNA (RAPD) are among the most widely used marker systems in DNA fingerprinting. Although the differentiation between allelic types is hampered in the output data from these molecular markers, many features have made them quite widespread, such as low costs and the possibility of generating a large amount of informative marker bands in a short time. Besides, there is no need for prior knowledge about DNA sequences of the studied organism when this kind of molecular markers is used (Weising et al., 2005).

As mentioned above, just a few analyses were carried out using dominant markers to access polymorphisms among castor bean accessions. Despite the great potential of ISSR and AFLP in characterizing the genetic diversity of several crops (Kumar et al., 2009; Weising et al., 2005), it is noteworthy that these powerful marker systems have been underused in genetic diversity analyses with this species. To the best of our knowledge, the only study in which AFLP markers were used to describe the genetic diversity of the species was performed by Allan et al. (2008). In a preliminary application of 16 AFLP primer combinations, these authors reported low levels of variability among 14 castor bean genotypes from different regions of the world. Thereafter, the authors selected the three most polymorphic primer combinations and applied them to a wider number of accessions (41 in total) that indicated weak geographically structured populations among germplasm collections of the five continents. These results were quite similar to those obtained with genomic SSR markers in the same work (discussed below), and this was the first indicative of a narrower genetic base than first thought for the species. However, due to the small number of generated marker bands (only 119), the low polymorphism sampled by Allan et al. (2008) could be an underestimation of the factual levels of information that this marker system might reach within the species.

Differences regarding polymorphism levels of AFLP markers have been reported for many other crops (Weising et al., 2005). The average percentage of polymorphic markers obtained by Tatikonda et al. (2009) for physic nut, for instance, was higher than that found by Pamidimarri et al. (2010) (88.2% and 61.2%, respectively). Additionally, even when the same primer combination (E-ACA + M-CAT) was used, different polymorphism levels were obtained [82.8% by Tatikonda et al. (2009) and 68.1% by Pamidimarri et al. (2010)]. This fact may be occurred due to differences in genetic diversity levels between the two sampled germplasm collections.

Results concerning lack of genetic structure through geographic distribution, as first indicated with AFLP markers, have been obtained by other research groups. Analyzing 32 castor bean lines from different countries using an association between RAPD markers and quantitative phenotypic traits (volume and weight of seeds, root length, time of germination and first flowering), Milani et al. (2009) obtained a certain degree of convergence between the resultant clusters regarding data from these two different approaches. Once again, accessions from different origins were put together, confirming the previously reported lack of geographically structured clusters among castor bean genotypes.

Afterwards, Gajera et al. (2010) have published a wider analysis with these low cost dominant markers, in which 200 RAPD primers and 21 ISSR primers were tested for generation of informative characters among Indian castor bean lines, and thus 30 and five, respectively, were selected for further polymorphism screening using 22 genotypes. Like in the previous RAPD analysis, the authors have found a remarkable level of polymorphism with both marker systems, in particular with RAPD analysis, in which 80.1% of the 256 marker bands were polymorphic. However, the lower variability obtained with ISSR markers by Gajera et al. (2010) may be occurred due to the smaller number of used primers (only five) compared to the RAPD approach. In general, ISSR markers tend to be more polymorphic because of its target site in the genome. Microsatellite sequences are known as one of the most variable and widespread types of repetitive DNA (Edwards et al., 1991; Weising et al., 2005). Thus, these results obtained by amplifying ISSR markers, which is a powerful tool that has been widely used to detect polymorphism either among crop cultivars or among wild populations of plants (Reddy et al. 2002), may not reveal the real polymorphism level in castor bean. For walnut (*Juglans regia*), for instance, Christopoulos et al. (2010) found a higher level of polymorphism with ISSR markers (82.8%) than the reported values by Nicese et al. (1998) for RAPD markers (25%).

Therefore, in order to increase the repertory of available ISSR markers for diversity analysis of castor bean, we have tested 60 primers from the UBC set (Table 1) for amplification among three genotypes of the species ('BRS Nordestina', 'BRS Paraguaçu' and 'Epaba 81'), using the protocol described by Bornet & Branchard (2001). PCR conditions of cycle intervals and annealing temperatures were used as described by Amorim (2009). Extraction and purification of genomic DNA were according to the methodologies described by Weising et al. (2005; CTAB protocol I) and Michaels et al. (1994), with minor modifications. Our results have revealed a preference for amplification of regions with AG/CT repeats in sampled primers and annealing temperatures used (Table 1). Most primers directed to AT microsatellite repeats have not amplified any DNA fragment with PCR conditions herein referred, although there is a high density of these regions in castor bean genome (as presented below). Possibly, the relatively high annealing temperatures [see, Gajera et al. (2010) and Tanya et al. (2011)], which were used to increase the PCR stringency, may have affected the amplification capability of the primers. However, the higher specificity of the DNA amplification, which was propitiated by this measure, ensures the validity of generated markers and the reproducibility of results.

Additionally, we carried out amplification tests with all 64 primer combinations from the previously cited AFLP kit (Table 2) with genomic DNA from the same used castor bean genotypes in the ISSR assay, following the protocol recommended by the manufacturer. All combinations of primers MseI-CAA, MseI-CAG, MseI-CAT and MseI-CTA successfully amplified fragments among used accessions and can be used for further analyses regarding genotyping and characterization of castor bean germplasm collections, complementing the possibilities of markers to be used in genetic diversity studies in the species. On the other hand, the primer MseI-CTG only worked when used with the primer EcoRI-AAC. These novel AFLP markers that can be used to access polymorphisms among and within castor bean germplasm collections will certainly be of great help during the process of genetic improvement of the species.

| Primer | SSR motif | Ta (°C) | Amp. | Primer | SSR motif | Ta (°C) | Amp. |
|--------|-----------|---------|------|--------|-----------|---------|------|
| UBC-801 | (AT)8T | 52.0 | – | UBC-845 | (CT)8RG | 54.0 | – |
| UBC-803 | (AT)8C | 52.0 | – | UBC-846 | (CA)8RT | 54.0 | + |
| UBC-804 | (AT)8A | 52.0 | – | UBC-847 | (CA)8RC | 54.0 | – |
| UBC-805 | (TA)8C | 52.0 | – | UBC-848 | (CA)8RG | 54.0 | + |
| UBC-806 | (TA)8G | 52.0 | – | UBC-849 | (GT)8YA | 52.0 | + |
| UBC-807 | (AG)8T | 50.4 | + | UBC-850 | (GT)8YC | 54.0 | + |
| UBC-808 | (AG)8C | 52.0 | + | UBC-851 | (GT)8YG | 54.0 | – |
| UBC-809 | (AG)8G | 54.0 | – | UBC-852 | (TC)8RA | 52.0 | – |
| UBC-810 | (GA)8T | 50.4 | + | UBC-853 | (TC)8RT | 52.0 | + |
| UBC-811 | (GA)8C | 54.0 | + | UBC-855 | (AC)8YT | 52.0 | + |
| UBC-812 | (GA)8A | 50.4 | + | UBC-856 | (AC)8YA | 52.0 | + |
| UBC-816 | (CA)8T | 50.0 | + | UBC-857 | (AC)8YG | 54.0 | + |
| UBC-817 | (CA)8A | 50.0 | + | UBC-858 | (TG)8RT | 52.0 | + |
| UBC-818 | (CA)8G | 54.0 | + | UBC-859 | (TG)8RC | 54.0 | – |
| UBC-819 | (GT)8A | 50.0 | – | UBC-860 | (TG)8RA | 52.0 | + |
| UBC-824 | (TC)8G | 52.0 | + | UBC-864 | (ATG)6 | 52.0 | + |
| UBC-825 | (AC)8T | 50.4 | + | UBC-868 | (GAA)6 | 50.0 | + |
| UBC-826 | (AC)8C | 52.8 | + | UBC-869 | (GTT)6 | 50.0 | – |
| UBC-827 | (AC)8G | 52.8 | + | UBC-873 | (GACA)4 | 52.0 | + |
| UBC-828 | (TG)8A | 52.0 | – | UBC-876 | (GATA)4 | 50.0 | – |
| UBC-830 | (TG)8G | 52.0 | – | UBC-878 | (GGAT)4 | 52.0 | + |
| UBC-831 | (AT)8YA | 54.0 | + | UBC-879 | (CTTCA)5 | 50.4 | + |
| UBC-834 | (AG)8YT | 52.0 | + | UBC-880 | (GGAGA)5 | 50.0 | + |
| UBC-835 | (AG)8YC | 54.0 | + | UBC-884 | HBH(AG)7 | 54.0 | + |
| UBC-836 | (AG)8YA | 54.0 | + | UBC-886 | VDV(CT)7 | 54.0 | + |
| UBC-840 | (GA)8YT | 52.0 | + | UBC-887 | DVD(TC)7 | 52.0 | + |
| UBC-841 | (GA)8YC | 52.0 | + | UBC-888 | BDB(CA)7 | 52.0 | + |
| UBC-842 | (GA)8YG | 54.0 | + | UBC-889 | DBD(AC)7 | 52.0 | + |
| UBC-843 | (CT)8RA | 54.0 | – | UBC-890 | VHV(GT)7 | 54.0 | + |
| UBC-844 | (CT)8RC | 54.0 | – | UBC-891 | HVH(GT)7 | 52.0 | + |

Table 1. ISSR primers from the University of British Columbia set tested for amplification with three castor bean accessions ('BRS Nordestina', 'BRS Paraguaçu' and 'Epaba 81'). Primers that were either successfully amplified (+) or not (–) using the annealing temperatures (Ta) indicated by Amorim (2009) for cowpea (*Vigna unguiculata*) are indicated.

## 2.2 Genome sequencing and the use of co-dominant markers

As mentioned above, sequencing castor bean genome (Chan et al., 2010) has just opened a wide range of possibilities in analyzing the genetic diversity of this economically important species. Thus, several approaches that demand previous information about DNA sequences can be readily used in large scale analyses regarding the genetic diversity characterization of germplasm collections. In general, co-dominant marker systems are the most reliable

| | | MseI primer | | | | | | | |
|---|---|---|---|---|---|---|---|---|---|
| | | CAA | CAC | CAG | CAT | CTA | CTC | CTG | CTT |
| EcoRI primer | AAC | + | + | + | + | + | + | + | + |
| | AAG | + | + | + | + | + | + | − | + |
| | ACA | + | + | + | + | + | − | − | + |
| | ACC | + | + | + | + | + | + | − | + |
| | ACG | + | + | + | + | + | + | − | + |
| | ACT | + | + | + | + | + | + | − | + |
| | AGC | + | + | + | + | + | − | − | + |
| | AGG | + | − | + | + | + | − | − | − |

Table 2. Amplification panel for primer combinations from AFLP Starter Primer Kit (Invitrogen, Carlsbad, USA) tested with three castor bean accessions ('BRS Nordestina', 'BRS Paraguaçu' and 'Epaba 81'). Primer combinations that were either successfully amplified (+) or not (–) are indicated.

markers for characterizing the genetic variability because of their capability to distinguish allelic types providing valuable information about the heterozygosity state of a given species (Kumar et al., 2009). However, there are factors that may restrict the use of these markers, as the high cost and the demanded time to make the DNA sequences available (Weising et al., 2005).

Among the most used co-dominant marker systems in evaluating plant diversity are microsatellite markers (or SSR) and single nucleotide polymorphisms (SNP) (Kumar et al., 2009). While the former have been widely employed since its publication in the early 1990s (see Morgante & Olivieri, 1993), SNP markers are becoming more popular as information about the genomes of plant species are increasing (e.g. Amar et al., 2011; Dong et al., 2010; Li et al., 2010).

In a worldwide-range germplasm characterization, Foster et al. (2010) evaluated the genetic diversity among 488 castor bean accessions from 45 countries using 48 SNPs, observing a molecular variance far higher within populations (74%) than among populations (22%) and countries (4%). These results also confirmed a very weak geographic structuration among castor bean populations, confirming previous results obtained with dominant markers (Allan et al., 2008; Milani et al., 2009).

Even within a minor geographic range, among 188 castor bean accessions from 13 wild populations from Florida (USA), distribution patterns of SNP alleles were not clear and indicated extensive homogenization either due to a high gene flow or because of multiple introductions (see Foster et al., 2010). Despite the great number and the wide distribution of sampled germplasm collections and better marker coverage compared to the previous report (Allan et al., 2008), the genomic coverage of the 48 SNPs described by Foster et al. (2010) was quite lower than the coverage of the soybean genome achieved by Li et al. (2010) who used 554 SNPs and 303 accessions.

Chan et al. (2010) estimated that more than half of the castor bean DNA consists of repetitive sequences, and SSR motifs are supposed to be widely spread through the species genome. In the last years, microsatellite markers have been increasingly employed to characterize genetic diversity within castor bean germplasm collections (Allan et al., 2008; Bajay et al.,

2009, 2011; Qiu et al., 2010) although still there is not an estimate of the extent of SSRs in the whole genome of the species. Qiu et al. (2010), analyzing microsatellite repeats associated to expressed sequence tags (ESTs), have reported a higher density of SSRs (excluding mono-repeats) in castor bean genic sequences (1/5.0 kbp) than the average described for other crops, such as maize (*Zea mays*) with 1/8.1 kbp, tomato (*Solanum lycopersicum*) with 1/11.1 kbp) and cotton (*Gossypium hirsutum*) with 1/20.0 kbp (Cardle et al., 2000), for instance. Qiu et al. (2010) suggested that such a high SSR density in ESTs may be associated to the small genome size of castor bean (~350 Mbp; Chan et al., 2010).

In order to search for occurrence and distribution of microsatellite repeats across the whole genome of castor bean, we have run the SciRoKo (Kofler et al., 2007) software, using the genome assembly available at http://castorbean.jcvi.org/downloads.php. Excluding 18,718 mononucleotide repeats (ca. 97% comprising poli-A SSRs), more than 95,000 SSR sequences were revealed in the analysis (Table 3), with one microsatellite occurring each 18.4 kbp, a density far below the described for EST-SSRs by Qiu et al. (2010), just as have been reported for plants in general (see Morgante et al., 2002).

| SSR motif | Counts | Counts/Mbp | AL [a] | AM [b] | AD [c] | PWG [d] |
|-----------|--------|------------|--------|--------|--------|---------|
| AT | 19286 | 55 | 29.32 | 0.48 | 18180.6 | 20.23% |
| AAAT | 12890 | 36.76 | 19.25 | 0.51 | 27201.79 | 13.52% |
| AAT | 12763 | 36.4 | 25.05 | 0.93 | 27472.46 | 13.39% |
| AAAAT | 7104 | 20.26 | 20.15 | 0.56 | 49356.84 | 7.45% |
| AG | 6748 | 19.25 | 24.72 | 0.46 | 51960.73 | 7.08% |
| AAG | 5820 | 16.6 | 22.27 | 0.74 | 60245.88 | 6.11% |
| AAAG | 3608 | 10.29 | 19.69 | 0.69 | 97181.54 | 3.79% |
| AAAAG | 3106 | 8.86 | 22.91 | 0.93 | 112888.29 | 3.26% |
| AATT | 2126 | 6.06 | 19.91 | 0.75 | 164925.22 | 2.23% |
| AAATT | 1979 | 5.64 | 18.99 | 0.5 | 177175.85 | 2.08% |
| AAAAAT | 1760 | 5.02 | 22.62 | 0.47 | 199222.17 | 1.85% |
| AAAAAG | 1426 | 4.07 | 27.38 | 0.94 | 245884.3 | 1.50% |
| ATC | 1426 | 4.07 | 19.53 | 0.5 | 245884.3 | 1.50% |
| AC | 1106 | 3.15 | 22.33 | 0.27 | 317026.23 | 1.16% |
| AGC | 855 | 2.44 | 18.07 | 0.32 | 410094.75 | 0.90% |
| ACC | 830 | 2.37 | 20.41 | 0.6 | 422447 | 0.87% |
| AATAT | 818 | 2.33 | 21.51 | 0.82 | 428644.27 | 0.86% |
| AAC | 626 | 1.79 | 18.89 | 0.44 | 560113.44 | 0.66% |
| AAAC | 614 | 1.75 | 17.53 | 0.29 | 571060.28 | 0.64% |
| AGG | 551 | 1.57 | 19.54 | 0.46 | 636353.93 | 0.58% |
| Others | 9871 | 0.09 | 21.75 | 0.36 | 43916350.09 | 10.36% |
| Average | – | 61.39 | 23.22 | 0.59 | 18393.66 | – |
| Total | 95313 | – | – | – | – | 100% |

Table 3. Composition, distribution and frequency of microsatellite motifs across the castor bean genome. Microsatellite statistics was performed with the software SciRoKo (Kofler et al., 2007), using default settings. Absolute SSR counts and weighted average counts per Mbp (10[5] base pairs) are presented in this panel. [a]AL weighted average length of the SSRs (bp); [b]AM weighted average number of mismatches; [c]AD weighted average distance between SSR repeats (bp); [d]PWG percentage of occurrence of the SSR through the whole genome.

Thus, we have found that the genomic SSRs consisted of 27,151 dinucleotide repeats (28.49%), 23,485 trinucleotide repeats (24.64%), 21,266 tetranucleotide repeats (22.31%), 15,988 pentanucleotide repeats (16.77%) and 7,423 hexanucleotide repeats (7.79%). On the other hand, the overall proportions of the microsatellite classes in EST-SSRs (Qiu et al., 2010) were quite different, in which the major proportion of the repeats were constituted by trinucleotide motifs (61.06%), followed by di- repeats (32.02%), tetra- repeats (3.63%), penta-repeats (1.01%) and hexa- repeats (2.28%). This higher percentage of trinucleotide repeats (as well as the hexa- repeats with a higher proportion than the penta- repeats) in EST-SSRs might be related to the function of this type repetitive DNA within transcribed regions. It is in agreement with results reported for other plants (Morgante et al., 2002). Due to the structure of tri- repeats and hexa- repeats, these types of SSR may change in the number of repetitions without affecting the reading frame of the gene (Metzgar et al., 2000). The general occurrence pattern of specific microsatellite motifs has diverged between EST-SSRs (Qiu et al., 2010) and genomic SSRs (Table 3). While Qiu et al. (2010) observed the AG-based dinucleotide repeats as the most frequent in the EST-SSRs (22.29%), our results showed that the AT mofit was the most abundant in genomic SSR sequences of castor bean (20.23%), as described by Morgante et al. (2002) for *Arabidopsis thaliana*. Likewise, we have found that AAT was the most prominent tri- repeat motif within genomic microsatellites (13.39%), as the AAG motif was more frequent in transcribed regions (14.35%; Qiu et al., 2010).

SSR motifs are also abundant in heterochromatic regions, which are quite difficult to sequence because of the extremely repetitive nature of this class of chromatin. In this case, cytogenetic tools as fluorescent *in situ* hybridization (FISH) with SSR-like probes in mitotic chromosomes may be helpful in characterizing distribution and polymorphisms of large microsatellite repeats along the chromosome set (Cuadrado & Jouve, 2010).

Cuadrado & Jouve (2007), for instance, analyzing the distribution pattern of trinucleotide repeats in barley (*Hordeum vulgare*) chromosomes by means of FISH, observed a preference of these large microsatellite clusters for heterochromatic regions, except for the ACT-based probe. In relation to the distribution of heterochromatin through castor bean chromosomes, large heterochromatic blocks have been related, including all pericentromeric regions (Jelenkovic & Harrington, 1973; Paris et al., 1978; Vasconcelos et al., 2010), in which the SSR sequences may be important constituents. Thus, in the present work an *in situ* hybridization was performed with the synthetic oligonucleotide (TGA)$_6$ as probe, aiming to test the potentiality of using large microsatellite clusters to characterize castor bean accessions. Cell preparations, image documentation and FISH conditions followed Vasconcelos et al. (2010); probe preparation was done according to Cuadrado & Jouve (2007).

In contrast to the observed for barley chromosomes, hybridization signals of the TGA-based probe, which is directed to the same target of the (CAT)$_5$ oligonucleotide used by Cuadrado & Jouve (2007), were observed in all chromosomes, mostly associated to GC-rich heterochromatin [evidenced by cromomicin A$_3$ (CMA)] (Fig. 3). While the chromosome E presented a non-heterochromatic site of the sampled repeat, the chromosomes B and D were the only without a pericentromeric site (Fig. 3). Taking into account the successful hybridization of the oligonucleotide in castor bean chromosomes, it is clear that the use of FISH to analyze microsatellite distribution through the species genome may be a very useful approach.

In characterization of castor bean germplasm collections through SSR markers, all studies conducted so far indicated congruent results with other marker systems (Allan et al., 2008; Bajay et al., 2009, 2011; Qiu et al., 2010). As mentioned above, sampling 41 genotypes from 35 countries, Allan et al. (2008) first indicated a relatively narrow genetic diversity in the species by using only nine genomic SSR markers and three AFLP primer combinations. Although SSR markers have yielded more polymorphism than AFLPs in the same analysis, both marker systems led to similar results of molecular variance indexes (Allan et al., 2008). Subsequently, Bajay et al. (2009, 2011) developed and tested a total of 23 SSR markers from a microsatellite-enriched library in two subsequent analyses of genetic diversity within two Brazilian castor bean germplasm collections. Similarly to previous results, these two studies have revealed relatively low heterozigozity levels among castor bean genotypes.

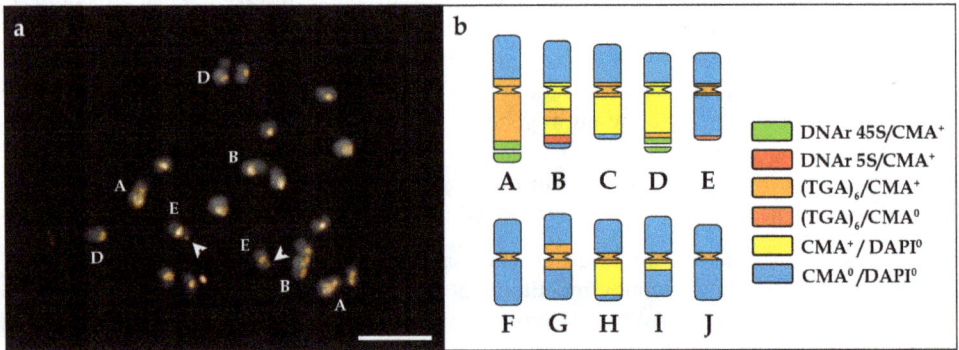

Fig. 3. Fluorescent *in situ* hybridization with the synthetic oligonucleotide (TGA)$_6$ in mitotic chromosomes of the castor bean cultivar 'BRS Energia'. In (**a**) a mitotic metaphase is showed evidencing the chromosomes bearing the main rDNA sites (A, B and D) and the only chromosome (E) with a terminal signal of hybridization (arrowheads). (**b**) Representative idiogram for castor bean showing the location of the (TGA)$_6$ sites in relation to the marks described by Vasconcelos et al. (2010).

After searching for SSR markers derived from the ESTs of castor bean, Qiu et al. (2010) selected 118 primer pairs (out of 379) that were used to estimate relationships among 24 accessions. The proportion of polymorphic amplicons (41.1%) generated in the analysis can be considered as satisfactory, taking into account that Raji et al. (2009) observed 50.6% of polymorphism using EST-SSR markers to analyze genetic diversity among cassava, a crop with a more recent history of cultivation, in comparison to castor bean. Likewise, PIC and heterozigozity values observed for castor bean EST-SSRs were relatively high and quite similar to those in cassava (Raji et al., 2009). In contrast to the results observed by Allan et al. (2008) and Foster et al. (2010), some degree of geographically structured clusters was observed among the accessions used by Qiu et al. (2010), although the authors recognized the small number of sampled genotypes, which may have still hindered the results.

It is clear that there is a great difference in the availability between EST-SSRs and genomic SSR to evaluate the extant genetic diversity in castor bean accessions. Albeit less frequent, the genomic microsatellites are less likely to suffer mutations with deleterious effects than the EST-SSRs, a fact that makes genomic SSRs more prone to polymorphisms (see Kalia et al., 2011;

Varshney et al., 2005; Weising et al., 2005). Thus, in order to provide a wider range of genomic microsatellite markers for castor bean genotyping, we have performed a data mining through the whole genome of the species by running the online software WebSat (Martins et al., 2009) to locate SSR motifs (minimum size of 30 nucleotides, excluding all microsatellites composed by mono- repeats, either simple or compound) and design primers using default settings.

Covering more than 11 Mbp of the castor bean genome (approximately 3%), a total of 134 primer pairs were herein designed (Table 4). Despite the low genomic coverage, especially if compared to the work carried out by Cavagnaro et al. (2010), in which the whole genome of cucumber (*Cucumis sativus*) was scanned, the analyzed fraction of the genome was close to

| Primer | Sequence 5' – 3' | T$m$ (°C) | SSR motif | EAS |
|---|---|---|---|---|
| RC9-V28828.2 | F: ATTACTTGGGTTCTGTGCCTGT<br>R: TTACGAGCTAAGAAAGTCGCTAAAG | 59.9 | $(TA)_{17}$ | 385 |
| RC19-V28828.6 | F: TCGAGCATAGCAAACATGAA<br>R: CGGCAGAACTGTGAGATAAAGT | 58.3 | $(TAA)_{16}$ | 305 |
| RC31-V29842.10 | F: AGCAGATTGTGAAGGGATTGTT<br>R: ACCCCTGAACTTTTGTGATTTG | 60.1 | $(AT)_{24}$ | 384 |
| RC35-V29842.14 | F: GGGATGACTCCAACACCAAT<br>R: TTTCCTCCACTCAAAGCAAAC | 59.5 | $(ATT)_{10}(TAG)_{10}$ | 391 |
| RC41-V29842.20 | F: AATTGTCCTCTCGCATGGTC<br>R: GGGGATTAGGGTTTGTGAGG | 60.3 | $(AAAAGG)_4(GT)_9$ | 365 |
| RC42-V29842.21 | F: TCCGACGCAATAGTACCTTGT<br>R: TGTCTCTTCTCCATTGTCTCCC | 60.1 | $(AT)_{18}$ | 182 |
| RC54-V29842.33 | F: ACATTACACGAACTCGAACCG<br>R: TGCCGGTAGGGCTATAAACA | 60.3 | $(TA)_{37}(GA)_8$ | 557 |
| RC55-V29842.34 | F: TTTCCCTCTGTGTCTTCGCT<br>R: TAGCGCGAGAAAGTCGAGAT | 60.1 | $(CTT)_{12}$ | 129 |
| RC58-V29842.37 | F: CGGAAGTTGATGACAAACGA<br>R: ATGGTTTTGCTGCTGTTGC | 59.8 | $(AT)_{21}$ | 303 |
| RC65-V29842.44 | F: AGCTCCTTGTCCTGTGCCTA<br>R: CTGGTGAAGGGTCTGGTCAT | 60.0 | $(CCACCT)_5$ | 389 |
| RC67-V29842.46 | F: ATACCCTCCACCAATCCCTC<br>R: AGTTGCGGAAGTTTCTTTGC | 59.8 | $(CCA)_7(CCA)_9$<br>$(TCCACC)_4$ | 390 |
| RC71-V29842.50 | F: CAGCTTATGGGAAGATGCTAAA<br>R: ATGCAGGATTCACAACAGGA | 58.8 | $(GT)_{25}$ | 385 |
| RC86-V29842.65 | F: ATATCCCAAAAGCACCCACA<br>R: CCAAAATCATCAGCTCGCTT | 60.3 | $(AT)_{17}$ | 494 |
| RC89-V29842.68 | F: AGCTATCCATTCATGCGGAG<br>R: AGCTCCTGTCAACATCCCTG | 60.2 | $(CAT)_{13}$ | 250 |
| RC94-V29842.73 | F: AGTAAGTTGCGTCAGTCCGC<br>R: GTGCTCGCCTTTGTTTGACT | 60.5 | $(AG)_{20}$ | 265 |
| RC98-V29842.77 | F: TGGTTGGAGAGAGGTTTGTTTT<br>R: GAGCTGTCTTTTGTAGCCCAGT | 60.0 | $(TA)_{15}$ | 228 |
| RC108-V29842.87 | F: GAATCTCACCTGCTATTATGCCT<br>R: CCAAAGCCAAAGAGGGTACTT | 59.2 | $(AT)_{20}$ | 210 |
| RC110-V29842.89 | F: CCCATCCAATGAAATAGGGA<br>R: TGTCCTGCATCAACTGGTTT | 59.4 | $(TA)_{16}$ | 287 |

| Primer | Sequence 5' – 3' | Tm (°C) | SSR motif | EAS |
|---|---|---|---|---|
| RC128-V29842.107 | F: AGAGAGTGAAAGGCGAGTCAGT<br>R: AATCAGTTTGGTCGCGTAGC | 60.0 | $(TA)_{20}$ | 260 |
| RC134-V29842.113 | F: GGTCCAGTTGCCTCTAACCA<br>R: AGATCAGCATACAAGGCGCT | 60.1 | $(TC)_{17}$ | 289 |
| RC136-V29842.115 | F: ATGCCCGAACCTATAACCCT<br>R: AACCCTATTTCTTCGTCATTGC | 59.6 | $(AT)_{15}$ | 208 |
| RC138-V29842.117 | F: TTGAGGCAAACACTTGCACTA<br>R: ATTTTAACTTTGGGCACGCT | 59.1 | $(AT)_{28}$ | 275 |
| RC140-V29842.119 | F: GGCTGTCTAACGCCTAGCAA<br>R: AAGGAAATGGTGTGGCAAAA | 60.0 | $(TA)_{21}$ | 315 |
| RC141-V29842.120 | F: CAATGGTGAATGATGAACGG<br>R: AGTTGGTGCCAGGAGAAGAA | 59.8 | $(AT)_{22}$ | 185 |
| RC146-V29842.125 | F: TTGGCAGTCATTGTTCTTCTG<br>R: GTTTTGGAGTGGCAGAGCT | 59.2 | $(CT)_{11}(TA)_{11}$ | 393 |
| RC151-V29842.130 | F: TTGTGTCCATACCAACATCG<br>R: GGATAGGAGCATCAAGAAGGTT | 58.3 | $(AT)_{31}$ | 389 |
| RC158-V29842.137 | F: AATTGGAGTGGGTAGAGAGGG<br>R: CAGGTCTCAAGTTTTCCACCA | 59.8 | $(AAT)_{10}$ | 212 |
| RC161-V29842.140 | F: CGCAAATAAACGGAGCATTA<br>R: ACAAGCCCACACCCATAAGT | 59.1 | $(GAAAAA)_{7}$ | 318 |
| RC165-V29842.144 | F: TCTGCAACAGGAACGCATAG<br>R: TGCAAGGATGTTTGTACTTTGG | 60.0 | $(CA)_{10}(TA)_{5}(AT)_{12}$ | 280 |
| RC176-V29842.155 | F: CAGAGAAAGAACCCTCCGAA<br>R: CCGGTCTGAAACTCTTCTGC | 59.7 | $(AT)_{7}(TA)_{9}$ | 385 |
| RC185-V29381.1 | F: AGCTGGTAATGGCTCCAAGA<br>R: AGAGTGGTTGCCTGATTTGC | 60.0 | $(TA)_{32}$ | 359 |
| RC186-V29381.2 | F: CCTTTGTCTCTCTGGCTCGT<br>R: GAAAATGGTCCCTCGTTTCA | 59.8 | $(TA)_{21}$ | 323 |
| RC196-V29849.7 | F: AGGAAATGAAATGGTGATGG<br>R: TGAAGATTGTATGGGGAGGA | 57.6 | $(TAT)_{14}(TTA)_{6}$ | 173 |
| RC205-V29849.16 | F: TGCGTGAGGGGTTAATTGAT<br>R: TGGTGATGGTAGAGATTCCCC | 60.7 | $(AAT)_{12}$ | 188 |
| RC207-V29849.18 | F: AGCCCCTCACTCCATTATCA<br>R: TGTCATATAGGCCCAAGTCG | 59.1 | $(TG)_{7}(TA)_{10}$ | 294 |
| RC208-V29849.19 | F: AAGGAAGCTCTCTCCAAGGG<br>R: ATCGGGGAAGAAGAAAGGAA | 60.0 | $(CT)_{6}(CT)_{9}(CT)_{9}$ | 308 |
| RC219-V28829.3 | F: TCCTCAAAGCTGTCCAAACA<br>R: GCGAGTGCTGTCTAAAGCCTA | 59.6 | $(TA)_{33}$ | 267 |
| RC235-V27471.12 | F: AAAGGGGTACAACCGGAAAG<br>R: CTTACGCAGCACTGGAAACA | 60.1 | $(TC)_{6}(TA)_{16}$ | 315 |
| RC244-V27471.21 | F: TGAACCAAGCTCGCATGTAG<br>R: GCCAAGAAGATGATTACCCG | 59.8 | $(AAT)_{9}$<br>$(GAAGGA)_{4}$ | 385 |
| RC252-V27471.29 | F: AGATAAACATGACAGCAGGGC<br>R: TTGACATAACTGGCTTTGGATG | 59.6 | $(AT)_{23}$ | 387 |
| RC253-V27471.30 | F: ATGCTGCTTTGTCGTTTCCT<br>R: TTCTAACTCCCCATTCGGTG | 59.9 | $(AT)_{12}(AT)_{10}$ | 569 |
| RC267-V29697.9 | F: ACCCTGTGAGGCGTACATTT | 59.2 | $(TA)_{28}$ | 261 |

| Primer | Sequence 5′ – 3′ | Tm (°C) | SSR motif | EAS |
|---|---|---|---|---|
| RC277-V28725.9 | R: CCACTCTTTTGGGGTTGTTT<br>F: ATGAAGGGTGATGGCAGAAG<br>R: ACGTCAGCCGAGTTCAAAAT | 59.9 | $(TC)_{15}$ | 165 |
| RC279-V28725.11 | F: TAGGAAGAGGGGTGGTTTCC<br>R: TGGTTAAAGATCAGGCGGTC | 60.2 | $(AG)_{11}(GA)_6$ | 338 |
| RC280-V28725.12 | F: TCGAGAAGTGGGAACAGTGA<br>R: TTGGGAGAGATTTGAAGGAAAG | 59.5 | $(TC)_7(CT)_{14}$ | 331 |
| RC283-V28725.15 | F: TTCAAGTATTGGGAATGGCAC<br>R: GAGTGGGTTTGAGCAGAAGC | 59.9 | $(CT)_{19}$ | 188 |
| RC291-V28725.23 | F: TGAGGGAGGAGTTGAAAGAAA<br>R: TCAAGTGTGGAGCACGTAAAA | 59.1 | $(ATA)_{12}$ | 336 |
| RC292-V30197.1 | F: CATACGCACACAGCCTTGC<br>R: GGCCAATCTGGATCATACTTTCT | 61.1 | $(TA)_{19}$ | 408 |
| RC302-V30197.11 | F: GTTTTGATTTGGATGCTGGG<br>R: ACCCATAAGGTGGCATGTTC | 60.0 | $(AAT)_{11}$ | 349 |
| RC308-V30197.17 | F: GAAATTGCTACAGAAACCGC<br>R: TCGATTCTTGGTGAAAGTTAGG | 57.9 | $(TA)_{17}$ | 313 |
| RC326-V28359.17 | F: GACATGATTTTGACAGGTCGG<br>R: GCGGTGGGTGTTTAATTTTG | 60.3 | $(TA)_{15}$ | 353 |
| RC328-V28359.19 | F: CTAAACCCAGAGAGCGATGC<br>R: AGGTCGCAAAAGCTGTGAGT | 60.0 | $(AT)_{23}$ | 375 |
| RC330-V28359.21 | F: GTCACAATTCAACGCTGCTG<br>R: GCTCTACTGATTGATCCGGC | 60.1 | $(TA)_{21}$ | 391 |
| RC331-V29476.1 | F: ATTGCTTGTTATCGCCGTTC<br>R: TTGAGGGACTAAGGTGAAGAAGA | 59.8 | $(TCT)_{18}$ | 276 |
| RC332-V29476.2 | F: TGGAAGTCGCTGTCCTACG<br>R: TTAAGGAAGTTGAGGGACCAAA | 60.0 | $(CT)_9(CT)_7$ | 285 |
| RC338-V29828.6 | F: GCCCTACTTCTAACCATGTGC<br>R: GTGGTCCTTATGCAACCCAT | 59.2 | $(AT)_{29}$ | 478 |
| RC344-V29828.12 | F: CCCAACAAGCTCACACCTTT<br>R: CCTGCTAGGTTTTGCCAGAG | 60.1 | $(GA)_6(CTCTCA)_5$ | 264 |
| RC354-V28212.4 | F: TGAGGGACTAAATGGTGAAGAA<br>R: ACCAGACAAGCCATGAACG | 59.2 | $(GAA)_9(GA)_6$ | 277 |
| RC357-V29605.2 | F: AGCTCACTGGAAAAGCCAAA<br>R: GGACCAAATGTCGATAGGAAGA | 60.2 | $(CT)_6(TTC)_8$ | 387 |
| RC379-V29690.8 | F: TTCACTCCCTTCTAGTCGCC<br>R: AAAGGCACCAAAGATCCAAA | 59.5 | $(TAT)_{12}$ | 514 |
| RC381-V29690.10 | F: AAGCACAGAAAAGAGCAAAGC<br>R: AGTCCGCCCTTACATATCCA | 59.2 | $(AT)_{22}$ | 384 |
| RC387-V29690.16 | F: GCTCGTGAAGCTCATAGGC<br>R: GGACCATTTTATTTGCTGAGG | 58.6 | $(TA)_{20}$ | 400 |
| RC401-V27770.5 | F: TTTGCTTTTGCATTGTGGAC<br>R: TGCTTCATAAATTGGCTTGG | 59.2 | $(TA)_{26}$ | 177 |
| RC403-V27770.7 | F: TTCATGCAAGTGAGGAATGC<br>R: AGGCTGGAAAATGCAAACTC | 59.6 | $(TA)_{23}$ | 165 |
| RC410-V29745.2 | F: TCTCTATCGCCACATCACCA<br>R: ATTTGATACCACCACCGCTC | 60.0 | $(TA)_{40}$ | 562 |

| Primer | Sequence 5′ – 3′ | T$m$ (°C) | SSR motif | EAS |
|---|---|---|---|---|
| RC413-V29745.5 | F: CCGAAAGCTACCAGATCGAG<br>R: TCCTCCCATCTTCTTCTTCTTCT | 59.9 | $(GGAA)_6(GAA)_{10}$ | 341 |
| RC414-V29745.6 | F: GGGAGGGAAGTGGAAATAGC<br>R: AGCCATCAGAATTGGGTTCA | 60.2 | $(CT)_{12}(TA)_{11}$ | 385 |
| RC417-V29745.9 | F: CGAAACAAATCGCTGGAAT<br>R: GAGATTGAACGGGAAGGATG | 59.1 | $(TA)_{30}$ | 600 |
| RC420-V29745.12 | F: CAACCCAGAAGGGCAGAAG<br>R: TGGATTCGGATTTGGAGAAG | 60.4 | $(GAA)_{11}$ | 190 |
| RC423-V29745.15 | F: ACAAGGACGACGAAGACGAC<br>R: CAACAGCAGAACAGTAGCCG | 60.0 | $(AG)_8(GA)_{12}$ | 356 |
| RC426-V29745.18 | F: GAGAAGCTAAAGCCCACAGG<br>R: TTTGAAGCCAACTCAACTCG | 59.1 | $(TA)_{17}$ | 321 |
| RC430-V29763.1 | F: GCATTGTCCAACTGATGAGC<br>R: TGTGAGAAGCTGCGGTTAAAT | 59.6 | $(TA)_{20}$ | 134 |
| RC431-V29763.2 | F: CTCTTGGCTTGGTGTCCACT<br>R: CGACTGCAATTCTTCCCTGT | 60.3 | $(TC)_{15}(AT)_6$ | 395 |
| RC434-V29763.5 | F: GAAAAGAGAAGCCCAGATGGT<br>R: TTGAAAGGGACACACCAAAA | 59.4 | $(AAT)_{11}$ | 366 |
| RC436-V29763.7 | F: AGTGTTTGCTTGATGGGTTGA<br>R: TGCAGGCTTTCCAAATCG | 60.7 | $(AT)_{26}$ | 385 |
| RC438-V29763.9 | F: GTCGTGAGGCTGTGGAAAAT<br>R: AAGTCTAAGCTAATGCTCGCTGA | 60.0 | $(TA)_{19}$ | 275 |
| RC441-V29763.12 | F: GAGGTGGCAGACTATTTCTTGA<br>R: CTCCACTCCTTCATGCTTTTAG | 58.3 | $(ATA)_{10}$ | 325 |
| RC446-V28448.1 | F: GCCAAGGCTCTTTCTCTTTT<br>R: ACTGTGGTGATCGGAGAGG | 58.2 | $(ATA)_{28}$ | 378 |
| RC447-V28448.2 | F: AGCTGACCACCTGAGTACAACA<br>R: GTATGTGACGCCAACCATCA | 60.1 | $(TA)_{23}$ | 369 |
| RC452-V28448.7 | F: GCTTTTGGGTTGGAGTTCAG<br>R: GAGGAGACCAAGAAAACTAAAGGA | 59.5 | $(TC)_{19}$ | 161 |
| RC453-V28448.8 | F: TTGCATTTGTGTGTGTGTCG<br>R: ATTTTAGTTGGCTCCCCACC | 60.2 | $(GA)_{14}(AG)_6$ | 385 |
| RC457-V28448.12 | F: CCGGATTATGTGGTTAGTAGTGG<br>R: ATCGTCTCATCGGTGGTTTT | 59.5 | $(TA)_{34}$ | 296 |
| RC461-V28448.16 | F: GCTGGTATTGGGGTTCTGTG<br>R: TCATGGGTTGTTCTGCTTCA | 60.3 | $(GAA)_{10}$ | 376 |
| RC462-V28448.17 | F: ATAATGGTAGGCTCGGGTGA<br>R: TGGAGGATAAAGTGATGTGAGG | 59.2 | $(CT)_{19}$ | 140 |
| RC463-V28448.18 | F: AACAGGCCAAACAGACAATCTT<br>R: CATTAGGCAATAGTAGCTCCCC | 59.6 | $(TA)_{11}(AG)_9$ | 348 |
| RC465-V37395.1 | F: ATACTTGTCACGGGCTTTGG<br>R: AGCGGGAGACGATTGTTATG | 60.0 | $(TCTTT)_6$ | 395 |
| RC472-V29984.7 | F: TGGCATCCTACATTCACACG<br>R: CCTCACATTTGGGTGGACAT | 60.6 | $(TA)_{20}$ | 504 |
| RC479-V30056.4 | F: AACCGAGCAGGAAATGAGAA<br>R: GTCGTGCAGAAAGGAAGGAG | 59.9 | $(AT)_7(AAT)_7$ | 355 |
| RC480-V30056.5 | F: GCTCTGCCTCATCCTCACAT | 60.7 | $(TA)_{17}$ | 279 |

| Primer | Sequence 5' – 3' | T$m$ (°C) | SSR motif | EAS |
|---|---|---|---|---|
| RC504-V30056.29 | R: AATCAAATGTTCCGCGAGTG<br>F: GCAAGCTCGTTTATATGCTCAA<br>R: ATCCAACACCGACACTCCA | 59.7 | $(AAT)_{19}$ | 586 |
| RC511-V30056.36 | F: TTCTTGCCTCCCTACTGATAATG<br>R: AGGTTCTGATGCTTGTTGTCC | 59.4 | $(AAT)_{10}$ | 201 |
| RC515-V30056.40 | F: ATCAAAGGGTCTTGAGCCAG<br>R: ATCCTGCATTTTGTTGTTGC | 59.0 | $(AT)_{15}$ | 252 |
| RC537-V29764.12 | F: TCGATTCCACCTCCTCATCA<br>R: CCAGAAGCCAATTCACATGC | 61.6 | $(AT)_{13}(TG)_{12}$ | 598 |
| RC539-V29764.14 | F: TGCTCTGCCTCATCATCATC<br>R: CACGAATGGTTGTGTTTGCT | 59.8 | $(AT)_{23}$ | 313 |
| RC545-V29764.20 | F: AGCTCTTTTGCCTTCCCACT<br>R: AGGGTTAGAGGTAGGGGTTTTG | 60.1 | $(AT)_{24}$ | 355 |
| RC556-V30156.3 | F: GCCATTGATGAGTTTGCTGA<br>R: GATACACTTGCCGGAACGAC | 60.2 | $(AG)_{17}$ | 326 |
| RC558-V30156.5 | F: GTAGCGACTCCCATCTTTGC<br>R: AAAACTCTCACCTTTTCTTCCG | 59.4 | $(ATT)_{15}$ | 561 |
| RC559-V30156.6 | F: ACACGGGTGATTTGGGAATA<br>R: AAAGGGGCTTCGAGAGAAAA | 60.2 | $(AC)_{14}(AT)_9$ | 214 |
| RC565-V30156.12 | F: CATTTCGAGCCTAAGAACGC<br>R: AAGTTCAGGGACCAAAAGTCAA | 60.0 | $(TCT)_{13}$ | 314 |
| RC567-V30156.14 | F: TGGTGACTCGAACGAAGATG<br>R: CCACGTCTTTTGGATGCTTT | 60.0 | $(TA)_{27}$ | 337 |
| RC568-V30156.15 | F: ACGAAAGGACTGATTGGCAT<br>R: ATTATGGGAGGTTCACCAGC | 59.2 | $(TA)_{21}$ | 589 |
| RC571-V30156.18 | F: CTAAGGATGGTAGCCGAGTG<br>R: AATGAGTAGCGAACACACTATCTG | 57.6 | $(AT)_{22}$ | 475 |
| RC572-V30156.19 | F: CAGATAGTGTGTTCGCTACTCATT<br>R: CGTACAATACCCGAATCCAA | 58.0 | $(AT)_{15}$ | 263 |
| RC574-V30156.21 | F: AGCTCTCTCTCCAAGGGGAC<br>R: AATTCTGCGGGATGAAATTG | 59.9 | $(CT)_6(CT)_{12}$ | 348 |
| RC575-V30156.22 | F: GTGAGAAAAGAGAGTGTGAAAGTG<br>R: TCATTACCAAACCCGCTAAA | 57.8 | $(AT)_{15}$ | 380 |
| RC579-V30156.26 | F: TGCCATGACCCAACTAAACA<br>R: ATTGCTTCTTGCGCTCTTTC | 59.9 | $(GA)_8(TA)_{13}$ | 370 |
| RC580-V30156.27 | F: TTGGATGCTACCCTCTTAACTCT<br>R: CTTAAATGACCCGATTTGGC | 59.0 | $(TA)_{18}$ | 345 |
| RC582-V30156.29 | F: GTTAGCGGAAACTCGGCA<br>R: CATCCCGAACCACATTTCA | 60.1 | $(TAT)_{23}$ | 400 |
| RC583-V30156.30 | F: TGGTTAAGGGGTTATGGCAG<br>R: CGAATTTTAGAAGGAACAAGGC | 59.6 | $(AT)_{25}$ | 577 |
| RC584-V30156.31 | F: GCCTTGTTCCTTCTAAAATTCG<br>R: GAGGGAGAGCTGTTGTTGGT | 59.3 | $(AT)_{27}(GT)_{10}$ | 568 |
| RC594-V30156.41 | F: TGTGAAAAGGGAGTTCGGAG<br>R: ATTGCGGGTAAAACTGAAGC | 59.7 | $(AT)_{32}(TA)_{12}$ | 357 |
| RC596-V30156.43 | F: TTGAAGGGTGGAGGATTCTTT<br>R: CTGCTTGATTGATTTCCCACT | 59.6 | $(TA)_{24}$ | 397 |

| Primer | Sequence 5′ – 3′ | Tm (°C) | SSR motif | EAS |
|--------|------------------|---------|-----------|-----|
| RC598-V30156.45 | F: AACAGCAGCGACAACAACAC<br>R: CAGCAGAACTCAAACCGACA | 60.0 | $(CAC)_6(AAC)_9$ | 395 |
| RC603-V30156.50 | F: CACCTTGCAGAGCTTACGAA<br>R: CCCCAGTTCACACCAATACC | 59.7 | $(TA)_6(CT)_9$ | 341 |
| RC604-V30156.51 | F: TCCTACTCTTGTGAGAACGGACT<br>R: GGGCATACTTCAACCGAAGA | 59.8 | $(TA)_{27}$ | 300 |
| RC612-V30156.59 | F: CAGGCCGTATTAGCGATGTT<br>R: GGTATGTCTTAGTTAGGATTGGGC | 59.7 | $(TA)_{29}$ | 229 |
| RC614-V30156.61 | F: GCGATCACTTACACGATATGC<br>R: TCACTCATCACACACATCACCT | 58.7 | $(AT)_{22}$ | 494 |
| RC621-V30156.68 | F: TCGCATGAGTGGAACCTGTA<br>R: AAGAAGAAGGAGACACGGCA | 60.1 | $(AAT)_{12}$ | 319 |
| RC630-V30156.77 | F: CCAATAGGTTCAGTGTTTTGCC<br>R: ACAAGCTGGGCATGATTGA | 60.5 | $(AT)_{29}$ | 363 |
| RC638-V30156.85 | F: TGGTTGATGGTGCCAGACTA<br>R: TATCTGAAAATGCGCTGCTG | 60.1 | $(TC)_{15}(CT)_8$ | 305 |
| RC640-V30156.87 | F: CGACGTAGGAGCAACTAAAGG<br>R: ATCGGACATGGTGCTTAAAA | 58.6 | $(TTA)_{10}$ | 272 |
| RC644-V30156.91 | F: TGCATCCTTGTTTCCATGTC<br>R: TTTGCCGATCCTCAGTCTCT | 59.7 | $(AT)_{24}$ | 569 |
| RC645-V30156.92 | F: CTTGAGGGTCGTAGGAGCAG<br>R: GGTAGCAACTCTCTTTCTTCGC | 59.8 | $(AT)_{41}$ | 327 |
| RC647-V30156.94 | F: GGCAAGCACAGAGCATACG<br>R: GCGCCATAGAATTTGCACTT | 60.4 | $(TA)_{19}$ | 386 |
| RC650-V30156.97 | F: ATAATTCCAGGGGCAAAATC<br>R: CAAATGGCACCCAATAAGAA | 58.2 | $(AT)_{36}$ | 310 |
| RC654-V29781.1 | F: ATAGAGAGAGAGGGAGGGAGAC<br>R: TGCCAATATGCAGTCACATC | 57.6 | $(AG)_{15}$ | 327 |
| RC655-V29781.2 | F: GGCTGGAAGAGGAAGAAACA<br>R: GTGCTGATCCAAACTTGATAGG | 59.1 | $(AG)_{15}$ | 244 |
| RC658-V47284.1 | F: GAGATGTCTAATCACACGCCAA<br>R: AATCGGGGCCATTTACTTTT | 59.9 | $(ATT)_{11}$ | 323 |
| RC661-V29346.2 | F: TGGTTCCTTCTTGTGTGAGC<br>R: ATTTGCCTCCCCTAGTTTCC | 59.1 | $(CCGCA)_6$ | 206 |
| RC662-V29346.3 | F: TGCTGCCCTTGATTTTCTTT<br>R: CCTTGTGCCCTTTGTATGCT | 60.0 | $(TA)_{18}(AG)_{14}$ | 252 |
| RC687-V30006.20 | F: GCTTATGGGTGGTTGTTTCC<br>R: GTTTATCATCGAGCAAGCCA | 59.1 | $(CT)_{30}$ | 213 |
| RC690-V30006.23 | F: TAGAACGCCTGGTGATTGTG<br>R: AAACCATTGCACCCTTGAAA | 60.0 | $(TC)_7(AC)_{14}$ | 226 |
| RC691-V30006.24 | F: TGATTCGATTACAACTCCAGC<br>R: CACTAAATGGTGGAATGACTGA | 57.7 | $(AT)_{11}(AAT)_9$ | 368 |
| RC693-V30006.26 | F: TCCCAACCCCAACAATAGAA<br>R: CACATGGGTGGCAAAGAAA | 60.3 | $(TA)_{15}$ | 365 |

Table 4. SSR primer pairs obtained through data mining in a fragment of the castor bean genome sequence. Sequences, melting temperature and estimated allele sizes (EAS) of primer pairs (F: forward; R: reverse) are indicated in the panel.

the value obtained by Qiu et al. (2010) for genic sequences (13.68 Mbp – approximately 4% of the genome). Moreover, our stringent criteria for selection of microsatellites to be used sharply reduced the final number of annotated primer pairs. Without the adopted restriction of 30 nucleotides and using default parameters of the software, the total number of scored microsatellites increased from 134 to 696 (data not shown). Therefore, due to the higher number of repetitions of the targeted microsatellites, these molecular markers may be more liable to polymorphisms than the smaller SSRs.

## 3. Conclusion

Despite the recent efforts to characterize castor bean germplasm collections, there are relatively few molecular markers available. Curiously, the use of widely spread and low-cost anonymous markers, as RAPD and ISSR, in genetic diversity analyses is still problematic and insufficient. Even the powerful and reliable AFLP marker system was poorly used to describe the extant polymorphism in castor bean germplasm collections. However, still there is a need for selection of robust molecular markers able to distinguish accessions and/or for association with phenotypic traits of interest such as oil production, resistance to abiotic stress and pathogens. Thus, our results, in addition to compiled data from literature, will be very useful for breeding programs by providing important information about genetic diversity of this important crop. Furthermore, our efforts in describing novel molecular markers certainly should help the development of the first genetic map for castor bean.

## 4. Acknowledgment

The authors thank the following Brazilian agencies for financial support: Banco do Nordeste do Brasil S/A (BNB), Escritório Técnico de Estudos Econômicos do Nordeste (ETENE), Fundo de Desenvolvimento Científico e Tecnológico (FUNDECI); Conselho Nacional de Desenvolvimento Científico e Tecnológico (CNPq); Coordenação de Aperfeiçoamento de Pessoal de Nível Superior (CAPES); and Fundação de Amparo à Ciência e Tecnologia do Estado de Pernambuco (FACEPE).

## 5. References

Allan, G.; Williams, A.; Rabinowicz, P.D.; Chan, A.P.; Ravel, J. & Keim, P. (2008). Worldwide genotyping of castor bean germplasm (Ricinus communis L.) using AFLPs and SSRs. *Genetic Resources and Crop Evolution*, Vol.55, No. 3, (July 2007), pp. 365-378, ISSN 0925-9864

Amar, M.H.; Biswas, M.K.; Zhang, Z. & Guo, W.W. (2011) Exploitation of SSR, SRAP and CAPS-SNP markers for genetic diversity of Citrus germplasm collection. *Scientia Horticulturae*, Vol.128, No.3, (April 2011) pp. 220-227, ISSN 0304-4238

Amorim L.L.B. (2009). *Construção de um mapa genético para feijão-caupi com marcadores moleculares ISSR, DAF e CAPS*, Master Dissertation (Biological Sciences), Federal University of Pernambuco, Recife, Brazil

Andrade, I.M.; Mayo, S.J.; Van Den Berg, C.; Fay, M.F.; Chester, M.; Lexer, C. & Kirkup, D. (2009). Genetic variation in natural populations of *Anthurium sinuatum* and *A. pentaphyllum* var. *pentaphyllum* (Araceae) from north-east Brazil using AFLP molecular markers. *Botanical Journal of the Linnean Society*, Vol.159, No.1, (January 2009), pp. 88-105, ISSN 1095-8339

Bajay, M.M.; Pinheiro, J.B.; Batista, C.E.A.; Nobrega, M.B.D. & Zucchi, M.I. (2009). Development and characterization of microsatellite markers for castor (*Ricinus communis* L.), an important oleaginous species for biodiesel production. *Conservation Genetics Resources*, Vol.1, No.1, (July 2009), pp. 237-239, ISSN 1877-7252

Bajay, M.M.; Zucchi, M.I.; Kiihl, T.A.M.; Batista, C.E.A.; Monteiro, M. & Pinheiro, J.B. (2011). Development of a novel set of microsatellite markers for Castor bean, *Ricinus communis* (Euphorbiaceae). *American Journal of Botany*, Vol.98, No.4, (April 2011), pp. E87-E89, ISSN 1537-2197

Berman, P.; Nizri, S. & Wiesman, Z. (2011). Castor oil biodiesel and its blends as alternative fuel. *Biomass and Bioenergy*, In Press (Online First), pp. 6, ISSN 0961-9534

Billotte, N.; Jourjon, M.; Marseillac, N.; Berger, A.; Flori, A.; Asmady, H.; Adon, B.; Singh, R.; Nouy, B.; Potier, F.; Cheah, S.; Rohde, W.; Ritter, E.; Courtois, B.; Charrier, A. & Mangin, B. (2010). QTL detection by multi-parent linkage mapping in oil palm (*Elaeis guineensis* Jacq.). *Theoretical and Applied Genetics*, Vol.120, No.8, (February 2010), pp. 1673-1687, ISSN 0040-5752

Cardle, L.; Ramsay, L.; Milbourne, D.; Macaulay, M.; Marshall, D. & Waugh, R. (2000). Computational and experimental characterization of physically clustered simple sequence repeats in plants. *Genetics*, Vol.156, No.2, (October 2000), pp. 847-854, ISSN 0016-6731

Cavagnaro, P.F.; Senalik, D.A.; Yang, L.; Simon, P.W.; Harkins, T.T.; Kodira, C. D.; Huang, S. & Weng, Y. (2010). Genome-wide characterization of simple sequence repeats in cucumber (*Cucumis sativus* L.). *BMC Genomics*, Vol.11, No. 569, (October 2010), pp. 1-18, ISSN 1471-2164

Chan, A.P.; Crabtree, J.; Zhao, Q.; Lorenzi, H.; Orvis, J.; Puiu, D.; Melake-Berhan, A.; Jones, K.M.; Redman, J.; Chen, G.; Cahoon, E.B.; Gedil, M.; Stanke, M.; Haas, B.J.; Wortman, J.R.; Fraser-Liggett, C.M.; Ravel, J. & Rabinowicz, P.D. (2010). Draft genome sequence of the oilseed species *Ricinus communis*. *Nature Biotechnology*, Vol.28, No.9, (August 2010), pp. 951-956, ISSN 1087-0156

Christopoulos, M.V.; Rouskas, D.; Tsantili, E. & Bebeli, P. J. (2010). Germplasm diversity and genetic relationships among walnut (*Juglans regia* L.) cultivars and Greek local selections revealed by Inter-Simple Sequence Repeat (ISSR) markers. *Scientia Horticulturae*, Vol.125, No.4, (July 2010), pp. 584-592, ISSN 0304-4238

Cuadrado, A. & Jouve, N. (2007). The nonrandom distribution of long clusters of all possible classes of trinucleotide repeats in barley chromosomes. *Chromosome Research*, Vol.15, No.6, (August 2007), pp. 711-720, ISSN 0967-3849

Cuadrado, A. & Jouve, N. (2010). Chromosomal detection of simple sequence repeats (SSRs) using nondenaturing FISH (ND-FISH). *Chromosoma*, Vol.119, No.5, (April 2010), pp. 495-503, ISSN 0009-5915

Demirbas, A. (2007). Importance of biodiesel as transportation fuel. *Energy Policy*, Vol.35, No.9, (September 2007), pp. 4661-4670, ISSN 0301-4215

Dong, Q.H.; Cao, X.; Yang, G.; Yu, H.P.; Nicholas, K.K.; Wang, C. & Fang, J.G. (2010). Discovery and characterization of SNPs in *Vitis vinifera* and genetic assessment of some grapevine cultivars. *Scientia Horticulturae*, Vol.125, No.3, (June 2010), pp. 233-238, ISSN 0304-4238

Dyer, J.M.; Stymne, S.; Green, A.G. & Carlsson, A.S. (2008). High-value oils from plants. *Plant Journal*, Vol.54, No.4, (May 2008), pp. 640-655, ISSN 1365-313X

Edwards, A.; Civitello, A.; Hammond, H.A. & Caskey C.T. (1991). DNA typing and genetic mapping with trimeric and tetrameric tandem repeats. *American Journal of Human Genetics*, Vol.49, No.4, (October 1991), pp. 746-756, ISSN 0002-9297

Feng, S.; Li, W.; Huang, H.; Wang, J. & Wu, Y. (2009). Development, characterization and cross-species/genera transferability of EST-SSR markers for rubber tree (*Hevea brasiliensis*). *Molecular Breeding*, Vol.23, No.1, (August 2008), pp. 85-97, ISSN 1380-3743

Foster, J.T.; Allan, G.J.; Chan, A.P.; Rabinowicz, P.D.; Ravel, J.; Jackson, P.J. & Keim, P. (2010). Single nucleotide polymorphisms for assessing genetic diversity in castor bean (*Ricinus communis*). *BMC Plant Biology*, Vol.10, No.13, (January 2010), pp. 1-11, ISSN 1471-2229

Gaginella, T.S.; Capasso, F.; Mascolo, N. & Perilli, S. (1998). Castor oil: new lessons from an ancient oil. *Phytotherapy Research*, Vol.12, No.S1, (December 2008), pp. S128-S130, ISSN 0951-418X

Gajera, B.B.; Kumar, N.; Singh, A.S.; Punvar, B.S.; Ravikiran, R.; Subhash, N. & Jadeja, G.C. (2010). Assessment of genetic diversity in castor (*Ricinus communis* L.) using RAPD and ISSR markers. *Industrial Crops and Products*, Vol.32, No.3, (November 2010), pp. 491-498, ISSN 0926-6690

Gepts, P. (2004). Crop domestication as a long-term selection experiment, In: *Plant Breeding Reviews*, J. Janick, (Ed.), Vol. 24, Part 2, 1-44, John Wiley & Sons, Inc., ISBN 0-471-46892-4, Oxford, United Kingdom

Jelenkovic, G. & Harrington, E. (1973). Chromosome complement of *Ricinus communis* at pachytene and early diplotene. *Journal of Heredity*, Vol.64, No.3, (May 1973), pp. 137-142, ISSN 0022-1503

Kalia, R.K.; Rai, M.K.; Kalia, S.; Singh, R. & Dhawan, A.K. (2011). Microsatellite markers: an overview of the recent progress in plants. *Euphytica*, Vol.177, No.3, (November 2010), pp. 309-334, ISSN 0014-2336

Kofler, R.; Schlotterer, C. & Lelley, T. (2007). SciRoKo: a new tool for whole genome microsatellite search and investigation. *Bioinformatics*, Vol.23, No.13, (July 2007), pp. 1683-1685, ISSN 1460-2059

Kumar, P.; Gupta, V.K.; Misra, A.K.; Modi, D.R. & Pandey, B. K. (2009). Potential of molecular markers in plant biotechnology. *Plant Omics*, Vol.2, No.4, (July 2009), pp. 141-162, ISSN 1836-0661

Kumar, R.S.; Parthiban, K.T. & Rao, M.G. (2009). Molecular characterization of *Jatropha* genetic resources through inter-simple sequence repeat (ISSR) markers. *Molecular Biology Reports*, Vol.36, No.7, (November 2008), pp. 1951-1956, ISSN 0301-4851

Li, Y.H.; Li, W.; Zhang, C.; Yang, L.A.; Chang, R.Z.; Gaut, B.S. & Qiu, L.J. (2010). Genetic diversity in domesticated soybean (*Glycine max*) and its wild progenitor (*Glycine soja*) for simple sequence repeat and single-nucleotide polymorphism loci. *New Phytologist*, Vol.188, No.1, (October 2010), pp. 242-253, ISSN 1469-8137

Martins, W.S.; Lucas, D.C.; Neves, K.F. & Bertioli, D.J. (2009). WebSat - a web software for microsatellite marker development. *Bioinformation*, Vol.3, No.6, (January 2009), pp. 282-283, ISSN 0973-2063

Metzgar, D.; Bytof, J. & Wills, C. (2000). Selection against frameshift mutations limits microsatellite expansion in coding DNA. *Genome Research*, Vol.10, No.1, (January 2000), pp. 72-80, ISSN 1088-9051

Michaels S.D.; John, M.C. & Amasino R.M. (1994). Removal of polysaccharides from plant DNA by ethanol precipitation. *Biotechniques*, Vol.17, No.2, (August 1994), pp. 274-276, ISSN 0736-6205

Milani, M.; Dantas, F.V. & Martins, W.F.S. (2009). Genetic divergence among castor bean genotypes by morphologic and molecular characters. *Revista Brasileira de Oleaginosas e Fibrosas*, Vol.13, No.2, (May 2009), pp. 61-71, ISSN 1980-4830

Morgante, M. & Olivieri, A.M. (1993). PCR-amplified microsatellites as markers in plant genetics. *The Plant journal*, Vol.3, No.1, (January 1993), pp. 175-182, ISSN 0960-7412

Morgante, M.; Hanafey, M. & Powell, W. (2002). Microsatellites are preferentially associated with nonrepetitive DNA in plant genomes. *Nature Genetics*, Vol.30, No.2, (February 2002), pp. 194-200, ISSN 1061-4036

Nicese, F.P.; Hormaza, J.I. & McGranahan, G.H. (1998). Molecular characterization and genetic relatedness among walnut (*Juglans regia* L.) genotypes based on RAPD markers. *Euphytica*, Vol.101, No.2, (November 2004), pp. 199-206, ISSN 0014-2336

Ogunniyi, D.S. (2006). Castor oil: a vital industrial raw material. *Bioresource Technology*, Vol.97, No.9, (June 2006), pp. 1086-1091, ISSN 0960-8524

Pamidimarri, D.V.N.S.; Mastan, S.G.; Rahman, H. & Reddy, M. (2010). Molecular characterization and genetic diversity analysis of *Jatropha curcas* L. in India using RAPD and AFLP analysis. *Molecular Biology Reports*, Vol.37, No.5, (August 2009), pp. 2249-2257, ISSN 0301-4851

Paris, H.S.; Shifriss, O. & Jelenkovic, G. (1978). Idiogram of *Ricinus communis* L. *Journal of Heredity*, Vol.69, No.3, (May 1978), pp. 191-196, ISSN 0022-1503

Pinzi, S.; Garcia, I.L.; Lopez-Gimenez, F.J.; Castro, M.D.L.; Dorado, G. & Dorado, M.P. (2009). The ideal vegetable oil-based biodiesel composition: a review of social, economical and technical implications. *Energy & Fuels*, Vol.23, No.5, (April 2009), pp. 2325-2341, ISSN 0887-0624

Popova, M.G. & Moshkin, V.A. (1986). Botanical classification, In: *Castor*, V.A. Moshkin, (Ed.), 11-27, Amerind Publishing, ISBN 90-6191-466-3, New Delhi, India

Qiu, L.; Yang, C.; Tian, B.; Yang, JB. & Liu, A. (2010). Exploiting EST databases for the development and characterization of EST-SSR markers in castor bean (Ricinus communis L.). *BMC Plant Biology*, Vol.10, No.278, (December 2010), pp. 1-10, ISSN 1471-2229

Raji, A.A.J.; Anderson, J.V.; Kolade, O.A.; Ugwu, C.D.; Dixon, A.G.O. & Ingelbrecht, I.L. (2009). Gene-based microsatellites for cassava (*Manihot esculenta* Crantz): prevalence, polymorphisms, and cross-taxa utility. *BMC Plant Biology*, Vol.9, No.118, (September 2009), pp. 1-11, ISSN 1471-2229

Reddy, M.P.; Sarla, N. & Siddiq, E.A. (2002). Inter simple sequence repeat (ISSR) polymorphism and its application in plant breeding. *Euphytica*, Vol.128, No.1, (November 2002), pp. 9-17, ISSN 0014-2336

Savy-Filho, A. (2005). Melhoramento da mamona, In: *Melhoramento de espécies cultivadas*, A. Borém, (Ed), Second Edition, 429-452, Editora UFV, ISBN 85-7269-206-1, Viçosa, Brazil

Sayama, T.; Hwang, T.Y.; Komatsu, K.; Takada, Y.; Takahashi, M.; Kato, S.; Sasama, H.; Higashi, A.; Nakamoto, Y.; Funatsuki, H. & Ishimoto, M. (2011). Development and application of a whole-genome simple sequence repeat panel for high-throughput genotyping in soybean. *DNA Research*, Vol.18, No.2, (April 2011), pp. 107-115, ISSN 1756-1663

Scholz, V. & Silva, J.N. (2008). Prospects and risks of the use of castor oil as a fuel. *Biomass and Bioenergy*, Vol.32, No.2, (February 2008), pp. 95-100, ISSN 0961-9534

Simon, M.V.; Benko-Iseppon, A.M.; Resende, L.V.; Winter, P. & Kahl, G. (2007). Genetic diversity and phylogenetic relationships in *Vigna* Savi germplasm revealed by DNA amplification fingerprinting. *Genome*, Vol.50, No.6, (June 2007), pp. 538-547, ISSN 0831-2796

Singh, A.K. (2011). Castor oil-based lubricant reduces smoke emission in two-stroke engines. *Industrial Crops and Products*, Vol.33, No.2, (March 2011), pp. 287-295, ISSN 0926-6690

Sraphet, S.; Boonchanawiwat, A.; Thanyasiriwat, T.; Boonseng, O.; Tabata, S.; Sasamoto, S.; Shirasawa, K.; Isobe, S.; Lightfoot, D.A.; Tangphatsornruang, S. & Triwitayakorn, K. (2011). SSR and EST-SSR-based genetic linkage map of cassava (*Manihot esculenta* Crantz). *Theoretical and Applied Genetics*, Vol.122, No.6, (April 2011), pp. 1161-1170, ISSN 1161-1170

Sujatha, M.; Reddy, T.P. & Mahasi, M.J. (2008). Role of biotechnological interventions in the improvement of castor (*Ricinus communis* L.) and *Jatropha curcas* L. *Biotechnology Advances*, Vol.26, No.5, (May 2008), pp. 424-435, ISSN 0734-9750

Talia, P.; Nishinakamasu, V.; Hopp, H.E.; Heinz, R. A. & Paniego, N. (2010). Genetic mapping of EST-SSRs, SSR and InDels to improve saturation of genomic regions in a previously developed sunflower map. *Electronic Journal of Biotechnology*, Vol.13, No.6, (November), pp. 6, ISSN 0717-3458

Tanya, P.; Taeprayoon, P.; Hadkam, Y. & Srinives, P. (2011). Genetic diversity among *Jatropha* and *Jatropha*-related species based on ISSR markers. *Plant Molecular Biology Reporter*, Vol.29, No.1, (July 2010), pp. 252-264, ISSN 0735-9640

Tatikonda, L.; Wani, S.P.; Kannan, S.; Beerelli, N.; Sreedevi, T.K.; Hoisington, D.A.; Devi, P. & Varshney, R.K. (2009). AFLP-based molecular characterization of an elite germplasm collection of *Jatropha curcas* L., a biofuel plant. *Plant Science*, Vol.176, No.4, (April 2009), pp. 505-513, ISSN 0168-9452

Varshney, R.K.; Graner, A. & Sorrells, M.E. (2005). Genic microsatellite markers in plants: features and applications. *Trends in Biotechnology*, Vol.23, No.1, (January 2005), pp. 48-55, ISSN 0167-7799

Vasconcelos, S.; Souza, A.A.; Gusmao, C.L.S.; Milani, M.; Benko-Iseppon, A.M. & Brasileiro-Vidal, A.C. (2010). Heterochromatin and rDNA 5S and 45S sites as reliable cytogenetic markers for castor bean (*Ricinus communis*, Euphorbiaceae). *Micron*, Vol.41, No.7, (October 2010), pp. 746-753, ISSN 0968-4328

Vos, P.; Hogers, R.; Bleeker, M.; Reijans, M.; Van De Lee, T.; Hornes, M.; Friters, A.; Pot, J.; Paleman, J.; Kuiper, M. & Zabeau, M. (1995). AFLP: a new technique for DNA fingerprinting. *Nucleic Acids Research*, Vol.23, No.21, (November 1995), pp. 4407-4414, ISSN 0305-1048

Webster, G.L. (1994). Classification of the Euphorbiaceae. *Annals of the Missouri Botanical Garden*, Vol.81, No.1, (January 1994), pp. 3-32, ISSN 0026-6493

Weising K.; Nybom, H.; Wolff, K. & Kahl, G. (2005). *DNA Fingerprinting in plants: principles, methods, and applications*, Second Edition, CRC Press, ISBN 978-0-8493-1488-9, Boca Raton, USA

Yi, C.; Zhang, S.; Liu, X.; Bui, H. & Hong, Y. (2010). Does epigenetic polymorphism contribute to phenotypic variances in *Jatropha curcas* L.? *BMC Plant Biology*, Vol.10, No.259, (November 2010), pp. 1-9, ISSN 1471-2229

# Part 4

# Breeding For Pest and Disease Resistance

# 11

# Olive – *Colletotrichum acutatum*: An Example of Fruit-Fungal Interaction

Sónia Gomes[1], Pilar Prieto[2], Teresa Carvalho[3],
Henrique Guedes-Pinto[1] and Paula Martins-Lopes[1]
[1]*Institute of Biotechnology and Bioengineering, Centre of Genomics and Biotechnology -
University of Trás-os-Montes and Alto Douro (IBB/CGB-UTAD),*
[2]*Department of Mejora Genética Vegetal, Instituto de Agricultura Sostenible (CSIC),*
[3]*National Station of Plant Breeding, Department of Oliviculture,*
[1,3]*Portugal*
[2]*Spain*

## 1. Introduction

Plants are continuously exposed to an extensive array of environmental biotic and abiotic stresses. Among the first ones, viruses, bacteria, fungi, nematodes and insects are the causal agents of the most serious plant diseases. In some cases the interactions between host and pathogen result in the loss of agricultural yield that is often linked to product quality decrease (Bhadauria et al., 2010; Bray et al., 2000; Montesano et al., 2003; Zipfel, 2008). The best and most effective approach for increasing crop yield is to enhance the production efficiency and to reduce agricultural yield losses due to various plant diseases and several other stress factors. However, plants are under strong evolutionary pressure to maintain surveillance against pathogens. Part of this success is because plants have evolved a variety of sophisticated responses that recognise compounds produced and/or are released by the pathogens (elicitors) and employ these, to trigger defence signalling, normally designated by innate immunity (Montesano et al., 2003; Parker, 2009; Zipfel, 2008). Moreover, to grow in their natural environment, some crop genotypes naturally defend themselves against pathogen infection through the development of a series of morphological, physiological and molecular changes, which are all controlled by functional genomic networks. For instance, the interaction between fungal cell wall and plant surface is the beginning of compatible interaction establishment. Plant cuticle is the region where fungal infection structures differentiated, and the plant/fruit invasion is initiated. The molecular recognition of fungal cell wall may cause stress reactions and activate host's defence mechanisms. Pathogen stresses are one of the most significant damaging factors that limit the development and consequently, decrease the yield and quality of many crops. Understanding the pathogen infection mechanisms is crucial in an integrated analysis in order to target specialized functions as signalling, defence responses, and cell death, among others. Nowadays, the use of model plants may help to target candidate genes in other crop species, decreasing, therefore, the amount of work required in unknown genomes. This chapter is a review of the state-of-the-art concerning plant-pathogen interaction focused on: (i) plant defence responses (e.g., levels of host defence, pathogen molecules), (ii) plant defence signalling (e.g., molecular recognition, gene expression), and (iii) *Olea europaea* L. and *Colletotrichum acutatum* strategies in susceptible and resistant olive cultivars.

## 1.1 Compatible and Incompatible Interactions

The resistance in plants has been defined as an incompatible interaction between plant and pathogen. In general, a incompatible interaction involves the plant recognition processes that prevent or retard the pathogen growth, and spread it through plant cells. On the other hand, a compatible interaction is when plant disease occurs, due to an inadequate defence response of the host against the pathogen in terms of timing and intensity (Casado Díaz et al., 2006; Mysore & Ryu, 2004). In many plant species one of the most typical symptoms of defence response is the rapid plant cell death at the infection site, the hypersensitive response (HR), which limits pathogen from spreading to other cells (Dangl et al., 1996; Glazebrook, 2005; Oh et al., 2006). Van Der Plank (1966) reported two categories of disease resistance in plants: vertical and horizontal resistance. Vertical resistance (monogenic or oligogenic) protects hosts against only one pathogen race. This complete resistance is conferred by a few genes, or even it can be based on a single gene pair, i.e., host *R*-gene activation by pathogen avirulence gene (*avr*). The horizontal resistance (polygenic) or incomplete resistance is conditioned by many genes with minor effect. In the review work Mysore & Ryu (2004) proposed that non-host resistance against bacteria, fungi, and oomycetes can be also classified into: (i) type I, which does not result in visible cell death and (ii) type II, in which a hypersensitive response occurs, resulting in cell death at the infection site. The host resistance has been studied intensively in different pathosystem, while non-host resistance remains poorly understood (Oh et al., 2006). However, many aspects related to the gene expression patterns in some host-pathogen interaction have not yet been cleared.

## 1.2 Functional genomic

Functional genomics in plant-pathogen interaction involves studies that reveal the complex networks of host stresses perception and signal transduction, leading to the multiple defensive responses to pathogens (Langridge et al., 2006; Sreenivasulu et al., 2007; Vij & Tyagi, 2007). Functional genomics involves the development of global experimental approaches in order to analyse gene function, in contrast to structural genomics, where the entire nucleotide sequence of an organism's genome is determined (Hieter & Boguski, 1997). The purpose of functional genomics is to understand the genes function, how cells work, how cells form organisms, what goes wrong in disease and how components work together to comprise functioning cells and organisms (Lockhart & Winzeler, 2000). Advances on functional analysis considering genome, proteome and metabolome of an organism, together with the potential of bioinformatics and microscopic tools, enable scientists to assess global gene and, protein expression, and metabolite profiles of some damaged tissues, allowing a better understanding of the plant response mechanisms. The knowledge of complete genomic sequence of different crops is, sometimes the only way to gain access to the entire set of genes. The genome sequencing of the first higher plant, thale cress (*Arabidopsis thaliana*), provides nowadays an excellent model species to study host plant stress responses and to identify target genes for biotechnology applications (Bevan & Walsh,

2005; Zhang et al., 2006). For instance, cDNA from strawberry differently expressed upon challenge with the pathogen *Colletotrichum acutatum* has recently been isolated, and showed similarity to (*At* WRKY75) defence genes in *Arabidopsis thaliana* (Casado-Díaz et al., 2006; Encinas-Villarejo et al., 2009). With the development of high-throughput sequencing technologies, the number of genomes sequenced has been increasing fast. The *Oryza sativa* L., *Triticum aestivum* L., *Zea mays* L., *Vitis vinifera* L., *Glycine max* L., and *Fragaria vesca* L. genome sequencing have been extremely successful in the discovery of new genes. Similarly to plants, the knowledge of the pathogens genome is very important to monitor global changes that occur during plant-fungal interactions. Several fungal genomes have been now sequenced, being some of them extremely important in terms of yield lost in key crops such as *Botrytis cinerea* (grape grey mould, and other host species), *Fusarium graminearum* (cereal head blight), *Fusarium verticillioides* (corn seed rot), *Magnaporthe oryzae* (rice blast), *Mycosphaerella fijiensis* (banana black leaf streak), *Septoria tritici* (wheat leaf blotch), *Puccinia graminis* (cereal rust), *Phytophthora ramorum* (sudden oak death) and *Phytophthora sojae* (soybean stem/root rot) (Bhadauria et al., 2009).

## 2. Plant defence responses

### 2.1 Pre and post-invasive levels in plant-pathogen interaction

Disease resistance in plant-pathogen interactions requires sensitive and specific recognition mechanisms for pathogen-derived signals in plants. Plants lack mobile defender cells (like animal antibodies) and a somatically adaptive immune system (Jones & Dangl, 2006; Palma et al., 2009). However, they are equipped, at least, by two levels of defence: a *pre-invasive*, which is expressed at the cell wall and apoplastic spaces level aiming to prevent pathogens penetration, and a *post-invasive*, which mediates resistance relatively to pathogen that have successfully penetrated plant cells, and often results in a localized cell death at the infection site.

The physical barriers are the first level of a general plant defence against pathogens invasion, and include waxy cuticular skin layers, the plants' cell wall and actin cytoskeleton that play a key role in penetration resistance (Dangl & Jones, 2001; Kobayashi & Kobayashi, 2007). The plants' cell wall is one of the sites where the changes, due to the defence response, can be observed. These can include cell-wall thickening and lignification, papilla formation, phenolic compounds accumulation, phytoalexins and other secondary metabolites, as well transcriptional activation of pathogenesis-related proteins (Anand et al., 2009; Bhadauria et al., 2010; Montesano et al., 2003; Salazar et al., 2007; Shan & Goodwin, 2005; Zipfel, 2008). The papilla structures were observed on blueberry fruits inoculated with *Colletotrichum acutatum,* and it was formed beneath subcuticular hyphae and represented one of the host defence responses (Wharton & Schilder, 2008). If passive defences, such as the cell wall, is overcome by the pathogen, active defence responses are triggered a long lasting systemic response (systemic acquired resistance, SAR) which confers to the plant resistance against a broad spectrum of pathogens (Thordal-Christensen, 2003).

Primary or basal immune defence response is induced by perception of molecules called pathogen-associated molecular patterns (PAMPs or MAMPs) (Dangl & Jones, 2001; Jeong et al., 2009). The pathogen-associated molecular patterns (PAMPs) are recognised by the plant innate immune systems through receptor proteins called pattern recognition receptors

(PRRs), and include cold shock protein, flagellin, lipopolysaccharides of Gram-negative bacteria, lipids, elongation factor (EF-Tu), enzyme superoxide dismutase, peptidoglycan, and chitin of fungi (Bent & Mackey, 2007; Chisholm et al., 2006; Jeong et al., 2009; McDowell & Simon, 2008; Rafiqi et al., 2009; Zipfel, 2008). Perception of PAMPs occurs through PRRs located on the cell surface which could activate a chain of intracellular defensive signalling pathways, including the activation of a mitogen-activated protein kinase (MAPK) signalling cascade. When PAMPs are recognised by PRRs the PAMP-triggered immunity (PTI) system can halt microbial growth. In general, the PRRs are constituted by an extracellular leucine-rich repeat (LRR) domain, and an intracellular kinase domain.

The second molecular level of plants defence (*post-invasive*) occurs after the effectors pathogen invasion. The effectors secreted by pathogens into host cells are recognised by intracellular nucleotide-binding (NB)-LRR receptors which induce effector-triggered immunity (ETI). Successful pathogens have evolved effector molecules that target virulence effector proteins (Avr) that can overcome host defensive pathways. In a dynamic co-evolution between plants and pathogens, some plants have evolved disease resistance proteins (R) to recognise these effectors directly or indirectly, and activate an effective immune response like activation of localised cell death at the pathogen infection sites, called the hypersensitive response (HR) (Kim et al., 2008; Lindeberg & Collmer, 2009; Zipfel, 2008). The gene-for-gene resistance model, proposed by Flor (1971), requires that Avr-protein recognise the corresponding R-protein, which is accompanied by localised cell death. This effect triggers a cascade of signal transduction events that include rapid ion fluxes, extracellular oxidative burst, changes of phosphorylation status, induction of salicylic acid, and localised transcription reprogramming at the infection site (Dangl & Jones, 2001; Dangl & McDowell, 2006; Kim et al., 2008; Palma et al., 2009). Although they have been documented in others pathosystem in the olive fruits infected by *Colletotrichum acutatum* no reports considering these events have been made. New perspectives on plant defence signal transduction mechanisms have been given by the discovery of novel intracellular perception of pathogen effector proteins. The events of pathogen effectors recognition are mediated by plant NB-LRR proteins and allows resistance defence response to pathogens (Dodds & Rathjen, 2010). The plant NB-LRR proteins are able to recognise pathogen effectors through diverse pathways: (i) direct recognition, (ii) guard and decoy models, or by (iii) bait-and-switch model, and translate these interactions into a defence response (Collier & Moffett, 2009; Dangl & Jones, 2001; Dodds & Rathjen, 2010; Rafiqi et al., 2009; Van der Hoorn & Kamour, 2008). All models illustrated the diversity of perception mechanisms employed by plants to detect the broad variety of pathogen effectors, and activating plant defence responses to infection. In pathogen and plant direct recognition, the effectors triggers immune signalling by physical binding to the NB-LRR receptor. The LRR domain is considered the major determinant of perception and recognition specificity (Rafiqi et al., 2009). The guard and decoy models report a modification on an accessory protein which is then recognised by NB-LRR receptor. In bait-and-switch model the effector interacts with an accessory protein associated with NB-LRR, and then a recognition mechanism occurs between the effector and NB-LRR protein in order to trigger signalling. However, none of these models are completely understood. Thus we face the need to expand our knowledge on how plant immune receptors are activated by effector recognition and how the resistance signal is triggered.

## 2.2 Plant defence signaling molecules

Various signalling molecules mediate the expression of pathogenesis-related proteins, which can interfere with the plant resistant to a pathogen attack. The defence protein products include peroxidase, polyphenol oxidase, which catalyzes the formation of lignin and phenylalanine ammonia-lyase, involved in phytoalexin and phenolics biosynthesis (Salazar et al., 2007). Some pathogenesis-related proteins such as chitinase, and $\beta$-1,3 glucanase have potential antifungal activity which degrade the fungal cell wall and cause fungal cells lyses. Co-induction of chitinases, peroxidases, $\gamma$-thionins and $\beta$-1,3-glucanases gene expression, during pathogen infection, has been described in several plants, including wheat, strawberry, potato, soybean, maize, tobacco, tomato, bean, and pea, among others (Bettini et al., 1998; Casado-Díaz et al., 2006; Cheong et al., 2000; Lambais & Mehdy, 1998; Li et al., 2001; Liu et al., 2010; Petruzzelli et al., 1999; Vogelsang & Barz, 1993). Defence responses can also be mediated by endogenous signalling molecules such as salicylic acid, jasmonic acid, and ethylene (Encinas-Villarejo et al., 2009; Mysore & Ryu, 2004). Plant hormones have been reported to have a role in induce plant defence responses, operating in two major defence pathways in plants, depending on salicylic acid or jasmonic acid and ethylene, and conferring resistance to different pathogens (de Vos et al., 2005; Dempsey et al., 1999; Jones & Dangl, 2006; Métraux, 2001). Salicylic acid confers resistance to host plants, especially against biotrophs and hemibiotrophs pathogens, whereas jasmonic acid and ethylene signalling contributes to resistance against necrotrophs pathogens (Ausubel, 2005; Chisholm et al., 2006; Glazebrook, 2005). An incompatible interaction has been reported (Lee et al., 2009), where salicylic acid protects unripe fruit of pepper against *Colletotrichum gloeosporioides* infection through the inhibition of appressorium development. Genes encoding ethylene and jasmonic acid biosyntheses and indole-3-acetic acid regulation were found to be highly induced in citrus flowers during *Colletotrichum acutatum* infection (Lahey et al., 2004; Li et al., 2003).

## 2.3 *Colletotrichum spp.* as pathogenic fungi of *Olea europaea L.*

Fungi are the causal agents of most serious disease and are one of the pathogens that are able to breach the intact surfaces of hosts, rapidly establishing infections that can result in significant agricultural yield loss (Bhadauria et al., 2010). *Colletotrichum* species comprises a diverse range of important plant pathogenic fungi that cause pre- and postharvest crop losses worldwide. *Colletotrichum acutatum* and *Colletotrichum gloeosporioides* have been reported as causal agents of olive anthracnose, which is a major disease of cultivated olive orchards. The *Olea europaea* L. was domesticated in the Mediterranean region where it has a huge eco-social role. However, due to its importance as a crop it has spread to other Mediterranean climates worldwide (Gutiérrez & Ponti, 2009). Like other crops, olive is susceptible to a large number of diseases some of which are causing considerable damage to the olive orchards worldwide. Without the support of pest-control chemicals, such as fungicides or herbicides, crops are increasingly exposed to a range of biotic attacks. As for *Olea europaea* L. one of the main problems, concerning long-term cultivation, is the fungal contamination such as *Colletotrichum acutatum, Colletotrichum gloeosporioides, Fusicladium oleagineum, Phytophthora megasperma, Rhizoctonia, Verticillium dahliae, Pseudomonas syringae* pv. *savastanoi,* and the pests such as *Bactrocera oleae,* and *Prays oleae* (Sergeeva & Spooner-Hart, 2009; Tsitsipis et al., 2009).

Olive anthracnose was reported for the first time in Portugal in 1899 by Almeida (1899) which classified the *Gloesporium olivarum* as the causal agent of olive disease. In 1957, Von

Arx reported that *Colletotrichum gloeosporioides* was the species responsible for olive anthracnose. Latter, Simmonds introduced *Colletotrichum acutatum* in 1965 as one of the pathogen responsible for olive anthracnose. Research focused in host–pathogen interactions developed in Greece, Italy, and Spain has provided new insights into olive disease identification, and *Colletotrichum gloeosporioides* (originally: *Gloeosporium olivarum*) was reported as the primary cause of olive anthracnose (Mateo-Sagasta, 1968; Zachos & Makris, 1963). It has also been very difficult to discriminate between *Colletotrichum acutatum* and *Colletotrichum gloeosporioides* by traditional taxonomical methods. Morphologically, *Colletotrichum acutatum* and *Colletotrichum gloeosporioides* are very similar because of their host range overlapping and the wide variability found among their pathotypes. Considerable progress has been made in *Colletotrichum* species identification. In Portugal, Talhinhas et al. (2005) using molecular approaches, and Carvalho et al. (2003) with Potato dextrose agar enzyme linked immunosorbent assay (PDA-ELISA) reported that *Colletotrichum acutatum* is the predominate species in olive orchards, and consequently responsible for olive anthracnose. Nowadays, *Colletotrichum acutatum* is the dominant species in both olive growing regions: Alentejo in Portugal, and Andalusia in Spain (Moral et al., 2009; Talhinhas et al., 2005, 2009). Recently, molecular tools such as random amplification of polymorphic DNA (RAPD), species-specific primers for *Colletotrichum acutatum* isolates using internal transcribed spacer (ITS) region of rDNA sequences, and other regions of the genome have been effectively used to clearly discriminate *Colletotrichum acutatum* from *Colletotrichum gloeosporioides* and to identify genetically distinct subgroups of *Colletotrichum acutatum* (Peres et al., 2005; Talhinhas et al., 2005, 2011). Other characters, such as growth rates and sensitivity to benomyl fungicide, have been helpful to differentiate between *Colletotrichum acutatum* and *Colletotrichum gloeosporioides* (Peres et al., 2005).

Besides olive crop, *Colletotrichum acutatum* is a major constraint in global food production as it causes many of the world's most devastating diseases in cereals, grasses, legumes, vegetables, perennial crops and a number of fruit trees (Bailey et al., 1992; Peres et al., 2005; Wharton & Diéguez-Uribeondo, 2004). This species can infect all plant surfaces, but favours the young leaves, small branches and fruits of herbaceous species growing in a humid microclimate (Peres et al., 2005; Wharton & Diéguez-Uribeondo, 2004). In different regions of Portugal the *Colletotrichum acutatum* severely affects olive orchards leading to a decrease in olive oil production and quality, which compromises the protected denomination of origin (PDO) of Portuguese olive oils (Fig. 1). The *Colletotrichum acutatum* can affect up to 100% of the fruit on a olive tree during humid (or rain) autumns where susceptible olive cultivars are grown (Casado-Díaz et al., 2006; Freeman et al., 2002; Garrido et al., 2008; Peres et al., 2005; Talhinhas et al., 2011; Trapero & Blanco, 2008). Moreover, a poor quality and low stability olive oil is obtained from olives harvested in areas affected by anthracnose, presenting alterations in oil color (red), high acidity and other typical organoleptic characteristics (Carvalho et al., 2006; Moral et al., 2008; Talhinhas et al., 2009, 2011) (Fig.1).

In Portugal the olive oil production employs more than 400 000 people and the average yield is about 42 000 ton/year. The area occupied by olive trees is more than 340 thousand of hectares. In the last ten years, an expansion of olive tree planting area by Portuguese farmers was observed. This was as a result of an increasing of 2.5% in the demand of olives in the world. During the same period of time, in Portugal, the consume of olive oil was recuperated from 3.3 Kg to 7.0 Kg per capita. Beyond the improvement of national production/consume of

Fig. 1. Olive fruits infected by *Colletotrichum acutatum*. (**a**) Olive fruits without pathogen infection; (**b**) olive fruits with *Colletotrichum acutatum* infection, and (**c**) 192 hours after inoculation olive fruit is completely destroyed by pathogen infection.

olive oil, there was also an increase in exportations of about 19%. Nationally, the prevalent olive cultivar is 'Galega' that gives to olive oil a good specificity when compared with olive oils from others cultivars. This characteristic is unique in the world and for that reason it is a national imperative to preserve 'this heritage' at all cost. However, this cultivar has a disease resistant problem, related to its high susceptibility to olive anthracnose caused by the *Colletotrichum acutatum*, known in Portugal as 'gafa' disease. This disease is very aggressive and is one of the main constraints affecting both the Portuguese olive oil production and unique characteristics. The *Colletotrichum* species are known to produce enzymes that degrade carbohydrates and thus dissolve plant cell walls (e.g., polygalacturonases, pectin lyases and proteases) and hydrolyze fruit cuticles (Wharton & Diéguez-Uribeondo, 2004). In a response, several plants have evolved inhibitor proteins (PGIPs) that specifically recognise and inhibit fungal polygalacturonases (Mehli et al., 2004). Recently, a new report concerning the reduction of *Colletotrichum acutatum* infection by a polygalacturonase inhibitor protein extracted from apple, has been provided (Gregori et al., 2008).

The penetration of the plant tissue is always a crucial event for plant-pathogen interaction and the success of colonization depends on the ability of the pathogen to retrieve the nutrients from the host. In some host-pathogen interactions, such as olive fruits and pepper, the fruit index maturation may compromise the success of colonization (Lee et al., 2009; Moral et al., 2008). The resistance of immature fruits to colonization by *Colletotrichum* species can be related with the sugar content which is a non-suitable substrate to fulfil the nutritional and energy requirements of the pathogen (Wharton & Diéguez-Uribeondo, 2004). Additionally, there are a large number of secondary metabolites such as alkaloids, tannins, phenols and resins, which create a hostile and toxic environment for pathogens growth due to their anti-microbial activity (Dixon, 2001). Preformed fungi toxic compounds in unripe avocado fruit were reported to inhibit the *Colletotrichum gloeosporioides* growth

(Prusky et al., 2000). Phytoalexins have also been identified in *Capsicum annum* L. anthracnose, caused by *Colletotrichum capsici* and *Glomerella cingulata*. There are few reports about phytoalexins production after *Colletotrichum* species infection. The most recently, an evidence of phytoalexins production was observed on unripe blueberry fruits inoculated with *Colletotrichum acutatum* (Wharton & Diéguez-Uribeondo, 2004).

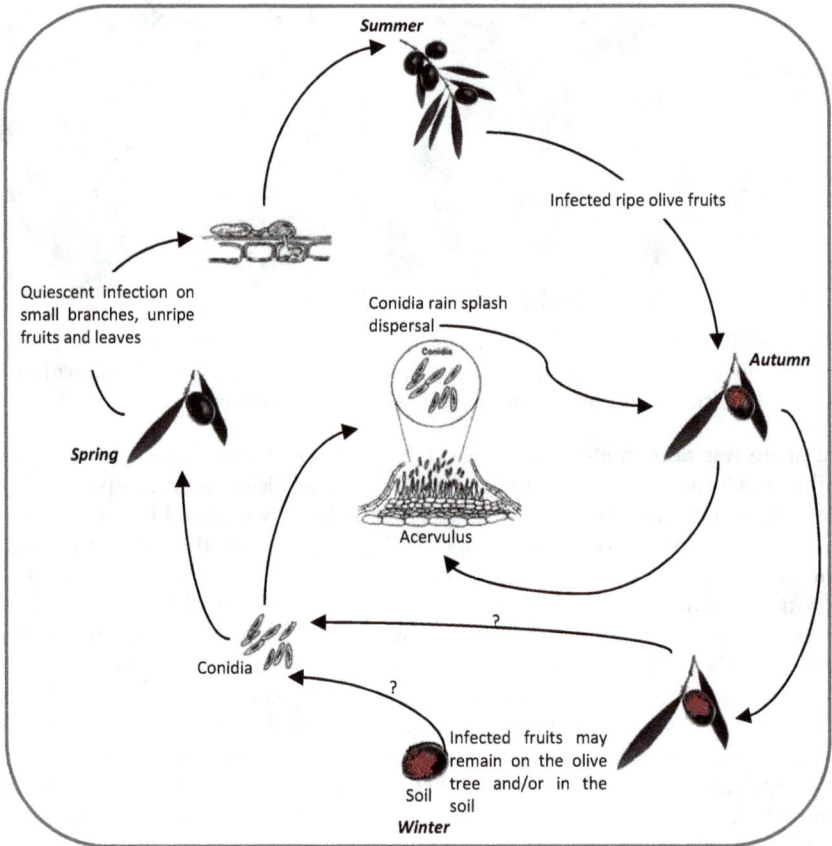

Fig. 2. The *Colletotrichum acutatum* life cycle on *Olea europaea* L. fruits. All *Colletotrichum* spp. produce acervuli with abundant spores that are rain-splashed and serve as an inoculum source for futures infections. Spore germinates to form appressoria, and quiescent infections can be established on fruits, small branches and/or leaves. Long biotrophic phase on unripe fruits are observed during spring, and summer. Necrotrophic phase appears primarily on ripe fruits and then on leaves and branches (adapted from Peres et al., 2005; Trapero & Blanco, 2008).

## 2.3.1 *Colletotrichum acutatum* life cycle on olive fruits

The early events of basic pathogenicity and susceptible interactions between olive cultivars and *Colletotrichum acutatum* as well as lifestyle of the *Colletotrichum* species are still not understood, and therefore needs to be explored. The *Colletotrichum* species life cycle comprises a teleomorph (sexual) and an anamorph (asexual) stages. However, in many plant-fungal systems asexual phase has still not been found, even on economically important crops like olive tree. The *Colletotrichum* species life cycle is summarized in Fig. 2. In general, the sexual stage accounts for the fungus genetic variability while the asexual is responsible for fungal spore formation and appressorium development. The *Colletotrichum* spores are essential for the fungus dispersal (Peres et al., 2005; Wharton & Diéguez-Uribeondo, 2004). Specific structures or developmental stages, such as acervuli formation, spore germination and appressorium formation are distinguished during the *Colletotrichum* species life cycle.

### 2.4 *Colletotrichum acutatum* and olive fruits interaction

To establish a compatible interaction with a host, the fungal pathogen does not only need to overcome host physical barriers, but also they have to establish a feeding relationship with the host. The early events in *Colletotrichum* spp. infection process are very similar among the different hosts. Frequently, the host-pathogen interaction includes (i) spore adhesion to the host surface, (ii) spore germination, (iii) appressorium development and (iv) fungal growth through the colonization of the hosts' tissues (Fig. 3). Spores are embedded in a matrix of moist hydrophilic mucilaginous material including glycoproteins, lipids and polysaccharides. These compounds are essential not only for spore adhesion, but also for their protection against desiccation, physical damage, and host defence responses. After spore germinate a short germ tube is generated which finally differentiate into an appressorium with a penetration pore (Fig. 3).

The appressorium with penetration pore are a penetration structure that is formed after germ tube germination, and allows many pathogens to enter into the host cells. Pathogens may force their way through plant surfaces by different means; some take advantage of natural doors, such as stomata or lenticels; some others enter through wounds or directly through cuticle (Bailey et al., 1992; Gomes et al., 2009; Jong & Ackerveken, 2009). In order to infect plants, fungal pathogens have developed a wide variety of infection strategies: (i) *Necrotrophs*, pathogen kill the host and feed from the cell contents; (ii) *Biotrophs*, require a living host to complete their life cycle; and (iii) *Hemibiotrophs*, act as both biotrophs and necrotrophs at different stages of infection. Most of the *Colletotrichum* species are hemibiotrophs with different spans of their biotrophic phase, and may also undergo a period of quiescence in order to overcome resistance mechanisms (Gomes et al., 2009; Peres et al., 2005). Two types of interaction between *Colletotrichum* species and their hosts have been reported: intercellular hemibiotrophy and subcuticular intramural necrotrophy (Gomes et al., 2009; O'Connell et al., 2000; Perfect et al., 1999) (Fig. 4).

In intercellular hemibiotrophic infections, a symptomless biotrophic phase is followed by a destructive necrotrophic one, during which symptoms become apparent. Examples of pathogens employing this infection strategy are *Colletotrichum graminicola* Politis & Wheeler, 1973). *Colletotrichum gloeosporioides* (Ogle et al., 1990), and *Colletotrichum lindemuthianum* (Mercer et al., 1975; O'Connell et al., 1985). In the subcuticular intramural infection strategy, rather than penetrating the epidermal cell wall, the fungus grows under the cuticle and within the periclinal and anticlinal walls of epidermal cells (Arroyo et al., 2007; O'Connell et al., 2000; Wharton & Diéguez-Uribeondo, 2004). *Colletotrichum phomoides* (Bailey et al., 1992), *Colletotrichum capsici* (Pring et al., 1995) and *Colletotrichum circinans* (Bailey et al., 1992) behave in this manner. On olive fruits, a combination of both strategies has been reported, and results in intra- and intercellular colonization of fruits (Gomes et al., 2009) (Fig. 4). In this pathosystem *Colletotrichum acutatum* used hemibiotrophy and subcuticular colonization strategies to colonize 'Galega' that is *Colletotrichum acutatum* susceptible olive cultivar and 'Picual' that is *Colletotrichum acutatum* resistant olive cultivar.

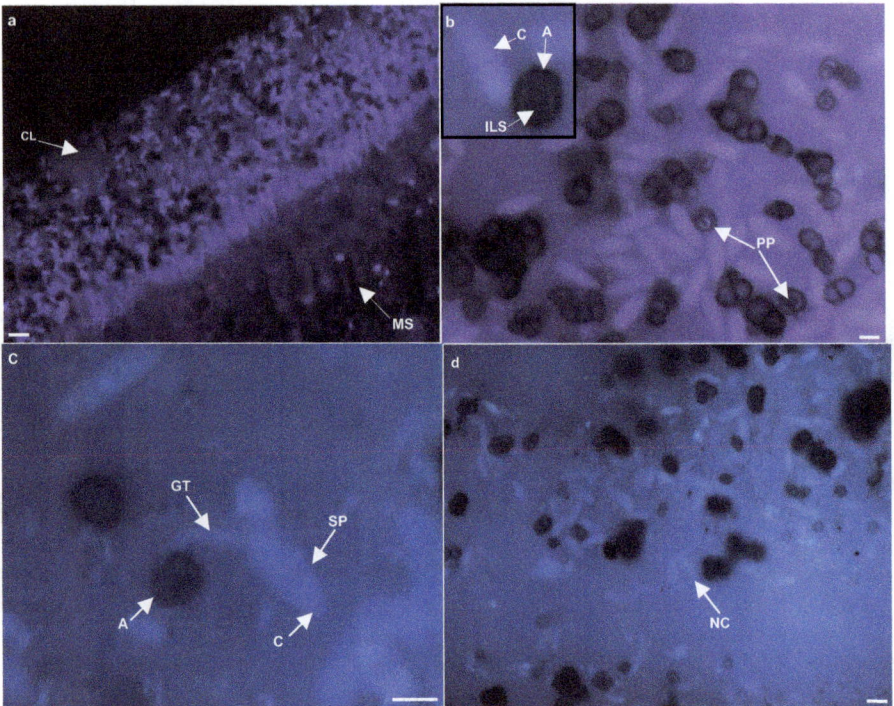

Fig. 3. *Colletotrichum acutatum* infection structures in olive fruits. (**a**) Massive adhesion of fusiform spore on the susceptible cultivar (Galega) was observed at 48 hours after inoculation (hai). (**b**) Germinated and ungerminated spore on the fruit surface 48 hai; a penetration pore has differentiated under a mature appressorium and internal light spot (*ILS*) was observed at 72 hai. (**c**) Conidia (with a septum at the equatorial zone (arrow)) developed a germ tube at the end of which a pigmented fluorescing appressorium differentiated within 72 hai. (**d**) The development of one or more secondary conidia simultaneously was observed at 48 hai. Scale bar represents 20 µm in panel (**a**) 10 µm in panels (**b**) and (**d**) and 5 µm in panel (**c**) CL - host cuticle; MS - mesocarp; C - conidium; A - appressorium; ILS - internal light spot; PP - penetration pore; GT - germ tubes; SP - septum; NC - new conidial formation.

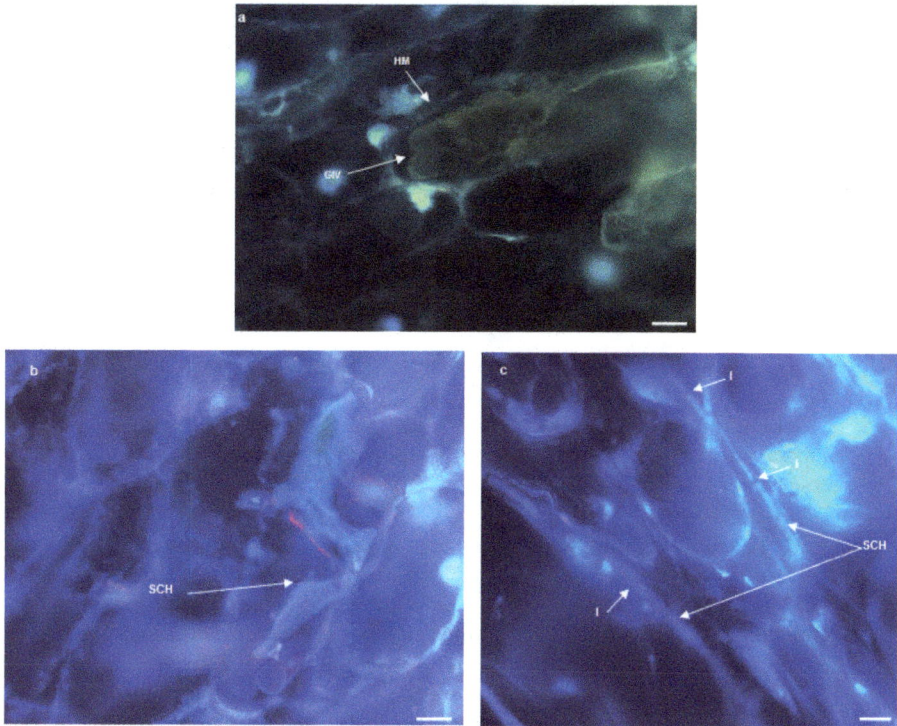

Fig. 4. Intracellular hemibiotrophic-like infection structures in infected olive fruit. Penetration of epidermal cells and fungal development inside the host were observed. (a) The host membrane (arrow) encloses a globose infection vesicle without damaging the host plasma membrane. (b) and (c) Subcuticular infection vesicles were observed in susceptible 'Galega' (b) and resistant 'Picual' (c) olive fruit. During the early events of infection, hyphae grow inside the host cell wall of 'Galega' without penetrating the lumen. (c) Development of hyphae with internodes are marked with arrows. Scale bar represents 10 μm in all panels. HM - host membrane; GIV - globose infection vesicle; I - internodes; SCH - subcuticular hypha.

## 3. Conclusion

Functional genomics studies in species like *Olea europaea* L., where no sequence data are available, has a major constraint, which limits the screening of potentially transcript-derived fragments (TDFs) and their function. The identification and the role of genes involved in olive polygenic resistance is complex and remain unknown. Current knowledge about *Colletotrichum acutatum*-olive interaction, in susceptible and resistant cultivars and its functional genomic impact on plant resistance is not yet sufficient to provide solid explanations. Data for model plants may direct strategies that are transferable to crop systems, even when there is a lack of sequencing information related to the crop under study, once they may provide guidelines to which direction studies may be conducted. Furthermore, *Colletotrichum acutatum*-strawberry (pathosystem model for *Colletotrichum acutatum* - plant interaction) interactions are still poorly understood making the task even more difficult in olive. The identification of resistance genes is essential not only to understand the basis of olive anthracnose disease but also to identify novel fungicide targets and, in the long term, environmental and human safe fungicides.

## 4. Acknowledgment

This research was funded by Fundação para a Ciência e Tecnologia, Portugal through SFRH/DB/25384/2005 grant and 'POCTI/AGR/57817/2004' project, and 'Acção Integrada Luso-Espanhola N.º E-97/09'.

## 5. References

Almeida, M.J.V. (1899). La Gaffa des Olives en Portugal. *Bulletin de la Societe Mycologique de France*, Vol. 15, pp. 90-94, ISSN 0395-7527

Anand, T.; Bhaskaran, R.; Raguchander. T.; Samiyappan, R.; Prakasam, V. & Gopalakrishnan, C. (2009). Defence Responses of Chilli Fruits to *Colletotrichum capsici* and *Alternaria alternate*. *Biologia Plantarum*, Vol.53, No.3, pp. 553-559, ISSN 0006-3134

Ausubel, F.M. (2005). Are Innate Immune Signaling Pathways in Plants and Animals Conserved? *Nature Immunology*, Vol.6, No.10, pp. 973–979, ISSN 1529-2908

Arroyo, F.T.; Moreno, J.; Daza, P.; Boianova, L. & Romero, F. (2007). Antifungal Activity of Strawberry Fruit Volatile Compounds against *Colletotrichum acutatum*. *Journal of Agricultural Food and Chemistry*, Vol.55, No.14, pp. 5701-5707, ISSN 00218561

Bailey, J.A.; O'Connell, R.J.; Pring, R.J. & Nash, C. (1992). Infection Strategies of *Colletotrichum* Species. In: *Colletotrichum: Biology, pathology and control*. In J. A. Bailey & M. J. Jeger, (Ed.), 88–120, CAB International, ISBN 10: 0851987567, Wallingford, UK.

Bent, A.F. & Mackey, D. (2007). Elicitors, Effectors, and *R* Genes: The New Paradigm and a Lifetime Supply of Questions. *Annual Review of Phytopathology*, Vol.45, (September 2007), pp. 399–436, ISSN 0066-4286

Bettini, P.; Cosi, E.; Pellegrini, M.G.; Turbanti, L.; Vendramin, G.G. & Buiatti, M. (1998). Modification of Competence for *in vitro* Response to *Fusarium oxysporum* in Tomato cells. III. PR-protein Gene Expression and Ethylene Evolution in Tomato Cell Lines Transgenic for Phytohormone-related Bacterial Genes. *Theoretical and Applied Genetics*, Vol.97, No.4, (April 1998), pp. 575–583, ISSN 1432-2242

Bevan, M. & Walsh, S. (2005). The *Arabidopsis* genome: a foundation for plant research. *Genome Research*, Vol.15, No.12, (December 2005), pp. 1632-1642, ISSN: 1532-2548

Bhadauria, V.; Banniza, S.; Wang, L.X.; Wei, Y.D. & Peng, Y.L. (2010). Proteomic Studies of Phytopathogenic Fungi, Oomycetes and their Interactions with Hosts. *European Journal of Plant Pathology*, Vol.126, No.1, (January 2010), pp. 81–95, ISSN 1573-8469

Bhadauria, V.; Banniza, S.; Wei, Y.D. & Peng, Y.L. (2009). Reverse Genetics for Functional Genomics of Phytopathogenic Fungi and Oomycetes. *Comparative and Functional Genomics*, Vol.2009, (July 2009), pp. 1-11, ISSN 1532-6268

Bray, E.A.; Bailey-Serres, J. & Weretilnyk, E. (2000). Responses to Abiotic Stresses. *In: Biochemistry and Molecular Biology of Plants*, Gruissem, W, Buchanan B. and Jones R. (Ed.), 1158-1249, American Society of Plant Physiologists, Rockville, MD

Carvalho, M.T.; Piteira, M.C.C. & Clara, M.I.E. (2003). Identificação de *Colletotrichum acutatum* em *Olea europaea* Afectada pela Doença da Gafa. *Proceeding of III Simpósio Nacional de Olivicultura- Revista de Ciências Agrárias*, pp. 54, Castelo Branco, Portugal, October 29-3, 2003

Carvalho, M.T.; Simões-Lopes, P.; Monteiro da silva, M.J.; Pires, S. & Gonçalves, M.J. (2006). The Effect of *Colletotrichum* Control on Getting High Quality Olive Oils. *Proceedings of Olivebioteq, Second International Seminar "Biotechnology and Quality of Olive Tree Products Around the Mediterranean Basin"*, pp. 239-242, Marzala del Vallo, Marsala, Italy, November 5-10, 2006

Casado-Díaz, A.; Encinas-Villarejo, S.; Santos, B.; Schiliro, E.; Yubero-Serrano, E.; Amil-Ruíz, F.; Pocovi, M.I.; Pliego-Alfaro, F.; Dorado, G.; Manuel Rey, M.; Romero, F.; Munoz-Blanco, J. & Caballero, J.L. (2006). Analysis of Strawberry Genes Differentially Expressed in Response to *Colletotrichum* Infection. *Physiologia Plantarum*, Vol.128, No.4, (December 2006), pp. 633–650, ISSN 1399-3054

Cheong, Y.H.; Kim, C.Y.; Chun, H.J.; Moon, B.C.; Park, H.C.; Kim, J.K.; Lee, S-H.; Han, C.D.; Lee, S.Y. & Cho, M.J. (2000). Molecular Cloning of a Soybean Class III beta-1,3-Glucanase Gene that is Regulated both Developmentally and in Response to Pathogen Infection. *Plant Science*, Vol.154, No.1, (May 2000), pp. 71–81, ISSN 0168-9452

Chisholm, S.T.; Coaker, G.; Day, B. & Staskawicz, B.J. (2006). Host-microbe Interactions: Shaping the Evolution of the Plant Immune Response. *Cell*, Vol.124, No.4 (February 2006), pp. 803-814, ISSN 0092-8674

Collier, S.M. & Moffett, P. (2009). NB-LRRs Work a 'bait and switch" on Pathogens. *Trends in Plant Science*, Vol.14, No.10, pp.521-529, ISSN 1360-1385

Dangl, J.L.; Dietric, R.A. & Richberg, M.H. (1996). Death Don't Have no Mercy: Cell Death Programs in Plant-Microbe Interactions. *The Plant Cell*, Vol.8, (October 1996), pp. 1793–1807, ISSN 1532-298X

Dangl, J.L. & Jones, J.D.G. (2001). Plant Pathogens and Integrated Defence Responses to Infection. *Nature*, Vol.411, (June 2001), pp. 826-833, ISSN 0028-0836

De Vos, M.; Van Oosten, V.R.; Van Poecke, R.M.P.; Van Pelt, J.A.; Pozo, M.J.; Mueller, M.J.; Buchala, A,J.; Métraux, J.P.; Van Loon, L.C.; Dicke, M. & Pieterse, C.M.J. (2005). Signal Signature and Transcriptome Changes of *Arabidopsis* During Pathogen and Insect Attack. *Molecular Plant-Microbe Interaction*, Vol.18, No.8, (May 2005), pp. 923-937, ISSN 0894-0282

Dempsey, D.A.; Shah, J. & Klessig, D.F. (1999). Salicylic Acid and Disease Resistance in Plants. *Critical Reviews in Plant Sciences*, Vol.18, pp. 547–575, ISSN 1549-7836

Dixon, R.A. (2001). Natural Products and Plant Disease Resistance. *Nature*, Vol.411, No.6839, (June 2001), pp. 843–847, ISSN 0028-0836

Encinas-Villarejo, S.; Maldonado, A.M.; Amil-Ruiz, F.; de los Santos, B.; Romero, F.; Pliego-Alfaro, F.; Muñoz-Blanco, J. & Caballero, J.L. (2009). Evidence for a Positive Regulatory Role of Strawberry (*Fragaria x ananassa*) *Fa* WRKY1 and *Arabidopsis At* WRKY75 Proteins in Resistance. *Journal of Experimental Botany*, Vol.60, No.11, (May 2009), pp. 3043-3065, ISSN 0022-0957

Dodds, P.N. & Rathjen J.P. (2010). Plant Immunity: Towards an Integrated View of Plant–Pathogen Interactions. *Nature Reviews Genetics*, Vol.11, (August 2010), pp. 539-548, ISSN : 1471-0056

Flor, H.H. (1971). Current Status of Gene-for-gene Concept. *Annual Review of Phytopathology*, Vol.9, No.1, (September 1971), pp. 275-296, ISSN 0066-4286

Freeman, S.; Shalev, Z. & Katan, J. (2002). Survival in Soil of *Colletotrichum acutatum* and *C. gloeosporioides* Pathogenic on Strawberry. *Plant Disease*, Vol.86, No.9, (September 2002), pp. 965-970, ISSN 0191-2917

Garrido, C.; Carbú, M.; Fernández-Acero, F.J.; Budge, G.; Vallejo, I.; Colyer, A. & Cantoral, J.M. (2008). Isolation and Pathogenicity of *Colletotrichum* spp. Causing Anthracnose of Strawberry in South West Spain. *European Journal of Plant Pathology*, Vol.120, No. 4, (October 2007), pp. 409-415, ISSN 0929-1873

Glazebrook, J. (2005). Contrasting Mechanisms of Defense Against Biotrophic and Necrotrophic Pathogens. *Annual Review of Phytopathology*, Vol.43, (March 2005), pp. 205-227, ISSN 0066-4286

Gregori. R.; Mari, M.; Bertolini, P.; Barajas, J.A.S.; Tian, J.B. & Labavitch, J.M. (2008). Reduction of *Colletotrichum acutatum* Infection by a Polygalacturonase Inhibitor Protein Extracted from Apple. *Postharvest Biology and Technology*, Vol.48, No.2, (May 2008), pp. 309-313, ISSN 0925-5214

Gomes, S.; Prieto, P.; Martins-Lopes, P.; Carvalho, T.; Martin, A. & Guedes-Pinto, H. (2009). Development of *Colletotrichum acutatum* on Tolerant and Susceptible *Olea europaea* L. cultivars: A Microscopic Analysis. *Mycopathologia*, Vol.168, No.4, (May 2009), pp. 203-211, ISSN 0301-486X

Gutiérrez, A.P. & Ponti, L. (2009). Can Climate Change Have an Influence on the Occurence and Management of Olive Pests and Diseases?. *Proceeding of 4th European Meeting of the IOBC/WPRS Working Group Integrated Protection of Olive Crops*, pp. 25, Córdoba, Spain, June 1-4, 2009

Hieter, P. & Boguski, M. (1997). Functional Genomics: it's All How You Read It. *Science*, Vol.278, No.5338, (October 1997), pp. 601-602, ISSN 1095-9203

Jeong, B.; Dijk, K. & Alfano, J. (2009). *Pseudomonas syringae* Type III-Secreted Proteins and Their Activities and Effects on Plant Innate Immunity. *Annual Plant Reviews*, Vol.34, (February 2009), pp. 48-76, ISSN 1756-9710

Jones, J.D. & Dangl, J.L. (2006). The Plant Immune System. *Nature*, Vol.444, (November 2006), pp. 323-329, ISSN 0028-0836

Jong, M. & Ackerceken, G. (2009). Fungal and Oomycete Biotrophy. *Annual Plant Reviews*, Vol.34, (February 2009), pp. 77-101, ISSN 1756-9710

Kim, M.G.; Kim, S.Y.; Kim, Y.W.; Mackey, D. & Lee, S.Y. (2008). Responses of *Arabidopsis thaliana* to Challenge by *Pseudomonas syringae*. *Molecules and Cells*, Vol.25, No.3, (May 2008), pp. 323-331, ISSN 0219-1032

Kobayashi, Y. & Issei Kobayashi, I. (2007). Depolymerization of the Actin Cytoskeleton Induces Defense Responses in Tobacco Plants. *Journal of General Plant Pathology*, Vol.73, No.5, (August 2007), pp. 360-364, ISSN 1610-739X

Lahey, K.A.; Yuan, R.; Burns, J.K.; Ueng, P.P.; Timmer, L.W. & Chung, KR. (2004). Induction of Phytohormones and Differential Gene Expression in Citrus Flowers Infected by the Fungus *Colletotrichum acutatum*. *Molecular Plant-Microbe Interaction*, Vol.17, No.12, (December 2004), pp. 1394-1401, ISSN 0894-0282

Lambais, M.R. & Mehdy, M.C. (1998). Spatial Distribution of Chitinases and beta-1,3-glucanase transcripts in bean arbuscular mycorrhizal roots under low and high soil phosphate Conditions. *New Phytologist*, Vol.140, No.1, pp. 33-42, ISSN 1469-8137

Langridge, P.; Paltridge, N. & Fincher, G. (2006). Functional Genomics of Abiotic Stress Tolerance in Cereals. *Briefings in Functional Genomics and Proteomics*, Vol.4, No.4, (February 2009), pp. 343-354, ISSN 1477-4062

Lee, S.; Hong, J.C.; Jeon, W.B.; Chung, Y.S.; Sung, S.; Choi, D.; Joung, Y.H. & Oh, BJ. (2009). The Salicylic Acid-Induced Protection of Non-Climacteric Unripe Pepper Fruit against *Colletotrichum gloeosporioides* is Similar to the Resistance of Ripe Fruit. *Plant Cell Report*, Vol.28, No.10, (August 2009), pp. 1573–1580, ISSN 1432-203X

Lindeberg, M. & Collmer, A. (2009). Gene Ontology for Type III Effectors: Capturing the Processes at the Host Pathogen Interface. *Trends in Microbiology*, Vol.17, No.7, (July 2009), pp. 304-311, ISSN 0966-842X

Li, W.; Yuan, R.; Burns, J.K.; Timmer, L.W. & Chung, K.R. (2003). Genes for Hormone Biosynthesis and Regulation are Highly Expressed in Citrus Flowers Infected with the Fungus *Colletotrichum acutatum*, the Causal Agent of Postbloom Fruit Drop. *Journal of American Society Horticulture Science*, Vol.128, No.4, (February 2003), pp. 578-583, ISSN 0003-1062

Li, W.L.; Faris, J.D.; Muthukrishnan, S.; Liu, D.J.; Chen, P.D. & Gill, BS. (2001). Isolation and Characterization of Novel cDNA Clones of Acidic Chitinases and beta-1,3-glucanases from Wheat Spikes Infected by *Fusarium graminearum*. *Theoretical and Applied Genetics*, Vol.102, No.2-3, (May 2000), pp. 353–362, ISSN 1432-2242

Liu, B.; Xue, X.D.; Cui, S.P.; Zhang, X.Y.; Han, Q.M.; Zhu, L.; Liang, X.F.; Wang, X.J.; Huang, L.L.; Chen, X.M. & Kang, Z.S. (2010). Cloning and Characterization of a Wheat beta-1,3-glucanase Gene Induced by the Stripe Rust Pathogen *Puccinia striiformis* f. sp *tritici*. *Molecular Biology Reports*, Vol.37, No.2, (February 2010), pp. 1045-1052, ISSN 1573-4978

Lockhart & Winzeler. (2000). Genomics, Gene Expression and DNA Arrays. *Nature*, Vol.405, No.6788, (June 2000), pp. 827-836, ISSN 0028-0836

Mateo-Sagasta, E. (1968). Estudios Básicos sobre *Gloeosporium olivarum* Alm. (*Deuteromiceto Melanconial*). *Boletin de Patologia Vegetal y Entomologia Agricola*, Vol.30, pp. 31-135.

McDowell, J.M. & Simon, S.A. (2008). Molecular Diversity at the Plant-Pathogen Interface. *Developmental and Comparative Immunology*, Vol.32, No.7, (December 2007), pp. 736–744, ISSN 0145-305X

Mehli, L.; Schaart, J.G.; Kjellsen, T.D.; Tran, D.H.; Salentijn, E.M.J.; Schouten, H.J. & Iversen, T.H. (2004). A Gene Encoding a Polygalacturonase-Inhibiting Protein (PGIP) Shows Developmental Regulation and Pathogen-Induced Expression in Strawberry. *New Phytologist Journal*, Vol.163, (February 2004), pp. 99–110, ISSN 0028-646X

Mercer, P.C.; Wood, R.K.S. & Greenwood, A.D. (1975). Ultrastructure of parasitism of *Phaseolus vulgaris* by *Colletotrichum lindemuthimum*. *Physiology Plant Pathology*, Vol.5, pp. 203-214.

Métraux, J.P. (2001). Systemic Acquired Resistance and Salicylic Acid: Current State of Knowledge. *European Journal of Plant Pathology*, Vol.107, No.1, (November 2000), pp. 13–18, ISSN 1573-8469

Montesano, M.; Brader, G. & Palva, T. (2003). Pathogen-Derived Elicitors: Searching for Receptors in Plants. *Molecular Plant Pathology*, Vol.4, No.1, pp. 73-79, ISSN 0885-5765

Moral, J.; Oliveira, R. & Trapero, A. (2009). Elucidation of the Disease Cycle of Olive Anthracnose caused by *Colletotrichum acutatum*. *Phytopathology*, Vol.99, No.5, (May 2009), pp. 548-556, ISSN 0031-949X

Moral, J.; Bouhmidi, K. & Trapero, A. (2008). Influence of Fruit Maturity, Cultivar Susceptibility, and Inoculation Method on Infection of Olive Fruit by *Colletotrichum acutatum*. *Plant Disease*, Vol.92, No.10, (October 2008), pp. 1421-1426, ISSN 0191-2917

Mysore, K.S. & Ryu, C.M. (2004). Nonhost Resistance: How Much do We Know?. *Trends in Plant Science*, Vol.9, No.2, (February 2004), pp. 97-104, ISSN 1360-1385

O'Connell, R.; Perfect, S.; Hughes, B.; Carzaniga, R.; Bailey, J. & Green, J. (2000). Dissecting the cell biology of *Colletotrichum* infection processes, In: *Colletotrichum: Host Specificity Pathology and Host-Parasite Interactions,* Prusky D, Freeman S, Dickman MD. (Ed.), pp. 57-77, The American Phytopathological Society, St. Paul, MN.

O'Connell, R.J.; Bailey, J.A. & Richmond, D.V. (1985). Cytology and physiology of infection of *Phaseolus vulgaris* by *Colletotrichum lindemuthianum*. *Physiological and Molecular Plant Pathology*, Vol.27, pp.75-98.

Ogle, H.J.; Gowanlock, D.H. & Irwin, J.A.G. (1990). Infection of *Stylosanthes guianensis* and *S. scabra* by *Colletotrichum gloeosporioides*. *Phytopathology*, Vol.80, No.9, pp.837-842, ISSN 0031-949X

Oh, S.K.; Lee, S.; Chung, E.; Park, J.M.; Yu, S.H.; Ryu, C.M. & Choi, D. (2006). Insight into Types I and II nonhost Rresistance using Expression Patterns of Defense-related Genes in Tobacco. *Planta*, Vol.223, No.5, (February 2006), pp. 1101–1107, ISSN 1432-2048

Palma, K.; Wiermer, M. & Li, X. (2009). Marshalling the Troops: Intracellular Dynamics in Plant Pathogen Defense. *Annual Plant Reviews*, Vol.34, (February 2009), pp. 177-219, ISSN 1756-9710

Parker, J. (2009). Molecular Aspects of Plant Disease Resistance, In: *Annual Plant Reviews*, Vol.34. eds. Wiley-Blackwell Publication Ltd, ISSN 1460-1494, Cologne, Germany

Peres, N.A.; Timmer, L.W.; Adaskaveg, J.E. & Correll, J.C. (2005). Lifestyles of *Colletotrichum acutatum*. *Plant Disease*, Vol.89, No.8, (August 2005), pp. 784-796, ISSN 0191-2917

Perfect, S.E.; Hughes, H.B.; O'Connell, R.J. & Green, J.R. (1999). *Colletotrichum*: A Model for Studies on Pathology and Fungal Plant Interactions. *Fungal Genetics and Biology*, Vol.27, No.2-3, (July-August 1999), pp. 186-198, ISSN 1096-0937

Petruzzelli, L.; Kunz, C.; Waldvogel, R.; Meins, F.Jr. & Leubner-Metzger, G. (1999). Distinct Ethylene- and Tissue-Specific Regulation of beta-1,3-glucanases and Chitinases during Pea Seed Germination. *Planta*, Vol.209, No.2, (August 1999), pp. 195–201, ISSN 1432-2048

Politis, D.J. & Wheeler, H. (1973). Ultrastructural study of penetration of maize leaves by *Colletotrichum graminicola*. *Physiological Plant Pathology*, Vol.3, pp.465-471.

Pring, R.J.; Nash, C.; Zakaria, M. & Bailey, J.A. (1995). Infection process and host range of *Colletotrichum capsici*. *Physiological and Molecular Plant Pathology*, Vol.46, No.2, pp.137-152, ISSN 0885-5765

Prusky, D.; Kobiler, I.; Ardi, R.; Beno-Moalem, D.; Yakoby, N. & Keen, N.T. (2000). Resistance Mechanisms of Subtropical Fruits to *Colletotrichum gloeosporioides*. In: *Colletotrichum: Host Specificity, Pathology, and Host-Pathogen Interaction*, Prusky D,

Freeman S, Dickman MB. (Ed.), 232-244, The American Phytopathological Society. St. Paul MN

Rafiqi, M.; Bernoux, M.; Ellis, J.G. & Dodds, P.N. (2009). In the Trenches of Plant Pathogen Recognition: Role of NB-LRR Proteins. *Seminars in Cell and Developmental Biology,* Vol.20, No.9, (May 2009), pp.1017-1024, ISSN 1084-9521

Salazar, S.M.; Castagnaro, A.P.; Arias, M.E.; Chalfoun, N.; Tonello, U. & Díaz Ricci, J.C. (2007). Induction of a Defense Response in Strawberry Mediated by an Avirulent Strain of *Colletotrichum. European Journal of Plant Pathology,* Vol.117, No.2, pp. 109–122, ISSN 1573-8469

Simmonds, J.H. (1965). A Study of the Species of *Colletotrichum* Causing Ripe Fruit Rots in Queensland. *Queensland Journal of Agricultural and Animal Science,* Vol.22, pp. 437-459, ISSN 0033-6173

Sergeeva, V. & Spooner-Hart, R. (2009). Olive Diseases and Disorders in Australia. *Proceeding of 4th European Meeting of the IOBC/WPRS Working Group Integrated Protection of Olive Crops,* pp. 30, Córdoba, Spain, June 1-4, 2009

Shan, X.C. & Goodwin, P.H. (2005). Reorganization of Filamentous Actin in *Nicotiana benthamiana* Leaf Epidermal Cells Inoculated with *Colletotrichum destructivum* and *Colletotrichum graminicola. International Journal of Plant Sciences,* Vol.166, No.1, (January 2005), pp. 31–39, ISSN 1058-5893

Sreenivasulu, N.; Sopory, S.K. & Kavi Kishor, P.B. (2007). Deciphering the Regulatory Mechanisms of Abiotic Stress Tolerance in Plants by Genomic Approaches. *Gene,* Vol.388, No.1-2, (February 2007), pp. 1-13, ISSN 0378-1119

Talhinhas, P.; Mota-Capitao, C.; Martins, S.; Ramos, A.P.; Neves-Martins, J.; Guerra-Guimarães, L.; Varzea, V.; Silva, M.C.; Sreenivasaprasad, S. & Oliveira, H. (2011). Epidemiology, Histopathology and Aetiology of Olive Anthracnose caused by *Colletotrichum acutatum* and *C. gloeosporioides* in Portugal. *Plant Pathology,* Vol.60, No.3, (June 2011), pp. 483-495, ISSN 1365-3059

Talhinhas, P.; Neves-Martins, J.; Oliveira, H. & Sreenivasaprasad, S. (2009). The Distinctive Population Structure of *Colletotrichum* Species Associated With Olive Anthracnose in the Algarve Region of Portugal Reflects a Host-Pathogen Diversity Hot Spot. *FEMS Microbiology Letters,* Vol.296, No.1, (May 2009), pp. 31–38, ISSN 1574-6968

Talhinhas, P.; Sreenivasaprasad, S.; Neves-Martins, J. & Oliveira, H. (2005). Molecular and Phenotypic Analyses Reveal Association of Diverse *Colletotrichum acutatum* Groups and a Low Level of *C. gloeosporioides* With Olive Anthracnose. *Applied and Environmental Microbiology,* Vol.71, No.6, (June 2005), pp. 2987-2998, ISSN 0099-2240

Thordal-Christensen, H. (2003). Fresh Insights into Processes of Nonhost Resistance. *Plant Biology,* Vol.6, No.4, (August 2003), pp. 351–357, ISSN 1435-8603

Trapero, A. & Blanco, M.A. (2008). Enfermedades. In: *El Cultivo de Olivo,* D. Barranco, R. Fernández-Escobar, and L. Rallo (Ed.), 595-656, Andalucía, Madrid, Spain

Tsitsipis, J.A.; Varikou, K.; Kalaitzaki, A.; Alexandrakis, V.; Margaritopoulos, J. & Skouras, P. (2009). Chemical Control of Olive Pests: Blessing or Curse?. *Proceeding of 4th European Meeting of the IOBC/WPRS Working Group Integrated Protection of Olive Crops,* pp. 31, Córdoba, Spain, June 1-4, 2009

van der Hoorn, R.A. & Kamoun, S. (2008). From Guard to Decoy: a New Model for Perception of Plant Pathogen Effectors. *Plant Cell,* Vol.20, pp. 2009–2017, ISSN 1532-298X van der Plank. (1966). Horizontal (polygenic) and Vertical (oligogenic) Resistance Against Blight. *American Journal of Potato Research,* Vol. 43, No.2, pp. 43-52, ISSN 1099-209X

Vij, S. & Tyagi, A.K. (2007). Emerging Trends in the Functional Genomics of the Abiotic Stress Response in Crop Plants. *Plant Biotechnology Journal,* Vol.5, No.3, (May 2007), pp. 1-20, ISSN 1467-7644

Von Arx, J.A. (1957). Die Arten der Gattung *Colletotrichum* Cda (in German). *Phytopathology, Z,* Vol. 29, pp. 413–468, ISSN 0031-949X

Vogelsang, R. & Barz, W. (1993). Purification, Characterization, and Differential Hormonal Regulation of a β-1,3-glucanase and Two Chitinases from Chick Pea (*Cicer arietinum* L.). *Planta,* Vol.189, No.1, (January 1993), pp. 60–69, ISSN 1432-2048

Wharton, P.S. & Diéguez-Uribeondo, J. (2004). The Biology of *Colletotrichum acutatum. Anales del Jardín Botánico de Madrid,* Vol.61, No.1, pp. 3-22, ISSN 1988-3196

Wharton, P.S. & Schilder, A.C. (2008). Novel Infection Strategies of *Colletotrichum acutatum* on Ripe Blueberry Fruit. *Plant Pathology,* Vol.57, (May 2007), pp. 122–134, ISSN 1365-3059

Zachos, D.G. & Makris, S.A. (1963). Studies on *Gloeosporium olivarum* in Greece II: Symptoms of the Disease. *Annals of Institute of Phytopathology, Benaki,* Vol.5, pp. 128-130, ISSN 1790-1480

Zhang, X.; Yazaki, J.; Sundaresan, A.; Cokus, S.; Chan, S.W.; Chen, H.; Henderson, I.R.; Shinn, P.; Pellegrini, M.; Jacobsen, S.E. & Ecker, J.R. (2006). Genome-wide high-resolution mapping and functional analysis of DNA methylation in *Arabidopsis. Cell,* Vol.126, No.6, (September 2006), pp. 1189–1201, ISSN: 0092-8674

Zipfel, C. (2008). Pattern-Recognition Receptors in Plant Innate Immunity. *Current Opinion in Immunology,* Vol.20, pp. 10–16, ISSN 0952-7915

# Part 5

# Plant Breeding Advances in Some Crops

# 12

# Breeding *Brassica napus* for Shatter Resistance

S. Hossain[1*], G.P. Kadkol[2], R. Raman[3], P.A. Salisbury[1,4] and H. Raman[3]
*[1]Department of Primary Industries,*
*[2]NSW Department of Primary Industries,*
*Tamworth Agricultural Institute*
*[3]EH Graham Centre for Agricultural Innovation,*
*an alliance between NSW Department of Primary*
*Industries and Charles Sturt University,*
*Wagga Wagga Agricultural Institute,*
*[4]Department of Agriculture and Food Systems,*
*Melbourne School of Land and Environment,*
*The University of Melbourne,*
*Australia*

## 1. Introduction

*Brassica napus* (canola or oilseed rape) has emerged as an important cultivated oilseed crop species grown in temperate climates of both the northern and southern hemispheres. In 2009, canola was sown to approximately 23.8 million hectares worldwide and production was approximately 53.3 million tonnes (FAOSTAT, 2011). The name "canola" identifies the "double low" oil and meal quality (low erucic acid content in the oil and low glucosinolate content in the meal) of the crop. Innovations such as herbicide resistance have enhanced the value of canola in weed management and crop rotations and improved its profitability. Further oil quality improvements have resulted in specialty canola varieties producing high oleic and low linolenic acid oils suitable for frying applications. However, one requirement that has persisted through the relatively short history of domestication of *B. napus* is the need for substantial improvement in shatter resistance to prevent significant seed loss especially under adverse harvest conditions.

Dehiscence of siliqua due to external forces at or after maturity leads to siliqua shatter (Kadkol et al., 1986a). Siliqua shatter can occur both prior to harvest due to adverse weather conditions and at harvest due to impact from combine harvesters. Dehiscence of ripe, dry fruit is a natural process by which many plant species disperse their seed in order to survive and spread in the wild. Whilst this mechanism is advantageous in nature, siliqua dehiscence in agriculture results in significant yield loss. Moreover, the dehisced seed can persist in the soil up to 10 years in winter *B. napus*, giving rise to volunteer plants or weeds in subsequent crops (Pekrun et al., 1996; Gulden et al., 2003). Typically yield losses are in the range of 10%-25% (Price et al. 1996). Seed losses of as much as 50% of expected yield have been reported when adverse climatic conditions delayed harvesting (MacLeod, 1981; Child & Evans, 1989). Current cultural practices to reduce siliqua shatter and to achieve better uniformity of ripening for harvest include windrowing (or swathing) and spraying desiccants. However, both these practices add to the cost of production and reduce flexibility in farm operations (Kadkol, 2009). Increased inherent shatter resistance could provide an option to delay harvesting to allow more even maturing of seeds and decrease the incidence of chlorophyll contamination from immature seeds in extracted oil (Morgan et al., 1998).

The fruits of *Brassicaceae* are botanically known as siliquae. Siliquae are derived from two carpels that form two locules separated by a thin, papery white replum. The fruit walls are valves that are attached to the replum forming a suture. The siliquae are attached to the raceme by a pedicel at the proximal end. At the distal end is the beak formed by the style (Fig 1). The suture is also known as dehiscence zone (DZ), where the valve margin is connected to the replum. Typically, a layer of thin parenchyma cells, that acts as a separation layer upon ripening, connects the valve margin to the replum. Dehiscence is usually initiated at the proximal end of the siliqua.

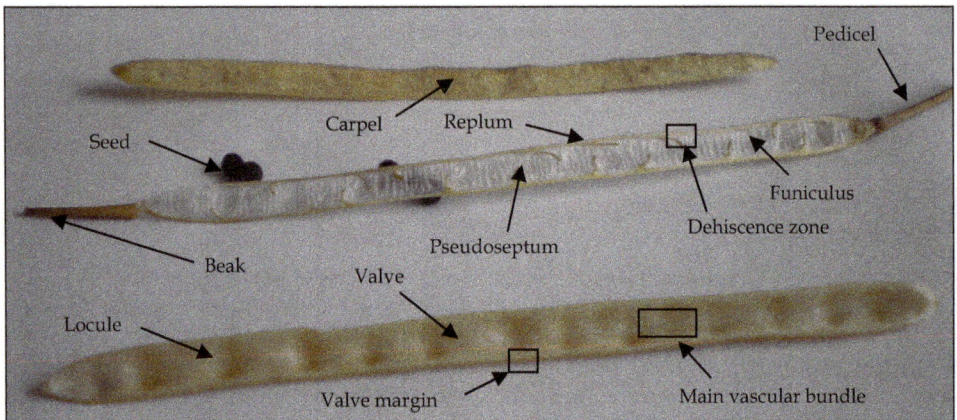

Fig. 1. The main structural features of a *Brassica napus* seed siliqua (from Kadkol, 2009)

Kadkol et al. (1986a) showed the presence of an abscission layer in the suture of siliquae of shatter susceptible *Brassica napus* and its absence in shatter-resistant *Brassica rapa* (Fig 2). They suggested that presence of an abscission layer is the basis of shatter susceptibility of *B. napus*. Differences in the vascular structure of siliquae and the width of the DZ has also been reported to be associated with variation for shatter resistance in a resynthesized *B. napus* line, 'DK142' in comparison with the shatter-susceptible winter *B. napus* line'Apex' (Child et al., 2003). The size of the main vascular bundle as it exited the valve and joined the vascular tissue of the replum was much larger in the resynthesized line.

Picart and Morgan (1984) investigated the physiological processes implicated in the control of siliqua dehiscence such as autolysis of the cells (degradation of pectic material in the middle lamella) of the DZ, senescence of the siliqua wall, water loss from thin walled cells, development of tensions resulting from different rates of drying of non-lignified and lignified cells of the valve and breakage of the vascular bundles at the base of the siliqua at the pedicel end. However, a study using polarizing microscopy by Kadkol et al. (1986a) suggested that development of tensions in the siliqua due to differential drying is unlikely.

A number of possible factors involved in the expression of the siliqua shatter resistance include morphological, anatomical and biochemical aspects of siliqua development and physiology. It may even encompass biotic and abiotic stress factors (Kadkol et al., 1986a; Morgan et al., 1998; Morgan et al., 2003; Summers et al., 2003). A summary of siliqua and plant characters as well as other factors reported to be involved in siliqua shatter are presented in Table 1.

| Source of trait | Trait | Trait type | Reference |
|---|---|---|---|
| Siliqua | Siliqua erectness | Morphological | Kadkol et al., 1984; Morgan et al., 2000 |
| | Siliqua size, shape and weight | Morphological | Morgan et al., 2000; Squires et al., 2003; Dinneny and Yanofsky 2004 |
| | Density of siliqua | Morphological | Kadkol et al., 1984 |
| | Pedicel length | Morphological | Morgan et al., 1998; Kadkol et al., 1984 |
| | Lignification of the suture/dehiscence zone | Anatomical | Kadkol et al., 1986a |
| | Lignification of the siliqua valves | Anatomical | Morgan et al., 1998 |
| | Size of main vascular bundle | Anatomical | Child et al., 2003; Kadkol et al., 1989; Morgan et al., 1998 |
| | Size of the dehiscence zone | Anatomical | Child et al., 2003 |
| | Enzymatic activity | Biochemical | Morgan et al., 1998; Child et al., 2003 |
| | Hormonal activity | Biochemical | Chauvaux et al., 1997; Child et al. 1998; Morgan et al., 1998 |
| Canopy structure | Interaction between plants | Morphological | Bowman, 1984; Kadkol et al., 1989; Summers et al., 2003 |
| Plant | Stem thickness | Morphological | Morgan et al., 1998 |
| | Uniformity of flowering | Physiological | Chandler et al., 2005; Morgan et al., 1998 |
| | Plant height | Morphological | Morgan et al., 1998; Morgan et al., 2000; Summers et al., 2003 |
| | Raceme structure | Physiological | Child & Huttly, 1999; Summers et al., 2003 |
| | Angle of the branches to the main stem | Morphological | Kadkol et al., 1984; Child & Huttly, 1999 |
| | Number of primary branches | Morphological | Kadkol et al., 1984 |
| Abiotic factors | Temperature | Environmental | Morgan et al., 2003; Summers et al., 2003 |
| | Rain and drought | Environmental | Morgan et al., 2003; Summers et al., 2003 |
| | Time of sowing | Environmental | Summers et al., 2003 |
| Biotic factors | Pests e.g. siliqua midge, aphids | Environmental | Meakin & Roberts , 1991; Summers et al., 2003 |
| | Pathogens e.g. alternaria | Environmental | Morgan et al., 2003 |

Table 1. Morphological, anatonomical, biochemical, physiological and environmental attributes implicated in siliqua shatter

## 2. Biochemical and molecular mechanisms underlying shatter resistance

Dehiscence of siliquae occurs as a result of highly coordinated and regulated events in growth and differentiation of the DZ and the degradation of the separation layer at ripening. This is due to triggering of enzymatic activity in DZ and cell separation predisposing siliqua to dehiscence from external forces. Several genes involved in growth and differentiation of the DZ have been identified and studied in *Arabidopsis* (e.g. Sorefan et al., 2009).

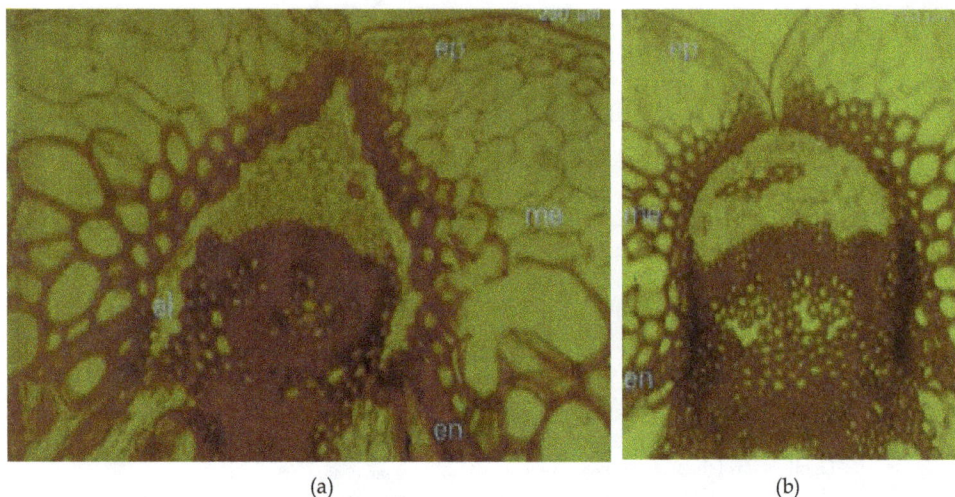

(a)                                                              (b)

Fig. 2. Transverse sections (x120) of fresh siliquae of *B. napus* (2a) and *B. rapa* cv. DS17D (2b) through dehiscence zones, stained with phloroglucine (al = abscission layer, en = endocarp, me = mesocarp, ep = epicarp) (from Kadkol et al., 1986a)

Many growth regulators such as abscicic acid (ABA), ethylene and auxin are well known for their role in abscission (Nemhauser et al., 2000; Ferrándiz, 2002; Sohlberg et al., 2006; Child et al., 1998; Meakin & Roberts, 1990*b*; Roberts et al., 2002). In *Brassica*, the role of an abscission cell layer in the siliquae dehiscence was first investigated by Kadkol et al. (1986a). Dehiscence is caused by the loss of cellular cohesion in the abscission layer, primarily attributable to the degradation of the middle lamella which appeared to result from an increased activity of hydrolytic enzyme cellulase leading to the cell separation process (Meakin & Roberts, 1990*a*; Meakin & Roberts, 1990*b*).

Johnson-Flanagan and Spencer (1994) found a climacteric of seed-produced ethylene preceding the pre-desiccation phase of *B. napus*. The evidence for ethylene acting as a regulator of dehiscence is unclear but it could still be a trigger for cellulase activity in DZ. Child et al. (1998) observed a correlation between delayed shattering and reduced ethylene production. The suppression of ethylene production by the treatment of siliquae with amino-ethoxyvinylglycine (AVG) delayed siliqua shatter. However, Roberts et al. (2002) reported *Arabidopsis* mutants that have nonfunctional ethylene receptors still exhibit a normal time-course of siliqua dehiscence and that the elevation in the cellulase $\beta$-1,4-glucanase in *B. napus* DZ occurs when the ethylene level in the siliqua is falling.

The activity of hydrolytic enzymes including $\beta$-1,4-glucanase and polygalacturonase involved in cell separation in the DZ appears to be regulated by auxin (Coupe et al., 1993). Chauvaux et al. (1997) observed that a decrease in auxin content in the DZ just prior to moisture loss in siliquae was correlated with a tissue specific increase in $\beta$-1,4-glucanase activity and hence with siliqua dehiscence. Auxin appears to have the opposite effect to ethylene and negatively regulates $\beta$-1,4-glucanase. Sorefan et al. (2009) demonstrated that formation of a local auxin minimum is required for specification of the valve margin abscission layer in *Arabidopsis* where dehiscence takes place. Thus, a low level of auxin seems to be a prerequisite for siliqua dehiscence and may allow for the induction of the activity of cell wall degrading enzymes.

In addition to cellulase activity, dissolution of the middle lamella in the DZ is another important process leading to cell separation. Jenkins et al. (1996) and Petersen et al. (1996) cloned and characterised two DNA fragments, SAC66 and RDPG1 associated with an endo-polygalacturonase (endo-PG). Both the DNA fragments related to a single *Arabidopsis* ortholog (called SAC70). Both transcriptional and post-translational control of PG activity has also been proposed (Roberts et al., 2002; Sander et al., 2001). However, in contrast to the activity of the cell wall degrading enzyme $\beta$-1,4-glucanase, polygalacturonase exhibits no correlation either temporally or spatially with siliqua dehiscence (Meakin & Roberts, 1990). This lack of siliqua DZ specificity of the endo-PG promoter has prevented the engineering of shatter resistance by silencing the endo-PG (Ostergaard et al., 2007).

## 3. Methods for screening germplasm for shatter resistance

Many of the early assessments used to evaluate siliqua shatter resistance have been based on imprecise, visual field observations (e.g., harvest yield and visual assessments) or manual tests (Table 2). These tests are somewhat subjective and are often not necessarily comparable due to the difference in maturity and moisture status of siliquae or differences in environmental conditions (Morgan et al., 1998).

| Approach | Type | Measure | Reference |
|---|---|---|---|
| Field observations | Visual scoring | Index | Josefsson, 1968 |
| | Direct harvesting vs. windrowing | Yield | Josefsson, 1968 |
| | Number of volunteer plants after harvest | Plants/area | Josefsson, 1968 |
| | Seed counting after harvest | % seed loss | Josefsson, 1968 |
| | Count shattered siliquae | % shattered siliquae | Tomaszewski & Koczowska, 1971 |
| Mechanical test | Compress plants between plates | % shattered siliquae | Jakubiec & Growchowski, 1963 |
| | Vibrate whole plants | % shattered siliquae | Voskerusa, 1971 |
| | Squeeze siliquae between fingers | Index | Tomaszewski & Koczowska 1971 |
| Anatomical test | Size of sclerenchymatic bridges between valves and replum | Thickness of sclerenchymatic bridge | Loof & Jonsson 1970 |

Table 2. Early tests used to identify shatter resistance in *Brassica* species

Kadkol et al. (1984) suggested that the methodology used to test siliqua shatter resistance should simulate shattering as it occurs in the field and during harvesting. They further suggested that it would be most appropriate to test the siliqua as a cantilever because most external forces acting on the siliqua would load it at the distal end whilst it is attached to the plant at the proximal end. However, many of the mechanical tests (Table 3) including the random impact test do not achieve this requirement. Another requirement of testing procedures is that they should be low cost, fast and efficient. This criterion is not met by tests of the DZ that involve considerable preparation of the sample and subsequent technical demanding analysis.

To date, several mechanical testing procedures have been employed to investigate shatter resistance (Table 3) which allowed for greater comparability, accuracy and repeatability across different lines and cultivars. Liu et al. (1994) developed a pendulum-based test (Fig 3) that was a further development from the quasi-static cantilever test developed by Kadkol et al. (1984). The use of a pendulum provided a dynamic cantilever test of the siliquae that simulates the natural process in the field and achieves rates of loading comparable to those in the field. Recently, Kadkol (2009) reported further refinements of computer software and the apparatus for the pendulum test (Fig 4) which have improved the efficiency of the process as a screening method for use in breeding.

| Name of the test | Purpose | Methodology | Reference |
| --- | --- | --- | --- |
| Manual bending test | Evaluate shatter resistance | Collected siliqua placed on flat surface with angles marked and with pedicel held firm. The siliqua is bent anticlockwise causing bending stress at which the angle is noted (this bending stress is similar to wind stress in field). | Roy, 1982 |
| Cantilever test | To measure the bending moment and energy required to cause siliqua fracture | Siliqua is clamped at the pedicel end in a Universal Testing Machine. A steel wedge fixed to the load cell was used to load the siliqua as a cantilever, the applied force is recorded on the chart. Shatter resistance was defined as the bending moment at the peak of the force displacement graph. Another measure of shatter resistance was energy measured as the area under the curve up to the peak. | Kadkol et al., 1984 |
| Microfracture test (MFT) | To establish the contribution of the main vascular bundle of the valve to the amount of energy needed to separate the valve from the replum. | Siliqua wall tissue is excised at the pedicel end of the valve or from the middle of the siliqua half-way between pedicel and the beak in order to isolate areas for testing that were ~1mm in length containing the septum and valve between which the DZ was intact. An L-shaped steel device is raised by a Universal Test Machine until fracture occurred. | Child et al., 2003 |

| Siliqua twisting (applying torque) | To determine the strength of the DZ by applying twisting force to the siliqua. Angle at which seed siliqua rupture occurs and the maximum torque required for siliqua rupture | Torque applied under twist of 180° in a holder using an INSTRON device. | Tys et al., 2007 |
|---|---|---|---|
| 'Ripping' method | To quantitatively determine siliqua dehiscence strength at 2.5 cm from pedicel | 6 siliqua per variety kept at 25°C and 50% RH for 2 weeks. A metallic thread laced around the siliqua 2.5 cm from pedicel and laced to the pedicel, siliqua glued to plate. An L-shaped probe of the texture analyser lifted thread and opened siliqua; probe recorded opening strength. | Tan et al., 2007 |
| Pendulum test | To measure energy absorbed by the pendulum in siliqua rupturing process | Siliqua is clamped vertically by its stalk at the bottom dead centre of the pendulum swing. An optical encoder is used to measures the loss of pendulum movement upon striking and shattering the siliqua which provides an estimate of the energy absorbed by the siliqua. | Kadkol et al., 1991; Liu et al., 1994 |
| Random Impact Test (RIT) | Measure breaking response of siliqua by mimicking conditions in the crop canopy caused by agitation during harvest or caused by poor weather conditions, fit a model and estimate half life of sample | Equilibrate siliqua in atmosphere of constant relative humidity (50%) and temperature (105°C) to achieve constant weight; 20 siliqua per sample (2 replications), Controlled agitation of sample in a receptacle (cylindrical of 20cm diameter, axis vertical) containing 6 steel balls (12.5mm diameter) and shaken in the horizontal plane, 17 seconds ; remove siliqua and classify them as shattered or intact. | Bruce *et al.* (2002); Morgan *et al.* (1998; 2003); Squires *et al.* (2003) |

Table 3. Recent attempts to evaluate siliqua shatter resistance.

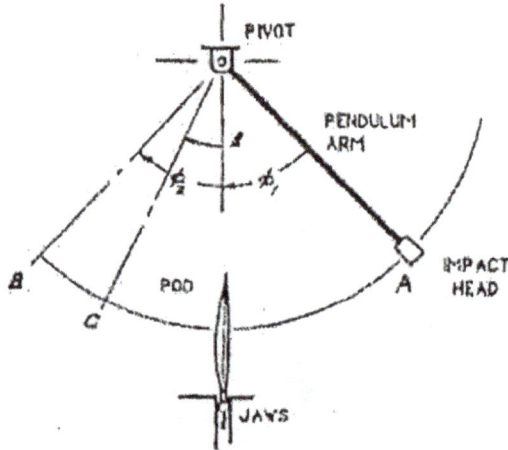

Fig. 3. Arrangement and analysis of pendulum (from Liu *et al.*, 1994).

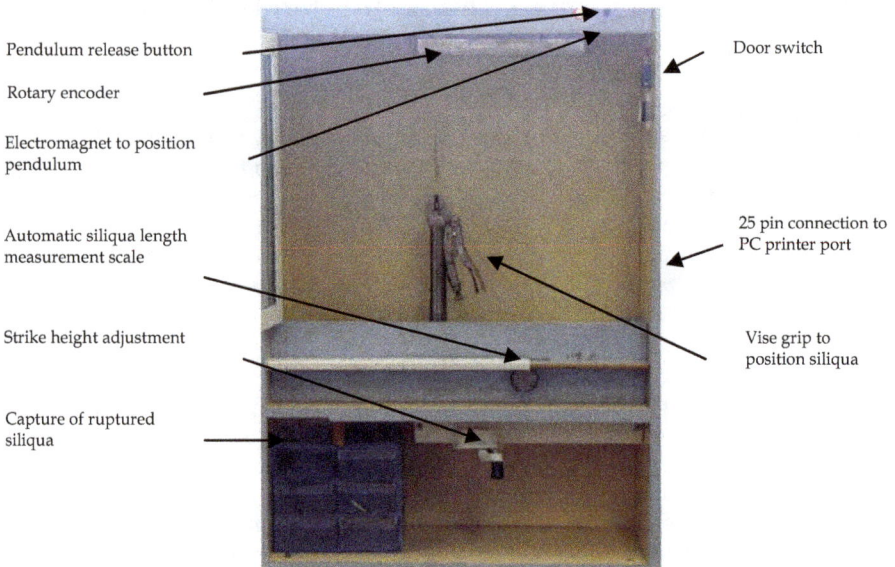

Pendulum release button

Rotary encoder

Electromagnet to position pendulum

Automatic siliqua length measurement scale

Strike height adjustment

Capture of ruptured siliqua

Door switch

25 pin connection to PC printer port

Vise grip to position siliqua

Fig. 4. The new pendulum machine for testing *Brassica* siliquae.

Morgan et al. (2003) cited the random impact test (RIT) as a good overall measure to compare the relative susceptibility of lines. The RIT involves agitation of 20 siliquae with ball bearings for 20 s and counting the number of intact siliquae. This test does not simulate the process of shatter as it happens in the field. These authors also quoted the tensile strength test as a useful test which correlates well to the RIT and field scores of shatter. However, the test appears to involve considerable sample preparation and hence is unsuitable for application in breeding programs. Wang et al. (2007) compared the degree of correlation between field data and results from pendulum test and RIT. Although it is difficult to accurately quantify harvest losses due to shatter in the field, in the study of Wang et al. (2007), RIT showed a lower level of association with field shatter than the pendulum test.

Morphological characters associated with shatter are more difficult to quantify. Delayed harvesting restricts the accuracy and effectiveness in discriminating between small differences in shattering affected by climatic and other environmental factors such as bird damage.

## 4. Genetic variation for shatter resistance

Genetic variation for shatter resistance exists both within Brassica species including B. rapa L., B. juncea L., B. hirta L. and within wild relatives of Brassica (Kadkol et al., 1985; Wang et al., 2007). Although there is some variation in B. napus, the level of resistance available is generally considered inadequate to avoid windrowing of crops on a routine basis (Raman et al., 2011). There have been a small number of reports characterizing genetic variation for shatter resistance in Brassica in germplasm collections. Wen et al. (2008) investigated the siliqua shattering resistance index of 229 accessions (mostly of Chinese origin) of B. napus using RIT. Most of the accessions (59.4%) were very susceptible to siliqua shatter. However, there were two lines considered to be shatter resistant which could potentially be used as parents to develop new varieties for improved this trait. Peng-Fei et al. (2011) evaluated 220 lines of B. napus for shatter resistance using 'ripping' method (Tan et al. 2007) and showed that ripping force ranged from 1.46N to 4.23N. The levels of pod strength reported in this study appear to be in general agreement with studies in Australia (Kadkol et al., 1984; Raman et al., 2011) indicating limited genetic variation in B. napus. Raman et al. (2011) evaluated 181 accessions of Brassica napus, one B. rapa, three B. juncea and two accessions of B. carinata, using a pendulum test (Kadkol, 2009) in two separate experiments. These accessions were collected from different parts of the world, representing contemporary cultivars and elite lines from Australian and international programs for shatter resistance. There was a moderate degree of correlation between the two sets of data. Siliqua strength (rupture energy - RE) values varied from 2.09 to 5.28 mJ and 2.34 to 5.58 mJ respectively, in the two experiments, indicating good correspondence between the two trials. These levels of RE are associated with intermediate shatter resistance which could prevent pod shatter in standing crops but insufficient to prevent harvest shatter (Kadkol, 2009). Genetic variation for higher levels of siliqua strength necessary for resistance to harvest shatter is present in B. rapa vars Yellow Sarson and Brown Sarson (Kadkol et al., 1984; Liu et al., 1994; Mongkolporn et al., 2003; Kadkol, 2009). Shatter resistance could be improved by introgressing the trait from these types and B. juncea (Kadkol 2009; Raman et al., 2011).

## 5. Inheritance of shatter resistance

Kadkol et al. (1986b) considered the genetic variation for shatter resistance within B. napus to be limited and studied inheritance of shatter resistance (measured as siliqua strength) in B. rapa in crosses between Brown Sarson (shatter resistant) and Torch (shatter susceptible) and Yellow Sarson (resistant) and Torch (susceptible). Segregation in the $F_2$ generation indicated the presence of 2 to 3 recessive genes which showed dominant epistatic interaction controlling shatter resistance. Further genetic analysis in one cross (Torch x DS-17-D) showed the presence of significant non-additive and additive genetic variances and a high broad sense heritability of shatter resistance (Kadkol et al., 1986c). The degree of dominance for shatter resistance was close to one supporting results from Mendelian analysis. In a subsequent study, Mongkolporn et al. (2003) confirmed a phenotypic segregation ratio of 12:3:1 (susceptible: intermediate: resistant) in an $F_2$ population derived from the Torch x DS-17-D, which indicated two recessive major genes (sh1 and sh2) with dominant epistasis conferring the resistance. This supports the earlier findings of Kadkol et al. (1986b).

Morgan et al. (2000) reported that shatter resistance in *B. napus* was recessive and mostly determined by additive genes. In their study, correlation of shatter resistance with important agronomic traits was low, suggesting that it would be feasible to introgress the shatter resistance trait into commercial breeding lines. They also noted the absence of genetic linkage of siliqua strength with other siliqua characters such as short/long siliqua or erect/horizontal siliqua. This suggested that it should also be possible to enhance shatter resistance by combining it these characters. Peng-Fei et al. (2011) investigated inheritance of shatter resistance in *B. napus* by mixed model analysis of parental lines, $F_1$, $BC_1$, $RBC_1$ and $F_2$ generations. They showed that two genes with additive–dominance–epistatic effects plus polygenes with additive–dominance-epistatic effects control shatter resistance. The heritability of two major genes in the $F_2$ and backcross generation ranged from 49.4% to 50%, suggesting that significant genetic gain can be made through conventional breeding.

Molecular studies of dehiscence zone specific mRNAs have led to isolation of genes which have been considered to be involved in production and regulation of enzymes involved in degeneration of the separation layer upon siliqua ripening (Coupe et al., 1993, 1994; Petersen et al., 1996; Whitelaw et al., 1999). In *Arabidopsis*, seed shattering is controlled by the several MADS-box and homeodomain genes. Screening of *Arabidopsis* enhancer or gene trap lines (Ferrándiz, 2002) identified genes involved in DZ differentiation. SHATTERPROOF (*SHP1*) and SHATTERPROOF (*SHP2*), previously called *AGL1* and *AGL5* respectively, are closely related MADS-box genes and are members of a monophyletic clade that also includes *AGAMOUS* and *AGL11* control and promote DZ differentiation at the valve-replum boundary in *Arabidopsis* (Liljegren et al., 2000). The expression of shp1 and shp2 is regulated by *AGAMOUS* (Savidge *et al.*, 1995), *FRUITFULL* (Ferrandiz et al., 2000) and *REPLUMLESS* genes (Roeder et al., 2003). Recently, *SHP1* and *SHP2* have been shown to play an important role in promoting stigma, style and medial tissue development (Colombo et al., 2010). Another indehiscent mutant gene, *ALCATRAZ* (ALC), corresponding to the bHLH transcription factor, has been isolated which is involved in the development of the abscission layer in the DZ and direct cell differentiation (Rajani & Sundaresan, 2001). Girin et al. (2010) reported that the *REPLUMPNESS* (*RPL*) gene which acts by limiting the expression of the valve margin identity genes; Shp-1 and Shp-2, INDEHISCENT and ALC to the narrow strips where wall margins will form. In the valves, the *FRUITFULL* gene is required for post-fertilization development and elongation of the fruit and it acts similarly to the *RPL* by repressing Shp1/Shp2 and IND gene activity.

## 6. Breeding *B. napus* for shatter resistance

Previous research on evaluation of *B. napus* germplasm have revealed that there is limited variation in siliqua shatter resistance among current cultivars (Bowman, 1984; Kadkol et al., 1985; Downey & Röbbelen, 1989; Roberts et al., 2002). Ostergaard et al. (2007) ascribed this to the narrow genetic base as a result of breeding focus on 'double-low' cultivars originated from two cultivars, Bronowski and Liho. Also, the recent studies of variation for shatter resistance in germplasm collections (Raman et al., 2011; Wen et al., 2008) support previous reports of a general lack of variation for high levels of shatter resistance in *B. napus*.

Tolerance to field shattering has been developed in some Australian breeding programs by direct heading of breeding trials and plots as an indirect selection method for shatter resistance. Although the varieties from Australian programs have not been properly characterized for shatter resistance, there appears to be significant improvement in field shatter tolerance in new lines relative to older varieties (Kadkol, 2009; Hossain et al., 2011a). However, further improvement in shatter resistance is required to allow direct heading of commercial crops. The conventional approach to breed *B. napus* for higher levels of shatter resistance has been based on interspecific hybridisation or resynthesis of *B. napus* using shatter-resistant species from the triangle of U. This approach requires several cycles of breeding and selection to overcome chromosomal imbalances and consequent impairment of meiosis and improve fertility of the shatter-resistant segregants. Often, malformation of the siliqua on partly sterile plants results in high siliqua strength.

Prakash and Chopra (1990) carried out interspecific hybridisation between *B. juncea* and *B. napus* and were able to isolate a reconstituted *B. napus* plant with complete nondehiscent fruits. This plant had normal meiosis and formed 19 bivalents. However, the seed fertility was very poor (23%) although pollen fertility appeared acceptable (84%) and this indicated significant chromosomal imbalances which might not have been apparent in meiosis studies. Agnihotri et al. (1990) attempted to transfer shatter resistance from *Raphanus* into *B. napus* using *Raphanobrassica* as the bridging material. This resulted in genetic material with variable fertility. In a Canadian study, lines derived from complex crosses made for development of yellow seeded *B. napus* showed better shatter resistance than standard Canadian *B. napus* varieties (Wang et al., 2007). Summers et al. (2003) resynthesised *B. napus* from crosses between *B. oleracea var. alboglabra* and *B. rapa var. chinensis* and developed DK142 that showed superior shatter resistance based on RIT assessments. However, the line turned out to have significantly lower levels of seed set relative to Apex, a commercial check variety. Recently, Banga et al. (2011) transferred shatter resistance from *B. carinata* to *B. napus*. Hybrid derivatives were characterized cytologically and further evaluated for shatter resistance using delayed harvesting. Anatomical analysis of shatter susceptible lines indicated the presence of DZ comprising thin-walled parenchymatous cell and showed dissolution in 40 days. Whilst shatter resistant genotypes displayed well defined DZ but remained intact and no sign of dissolution of cells or change that could lead to separation of siliquae valve margins from replum. The degree of improvement achieved in siliqua strength in this work is unclear.

Interspecific hybridisation of *B. napus* with *B. rapa var. Brown Sarson* and *var. Yellow Sarson* (Kadkol et al., 1991; Hossain et al., 2011b) has provided promising initial results. Stable segregants with high levels of siliqua strength have been produced with potential to provide harvest shatter resistance. However, further work is required to fully characterise and assess the shatter-resistant selections for meiotic stability, seed set and agronomic traits.

There have been a few reports of genetic transformation for improving shatter resistance in *Brassicas*. Chandler et al. (2005) over-expressed *Sinapis alba MADSB* gene, a close homologue to *FRUITFULL* in *Arabidopsis,* using a transgenic approach in winter and summer oilseed rape plants. The expression of the *MADSB* transgene modified the dehiscence zone differentiation and produced indehiscent plants. Ostergaard et al. (2006) showed that ectopic expression of the *Arabidopsis FRUITFULL* gene in *B. juncea* is sufficient to produce shatter-resistant *Brassica* fruit and that the genetic pathway leading to valve margin specification is conserved between *Arabidopsis* and *Brassica.* Studies have shown that transgenic fruit produced this way were completely shatter-resistant and were too tough for a combine harvester to thrash (Ferrandiz et al., 2000; Vancanneyt et al., 2003; Ostergaard et al., 2006). This is possibly because of the loss of the basic siliqua structure with valves and sutures that facilitates siliqua rupture. Authors suggested that the use of mutated forms of *FUL* or RNAi techniques to inactivate valve margin identity genes will probably prove useful in the fine–tuning of the degree of shatter resistance. Although these studies have been unsuccessful in producing the correct anatomical phenotype, they demonstrate a genetic strategy that can be used for improving shatter resistance.

## 6.1 Targeting Induced Local Lesions IN Genomes (TILLING)

The TILLING approach has been utilized for a large number of plants such as in *Arabidopsis,* wheat, barley, maize, lotus and *B. napus* (Comai et al., 2004; Slade et al., 2005; Slade & Knauf, 2005; Dreyer et al., 2007). The major advantage of this approach is the identification of mutants in target genes without genetic transformation. It allows the identification of single base-pair allelic variation in a target gene in a high throughput manner and may offer an alternative approach to identifying variation in shatter resistance among *B. napus* cultivars. Using this approach, Laga et al. (2011) achieved down-regulation of *IND* (indehiscent) gene which led to an indehiscence in *B. napus,* however, siliquae had a tube –like phenotype and did not rupture during mechanical harvesting obviously due to the loss of the valve and DZ structure similar to the transgenic canola discussed above. Use of a reverse genetics approach has produced an agronomically desirable phenotype that has optimal levels of seed shatter reduction. This study isolated and combined a set of mutant (null, weak and dominant negative) *IND* allele combinations that generated a range of seed shattering levels from natural shattering to pods that were shatter-resistant. Mutant plants displayed a range of reduction in shattering (5 to 15%) depending upon the combination of mutations used. This variation is being utilized for variety development. However, the method of screening for shatter resistance is unclear.

## 6.2 Molecular marker assisted breeding for shatter resistance

Identification of markers for shatter resistance in *B. napus* has not been reported extensively in the published literature due to lack of 'useful' variation for this trait in *B. napus* germplasm. Mongkolporn et al. (2003) utilized bulk segregant analysis (BSA) and identified three RAPD markers in an $F_2$ population derived from Torch X DS17D. Two of these markers (RAC-$3_{900}$ and RX-$7_{1000}$) were linked to the *sh1* and *sh2* major genes for shatter resistance. RAPD marker SAC-$20_{1300}$ showed a complete linkage with dominant alleles *SH1* and *SH2* for shatter susceptibility. The authors suggested it is likely that the recessive alleles, *sh1* and *sh2*, could have originated from independent mutations at two duplicate loci during the evolution of *B. rapa.* The marker linked to the dominant alleles can be used for marker-assisted selection (MAS), once validated in genetically diverse backgrounds. Mongkolporn et al. (2003) did not determine the chromosomal location of the loci associated with shatter resistance and this will require further research.

Association mapping (AM) is a promising new strategy for identification of markers for shatter resistance. AM approach is based upon the principle that linkage disequilibrium is maintained between loci over many generations in a given gene pool. Association mapping has been used for discovery and validation of trait-marker associations identified in the classical quantitative trait loci (QTL) mapping for loci associated with blackleg resistance, flowering time, leaf traits, seed phytate content in rapeseed (Jestin et al., 2010; Raman et al., 2010; Raman et al., 2011; Zhao et al., 2007). AM overcomes the major limitations of the QTL analysis that utilize bi-parental populations such as doubled haploids and recombinant inbred lines, as it surveys a large number of alleles at one locus and saves resources and time required to construct purpose designed 'mapping and validation' populations derived from structured biparental crosses.

Association mapping was used by Raman et al. (2011) to identify loci for shatter resistance in 188 genotypes of Brassica napus, B. rapa, B. juncea and B. carinata. These lines were phenotyped for siliqua strength using the pendulum method (Kadkol 2009). All accessions were genotyped with 1513 markers based upon Diversity Array Technology (DArT), Simple Sequence Repeats (SSR) and candidate genes that are reported to be involved in shatter resistance. Association analysis revealed that 150 markers were significantly associated (P<0.05) by the mixed linear model whereas the generalized linear model detected a total of 266 markers showing significant associations with rupture energy. Significantly associated markers were located on chromosomes A1, A2, A4, A6, A7, A8, A10, C2, C3, C5, C8 and C9. These results are consistent with the findings of a comprehensive transcriptome analysis of silique development and dehiscence in Arabidopsis and Brassica (Jaradat et al 2010). This study identified 131 cell wall related genes and 112 transcription factors that may be involved in silique dehiscence. Raman et al (2011) utilized markers that were largely based upon Diversity Array Technology. The majority of these markers have not been genetically mapped yet on the linkage maps of Brassica napus. It is possible that many DArT markers may be cosegregating and therefore map on the same loci. Previous studies have shown that B. napus genome has several chromosome rearrangements and therefore some of the DArT markers may represent to multiple copies of the same gene. Validation and fine mapping of these genomic regions, utilizing structured (doubled haploid or intercross) populations, will allow identification of candidate genes and/or their pathways associated with shatter resistance. The genes/QTLs identified in this work would mainly include loci that influence biochemical processes leading to formation of a separation layer in the DZ and its degradation at ripening.

## 7. Conclusions

Resistance to shatter is an important trait for B. napus improvement. It is a difficult trait to measure and breed into adapted germplasm and requires multiple years of selection and screening. To date, various breeding approaches have been attempted for improving shatter resistance of B. napus, mainly through interspecific hybridization or resynthesis of B. napus using shatter-resistant species from the triangle of U. In recent years, the power of high-density genetic maps and candidate gene studies in Brassica crops have demonstrated that an understanding of the number of genes underpinning the trait and their mode(s) of inheritance is important for further progress. In addition, an understanding of the potential of environment to impact on genetics is also required for the successful introduction of this trait into commercial oilseed rape. This information will greatly enhance breeding efficiency by identifying associated QTL or development and use of molecular markers for marker assisted selection.

Field screening using delayed harvest and visual assessments has been widely used to evaluate pod shatter resistance. It still remains the simplest way to get an approximate understanding of the shatter susceptibility of a large subset of lines. However, such assessments may be somewhat unreliable due to large environmental influences and subjective due to the vague boundaries of the assessment criteria. The development of pendulum method and the associated software facilitating rapid tests of shatter resistance have made it possible to characterize large germplasm collections in an objective way.

## 8. Future approaches in incorporating shatter resistance

To date, various approaches have been attempted for improving shatter resistance of *B. napus*, including indirect selection in breeding programs by direct heading, interspecific hybridization and also transformation with genes from other species. Breeding and selection within the species has limited potential due to the low genetic variation for the trait but could still result in development of varieties that are tolerant to field shattering.

To achieve higher levels of shatter resistance, it would be necessary to obtain siliqua strength levels that are available only in other species such as *B. rapa* and *B. juncea*. Although the reported attempts have generally not demonstrated complete success, interspecific hybridization could still achieve transfer of shatter resistance into *B. napus* combined with genetic stability, normal meiosis, and complete fertility together with absence of association with yield negative traits. The power of interspecific hybridization as a means of incorporating useful traits has been demonstrated in *B. napus* notably for blackleg resistance (Crouch et al., 1994) and yellow seed colour (Relf-Eckstein et al., 2007) but such work often needs a consistent, targeted breeding program over several generations after the initial isolation of the segregates to improve genetic stability and fertility. Molecular marker technology, such as marker assisted backcrossing would be very important for efficient development of shatter-resistant commercial cultivars upon achievement of successful incorporation of shatter resistance into *B. napus*.

Genetic engineering offers a promising a alternative approach for developing shatter-resistant *B. napus* in view of the advances in research on biology of shattering in *Arabidopsis*. However, shatter-resistant transgenics developed to date appear to have radically altered siliqua anatomy such that valve differentiation, DZ structure and consequently threshability are lost. Further research could develop mutations that retain valve differentiation and siliqua DZ structure whilst eliminating the separation layer similar to the anatomical phenotype of the *Brown* and *Yellow Sarson* varieties.

## 9. References

Agnihotri, A., Shivanna, K.R., Raina, S.N., Lakshmikumaran, M., Prakash, S. and Jagannathan, V., 1990. Production of *Brassica napus* x *Raphanobrassica* hybrids by embryo rescue: an attempt to introduce shattering resistance into *B. napus*. *Plant Breeding*, 105: 292-299.

Banga, S, Kaur, G, Grewal, N, Salisbury, P.A and Banga, S. S. 2011. Transfer of resistance to seed shattering from *Brassica carinata* to *B. napus*. *Proceeding of the 13th International Rapeseed Congress, Prague, Czach Republic*, pp 863-866.

Bowman, J.G., 1984. Commercial oilseed rape breeding. *Aspects of Applied Biology*, 6: 31-36.

Brown, J., Brown, A.P., Davies, J.B. and Erickson, D., 1997. Intergeneric hybridization between *Sinapis alba* and *Brassica napus*. *Euphytica*, 93: 163-168.

Bruce, D.M., Farrent, J.W., Morgan, C.L. and Child, R.D., 2002. Determining the oilseed rape pod strength needed to reduce seed loss due to pod shatter. *Biosystems Engineering*, 81 (2): 179-184.

Chandler, J., Corbesier, L., Spielmann, P., Dettendorfer, J., Stahl, D., Apel, K. and Melzer, S., 2005. Modulating flowering time and prevention of pod shatter in oilseed rape. *Molecular Breeding*, 15: 87-94.

Chauvaux, N., Child, R., John, K., Ulvskov, P., Borkhardt, B., Prinsen, E. and Van Onckelen, H.A., 1997. The role of auxin in cell separation in the dehiscence zone of oilseed rape pods. *Journal of Experimental Botany*, 48: 1423-1429.

Child, R.D., Chauvaux, N., John, K., Ulvskov, P. and Van Onckelen, H.A., 1998. Ethylene biosynthesis in oilseed rape pods in relation to pod shatter. *Journal of Experimental Botany*, 49: 829-838.

Child, R.D. and Evans, D.E., 1989. Improvement of recoverable yields in oil seed rape (*Brassica napus*) with growth retardants. *Aspects of Biology*, 23: 135-143.

Child, R.D. and Huttly, A., 1999. Anatomical variation in the dehiscence zone of oilseed rape pods and its relevance to pod shatter. In: *Proceedings of the 10th International Rapeseed Congress, Canberra, Australia. 1999*.

Child, R.D., Summers, J.E., Babij, J., Farrent, J.W. and Bruce, D.M., 2003. Increased resistance to pod shatter is associated with changes in the vascular structure in pods of a resynthesised *Brassica napus* line. *Journal of Experimental Botany*, 54: 1919-1930.

Child, R.D., Summers, J.E., Farrent, J.W., Babij, J. and Bruce, D.M., 2003. Variation in resistance to pod shatter and underlying mechanism in the resynthesised *B. napus* line DK142. In: *Proceedings of the 11th International Rapeseed Congress, Copenhagen, Denmark. 6-10 July 2003*, pp. 402-404.

Colombo, M., Brambilla, V., Marcheselli, R., Caporali, E., Kater, M.M. and Colombo, L., 2010. A new role for the SHATTERPROOF genes during Arabidopsis gynoecium development. *Developmental Biology*, 337: 294-302.

Comai, L., Young, K., Till, B.J., Reynolds, S.H., Greene, E.A., Codomo, C.A., Enns, L.C., Johnson, J.E., Burtner, C., Odden, A.R. and Henikoff, S., 2004. Efficient discovery of DNA polymorphisms in natural populations by ecotilling. *Plant Journal.*, 37: 778-786.

Coupe, S.A., Taylor, J.E., Isaac, P.G. and Roberts, J.A., 1993. Identification and characterisation of a proline-rich mRNA that accumulates during pod development in oilseed rape (*Brassica napus* L.). *Plant Molecular Biology*, 23: 1223-1232.

Coupe, S.A., Taylor, J.E., Isaac, P.G. and Roberts, J.A., 1994. Characterisation of a mRNA that accumulates during development of oilseed rape pods. *Plant Molecular Biology*, 24: 223-227.

Crouch, J. H., Lewis, B. G. and Mithen, R. F., 1994. The effect of A genome substitution on the resistance of *Brassica napus* to infection by *Leptosphaeria maculans*. *Plant Breeding* 112: 265 – 278.

Dinneny, J.R. and Yanofsky, M.F., 2004. Drawing lines and borders: how the dehiscent fruit of *Arabidopsis* is patterned. *BioEssays*, 27: 42-49.

Downey, R.K. and Röbbelen, G. (1989). *Oilseed crops of the world: their breeding and utilisation*, U.S.A: McGraw-Hill Publishing Company.

Dreyer, F., Frauen, M., Leckband, G., Milkowski, C. and Jung, C. 2007. An EMS population of *Brassica napus* L. for TILLING. The *12th International Rapeseed Congress, Wuhan, China*, 26-3003 Presentation BO-2-8.

FAOSTAT (2011) Available at http//:apps.fao.org/defalt.htm.

Ferrándiz, C., 2002. Regulation of fruit dehiscence in *Arabidopsis*. *Journal of Experimental Botany*, 53 (1): 2031-2038.

Ferrándiz, C., Liljegren, S.J. and Yanofsky, M.F., 2000. Negative regulation of the *SHATTERPROOF* genes by FRUITFULL during *Arabidopsis* fruit development. *Science*, 289: 436-438.

Girin, T., Stephenson, P., Goldsack, C.M.P., Kempin, S.A., Perez, A., Pires, N., Sparrow, P.A., Wood, T.A., Yanofsky, M.F., and Oestergaard, L., 2010. Brassicaceae INDEHISCENT genes specify valve margin cell fate and repress replum formation. *The Plant Journal*, 63: 329-338.

Gulden, R.H., Shirtliffe, S.J. and Thomas, A.G., 2003. Secondary seed dormancy prolongs persistence of volunteer canola in western Canada. *Weed Science.*, 51: 904-913.

Hossain, S., Kadkol G. P. and Salisbury P. A., 2011a. Pod shatter resistance evaluation in Australian cultivars of *Brassica napus*. In: *Proceedings of the 13th International Rapeseed Congress, Prague, Czech Republic. 05-09 June 2011*, pp 280.

Hossain, S., Kadkol G. P. and Salisbury P. A., 2011b. Shatter-resistant canola germplasm from interspecific hybridization – a progress report. In: *Proceedings of the 17th Australian Research Assembly on Brassicas Conference, Wagga Wagga NSW Australia, August 2011.* pp 42-44.

Jakubiec, J. and Growchowski, L., 1963. Polowa i laboratoryjna ocena odporuosci dwoch odmian rzepaku jarego na pekanie tuszczyn. Rolnictwo, *Warszawa*, 7: 49-65.

Jenkins, E.S., Paul, W., Coupe, S.A., Bell, S.J., Davies, E.C. and Roberts, J.A., 1996. Characterisation of an mRNA encoding a polygalacturonase expressed during pod development in oilseed rape (*Brassica napus* L.). *Journal of Experimental Botany*, 47: 111-115.

Jaradat, M. R., Ruegger, M., Bowling, A., Butler, H., Sun, Y., Skokut, T., and cutler, A. 2010. A comparative transcriptome analysis of silique development and dehiscence in Arabidopsis and Brassica integerating genotypic, interspecies and developmental comparisions. Proceeding of the 17th Crucifer Genetics Workshop, Sep 5-9th , 2010, Saskatoon, Canada, pp 103

Jestin, C., Lodé, M., Vallée, P., Domin, C., Falentin, C., Horvais, R., Coedel, S., Manzanares-Dauleux, M. and Delourme, R., 2010. Association mapping of quantitative resistance for *Leptosphaeria maculans* in oilseed rape (*Brassica napus* L.). *Molecular Breeding*, 1-17.

Josefsson, E., 1968. Investigations on shattering resistance of cruciferous oil crops. Z . *Pflanzenzuchtg*, 59: 384-396.

Johnson-Flanagan, A.M., and Spencer, M.S., 1994. Ethylene production during the development of mustard (*Brassica juncea*) and canola (*Brassica napus*) seed. *Plant Physiology*, 106: 601-606.

Kadkol, G.P., 2009. Brassica shatter-resistance research update. In: *Proceedings of the 16th Australian Research Assembly on Brassicas Conference, Ballarat Victoria. 14-16 September 2009*, pp. 104-109.

Kadkol, G.P., MacMillan, R.H., Burrow, R.P. and Halloran, G.M., 1984. Evaluation of *Brassica* genotypes for resistance to shatter. I. Development of a laboratory test. *Euphytica*, 33: 63-73.

Kadkol, G.P., Halloran, G.M. and MacMillan, R.H., 1985. Evaluation of *Brassica* genotypes for resistance to shatter. II. Variation in siliqua strength within and between accessions. *Euphytica*, 34: 915-924.

Kadkol, G.P., Beilharz, V.C., Halloran, G.M. and MacMillan, R.H., 1986a. Anatomical basis of shatter resistance in the Oilseed Brassicas. *Australian Journal of Botany*, 34: 595-601.

Kadkol, G.P., Halloran, G.M. and MacMillan, R.H., 1986b. Inheritance of siliqua strength in *Brassica campestris* L. I. Studies of $F_2$ and backcross populations. *Canadian Journal of Genetical Cytology*, 28: 365-373.

Kadkol, G.P., Halloran, G.M. and MacMillan, R.H., 1986c. Inheritance of siliqua strength in *Brassica campestris* L. II. Quantitative genetic analysis. *Canadian Journal of Genetical Cytology*, 28: 563-567.

Kadkol, G.P., Halloran, G.M. and MacMillan, R.H., 1989. Shatter resistance in crop plants. *Critical Reviews in Plant Science*, 8: 169-188.

Kadkol, G.P., MacMillan, R.H. and Halloran, G.M., 1991. Breeding canola for shatter resistance – a progress report. In: *Proceedings of the 8th Australian Research Assembly on Brassicas, Horsham, Victoria. 1-3 October 1991*, pp. 135-141.

Laga, B., Stevens, M., Haesendonckx, B., Standaert, E. and Crommar, K., 2011. A reverse genetics approach allows fine-tuning of seed shatter reduction in canola (*Brassica napus*) to optimal levels. *Proceeding of the 13th International Rapeseed Congress, Prague, Czach Republic*, pp 905

Liljegren, S.J., Ditta, G.S., Eshed, Y., Savidge, B., Bowman, J.L. and Yanofsky, M.F., 2000. *SHATTERPROOF* MADS-box genes control seed dispersal in *Arabidopsis*. *Nature*, 404: 766-770.

Liu, X-Y., MacMillan, R.H. and Burrow, R.P., 1994. Pendulum test for evaluation of the rupture strength of seed pods. *Journal of Texture Studies*, 25 (2): 179-189.

Loof, B . and Jonsson, R., 1970 . Resultat av undersokningar rorande drasfastheten hos raps. Sverig. Utsadesforen. *Tidskr*, 80: 193-205.

MacLeod, J. (1981). 'Harvesting' in *Oilseed Rape*, Cambridge: Agricultural Publishing, pp 107–119.

Meakin, P.J. and Roberts, J.A., 1990a. Dehiscence of fruit in oilseed rape (*Brassica napus* L.), I. Anatomy of pod dehiscence. *Journal of Experimental Botany*, 41: 995-1002.

Meakin, P.J. and Roberts, J.A., 1990b. Dehiscence of fruit in oilseed rape (*Brassica napus* L.), II. The role of cell wall degrading enzymes and ethylene. *Journal of Experimental Botany*, 41: 1003-1011.

Meakin, P.J. and Roberts, J.A., 1991. Anatomical and biochemical changes associated with the induction of oilseed rape (*Brassica napus*) pod dehiscence by *Dasineura brassicae* (Winn.). *Annals of Botany*, 67: 193-197.

Mongkolporn, O., Kadkol, G.P., Pang, E.C.K. and Taylor, P.W.J., 2003. Identification of RAPD markers linked to recessive genes conferring siliqua shatter resistance in *Brassica rapa*. *Plant Breeding*, 122: 479-484.

Morgan, C.L., Bruce, D.M.., Child, R., Ladbrooke, Z.L. and Arthur, A.E., 1998. Genetic variation for pod shatter resistance among lines of oilseed rape developed from synthetic *B. napus*. *Field Crops Research*, 58: 153-165.

Morgan, C.L., Ladbrooke, Z.L., Bruce, D.M., Child, R., and Arthur, A.E., 2000. Breeding oilseed rape for pod shattering resistance. *Journal of Agricultural Science, Cambridge*, 135: 347-359.

Morgan, C., Bavage, A., Bancroft, I., Bruce, D., Child, R., Chinoy, C., Summers, J. and Arthur, E., 2003. Using novel variation in *Brassica* species to reduce agricultural inputs and improve agronomy of oilseed rape – a case study in pod shatter resistance. *Plant Genetic Resources*, 1: 59-65.

Nemhauser, J., feldmann, L.J. and Zambryski, P.C., 2000. Auxin and ETTIN in *Arabidopsis* gynoecium morphogenesis. *Development*, 127: 3877-3888.

Ostergaard, L., Borkhardt, B. and Ulvskov, P. (2007). 'Dehiscence' in *Plant Cell Separation and Adhesion*, eds. Roberts, J.A. and Gonzalez-Carranza, Z.H., Victoria: Blackwell Publishing, pp. 137–163.

Ostergaard, L., Kempin, S.A., Bies, D., Klee, H.J. and Yanofsky, M.F. 2006. Pod shatter-resistant Brassica fruit produced by ectopic expression of the FRUITFULL gene. *Plant Biotechnology Journal*, 4: 45-51.

Pekrun, C., Lutman, P.J.W. and Baeumer, K., 1996. Introduction of secondary dormancy in rape seeds (*Brassica napus* L.) by prolonged imbibition under conditions of water stress or oxygen deficiency in darkness. *Eur. J. Agron.*, 6: 245-255.

Peng, P. Li, Y., Mei, D., Li, Y., Xu, Y. and Hu, Q., 2011. Evaluation and genetic analysis of pod shattering resistance in *Brassica napus*. In: *Proceedings of the 13th International Rapeseed Congress, Prague, Czech Republic. 05-09 June 2011*, pp. 185.

Petersen, M., Sander, L., Child, R. van Onckelen, H., Ulvskov, P. and Borkhardt, B., 1996. Isolation and characterisation of a pod dehiscence zone-specific polygalacturonase from *Brassica napus*. *Plant Molecular Biology*, 31: 517-527.

Picart, J.A. and Morgan, D.G., 1984. Pod development in relation to pod shattering. *Aspects of Applied Biology*, 6: 101-110.

Prakash, S. and Chopra, V.L., 1990. Reconstruction of allopolyploid Brassicas through non-homologous recombination: introgression of resistance to pod shatter in *Brassica napus*. *Genetical Research, Cambridge*, 56: 1-2.

Price, J.S., Hobson, R.N., Neale, M.A. and Bruce, D.M., 1996. Seed losses in commercial harvesting of oilseed rape. *Journal of Agricultural Engineering Research*, 65: 183-191.

Rajani, S. and Sundaresan, V., 2001. The *Arabidopsis* myc/bHLH gene *ALCATRAZ* enables cell separation in fruit dehiscence. *Current Biology*, 11: 1914-1922.

Raman, R., Raman, H., Kadkol, G.P., Coombes, N., Taylor, B. and Luckett, D., 2011. Genome-wide association analyses of loci for shatter resistance in Brassicas. In: *Proceedings of the Australian Research Assembly on Brassicas*, Wagga Wagga, NSW, Australia, pp 36-41.

Raman, H., Stodart, B., Ryan, P., Delhaize, E., Emberi, L., Raman, R., Coombes, N. and Milgate, A., 2010. Genome wide association analyses of common wheat (*Triticum aestivum* L) germplasm identifies multiple loci for aluminium resistance. *Genome*, 53:957-966.

Relf-Eckstein, J., Rakow, G.F.W., Rode, D.A. and Gugel, R.K., 2007. Agronomic performance and blackleg disease reactions of yellow-seeded Brassica napus canola., GCIRC *12th International Rapeseed Congress, Wuhan City, Hubei Province, China*, March 26-30, 2007, Vol. IV, pp. 133-136.

Roberts, J.A., Elliott, K.A. and Gonzalez-Carranza, Z.H., 2002. Abscission, Dehiscence and other cell separation processes. *Annual Reviews in Plant Biology*, 53: 131-158.

Roeder, A.H.K., Ferrándiz, C. and Yanofsky, M.F., 2003. The role of the REPLUMLESS homeodomain protein in patterning the Arabidopsis fruit. *Current Biology*, 13: 1630-1635.

Sander, L., Child, R., Ulvskov, P., Albrechtsen, M. and Borkhardt, B., 2001. Analysis of a dehiscence zone endo-polygalacturonase in oilseed rape (*Brassica napus*) and *Arabiopsis thaliana*: evidence for roles in cell separation in dehiscence and abscission zones, and in stylar tissues during pollen tube growth. *Plant Molecular Biology*, 46: 469-479.

Slade, A.J., Fuerstenberg, S.I., Loeffler, D., Steine, M.N. and Facciotti, D., 2005. A reverse genetic, nontransgenic approach to wheat crop improvement by TILLING. *Nature Biotechnology*, 23: 75-81.

Slade, A.J. and Knauf, V.C., 2005. TILLING moves beyond functional genomics into crop improvement. *Transgenic Research*, 14:109-115.

Sohlberg, J., Myrenas, M., Kuusk, S., Lagercrantz, U., Kowalczyk, M., Sandberg, G. and Sundberg, E., 2006. STY1 regulates auxin homeostasis and affects apical-basal patterning of the Arabidopsis gynoecium. *Plant Journal*, 47: 112-123.

Sorefan, K., Girin, T., Liljegren, S.J., Ljung, K., Robles, P., Galvan-Ampudia, C.S., Offringa, R., Friml, J., Yanofsky, M.F. and Ostergaard, L., 2009. A regulated auxin minimum is required for seed dispersal in *Arabidopsis*. *Nature*, 459: 583-587.

Squires, T.M., Gruwel, M.L.H., Zhow, R., Sokhansanj, S., Abrams, S.R. and Cutler, A.J., 2003. Dehydration and dehiscence in siliques of *Brassica napus* and *Brassica rapa*. *Canadian Journal of Botany*, 81: 248-254.

Summers, J.E., Bruce, D.M., Vancanneyt, G., Redig, P., Werner, C.P., Morgan, C. and Child, R.D., 2003. Pod shatter resistance in the resynthesised *Brassica napus* line DK142. *Journal of Agricultural Science*, 140: 43-52.

Tan, X., Zhang, J., Zhang, Z., Zhou, J., Jiang, S., Qi, C. and Li, J., 2007. Quantitative determination of the strength of rapeseed pod dehiscence. In: *Proceedings of the 12th International Rapeseed Congress, Wuhan, China. 26-30 March 2007*, pp. 280-283.

Tomaszewski, Z. and I . Koczowska, I., 1971. Metoda hodowli rzepiku ozimego TK-67. *Biuletyn Instytutu Hodowli i Aklimatyzacji Roslin*, 5: 73-75.

Tys, J., Stasiak, H., Borychowski, A. and Rybacki, R., 2007. Crack resistance of pods in some varieties of winter rapeseed. In: *Proceedings of the 12th International Rapeseed Congress, Wuhan, China. 26-30 March 2007*, pp. 420-422.

Vancanneyt, G., Redig, P., Child, R., Yanofsky, M. and Botterman, J., 2003. Podshatter resistance: from gene function validation in *Arabidopsis* towards a productivity trait in oilseed rape. In: *Proceedings of the 11th International Rapeseed Congress, Copenhagen, Denmark. 6-10 July 2003*, pp. 79-81.

Voskerusa, J., 1971. Z vyfesenych vyzkumnych ukolu v odvetvi olejnin. *Vestnik Ceskoslovenske Akademie Zemedelskie* 18: 538-541.

Fan., 2011. Production and characterization of interspecific somatic hybrids between *Brassica oleracea* var. *botrytis* and *B. nigra* and their progenies for the selection of advanced pre-breeding materials. *Plant Cell Rep*, 1-11. doi:10.1007/s00299-011-1088-9

Wang, R., Ripley, V.L. and Rakow, G., 2007. Pod shatter resistance evaluation in cultivars and breeding lines of *Brassica napus*, *B. juncea* and *Sinapis alba*. *Plant Breeding*, 126: 588-595.

Wen, Y.C., Fu, T.D., Tu, J.X., Ma, C.Z., Shen, J.X. and Zhang, S.F., 2008. Screening and analysis of resistance to siliquae shattering in rape (*Brassica napus* L.). *Acta Agronomic Sinica*, 34:163-166.

Whitelaw, C.A., Paul, W., Jenkins, E.S., Taylor, V.M. and Roberts, J.A., 1999. An mRNA encoding a response regulator protein from *Brassica napus* is up-regulated during pod development. *Journal of Experimental Botany*, 50: 335-341.

Zhao, J., Paulo, M-J., Jamar, D., Lou, P., van Eeuwijk, F., Bonnema, G., Vreugdenhil, D. and Koornneef, M., 2007. Association mapping of leaf traits, flowering time, and phytate constent in *Brassica rapa*. *Genome*, 50:963-973.

# 13

# Genetic Variability Evaluation and Selection in Ancient Grapevine Varieties

Elsa Gonçalves and Antero Martins
*Instituto Superior de Agronomia/Technical University of Lisbon*
*Portugal*

## 1. Introduction

Contrary to what usually occurs with other crops, grapevine varieties are mostly landraces that were domesticated by humans centuries or millennia ago from populations of wild grapevines (*Vitis vinifera* ssp. *sylvestris*). It is logical to assume that domestication was not been a single, instantaneous act by any of the first farmers but was rather a long succession of negative mass selections of wild plants, followed by more stringent positive selections and a final selection of a single initial plant that was destined to become the source of a new variety.

Certainly, from early times, the first vines were multiplied through vegetative propagation because of two main reasons: (1) the high heterozygosity of the plants causes strong segregation during sexual reproduction, with a consequent loss of features that were previously selected for, and (2) the progeny plants take a long time to reach sexual maturity, resulting in a loss of productive capacity for several years. In contrast to sexual reproduction, vegetative propagation allows for the fast development of new, productive plants from an initial plant and ensures that previously selected traits are maintained.

Selecting a single plant and vegetatively multiplying it creates a clone that consists of all of the genetically identical plants. However, although vegetative propagation ensures relatively homogeneous descendants, this homogeneity is not absolute. Indeed, the propagation depends on the growth of the tissues, on the cell division (mitosis) and on the DNA replication that occurs before each cell division. DNA replication is the source of gene mutations, that is, genetic variation. As a result, an initially homogeneous clone becomes genetically variable (a set of clones) over the course of time following successive multiplications. This set of genetically different clones corresponds to the variety that is presently cultivated.

Many of the genetic variations mentioned above are present in grapevines that are currently cultivated, and occasionally, we can directly observe the emergence of new features in vine plants. Typical examples are variations in the colour of the berry (for example, Pinot Blanc/Pinot Gris/Pinot Noir), the leaf's shape (for example, Chasselas/Chasselas Cioutat) and hairy leafs (for example, Garnacha/Garnacha Peluda). These types of features are determined by a single gene that has a strong genetic expression (major genes or macrogenes), and alleles of these genes came from mutations. Such features that have discrete distributions in heterogeneous populations are called "qualitative traits". The fact that these traits have discrete distributions has important consequences; the boundaries between different classes become evident to the point of justifying plants with different traits being considered to belong to distinct varieties. That is why Chasselas and Chasselas Cioutat are considered two different varieties. There are many other similar cases of varieties that differ exclusively at one macrogene.

However, most of the characteristics of the vine, including those of greatest economic importance (yield, soluble solids, acidity, anthocyanins and many others), are from another completely different type of trait. These traits have continuous and symmetric distributions, i.e., normal distributions, in heterogeneous populations and are therefore known as "quantitative traits". The normal distribution of quantitative traits can be explained by genetic and mathematical deduction under the assumption that they are determined by many genes (an undefined number) that exhibit a low and cumulative action (called microgenes or polygenes) and by strong environmental deviations. Although quantitative traits may also show a wide range of variation in some populations, this is not a sufficient criterion for their sub-division into distinct varieties. From this set of circumstances arises the fact that the yield can vary within the ancient variety up to tenfold, while some important characteristics of quality (soluble solids, acidity and anthocyanins) can vary up to twofold (Martins et al., 2006; Martins, 2007; Martins, 2009). All this variability is useful for important theoretical analysis regarding evolution and other topics as well as for the practical purposes of mass and clonal selection.

The analyses of qualitative and quantitative traits have traditionally been performed based on the principles and methods of Mendelian genetics and quantitative genetics, respectively. In recent years, powerful new molecular methods have contributed to major advances in our understanding of the genetic variation within varieties and its practical uses. The discovery and widespread use of microsatellite and other genetic markers has led to radical advances in the identification of varieties, the quantification of variability and our understanding of phylogeny and evolution (Arroyo-García et al., 2006; Sefc et al., 2001; Pelsy et al., 2010; Laucou et al., 2011; Myles et al., 2011). Genome-wide assessments of genetic diversity through techniques based on single nucleotide polymorphism (SNPs) are new, promising approaches that allow for the selection of varieties (Myles et al., 2011).

Nevertheless, the possibilities of the classical methods for the analysis of diversity are not dead, and many complementarities between the traditional methods and modern molecular techniques can be found. In particular, quantitative genetics is a well-cemented theory with a strong mathematical foundation that is directed at the study of quantitative traits and that continues to have great potential (Falconer & Mackay, 1996; Lynch & Walsh, 1998). This theory continues to show renewed efficiency, particularly when advanced data analysis based on mixed models and powerful computational methods are applied.

The intravarietal genetic variability of quantitative traits within ancient varieties is of great importance in the vine and wine industry. First, this diversity is the "raw material" with which to carry out selection; therefore, selection can only be successful if there is enough variability in the target traits. Selection of ancient varieties has been underway in some European countries since the beginning of the 20th century and became widespread in Europe at the middle of that century. Today, the genetic selection of varieties is a common practice in all developed viticultural countries and supports the large genetic gains of the main vine traits and contributes to the improvement of the competitiveness of the vine and wine industry in those countries (Martins et al., 1990).

In addition to its direct use for selection, the intravarietal variability is also interesting for other reasons. As the accumulation of variability is a function of time, it becomes possible to know the evolutionary age of the variety (the time elapsed since its domestication) by quantifying its intravarietal variability. Moreover, often the evolution of the variety was not confined to the region where it was created (or domesticated). On the contrary, it may have been imported into one or more other regions at different points in its history. These imports have generally been made through a single or few plants, which is an insufficiently sized set to represent the pre-existing intravarietal variability. That is, the imported variety will have returned to a state of near genetic homogeneity (as occurred at the time of domestication) and from there re-initiated the creation and accumulation of new variability. Accordingly, the quantification of intravarietal variability of a variety in different growing regions allows us to suggest hypotheses about the region where it was domesticated and to trace the path and timing of its expansion to other regions (Martins et al., 2006). This information can then be compared with the written historical information and other sources of information to contribute to our understanding of the histories of the varieties and the history of viticulture in general.

A third advantage of quantifying the intravarietal variability, both overall and in each of the growing regions of the variety, arises from the urgent need to halt the genetic erosion to which varieties are currently subject to. Variability has been gradually created over the time since domestication and has continued to gradually accumulate in the vineyards for centuries and millennia. However, during the 20th century, vine and wine technology and many political and administrative processes in viticulture changed deeply, resulting in dramatic genetic erosion. Today, new plantations are made with a small number of varieties and a very small number of clones. The variability continues to be created by the same genetic mechanisms, but the resulting plants are no longer kept in the vineyards. This new situation requires the use of a new strategy to preserve the variability that still exists: plantation of vineyards that are dedicated to the purpose of conserving the intravarietal variability.

Given the great importance of the viticulture, both in Europe and worldwide, and the interest in the intravarietal variability of the ancient varieties (for selection, for reconstructing history, to halt genetic erosion), the present work aims to contribute to a better understanding and use of that variability and will address the issues listed below.

Section 2.1 addresses the question of how to obtain a representative sample of the variability within a variety. The variability corresponds to the differences among all of the genotypes in all of the vineyards with a variety, but it is obviously impossible to study everything. Therefore, the solution is to work with a representative sample of the variability and to then make inferences regarding the entire variety. Section 2.2 describes experimental designs that are suitable for large field trials that contain more than 100 genotypes. Section 2.3 concerns the study of some mixed models, which can be used to analyse data from large field trials of grapevine varieties. In section 2.4, some results on the intravarietal genetic variability of two autochthonous Portuguese varieties are provided. Section 2.5 concerns the quantification of the genetic variability of two of the most important traits (yield and °Brix) in two other autochthonous Portuguese varieties. In addition, mass genotypic selection regarding those traits is carried out to demonstrate the potential of genetic variability and the advantages of the mass genotypic selection over clonal selection.

## 2. Quantification of genetic variability and mass genotypic selection within ancient varieties

### 2.1 How to obtain a representative sample of an ancient grapevine variety

To answer this question, two points should be considered: (1) the minimum sample size needed to represent the variety in a growing region; (2) the rules for sampling among old vineyards that contain the variety in order to ensure that the sample is random. With regard to the first point, a previous study related to this subject was performed by Martins et al. (1990) through a simulation study with the Touriga Nacional variety. The results suggested that 40 genotypes is the minimum number that should be used for representing a region. To reinforce the conclusions of that work, a new experimental approach to this problem will be presented below.

The experimental strategy to determine the minimum sample size (to ensure its representativeness) is to plant an oversized sample and then simulate several samples of smaller size by extracting random subsets from the first larger sample. Further evaluations and analysis of a given trait in samples of different sizes will allow us to determine the lowest number of genotypes that is needed for an accurate and precise estimate of diversity. Several characteristics of the vine can be used for this analysis. The yield per plant is generally the preferred trait because it is of general interest in all selection programs, it is easy to evaluate, and has a greater range of variation than other traits.

Using this strategy, a study was performed that took into account 4 examples of Portuguese autochthonous varieties from 4 regions of Portugal. The varieties were Tinta Miúda, Viosinho, Antão Vaz and Negra Mole. The initial trials of these varieties included 100, 199, 210 and 186 genotypes, respectively, and a randomised complete block design with 4 replicates for Tinta Miúda and 5 replicates for the other varieties. The data analysis was based on the mean yield obtained over several years. The estimates obtained for the genotypic variance component and the broad-sense heritability resulting from fitting a mixed model to the mean yield data for all of the genotypes in the trial are indicated in Table 1. In the model the block effects were assumed as fixed, and the genotypic effects and random errors were assumed as independent and identically distributed random normal variables.

| Variety | No. of Genotypes | $\hat{\sigma}_g^2$ | $\hat{H}^2$ |
|---------|------------------|--------------------|-------------|
| Tinta Miúda | 100 | 0.537 | 0.956 |
| Viosinho | 199 | 0.332 | 0.816 |
| Antão Vaz | 210 | 0.378 | 0.777 |
| Negra Mole | 186 | 0.115 | 0.644 |

Table 1. The genotypic variance and broad-sense heritability estimates ($\hat{\sigma}_g^2$ and $\hat{H}^2$, respectively) obtained from the yield data analysis of all of the genotypes in the trials of 4 grapevine varieties.

From these examples, the strategy is to verify if the same results could have been obtained with smaller sample sizes. Thus, samples of 10, 20, 30, 40, 50, 60, 70, 80, 90 and 100 genotypes from the entire set of the genotypes in the trial were obtained. For each number of genotypes, 1000 random extractions were made, and the consequent fitting of the mixed model to the data in order to estimate the genotypic variance and the broad-sense heritability was performed. SAS code, version 9.2 (SAS Institute, 2008) was used: PROC SURVEYSELECT for sampling and PROC MIXED for data analysis.

The interpretation of the results was based on the quality of genotypic variance and broad-sense heritability estimates compared with those obtained using all of the genotypes in the trial. That is, the estimates for the bias ( $\hat{\text{Bias}}$ ) and the mean square error ( $\hat{\text{MSE}}$ ) were computed as

$$\hat{\text{MSE}}\left(\hat{\theta}\right) = \left[\hat{\text{Bias}}\left(\hat{\theta}\right)\right]^2 + \hat{\text{Var}}\left(\hat{\theta}\right)$$

$$\hat{\text{Bias}}\left(\hat{\theta}\right) = \overline{\hat{\theta}} - \theta$$

where:

$\theta$ is the parameter estimate that is obtained when a model using all of the genotypes was fitted,

$$\overline{\hat{\theta}} = \frac{1}{1000}\sum\nolimits_{d=1}^{1000}\hat{\theta}(d) ,$$

$$\hat{\text{Var}}\left(\hat{\theta}\right) = \frac{1}{999}\sum\nolimits_{d=1}^{1000}\left(\hat{\theta}(d) - \overline{\hat{\theta}}\right)^2 ,$$

$\hat{\theta}(d)$ is the parameter estimate obtained for the $d^{th}$ simulation, where $d = 1,\cdots,1000$ .

To compare the results among the different cases studied we used two relative measures, the relative bias, which is defined as

$$\text{RB (\%)} = \left[\hat{\text{Bias}}\left(\hat{\theta}\right)\Big/\theta\right] \times 100$$

and the relative MSE (RMSE), which is defined as

$$\text{RMSE (\%)} = \left[\hat{\text{MSE}}\left(\hat{\theta}\right)\Big/\theta^2\right] \times 100 .$$

The results for the estimates of genotypic variance are shown in Figure 1. As indicated, the RB values were all close to zero for all of the sample sizes and the varieties studied. The RMSE values were smaller for the trials with a higher heritability and decreased as the sample size increased. These values increase from Fig. 1A to Fig. 1D (Tinta Miúda < Viosinho < Antão Vaz < Negra Mole). With a sample size of 10 genotypes, the RMSE ranged from 20% in the trial with Tinta Miúda to 45% in the trial with Negra Mole. When taking a sample size of 40 genotypes, the RMSE ranged from 3.1% for Tinta Miúda to 8.5% for Negra Mole. At a sample size of 60, the RMSE ranged from 1.4% for Tinta Miúda to

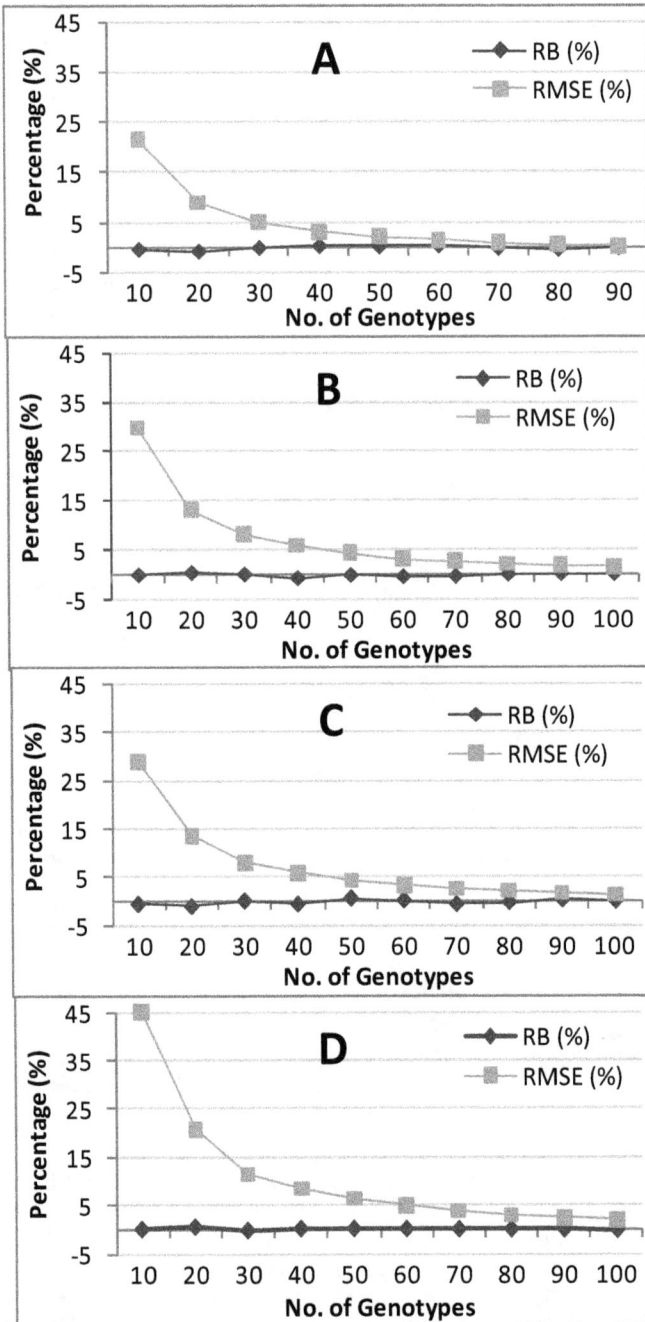

Fig. 1. Relative bias (RB) and relative mean square error (RMSE) for the genotypic variance estimates. A – Tinta Miúda; B – Viosinho; C – Antão Vaz; D – Negra Mole.

3.8% for Negra Mole. The differences between the RMSE values start to get very small with a sample size of 40 genotypes. When examining 60-70 genotypes, the differences are close to zero. Thus, the results obtained using 60-70 genotypes are nearly as good as those obtained using all of the genotypes in the trial.

The results for the estimates of broad-sense heritability are shown in Figure 2. As with the genotypic variance estimates, lower RMSE values were observed in the Tinta Miúda variety, and higher RMSE values were observed in the Negra Mole variety. The RMSE decreased as the sample size increased; however, this decrease became less marked with a sample size of 40 clones for all the studied varieties. Looking at the RB values, one can see that for sample sizes less than 40 the broad-sense heritability is underestimated, especially in the case of Negra Mole. From Figure 2, it is apparent that broad-sense heritability estimates obtained from samples with approximately 40 genotypes are close to those obtained with all of the genotypes in the trial; that is, the values of RB and RMSE obtained for the broad-sense heritability estimates are very close to zero.

In summary, the results for the estimates of broad-sense heritability indicated that estimates based on 40 clones showed approximately the same results as using all of the clones in the trial. However, the results obtained for the component of genotypic variance analysis are not so clear.

As this study is based on actual field trials, the quality of the estimates of the genotypic variance of the yield varied with the trial. The higher the heritability measurements obtained for the trial, the lower the number of genotypes that were required to obtain accurate estimates of the genetic variance. The results showed that the minimum number of genotypes needed to adequately represent the genetic variability of a variety ranged from 40 to 50 genotypes per growing region. However, at a sample size of approximately 70 genotypes, the quality of the estimates of genotypic variance started to become independent from the quality of the trial. From this number, the results obtained with all trials are the same, and therefore, a sample size of 70 will protect the analysis from less than favourable experimental conditions that may arise.

Now that we know the minimum number of genotypes, or parental plants, to integrate into the representative sample, the question of how to mark the plants in the vineyards and to ensure its representativeness remains. First, the set of marked plants must have a geographic distribution that is similar to the density distribution of old vines in the region that they are intended to represent. The restriction to the old vines means to prospect plants in vineyards that were planted prior to the existence of selection and nursery activities because only those preserve the diversity that was created in the past. The vineyards explored should be as geographically distant as possible and should not have be related (meaning the vineyards should have different owners, different years of planting, etc.). As a consequence, the total number of plants should come from the largest possible number of vineyards (20 or more), and only a few plants from each vineyard should be sampled (5 or less). Within each vineyard, the plants should be separated and must be marked in a casual way (except in cases of serious diseases of a systemic type).

## 2.2 Experimental designs suitable for large field trials

To quantify the genetic variation within a variety and to perform efficient selection, it is necessary to plant a very large field trial (normally between 100 and 400 clones), which will

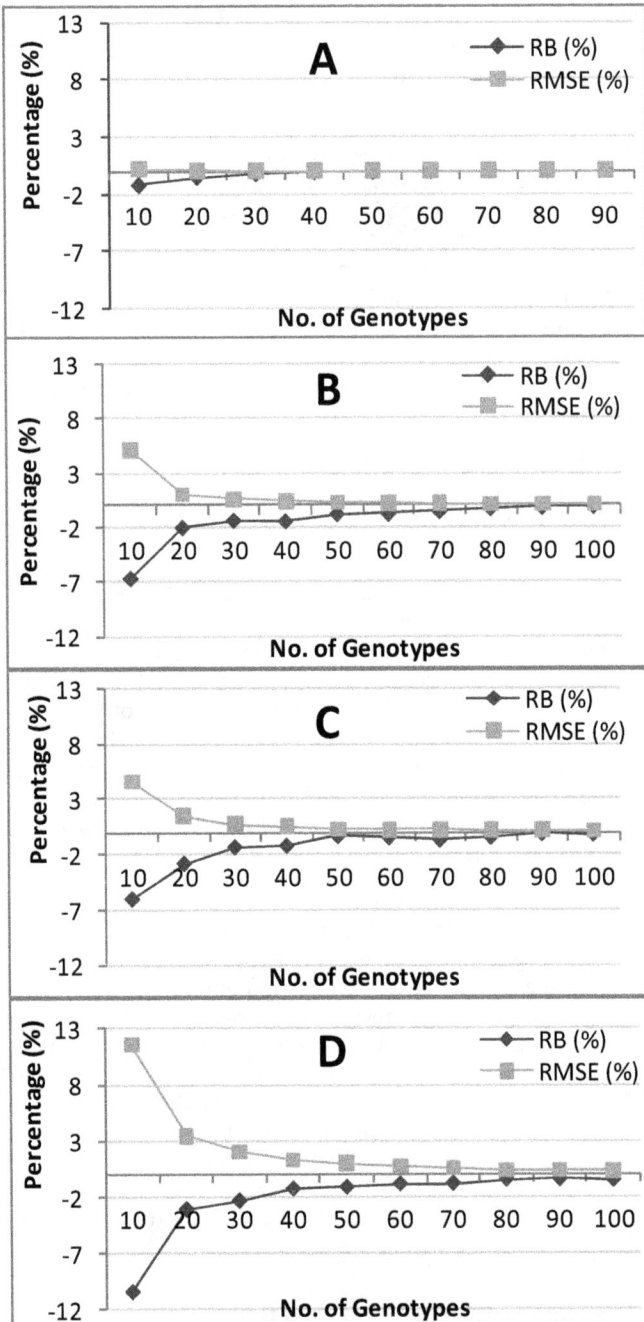

Fig. 2. Relative bias (RB) and relative mean square error (RMSE) for the broad-sense heritability estimates. A – Tinta Miúda; B – Viosinho; C – Antão Vaz; D – Negra Mole.

contain a representative sample of the variability within the variety across the different regions in which it is grown. Thus, the initial field trials for a grapevine variety would cover an unusually large area (from 0.75 to 1.5 ha), which by itself can cause a large amount of environmental variation. Therefore, the importance of experimental design in this type of trial is crucial to quantify the genetic variability and successfully select a superior group of clones.

The most relevant experimental designs for working with a high number of treatments (greater than 100 genotypes) are the alpha designs (Patterson & Williams, 1976), the row-column designs (Williams & John, 1989), the t-latinised designs (John & Williams, 1998) and the resolvable spatial row-column designs (Williams et al., 2006). The use of these designs in initial trials of grapevines was studied by Gonçalves et al. (2010). In that work, the authors compared several experimental designs via simulations, including randomised complete block, alpha and row-column designs, with the aim of identifying the designs that are most suitable for quantifying and utilising the genetic variability. For these purposes, they concluded that the alpha and row-column designs were better than the randomised complete block (RCB) design.

## 2.3 A review of several mixed models that are used in data analysis from large grapevine field trials

The theory of mixed models was developed in recent years, has been applied to a wide scope of sciences (Searle et al., 1992; Pinheiro & Bates, 2000; Verbeke & Molenberghs, 2000; McCulloch & Searle 2001; Giesbrecht, & Gumpertz, 2004; Littell et al., 2006; Butler et al., 2009; Lawson, 2010) and forms the basis for data analysis from grapevine selection trials.

Generally in these models, the genotypic effects are considered to be random effects because a random sample of genotypes of the cultivated variety is studied. The spatial control is done with the factors of the experimental design and, when necessary, through the variance-covariance matrix of the vector of the errors. Examples of mixed spatial models that are applied to grapevine initial field trials are described in Gonçalves et al. (2007). Models for data analysis of different experimental designs, namely, models for the analysis of randomised complete block, alpha, row-column and latinised designs, are described in Gonçalves et al. (2010).

The general linear mixed model can be written as

$$y = X\beta + Zu + e$$

where $y_{(n\times1)}$ is the vector of observations, $X_{(n\times p)}$ is the design matrix of fixed effects, $\beta_{(p\times1)}$ is the vector of fixed effects, $Z_{(n\times q)}$ is the design matrix of random effects, $u_{(q\times1)}$ is the vector of random effects and $e_{(n\times1)}$ is the vector of random errors.

The vectors, $u$ and $e$, are assumed to be independent with a multivariate normal distribution ($MVN$) with mean vector $0$ and variance-covariance matrices, $G_{(q\times q)}$ and $R_{(n\times n)}$, respectively. The distribution of $y$ is then multivariate normal with mean vector $X\beta$ and a variance-covariance matrix $V$,

$$V = ZGZ^T + R,$$

where $Z^T$ designates the transposition of $Z$.

When several traits, generally uncorrelated ones, are evaluated, the vector of observations, $y_{(n\times1)}$, has the form

$$y = \left(y_1^T, y_2^T, \cdots, y_t^T\right)^T,$$

and the vector of the random errors has the form

$$e = \left(e_1^T, e_2^T, \cdots, e_t^T\right)^T,$$

where $t$ represents the number of the evaluated traits and, for example, $y_1$ is a vector with $n_1$ observations for yield, $y_2$ is a vector with $n_2$ observations for °Brix, etc., and $e_1$, $e_2$, etc., are the correspondent vectors of random errors.

The vector $\beta$ contains the overall mean and other effects such as effects associated with experimental design, effects associated with the different growing regions of the variety when several regions are considered, the effects of different traits when several traits are studied, and so forth.

The vector $u$ usually consists of $k$ sub-vectors, such that

$$u = \left(u_1^T, \cdots, u_k^T\right)^T,$$

and the design matrix associated with the vector $u$, is given by

$$Z = \begin{bmatrix} Z_1 & Z_2 & \cdots & Z_k \end{bmatrix},$$

where $Z_1, Z_2, \cdots, Z_k$ are the design matrices associated to random effects vectors $u_1, u_2, \cdots, u_k$, respectively.

Therefore, to generalise to $k$ random effect factors, $Zu$ is decomposed as

$$Zu = \begin{bmatrix} Z_1 & \cdots & Z_k \end{bmatrix} \begin{bmatrix} u_1 \\ \vdots \\ u_k \end{bmatrix} = \sum_{i=1}^{k} Z_i u_i.$$

Each random effects factor, represented by $u_i$, may represent the genotypic effects of a trait, the effects associated with the experimental design of the trial, and so forth and has the properties

$$E[u_i] = 0,$$

$$Var[u_i] = \sigma_i^2 I_{q_i} = G_i,$$

$$Cov[u_i, u_{i'}] = 0 \text{ for } i \neq i'.$$

Consequently,

$$Var[u] = \bigoplus_{i=1}^{k} G_i = G \,,$$

where $\sigma_i^2$ is the variance of the random effects factor $i$, $I_{q_i}$ is the $q_i \times q_i$ identity matrix, and $G$ is the direct sum of matrices $G_i$.

In the simplest formulation, it is assumed that the elements of the vector $e$ are independent and identically distributed (iid) normal random variables, which leads to a variance-covariance matrix that is defined as

$$R = \sigma_e^2 I_n \,,$$

where $\sigma_e^2$ is the variance of the error and $I_n$ is the $n \times n$ identity matrix.

However, according to studies that have already been conducted in initial field trials with the grapevine (Gonçalves et al., 2007), the vector $e$ often represents the sum of two vectors, $\boldsymbol{\varepsilon} + \boldsymbol{\eta}$. The components of the vector $\boldsymbol{\varepsilon}$ are dependent from space, and it is assumed that $\boldsymbol{\varepsilon} \sim MVN(0, \sigma^2 \Sigma)$, where $\sigma^2$ is the spatially dependent variance and $\sigma^2 > 0$, and that $\Sigma$ is a $n \times n$ spatial correlation matrix, whose nondiagonal elements will be given by an anisotropic power correlation function. The components of the vector $\boldsymbol{\eta}$ are iid random variables and $\boldsymbol{\eta} \sim MVN(0, \sigma_\eta^2 I_n)$, where $\sigma_\eta^2$ is the nugget effect and $\sigma_\eta^2 \geq 0$ and $I_n$ is the $n \times n$ identity matrix. Consequently, the variance-covariance matrix for the vector of the errors is defined as

$$R = \sigma^2 \Sigma + \sigma_\eta^2 I_n \,.$$

When block effects are assumed to be fixed, the spatial modelling is made block by block, and it is usually assumed to be equal for all blocks.

When several traits are considered to be uncorrelated, the matrix $R$ takes the form

$$R = \bigoplus_{i=1}^{t} R_i \,,$$

where $R_i$ is the error variance-covariance matrix for the trait $i$.

Obviously, this is only a short review of the possible models that can be applied to data analysis from grapevine initial selection trials, and many other models could be addressed. However, our objective was to introduce the methodology that will be applied in the examples that follow in sections 2.4 and 2.5. One of these is a model that incorporates a nested structure for the genotypic effects to quantify the genetic variability by the growing region. The other is a model that analyses several traits to quantify the genetic variability per trait. In both situations, an RCB design will be considered.

The first approach will be supported by a mixed model, where several genotypic variance components are estimated and each one corresponds to a growing region. When considering an RCB design, the following model can be used

$$y_{ijk} = \mu + region_i + block_j + clone(region)_{ik} + e_{ijk} \tag{1}$$

for $i = 1, \ldots, s$, $j = 1, \ldots, r$ and $k = 1, \ldots, q_i$, where $q_i$ is the number of genotypes in the region $i$. In the model, $y_{ijk}$ represents the observation of the clone $k$ of the region $i$ in the block $j$, $\mu$ represents the overall mean, $region_i$ represents the fixed effect of the region, $block_j$ represents the random complete block effect or, depending on the trial, the fixed complete block effect, $clone(region)_{ik}$ represents the random genotypic effect within the region and $e_{ijk}$ represents the random error associated to the observation $y_{ijk}$.

With regard to the second approach, the model simultaneously analyses several traits. It uses a mixed model approach in which several genotypic variance components are estimated, and each one corresponds to a trait. When considering an RCB design, the model for this analysis can be written as

$$y_{ijk} = \mu + trait_i + block(trait)_{ij} + clone(trait)_{ik} + e_{ijk} \tag{2}$$

for $i = 1, \ldots, t$; $j = 1, \ldots, r_i$, where $r_i$ is the number of blocks for the trait $i$; and $k = 1, \ldots, q_i$, where $q_i$ is the number of genotypes evaluated for the trait $i$. In the model, $y_{ijk}$ represents the observation of the trait $i$ in the clone $k$ in the block $j$, $\mu$ represents the overall mean, $trait_i$ represents the fixed effect of the trait, $block(trait)_{ij}$ represents the random complete block effect for the trait or, depending on the trial, the fixed complete block effect for the trait, $clone(trait)_{ik}$ represents the random genotypic effect of the trait and $e_{ijk}$ represents the random error associated to the observation $y_{ijk}$.

## 2.4 Examples of quantification of genetic variability within ancient varieties

To study the intravarietal genetic variability, the cases of two Portuguese autochthonous grapevine varieties, Trincadeira and Síria, are given as example. The genotypes of Tricandeira were sampled in 4 regions (Alentejo, Oeste, Dão and Pinhel). The genotypes of Síria were sampled in 3 regions (Algarve, Alentejo and Pinhel). The field trials of Trincadeira and Síria were planted in Ribatejo and Pinhel, respectively, and both were laid out according to a randomised complete block design with 5 resolvable replicates and 4 plants per plot.

Data analysis was based on the average yield values observed over several years (1988, 1989 and 1990 for Trincadeira and 1988 and 1989 for Síria) and was performed using PROC MIXED of SAS version 9.2 (SAS Institute, 2008). For each variety, several mixed models were fitted to the yield data in order to address several relevant questions. The parameters involved in the model were estimated by residual maximum likelihood (REML) (Patterson & Thompson, 1971), using the Fisher Scores algorithm (Jennrich & Sampson, 1976).

The first question is to clarify if the varieties have significant genetic variability in the yield in Portugal. Thus, the first model that was fitted to the yield data was a model that considered all of the genotypes to be a sample from a single origin, Portugal. Additionally, the model assumed random block effects, used an anisotropic power function for the spatial correlated errors and a nugget effect (later called model A). A residual likelihood ratio test (REMLRT) was used to test the null hypothesis that the genotypic variance component was equal to zero. Since the null hypothesis, which involves a variance component, was on the boundary of the parameter space, the $p$-value of the test was half of the reported $p$-value from the chi-squared distribution with one degree of freedom (Self & Liang, 1987; Stram & Lee, 1994).

However, as these varieties exist in different regions of Portugal, another important question is whether this genetic variability is equal for all regions or whether, on the contrary, it differs according to region. To answer this question, two models that considered the origin of the genotypes by region of Portugal were fitted. One model assumed an equal genotypic variances for all of the regions and was later referred to as model B. The other assumed an unequal genotypic variances among the regions and was later referred to as model C. In both models, growing region effects were considered as fixed, block effects were considered as random and, for the random errors, an anisotropic power function for the spatial correlated errors and a nugget effect were considered.

A REMLRT was used to compare the fit of model B with the fit of model C. The distribution of the REMLRT statistic was considered to be a chi-squared with three degrees of freedom to compare models B and C in Trincadeira and with two degrees of freedom when comparing the models in Síria. These models were also compared using the Akaike's Information Criterion (AIC) and the Bayesian Information Criterion (BIC), and for these criteria, smaller values indicate a better fit.

To develop a better understanding of the amount of genetic variability between regions of the variety and between the varieties themselves, the coefficient of genotypic variation (the ratio between the estimate of the genotypic standard deviation and the estimate of the overall mean) was also computed.

The results for the quantification of the genetic variability of yield without taking into account the factor region (model A) are illustrated in Table 2. The genotypic variance was highly significant for both varieties. The REMLRT statistics ($(-2l_{R0}) - (- 2l_R)$) were 174.6 with a $p$-value<0.0001 for Trincadeira, and 352.2 with a $p$-value<0.0001 for Síria. When comparing the genetic variability of the yield between varieties under the experimental conditions studied, it was noted that Síria had a higher degree of genetic variability than Trincadeira. This conclusion can be more easily perceived through the values for $CV_G$, which were 26.0% for Síria and 15.3% for Trincadeira. These indicators point to a greater antiquity of Síria in Portugal compared with Trincadeira.

To compare the genetic variability within the variety in different growing regions, models B and C were fitted, and the results for minus twice the residual log-likelihood, AIC and BIC, are listed in Table 3.

| Variety | No. of Genotypes | $\hat{\sigma}_g^2$ | $-2l_R$ | $-2l_{R0}$ | $\hat{\mu}$ | $CV_G$ (%) |
|---|---|---|---|---|---|---|
| Trincadeira | 246 | 0.025 | 211.7 | 386.3 | 1.030 | 15.3 |
| Síria | 210 | 0.251 | 1955.4 | 2307.6 | 1.930 | 26.0 |

Table 2. Estimates of the overall mean ( $\hat{\mu}$ ), the genotypic variance ( $\hat{\sigma}_g^2$ ) and the coefficient of genotypic variation ($CV_G$), the minus twice the residual log-likelihood ($-2l_R$) obtained with the fitting of model A and the minus twice the residual log-likelihood obtained with a variant of the model A without genotypic effects ($-2l_{R0}$).

| Variety | Model | $-2l_R$ | AIC | BIC |
|---------|-------|---------|-----|-----|
| Trincadeira | B | 207.4 | 219.4 | 240.4 |
|  | C | 203.3 | 221.3 | 252.9 |
| Síria | B | 1929.5 | 1939.5 | 1956.2 |
|  | C | 1912.5 | 1926.5 | 1950.0 |

Table 3. Minus twice the residual log-likelihood ($-2l_R$), Akaike's Information Criterion (AIC) and Bayesian Information Criterion (BIC) obtained from the fitting of models B and C.

For Trincadeira, the observed value for the REMLRT statistic $((-2l_{RB})-(-2l_{RC}))$ was 4.1 with a $p$-value of 0.2508. Consequently, the null hypothesis of equal genotypic variances of yield was not rejected, indicating that the equal variance model (model B) is adequate to describe the data. The superiority of model B was also confirmed by the lower values obtained for the AIC and the BIC (219.4 and 240.4, respectively).

On the contrary, the null hypothesis of equal genotypic variances of yield among the regions of Alentejo, Algarve and Pinhel was rejected for Síria. In fact, the observed value for the REMLRT statistic $((-2l_{RB})-(-2l_{RC}))$ was 17 with a $p$-value of 0.0002. Additionally, on the basis of the AIC and the BIC, the unequal variance model (model C) was better than model B. The values obtained for these criteria were, respectively, 1926.5 and 1950.0 for model C, and 1939.5 and 1956.2 for model B.

In fact, observing the differences between the genotypic variance estimates for the different regions and the corresponding values of the $CV_G$, the differences were more marked in Síria (Table 4). The yield genotypic variances are quite different, ranging from 0.060 in Alentejo to 0.291 in Pinhel, which corresponds to values of 11.2% and 32% for the $CV_G$, respectively.

| Variety | Region | No. of Genotypes | $\hat{\mu}$ | $\hat{\sigma}_g^2$ | $CV_G$ (%) |
|---------|--------|------------------|-------------|--------------------|------------|
| Trincadeira | Alentejo | 55 | 1.058 | 0.025 | 14.9 |
|  | Oeste | 61 | 0.961 | 0.030 | 18.0 |
|  | Dão | 87 | 1.083 | 0.020 | 13.1 |
|  | Pinhel | 43 | 0.984 | 0.013 | 11.6 |
| Síria | Algarve | 78 | 2.031 | 0.217 | 23.0 |
|  | Alentejo | 49 | 2.184 | 0.060 | 11.2 |
|  | Pinhel | 83 | 1.683 | 0.291 | 32.0 |

Table 4. Estimates of the overall mean ($\hat{\mu}$), genotypic variance ($\hat{\sigma}_g^2$) and coefficient of genotypic variation ($CV_G$) obtained with the fitting of model C to the yield data (kg/plant).

To compare the antiquity of the varieties, it is more important to understand the variability of each one on its more heterogeneous region than to know the total average variability. On the basis of this criterion, the higher genetic variability of the Síria variety is also confirmed if we look to the $CV_G$ obtained for the regions with a higher genetic variability (Table 4). That is, Síria in the Pinhel region had a $CV_G$ of 32% and Trincadeira in the Oeste Region had a $CV_G$ of 18%. Once again, these indicators reinforce the greater antiquity of Síria compared with Trincadeira in Portugal.

To summarise, for Trincadeira, the genetic variability of the yield was equal in all regions. Trincadeira is a greatly expanded variety in Portugal, and it is likely that the constant exchange of material among the regions homogenised this variability.

For Síria, the greatest genetic variability was was found in the Pinhel region. This finding may indicate that the variety originated in Pinhel. Then, it was exported to the other regions, likely through selected material. This is logical, as the region with the highest genetic variability is the one that shows the lowest mean yield. When performing multiple pairwise comparisons of the means, the adjusted $p$-values indicated that the mean yield for Pinhel is lower than the means for Algarve and Alentejo ($p$-values<0.05). Likely, the exportation of the variety to other regions occurred only once. If they were several exports then the variability would tend to be equal in all regions as in the case of Trincadeira.

## 2.5 Examples of mass genotypic selection for important traits

The examples for mass genotypic selection will be conducted in another two Portuguese varieties, Arinto and Vital. The trial of Arinto was located in Setúbal and was laid out in a randomised complete block design with 247 genotypes, 4 plants per plot and 4 complete blocks. The traits considered for this study were yield (kg/plant) in 1995, 1998, 1999 and 2000 and soluble solids (°Brix) in 2005 and 2006. The trial with Vital was located in Caldas da Rainha and was laid out in a randomised complete block design with 232 genotypes, 4 plants per plot and 4 complete blocks. The traits considered were yield (kg/plant) in 1990, 1991 and 1992 and soluble solids (°Brix) in 1992. Because of the feasibility of evaluating the soluble solids, samples of 60 berries were collected by plot in the 3 most homogenous blocks. The analysis of the grape berries for evaluation of soluble solids was made following standard laboratory techniques.

Once again, the theory of mixed models and REML estimation were used and mixed model 2 of section 2.3, which assumed the block was fixed, was fitted to the data using PROC MIXED of SAS version 9.2 (SAS Institute, 2008). For Arinto, an anisotropic power function for correlated errors and a nugget effect were considered. For Vital, only independent and identically distributed errors were assumed. For both varieties the traits analysed were assumed to be uncorrelated. This decision was supported by previously descriptive analysis, which provided a Pearson's correlation coefficient between the traits of 0.11 for Arinto, and -0.17 for Vital.

After fitting the mixed models to the yield and °Brix data, the empirical best linear unbiased predictors (EBLUPs) of genotypic effects of those traits were obtained through the mixed model equations (Henderson, 1975; Searle et al., 1992). A generalised t-test (McLean & Sanders, 1988; Kenward & Roger, 1997) was performed to test the null hypothesis that the genotypic effect of a trait in a clone was equal to zero. The prediction standard errors were adjusted according to Prasad & Rao (1990) and Harville & Jeske (1992).

The EBLUPS of the genotypic effects were used to select clones and two distinct groups of clones from each variety were selected according to the trait. That is, two types of mass genotypic selections were made, one with the best clones for yield and the other with the best clones for °Brix.

The predicted genetic gain to be obtained through the selection of those groups was computed as the average of the EBLUPs of the respective genotypic effects. To better interpret the results, the predicted genetic gain was expressed as a percentage of the overall mean.

The results for the mass genotypic selection are shown in Table 5. The genotypic variance component was statistically significant ($p$-value<0.0001) for the two analysed traits for both varieties (the statistical test was the same as described in the previous section). This result indicates that there is sufficient "raw material" (genetic variability) to apply selection. For a selection of top 15% of the genotypes, the predicted genetic gains of yield were 32.1% and 43.1% for Arinto and Vital, respectively. For the °Brix, the predicted genetic gains were 10% for Arinto and 10.8%, for Vital. For Arinto, all of the genotypes in the group selected for yield showed a significant genotypic effect (p-value<0.05), and the genotypic effect of the last clone of the group was significant with a $p$-value of 0.0167. The °Brix group also had significant genotypic effects for all of the clones (p-value<0.05). The genotypic effect of the last clone of the group was significant with a $p$-value of 0.0024. For Vital, all of the genotypes in the group selected for yield showed a significant genotypic effect (p-value<0.05), and the genotypic effect of the last clone of the group was significant ($p$-value=0.0485). However, in the group selected for °Brix, four of the clones did not reveal a significant genotypic effect ($p$-value>0.05). This result indicates that the efficiency of selection is not equal in the two varieties and is higher in the Arinto variety.

| Variety | Trait | $\hat{\sigma}_g^2$ | $p$-value | Predicted genetic gain (%) obtained by selection of 15% of the superior genotypes |
|---------|-------|--------|---------|------------------------------------------------------------------|
| Arinto | Yield (kg/plant) | 0.1283 | <0.0001 | 32.1% |
|  | Soluble solids (°Brix) | 2.7497 | <0.0001 | 10.0% |
| Vital | Yield (kg/plant) | 0.3361 | <0.0001 | 43.1% |
|  | Soluble solids (°Brix) | 2.4941 | <0.0001 | 10.8% |

Table 5. The genotypic variance estimates for yield and °Brix ( $\hat{\sigma}_g^2$ ) and the predicted genetic gains obtained with selection proportion of 15% for the two studied varieties.

One issue remains to be clarified: why perform mass genotypic selection instead of using clonal selection?

To answer this question, we use a short demonstration. A group with the 30 top genotypes for yield was selected for each variety on the basis of the analysis of the mean yield for several years. The predicted genetic gain for this group of genotypes was computed separately year by year. In parallel, three individual top genotypes were selected for yield, also based on the mean yield of several years. Again, their behaviour was evaluated for individual years. The results are reported in Table 6.

For Arinto, the predicted genetic gain in the group of 30 clones is more stable over the years, ranging from 31.8% to 32.2%, than the behaviour of the individual clones. The clone, AR4108, always remained above the group of 30 genotypes and the genotypic effects of yield were always significantly different from zero (p-value<0.05). However, other genotypes did not always show a yield above the selected group, and their yield genotypic effects were not significant ($p$-value>0.05) for some years. The AR3605 clone in 1998 and the AR3903 clone in 2000 are two examples of this occurrence.

| Variety/clone | Year | EBLUP | *p*-value | % of yield above the overall mean |
|---|---|---|---|---|
| Arinto/AR4108 | 1998 | 0.776 | 0.0009 | 46.9 |
| | 1999 | 0.872 | <0.0001 | 51.6 |
| | 2000 | 0.847 | <0.0001 | 68.3 |
| Arinto/AR3605 | 1998 | 0.403 | 0.0839 | 24.4 |
| | 1999 | 0.787 | 0.0002 | 46.5 |
| | 2000 | 1.190 | <0.0001 | 96.0 |
| Arinto/AR3903 | 1998 | 0.686 | 0.0034 | 41.5 |
| | 1999 | 0.663 | 0.0018 | 39.2 |
| | 2000 | 0.288 | 0.1261 | 23.2 |
| **Arinto/Group** | **1998** | **0.529** | | **32.0** |
| | **1999** | **0.537** | | **31.8** |
| | **2000** | **0.400** | | **32.2** |
| Vital/VT1402 | 1990 | 0.441 | 0.0067 | 83.0 |
| | 1991 | 0.373 | 0.4336 | 16.3 |
| | 1992 | 1.653 | <0.0001 | 70.7 |
| Vital/VT1218 | 1990 | -0.061 | 0.7055 | -11.5 |
| | 1991 | 1.297 | 0.0067 | 56.7 |
| | 1992 | 1.251 | 0.0026 | 53.5 |
| Vital/VT1208 | 1990 | 0.152 | 0.3475 | 28.6 |
| | 1991 | 1.352 | 0.0047 | 59.1 |
| | 1992 | 1.027 | 0.0133 | 43.9 |
| **Vital/Group** | **1990** | **0.200** | | **37.7** |
| | **1991** | **0.818** | | **35.7** |
| | **1992** | **0.929** | | **39.7** |

Table 6. Behaviour of the group of clones and individual clones with respect to yield over the years compared on the basis of the empirical best linear unbiased predictors (EBLUPs) of the genotypic effects of yield and of the percentage of the yield above the overall mean. For the group, EBLUP is the average of the EBLUPS of the genotypic effects of the yield of the correspondent clones.

For Vital, the predicted genetic gain with the group of 30 clones was also more stable, ranging from 35.7% to 39.7%, than the behaviour of individual clones. The genotypic effects on the yield of the VT1218 and VT1208 clones were not significant ($p$-value>0.05) in 1990. In that year, these genotypes revealed a yield below the group of clones. In 1991, the VT1402 clone had a lower yield than the group of clones, and the yield genotypic effect was not significant ($p$-value>0.05).

To summarise, the results clearly indicate the difference between the behaviour of a group of clones and an individual clone. With a group of clones or with mass genotypic selection, the predicted genetic gains of yield remain nearly constant over the years considered. In contrast, the individual clones showed a more unstable behaviour. There were clones that appeared to be more stable, for example, the AR4108 clone. However, the sample of the evaluated environments is not sufficiently representative to make a more objective interpretation. This phenomenon that we are referring to is nothing more than the genotype×environment interaction. Because of this phenomenon, and to minimise its negative effects, reliable clonal selection requires the installation of numerous regional trials and data collection for several traits over a number of years. This process has at least two highly negative consequences, the costs become very cumbersome and may derail the selection work and the objectives formulated at the beginning of selection may already be outdated when it is completed 10-15 years later. Clonal selection is more time consuming and more expensive than mass genotypic selection. The option of selecting groups of clones from the initial field trial is cheaper than clonal selection, preserves some genetic variability of the variety in vineyards and it is a good strategy for overcoming genotype×environment interactions.

## 3. Conclusion

This chapter focused on the study of ancient grapevine varieties. Indeed, only ancient varieties can have enough intravarietal genetic variability to ensure the success of the presented methodologies.

The analyses were directed to situations in which we do not have any prior knowledge on the genetic variability within the varieties. Therefore, the approach was based on methods that, when properly articulated, can simultaneously respond to three key issues: (1) quantification and knowledge on geographical distribution of intravarietal variability, (2) selection of a high performance group of genotypes and (3) conservation of genetic variability and halting genetic erosion.

The main conclusions cover 3 important points.

1. The results on sampling of variability showed that the minimum number of genotypes needed to adequately represent the genetic variability of a variety ranges from 40 to 50 genotypes per growing region.
2. It was shown that there are high levels of genetic variability in two of the most important traits of grapevine and that those levels are different in different varieties and in different growing regions of the same variety.
3. Mass genotypic selections for two of the most important traits were successfully performed. This methodology is defended as a cheap, fast and efficient selection procedure and has the additional advantage of minimizing the genotype×environment interaction.

## 4. Acknowledgment

We are grateful to our colleagues of "National Network for Grapevine Selection" for their help in data collection. This work was funded by "Fundação para a Ciência e Tecnologia, Portugal" (BPD/43218/2008; PEst-OE/AGR/UI0240/2011).

## 5. References

Arroyo-García, R.; Ruiz-García, I.; Bolling, L.; Ocete, R.; López, M.A.; Arnold, C.; Ergul, A.; Söylemezoglu, G.; Uzun, H.I.; Cabello, F.; Ibáñez, J.; Aradhya, M.K.; Atanassov, A.; Atanassov, I.; Balint, S.; Cenis, J.L.; Costantini, L.; Gorislavets, S.; Grando, M.S.; Klein,

B.Y.; McGovern, P.E.; Merdinoglu, D.; Pejic, I.; Pelsy, F.; Primikirios, N.; Risovannaya, V.; Roubelakis-Angelakis, K.A.; Snoussi, H., Sotiri, P.; Tamhankar, S.; This, P.; Troshin, L.; Malpica, J.M.; Lefort, F. & Martinez-Zapater, J. M. (2006). Multiple origins of cultivated grapevine (*Vitis vinifera* L. ssp. *sativa*) based on chloroplast DNA polymorphisms. *Molecular Ecology*, Vol.15, No.12, pp. 3707-3714, ISSN 1365-294X

Butler, D.; Cullis, B.; Gilmour, A. & Gogel, B. (2009). *Mixed models for S language environments, ASReml-R reference manual.* The Department of Primary Industries and Fisheries (DPI&F), ISSN 0812-0005, Queensland, Australia

Falconer, D. & Mackay, T. (1996). *An introduction to quantitative genetics.* 4th edn.. Prentice Hall, ISBN 0582-24302-5, London, United Kingdom

Giesbrecht, F. & Gumpertz, M. (2004). *Planning, construction, and statistical analysis of comparative experiments.* John Wiley & Sons Inc., ISBN 0-471-21395-0, Hoboken, New Jersey, USA

Gonçalves, E.; St.Aubyn, A. & Martins, A. (2007). Mixed spatial models for data analysis of yield on large grapevine selection field trials. *Theor Appl Genet*, Vol.115, No.5, pp.653-663, ISSN 1432-2242

Gonçalves, E.; St.Aubyn, A. & Martins, A. (2010). Experimental designs for evaluation of genetic variability and selection of ancient grapevine varieties: a simulation study. *Heredity*, Vol.104, No. 6, pp. 552-562, ISSN 0018-067X

Harville, D. & Jeske, D. (1992). Mean squared error of estimation or prediction under a general linear model. *J Am Stat Assoc*, Vol.87, No.419, pp. 724-731, ISSN 0162-1459

Henderson, C. (1975). Best linear unbiased estimation and prediction under a selection model. *Biometrics*, Vol.31, No.2, pp. 423-447, ISSN 1541-0420

Jennrich, R. & Sampson, P. (1976). Newton-Raphson and related algorithms for maximum likelihood variance component estimation. *Technometrics*, Vol.18, No.1, pp. 11-17, ISSN 1537-2723

John, J. & Williams, E. (1998). t-Latinized designs. *Aust Nz J Stat*, Vol.40, No.1, pp. 111-118, ISSN 1467-842X

Kenward, M. & Roger, J. (1997). Small sample inference for fixed effects from restricted maximum likelihood. *Biometrics*, Vol.53, No.3, pp. 983-997, ISSN 1541-0420

Laucou, V.; Lacombe, T.; Dechesne, F.; Siret, R.; Bruno, J.-P.; Dessup, M.; Dessup, T.; Ortigosa, P.; Parra, P.; Roux, C.; Santoni, S.; Vares, D.; Peros, J.-P.; Boursiquot, J.-M. & This, P. (2011). High throughput analysis of grape genetic diversity as a tool for germplasm collection management. *Theor Appl Genet*, Vol.122, No.6, pp. 1233-1245, ISSN 1432-2242

Lawson, J. (2010). *Design and Analysis of Experiments with SAS.* Champman & Hall, CRC, ISBN 978-1-4200-6060-7, New York, USA

Littell, R.; Milliken, G.; Stroup, W.; Wolfinger, R. & Schabenberger, O. (2006). *SAS system for mixed models.* 2nd edn.. SAS Institute, ISBN 978-1-59047-500-3, Cary, NC, USA

Lynch, M. & Walsh, B. (1998). *Genetics and analysis of quantitative traits.* Sinauer Associates, Inc., ISBN 0-87893-481-2, Sunderland, United Kingdom

Martins, A. (2007). Variabilidade genética intravarietal das castas. In: *Portugal vitícola, o grande livro das castas*, J. Böhm, (Ed.), 53-56, Chaves Ferreira Publicações, ISBN 9728987102, Lisboa, Portugal

Martins, A. (2009). Genetic diversity of portuguese grapevines: methods and strategies for its conservation, evaluation and conservation. *Acenología*, No. 112, Available from http://www.acenologia.com/cienciaytecnologia/variedades_portuguesas_cien1209.htm

Martins, A.; Carneiro, L. & Castro, R. (1990). Progress in mass and clonal selection of grapevine varieties in Portugal. *Vitis, special issue*, pp. 485-489, ISSN 0042-7500

Martins, A.; Carneiro, L.; Gonçalves, E. & Eiras-Dias, J. (2006). Methodology for the analysis and conservation of intravarietal variability: the example of grapevine cv Aragonez, *Proceedings of XXIX Congress of OIV*, ISSN: 251-06-001-9, Logroño, Spain, June 25-30, 2006

McCulloch, C. & Searle, S. (2001). *Generalized linear and mixed models*. John Wiley & Sons Inc., ISBN 0-471-19364-x, New York, USA

McLean, R. & Sanders, W. (1988). Approximating degrees of freedom for standards errors in mixed linear models, *Proceedings of the Statistical Computing Section*, American Statistical Association, pp. 50-59, New Orleans, USA

Myles, S.; Boyko, A.; Owens, C.; Brown, P.; Grassi, F.; Aradhya, B.; Reynolds, A.; Chia, J.; Ware, D.; Bustamante, C. & Buckler, E. (2011). Genetic structure and domestication history of the grape. *PNAS*, Vol.108, No.9, pp. 3530-3535., ISSN 1091-6490

Patterson, H. & Thompson, R. (1971). Recovery of inter-block information when block sizes are unequal. *Biometrika*, Vol.58, No.3, pp. 545-554, ISSN 1464-3510

Patterson, H. & Williams, E. (1976). A new class of resolvable incomplete block designs. *Biometrika*, Vol.63, No.1 , pp. 83-92, ISSN 1464-3510

Pelsy, F.; Hocquigny, S.; Moncada, X.; Barbeau, G.; Forget, D.; Hinrichsen, P. & Merdinoglu, D. (2010). An extensive study of the genetic diversity within seven French wine grape variety collections. *Theor Appl Genet*, Vol.120, No.6, pp. 1219-1231, ISSN 1432-2242

Pinheiro, J. & Bates, D. (2000). *Mixed-effects models in S and S-plus*. Springer-Verlag, ISBN 978-1-4419-0317-4, New York, USA

Prasad, N. & Rao, J. (1990). The estimation of the mean squared error of small-area estimators. *J Am Stat Assoc*, Vol.85, No.409, pp. 163-171, ISSN 0162-1459

SAS Institute Inc. (2008). *SAS/STAT® 9.2 User's Guide*. SAS Institute Inc., ISBN 978-1-59047-949-0, Cary, NC, USA

Searle, S.; Casella, G. & McCulloch, C. (1992). *Variance components*. John Wiley & Sons Inc., ISBN 13-978-0-470-00959-8, Hoboken, New Jersey, USA

Sefc, K.M.; Lefort, F.; Grando, M.; Steinkellner, H. & Thomas, M. (2001). Microsatellite markers for grapevine: a state of the art. In: *Molecular biology biotechnology of grapevine*, K.A. Roubelakis-Angelakis, (Ed.), 433-463, Kluwer Publishers, ISBN 0-7923-6949-1, Amsterdam, Netherlands

Self, S. & Liang, K. (1987). Asymptotic properties of maximum likelihood estimators and likelihood ratio tests under nonstandard conditions. *J Am Stat Assoc*, Vol.82, No.398, pp. 605-610, ISSN 0162-1459

Stram, D. & Lee, J. (1994). Variance components testing in the longitudinal mixed effects model. *Biometrics*, Vol.50, No.4 , pp. 1171–1177, ISSN 1541-0420

Verbeke, G. & Molenberghs, G. (2000). *Linear mixed models for longitudinal data*. Springer-Verlag, ISBN 0-387-95027-3, New York, USA

Wiliams, E. & John, J. (1989). Construction of row and column designs with contiguous replicates. *Appl Stat*, Vol.38, No.1, pp. 149-154, ISSN 1467-9876

Williams, E.; John, J. & Whitaker, D. (2006). Construction of resolvable spatial row-column designs. *Biometrics*, Vol.62, No.1 , pp. 103-108, ISSN 1541-0420

# 14

# Heritability of Cold Tolerance (Winter Hardiness) in *Gladiolus xgrandiflorus*

Neil O. Anderson[1,3], Janelle Frick[1,3],
Adnan Younis[1,2] and Christopher Currey[1,4]
[1]*University of Minnesota*
[2]*University of Agriculture*
[3]*University of Arkansas*
[4]*Purdue University*
[1,3,4]*USA*
[2]*Pakistan*

## 1. Introduction

Gladiolus(-i) are herbaceous perennials with long, sword-like leaves and tall spikes of showy, colorful flowers (Goldblatt et al., 1998). Numerous cultivars (>10,000) have been bred (Sinha & Roy, 2002) with extended vase life, floral novelty, or extended flowering periods (Kumar et al., 1999; Takatsu et al., 2002). Recent focus has included transformation for potential creation of a genetically modified organism (GMO) cultivar (Kamo, 2008).

Most of the 180 *Gladiolus spp.* are originally from South Africa (Duncan, 1996, 2000; Goldblatt, 1996; Goldblatt & Manning, 1998; Goldblatt et al., 1993; Manning et al., 2002), although they are widely distributed to as far north as Russia and into the Mediterranean. In the Cape Province region alone, as many as 72 species have been identified as being native (Kidd, 1996). Winter-hardy species from Russia include *G. imbricatus* and *G. palustris*. It has been reported that several species are adaptable to cultivation, including *G. alatus, G. angustus, G. cardinalis, G. carmineus, G. carneus, G. dalenii, G. ochroleucus, G. pritzelii, G. saundersii,* and *G. sempervirens* (Duncan, 1982).

Most modern cultivated gladioli, *Gladiolus xgrandiflorus* Hort. (=*G. hortulanus*), are tetraploid ($2n=4x=60$) interspecific hybrids and have been cultivated for >260 years (Goldblatt, 1996). Modern *G. xgrandiflorus* hybrids are derived from n=6-12 S. African species (Barnard, 1972). Modern gladioli are primarily grown as summer-growing cut flowers and tender annuals. They are derived from summer-growing species, including *G. dalenii, G. oppositiflorus, G. papilio,* and *G. saundersii. Gladiolus cardinalis,* a winter growing species (winter rainfall region), has also been used in hybridization. The dwarf modern gladioli, 'Nanus' hybrids' (=*G. nanus*), are also interspecific hybrids derived from *G. tristis, G. carneus, G. primulinus,* and *G. cardinalis* (Goldblatt, 1996; Goldblatt & Manning, 1998). A new series of miniature gladiolus has also been derived from crossing the cultivated hybrids with wild species, i.e. [*G. xgrandiflorus*] *x G. tristis* (Cohen & Barzilay, 1991).

There are several environmental factors that affect the winter hardiness trait, including low temperatures, variable snow/ice cover, low light periods, and secondary invasion by pathogens (Blum, 1988; Tcacenco et al., 1989; Walker et al., 1995). Winter hardiness is a necessary trait for herbaceous perennials growing in northern climates and is important for floriculture crops as well as consumers (Kim & Anderson, 2006). Underground storage organs in geophytes, e.g., corms, bulbs, tubers, rhizomes, etc., allow herbaceous perennials to survive cold winters. The underground structure of perennial *Gladiolus* is a corm or fleshy storage stem from which shoots and roots grow. *Gladiolus* is a genus that has not been studied to any great extent in the area of winter hardiness. Bettaieb et al., (2007) found that low temperature stress of 8°C caused increased catalase (CAT) activity and lower hydrogen peroxide ($H_2O_2$) levels in gladiolus, but such information has not resulted in breeding for winter-hardy gladioli. Most or all cultivars are 'non-hardy' in Minnesota and other northern latitudes (Anderson, unpublished data).

The gladiolus breeding program at the University of Minnesota is part of a larger project in the Herbaceous Perennial Breeding Program to revolutionize geophytes. Gladiolus which are winter hardy or cold tolerant in USDA Zones 3-4 would allow this crop to overwinter in northern growing conditions and eliminate the need to dig the corms each fall and replant the subsequent spring. In recent years, due to the lack of adequate snow cover and cold temperatures the breeding program has had to supplement field overwintering over successive winters with laboratory freezing tests—a routine procedure widely used for woody and herbaceous perennial plants (Kim & Anderson, 2006). Our studies with USDA Z4 winter hardy chrysanthemums and gaura have shown that herbaceous perennial crowns and root systems must tolerate temperatures of –10°C (Z4) or –12°C (Z3) to survive Minnesota winters (Kim & Anderson, 2006; Pietsch et al., 2009). Most likely this is the case for gladiolus, although to the best of our knowledge, there have not been prior laboratory freezing tests of gladiolus corms for this purpose.

There are three inter-related research objectives for this study. First, selected cultivars and selections of *Gladiolus xgrandiflorus* will be tested to determine the range of cold tolerance at temperatures of 0°C to –10°C for all corm tissues and their subsequent regrowth potential. The second objective will be to determine the nuclear DNA genetic variation (using inter-simple sequence repeats, ISSRs) of the tested genotypes in comparison with wild species and other hybrids. Third, the heritability ($h^2$) of cold tolerance will be assessed in hybrids derived from crossing tested winter hardy and non-hardy parents.

## 2. Materials and methods

### 2.1 Experiment 1: initial freezing tests to determine cold tolerance response

Six cultivars or numbered breeding selections of *G. xgrandiflorus* were used in this study: 'Gemini', 'Great Lakes', 'King's Gold', 'Lady Lucille', Sel'n. 95-49 and 98-29 (Don Selinger, Minnesota Gladiolus Society, Woodbury, MN, USA). Prior to receipt, plants were field-grown in 2005 (45°N lat.) under a natural progression of short days prior to June (June 1=15.2 h), long days in mid-summer, to short days in September (Sep. 1=13.25 h; United States Nautical Almanac Office, 1977); the soil type was a Waukegan silt loam. Mean temperatures (monthly highs, lows) during the 2005 growing season (May 1 – Sep. 30) ranged from 24.5±8.1/10.9±3.8°C (day/night) (University of Minnesota, Department of Soil,

Water & Climate, 2005). Mature corms from plants which had flowered were dug in weeks 40-41 (2005), dried down (~21°C) for 2-3 weeks and shipped. Upon receipt, corms were cooled at 2°C in darkness (Widmer, 1958) for >1,000 h. In week 2 (2006), mature corms were planted into 11.4 cm² square deep PF Landmark containers (Belden Plastics, St. Paul, MN, USA) in trays filled with Sunshine #8/LC8 Professional Growing Mix (SunGro Horticulture, Bellevue, WA USA). Potted corms returned to 2°C (dark) until week 4 (2006) when the programmed freezing tests began.

### 2.1.1 Programmed freezing tests

There were 5 reps/treatment (corms) x 4 treatments (freezing temperatures) for a total of 20 experimental units for each genotype included in the programmed laboratory freezing tests (Kim & Anderson, 2006). Freezing tests were conducted using a Tenney environmental growth chamber (Model No. T20S, Series 942; Tenney Environmental Lunaire Co., Williamsport, PA, USA) with a programmable Series 942 Ramping Controller (Watlow Controls, Winona, MN). This created precision in the "profile control" of pre-programmed multi-step ramp (linear temperature change) and soak (constant temperature) times (Waldron, 1997). Precision temperature data loggers (Veriteq SP-1000; Veriteq Instruments, Inc., Richmond, British Columbia, Canada) were buried in the center (adjacent to the planted corms) of randomly selected pots throughout the chamber to ensure the ramp and soak temperatures were actually obtained. Once the pots were placed in the growth chamber (+2°C) in a randomized complete block (RCB) design, 15.25 h elapsed after transfer for the soilless medium to return to +2°C. Cold tolerance was assessed at 0°C, -3°C, -6°C, and -10°C with 5 h ramp time periods (5 h each for +2°C to 0°C, 0°C to -3°C, -3°C to -6°C, -6°C to -10°C) and a 13.25 h soak time at each temperature treatment (0°C, -3°C, -6°C, -10°C) following completion of each ramp down period (Kim & Anderson, 2006). After each treatment, samples were removed for cold damage assessment upon completion of each soak time and placed into a cooling chamber at 2°C for 2-3 d (dark) until the soilless medium was completely thawed (>0°C).

### 2.1.2 Regrowth assay

Once thawed, potted corms for each temperature treatment were moved to glasshouse conditions of 20/17 °C (day/night) temperature with a 16 h photoperiod (0800-2000 HR) to assay freezing damage to corms and regrowth potential of roots from the basal plate and shoots from the apical meristem. Three weeks later, plant and corms samples were harvested; soilless medium was washed away from the roots. Root number, root lengths (cm), shoot number, and shoot lengths (cm) were recorded. Root and basal plate cell damage (live=white, dead=brown to necrotic) was also determined. Corm damage was assessed after cutting each corm exactly in half with a vertical cut. Tissues were scored on a 1-5 scale, with 1=dead (completely brown or black), 2=brown coloration, 3=green or yellow coloration, 4=slight discoloration, and 5=completely undamaged (Fuller & Eagles, 1978; Perry & Herrick, 1996; Kim & Anderson, 2006; Waldron, 1997).

### 2.1.3 Data analyses

The determination of $LT_{50}$ (the temperature at which 50% of the population sample is killed) was calculated for each genotype (Pomeroy & Fowler, 1973; Tcacenco et al., 1989).

Quantitative data were analyzed by Analysis of Variance (ANOVA) and mean separations using Tukey's Honestly Significant Difference (HSD) test at α=0.05 using the Statistical Package for the Social Sciences (SPSS), Ver. 9.0 (SPSS, Chicago, IL, USA).

## 2.2 Experiment 2: molecular analysis of potential parents

Twenty genotypes of *Gladiolus species* (wild species accessions, hybrid cultivars, and numbered selections), including those used in Experiment 1, were used in this study. These represented a diversity of non-winter-hardy and winter-hardy accessions pre-screened (2005-2008) by the University of Minnesota Gladiolus Breeding Program: ten wild *Gladiolus spp.* (*G. callianthus* [Code: Acidanthera], *G. byzantinus*, *G. carinatus* [1TRB, 2TRB, 3TRB], *G. dalenii* [GL-DAHL], *G. geardii* [GL-Gear], *G. gracilis* [GL-GRAC], *G. saundersii* [03 Gsaun-SS 03-61], *G. tristis* var. *concolor* [GL-TRIS]), five named cultivars (*G. nanus* 'Impressive', *G. xgrandiflorus* 'Gemini', 'Great Lakes', 'King's Gold', 'Lady Lucille'), and five numbered selections (*G. xgrandiflorus* 95-49, 98-29, VT03, 04GL-27-6, 04GL-57-7). Seeds of wild species were obtained from Silverhill Seeds (Capetown, South Africa), numbered selections from Don Selinger (Sel'ns. 95-49, 98-29; Woodbury, MN, USA) and the University of Minnesota Gladiolus Breeding Program (VT03, 04GL-27-6, 04GL-57-7), while named cultivars were sourced from Don Selinger (Woodbury, MN, USA) and Holland Bulb Farms (Milwaukee, WI, USA). DNA was extracted from young leaf tissue of plants grown in a glasshouse (20/17 °C day/night, 16 h photoperiod 0800-2000 HR) using a cetyltrimethyl ammonium bromide (CTAB) protocol based on Doyle & Doyle (1987).

### 2.2.1 Polymerase Chain Reaction (PCR) conditions

Six UBC primers (University of British Columbia Vancouver, Canada, http://www.biotech.ubc.ca/frameset.html) with a high level of polymorphism and scorability (UBC 808, 810, 811, 813, 814, 818) — which have been successfully used with another monocot and geophyte (*Lilium longiflorum*; Anderson et al., 2010) and modifications of Roy et al. (2006) were used for PCR amplification and ISSR analysis. Each PCR contained 0.25 units Flexitaq™ DNA polymerase, 20µM of a single primer, 10mM dNTP, 1.0mM MgCl$_2$, 2µL diluted DNA solution and 5x Flexibuffer™, which was supplied with the polymerase, for a total volume of 25 µL in each reaction (Yamagishi et al., 2002). Amplifications were carried out in a thermocycler (PTC-100, MJ research Inc., Hayward, CA 94545, U.S.A.). Amplification conditions were 7 minutes of denaturation at 94°C followed by 50 cycles of temperatures [94°C for 30 seconds, 43°C for 70 seconds, 72°C for 120 seconds and a final 7 minute extension at 72°C]. Each sample was replicated thrice.

### 2.2.2 Gel electrophoresis

A 1.5 % agarose (mixed with 100 ml of electrophoresis, 1XTris Acetate) buffer was made and heated in a microwave at high power for ~2 mins. until completely melted. Ethidium bromide was added to the gel (final concentration 0.5 ug/ml) at this point to facilitate visualization of DNA after electrophoresis. Electrophoresis chambers were filled with 1XTris Acetate EDTA plus 5 µg/ml ethidium bromide (Sambrook et al., 1989). Sample DNA volumes (12 ml) were then loaded and a current of 75 Volts for 2~2.5 hours was applied. Gels were visualised under UV light and recorded using a Fluro Chem 500 camera (Alpha Innotech Corp., San Leandro, CA, USA).

## 2.2.3 Data analyses

Unequivocally scorable and consistently reproducible amplified DNA fragments for the primers (UBC 808, 810, 811, 813, 814, 818; Anderson, et al., 2010) across all three replications were transformed into binary character matrices (1 = presence, 0 = absence). Cross Checker was used to score and analyze binary interpretation of DNA fingerprints (http://www.spg.wau.nl/pv/pub/CrossCheck/). Variation was measured using Hedrick's Index ($I$) (Hancock & Bringhurst, 1979). To obtain interpopulation variance estimates, Russell & Rao (1940) coefficients of similarity were calculated. Clustering, dendrograms were created by the unweighted pair group method (UPGMA) using the SAHN-clustering and TREE programs from NTSys version 2.02 (Computer Exeter Software, Setauket, NY, USA).

## 2.3 Experiment 3: heritability of cold hardiness

In this experiment laboratory freezing tests were used to determine the level of heritability ($h^2$) (Kim & Anderson, 2006). Three population groups of segregating interspecific $F_1$ hybrids, derived from crossing G. xgrandiflorus hardy (H) or non-hardy (NH) parents (H x H, H x NH, NH x NH), were analyzed for cold hardiness. Since most of the $F_1$ hybrids are derived from interspecific hybrid parents, the $F_1$ generation is most likely to segregate (rather than the $F_2$; Anderson & Ascher, 1996). Seeds from 122 crosses or self pollinations (8,833 seeds) between the G. xgrandiflorus genotypes tested in Experiment 1 and an additional hardy numbered selection (VT03) or its inbred seedlings (VT03-2, -3, -5) made in 2006 were germinated. Due to low germination rates and seedling numbers, a total of 375 genotypes/population group were analyzed (Table 1).

The $F_1$ hybrids were grown in the glasshouse in 7.6 $cm^2$ containers filled with pasteurized gladiolus soilless mix (40% peat moss, 60% sand) under long day photoperiods (0600-2200 HR light) at 19/17°C (day/night) temperatures for 6 wk., followed by a dry down period of 4 wk at 21°C and darkness (to mimic summer dormancy). At the end of the dry down period, the corms were repotted in the gladiolus soilless mix and acclimated at 2°C for 1 wk prior to freezing tests.

A total of 375 corms/group were selected for the study (Table 1). The experimental design was a completely randomized design (CRD), consisting of 5 temperatures x 3 population groups x 5 freezing runs x 15 corms/population group/temperature, for a total of 1125 experimental units (corms). The five temperatures tested (0°C, -3°C, -6°C, –10°C, -12°C) included the four from Experiment 1 with an additional lower temperature (-12°C) in case the H x H genotypes were heterotic and tolerated lower temperatures than their parents. Corms from each of these three population groups were moved to the Tenney environmental growth chamber and re-acclimated at 2°C for 15.25 hours (cf. Experiment 1). Since the containers were smaller than those used in Experiment 1, the ramp times were shortened to 5 hr with a 2 hr soak time at each test temperature. Otherwise, the experiment protocols for laboratory freezing, thawing, reforcing in the glasshouse, as well as data collection, were the same as Experiment 1. Broad sense heritability ($h^2$) estimates and confidence intervals were also calculated (Knapp et al., 1985; Fehr, 1987). These $h^2$ estimates are a ratio of the total genotypic variance (additive, dominance, and epistasis) to the phenotypic variance.

| Population Group | Cross No. | Female Parent | x | Male Parent | No. Seedlings tested |
|---|---|---|---|---|---|
| H x H | 06GL-62 | 98-29 | x | VT03 | 133 |
| | 06GL-63 | 98-29 | x | VT03-5 | 14 |
| | 06GL-65 | VT03-2 | x | VT03-3 | 50 |
| | 06GL-66 | VT03-3 | x | VT03-2 | 14 |
| | 06GL-78 | Lady Lucille | x | 95-49 | 50 |
| | 06GL-84 | VT03 | x | VT03-3 | 60 |
| | 06GL-117 | 98-29 | x | Self | 54 |
| H x NH | 06GL-29 | 04GL-57-16 | x | 98-29 | 35 |
| | 06GL-31 | 04GL-57-23 | x | VT03-5 | 38 |
| | 06GL-32 | 04GL-57-24 | x | 98-29 | 66 |
| | 06GL-35 | 04GL-57-5 | x | VT03 | 69 |
| | 06GL-61 | 98-29 | x | 95-49 | 89 |
| | 06GL-72 | Great Lakes | x | VT03-5 | 78 |
| NH x NH | 06GL-37 | 04GL-57-7 | x | Great Lakes | 33 |
| | 06GL-73 | King's Gold | x | 95-49 | 61 |
| | 06GL-93 | 04GL-27-14 | x | Self | 52 |
| | 06GL-99 | 04GL-27-8 | x | Self | 49 |
| | 06GL-102 | 04GL-57-12 | x | Self | 38 |
| | 06GL-103 | 04GL-57-12 | x | Self | 41 |
| | 06GL-107 | 04GL-57-23 | x | Self | 5 |
| | 06GL-111 | 04GL-57-5 | x | Self | 96 |

Table 1. Population group (hardy x hardy [H x H], hardy x non-hardy [H x NH], non-hardy x non-hardy [NX x NH]), cross number, female/male parents of $F_1$ hybrid or selfed gladiolus crosses and the number of seedlings tested in Experiment 3 for heritability of winter hardiness. There were 375 seedlings tested within each population group for a total of 1,123 genotypes.

## 3. Results and discussion

### 3.1 Experiment 1: initial freezing tests to determine cold tolerance response

Mean root numbers ranged from 0.0 (King's Gold at 0, -3, -6, -10°C; 'Lady Lucille', 'Gemini', 'Great Lakes', Sel'n. 95-49 at -10°C; Sel'n. 98-29 at -6, -10°C) to 29.6 ('Great Lakes', 0°C) (Table 2). In the ANOVA, both genotypes and freezing temperatures as well as their interaction were very highly significant ($p \leq 0.001$) for root number and data could not be pooled. 'King's Gold' had no roots after any of the freezing temperature treatments, including 0°C, indicating that all root initials were killed (Table 2). In most instances, dead root initials occurred when the basal plate was completely dead (black). The significantly greatest mean number of roots occurred with 'Lady Lucille', 'Great Lakes' at 0, -3°C; Sel'n. 98-29 at -3°C (Table 2).

'King's Gold' had no living apical meristem (shoot) or root initial tissues with blackened basal plates (Fig. 1A). Mean root lengths showed similar responses to mean root numbers, ranging from 0.0 cm long (King's Gold at 0, -3, -6, -10°C; 'Lady Lucille', 'Gemini', 'Great Lakes', Sel'n. 95-49 at -10°C; Sel'n. 98-29 at -6, -10°C) to 6.4 cm (Sel'n. 98-29, Table 2). If roots existed, most genotypes and freezing temperatures averaged ~1.0 cm root lengths with only 'Lady Lucille', 'Great Lakes' and Sel'n. 98-29 (-3°C) having the significantly longest roots (Table 2). Oddly enough these genotypes, which also had significantly higher root numbers at 0°C, had significantly shorter roots than at -3°C (Table 2). Both genotypes and freezing temperatures, as well as their interaction, were very highly significant ($p \leq 0.001$) for root number and data could not be pooled.

Fig. 1. Corm damage scoring examples of phenotypic responses in gladiolus cultivars and selections after programmed laboratory freezing tests at 0°C, -3°C, -6°C, and -10°C: 'King's Gold' (A), 'Great Lakes' (B), 'Lady Lucille' (C). Regrowth of shoots at 0°C, -3°C, and -6°C (D) for 'Lady Lucille' (the -10°C tretment was completely dead, *cf.* Fig. 1C).

The range in response among genotypes for the corm damage scoring (Fig. 1) is demonstrated by 'King's Gold' (Fig. 1A) with all root and shoot tissues dead, to 'Great Lakes' with living root/shoot tissues at 0, -3°C only (Fig. 1B), and finally 'Lady Lucille' with living root (0 to -6°C) and apical meristem (-3°C) tissues (Fig. 1C). Slight variations between replications occurred (Table 2). Corm damage values did not differ significantly for genotypes ($p=0.132$) or the interaction of genotype x freezing temperature ($p=0.272$), although freezing temperatures were highly significantly different ($p \leq 0.001$). Thus, genotypes could be pooled within temperature treatments (Table 2). The most severely damaged corms (1.0) occurred at -10°C which differed significantly from -3°C (4.23) (Table 2). Corm damage scores at 0, -3°C did not differ from each other and were the best and healthiest corms possible (nearly 5.0 on the 1-5 scale; Table 2). Since stem storage tissue in the corms was the primary score for corm damage, stem tissue is quite resilient to freezing (0, -3, -6°C all had significantly higher scores than -10°C; Figs. 1A, 1B, 1C, Table 2) although not enough to confer winter survival in USDA Z3-4 (Kim & Anderson, 2006; Pietsch et al., 2009). Significantly increased cold tolerance would be required to survive winter conditions in northern latitudes.

Most mean shoot lengths were at 0.0 cm, three weeks after the freezing tests, due to complete death of the apical meristem (Table 2). The significantly longest shoots occurred at the -3°C freezing temperature for 'Great Lakes' (0.84 cm) and Sel'n. 98-29 (1.74 cm) (Table 2). Example shoot regrowth (Fig. 1D) later demonstrated that some stems had floral meristems and would flower (Fig. 1D, 0°C and -3°C) while others were thinner-stemmed and vegetative (Fig. 1D, -6°C). Genotypes were significantly different (p=0.026) whereas both freezing temperature treatments and genotype x freezing temperature interaction were highly significant (p≤0.001).

| Genotype (cultivar, Sel'n.) | Temp. Trmt. | Root No. | Root length (cm) | Corm damage | Shoot Length (cm) |
|---|---|---|---|---|---|
| King's Gold | 0°C | 0.0±0.0a | 0.0±0.0a | **4.8±0.4c** | 0.0±0.0a |
| | -3°C | 0.0±0.0a | 0.0±0.0a | **4.97±0.0c** | 0.0±0.0a |
| | -6°C | 0.0±0.0a | 0.0±0.0a | **4.23±0.0b** | 0.0±0.0a |
| | -10°C | 0.0±0.0a | 0.0±0.0a | **1.0±0.0a** | 0.0±0.0a |
| Lady Lucille | 0°C | 25.0±10.5c | 1.1±0.8ab | | 0.0±0.0a |
| | -3°C | 23.2±16.4c | 2.5±2.3bc | | 0.3±0.5ab |
| | -6°C | 16.6±10.8b | 1.4±1.0b | | 0.0±0.0a |
| | -10°C | 0.0±0.0a | 0.0±0.0a | | 0.0±0.0a |
| Gemini | 0°C | 16.6±12.3b | 1.0±0.8ab | | 0.0±0.0a |
| | -3°C | 13.6±13.1b | 1.4±1.8b | | 0.3±0.7ab |
| | -6°C | 4.4±6.4a | 0.75±1.3ab | | 0.0±0.0a |
| | -10°C | 0.0±0.0a | 0.0±0.0a | | 0.0±0.0a |
| Great Lakes | 0°C | 20.4±5.2c | 1.8±0.7b | | 0.0±0.0a |
| | -3°C | 29.6±5.8c | 4.1±1.5c | | 0.8±1.0c |
| | -6°C | 4.4±9.8a | 0.3±0.6a | | 0.0±0.0a |
| | -10°C | 0.0±0.0a | 0.0±0.0a | | 0.0±0.0a |
| Sel'n. 98-29 | 0°C | 9.1±7.8a | 1.3±1.2b | | 0.0±0.0a |
| | -3°C | 20.0±6.9c | 6.4±3.7c | | 1.7±1.2c |
| | -6°C | 0.0±0.0a | 0.0±0.0a | | 0.0±0.0a |
| | -10°C | 0.0±0.0a | 0.0±0.0a | | 0.0±0.0a |
| Sel'n. 95-49 | 0°C | 9.4±8.1a | 1.7±2.6b | | 0.3±0.7ab |
| | -3°C | 12.5±8.7b | 1.8±1.4b | | 0.4±0.7b |
| | -6°C | 1.0±2.2a | 0.5±1.1ab | | 0.0±0.0a |
| | -10°C | 0.0±0.0a | 0.0±0.0a | | 0.0±0.0a |

Table 2. Mean ±S.D. number of roots, root length (cm), corm damage (1-5 scale, with 1=dead (completely brown or black), 2=brown coloration, 3=green or yellow coloration, 4=slight discoloration, and 5=completely undamaged (Fuller & Eagles, 1978; Perry & Herrick, 1996; Kim & Anderson, 2006; Waldron, 1997) and shoot length of six gladiolus tested for across four freezing temperatures (0°C, -3°C, -6°C, -10°C). Pooled values (corm damage) are shown in **bold typeface.** Mean separations (5% H.S.D.) were performed post-ANOVA.

Lethal temperature ($LT_{50}$) values was -10°C for corm damage (stem tissue) with the exception of Sel'n. 95-49 (Table 3). Root number and lengths had $LT_{50}$ values of -6°C for all tested genotypes with the exception of 'Lady Lucille' ($LT_{50}$ = -10°C) and 'King's Gold' ($LT_{50}$ >0°C) (Table 3). Shoot apical meristem $LT_{50}$s (shoot length measurements) were significantly higher at -3°C for 'Great Lakes' and >0°C for all other genotypes (Table 3). Overall, 'King's Gold' had significantly shorter shoots or fewer living apical meristems than Sel'n. 98-29 (which was the best). These $LT_{50}$s are similar in response to cold tolerant ($LT_{50}$=-10°C) to non-cold-tolerant ($LT_{50}\leq$-6°C) garden chrysanthemum and gaura rhizomes (Kim and Anderson, 2006; Pietsch et al., 2009). Since there were far fewer treatment and genotype combinations with shoots, compared to roots, apical meristems are more sensitive to cold and freezing than root initials. Thus, for the six tested gladiolus genotypes, the apical meristem (shoots) is more sensitive to freezing than roots which are likewise more sensitive than stem tissue (corms) to freezing temperatures. One could expect corms to remain in the ground after a freeze/thaw cycle until they have respired to death and/or been invaded by pathogens.

| Genotype tested (cultivar, sel'n.) | Root No. | Root length (cm) | Corm damage | Shoots (length; cm) |
|---|---|---|---|---|
| King's Gold | >0°C | >0°C | -10°C | >0°C |
| Lady Lucille | -10°C | -10°C | -10°C | >0°C |
| Gemini | -6°C | -6°C | -10°C | >0°C |
| Great Lakes | -6°C | -6°C | -10°C | -3°C |
| Sel'n. 98-29 | -6°C | -6°C | -10°C | >0°C |
| Sel'n. 95-49 | -6°C | -6°C | -6°C | >0°C |

Table 3. Lethal temperatures for root number, root lengths (cm), corm damage, and shoot lengths (cm) at which 50% ($LT_{50}$) of the population of gladiolus corm tissues for six genotypes were killed in the chamber freezing tests.

### 3.2 Experiment 2: molecular analysis of potential parents

Previous researchers only used two primers to determine clonal fidelity in gladiolus (Roy et al., 2006). Four (UBC 808, 810, 811, 818) out of the six primers tested produced scorable, unequivocal bands, giving increased stringency. In the present study, 62 well-defined (clear and unequivocal bands) and scorable markers across replications were obtained. A previous gladiolus ISSR study (Roy, et al., 2006) had its own synthesized primers, rather than those readily available from the University of British Columbia (UBC). Several of these UBC primers were previously used in our laboratory to test genetic variation in clonal *Lilium longiflorum* (Anderson et al., 2010). A total of 38 (61.29%) primer pairs were polymorphic. Replications did not differ significantly in their banding patterns. Total numbers of scorable loci/primer ranged from 16 (Primer pair 808) to 19 (Primer pair 811) for all tested gladioli whereas the number of scorable, polymorphic loci/primer ranged from 8 (810, 818) to 12 (808).

The genetic variation for the tested genotypes (Experiment 1) range across a wide spectrum of the gladiolus genome (Fig. 2). A constructed dendrogram for the tested gladiolus nuclear DNA genetic variation delineates two principle groups separating at Nei's (1972) genetic distance (GD) of 0.53 (Fig. 2) with only one subsequent monophyletic singleton (a set with exactly one genotype) found ('Great Lakes', GD=0.638). Sel'n. 95-49 forms a distinct group at GD=0.74 with 'Lady Lucille' and 'Gemini' (Fig. 2); the latter two have a GD=0.718. Each principle group further separated into additional overlapping, contiguous clades (contigs) (Fig. 2). The remaining genotypes tested for freezing tolerance (Sel'n. 98-29, 'King's Gold'; Experiment 1) were distributed within a clade at GD=0.75 (Fig. 2). Several pairs of genotypes

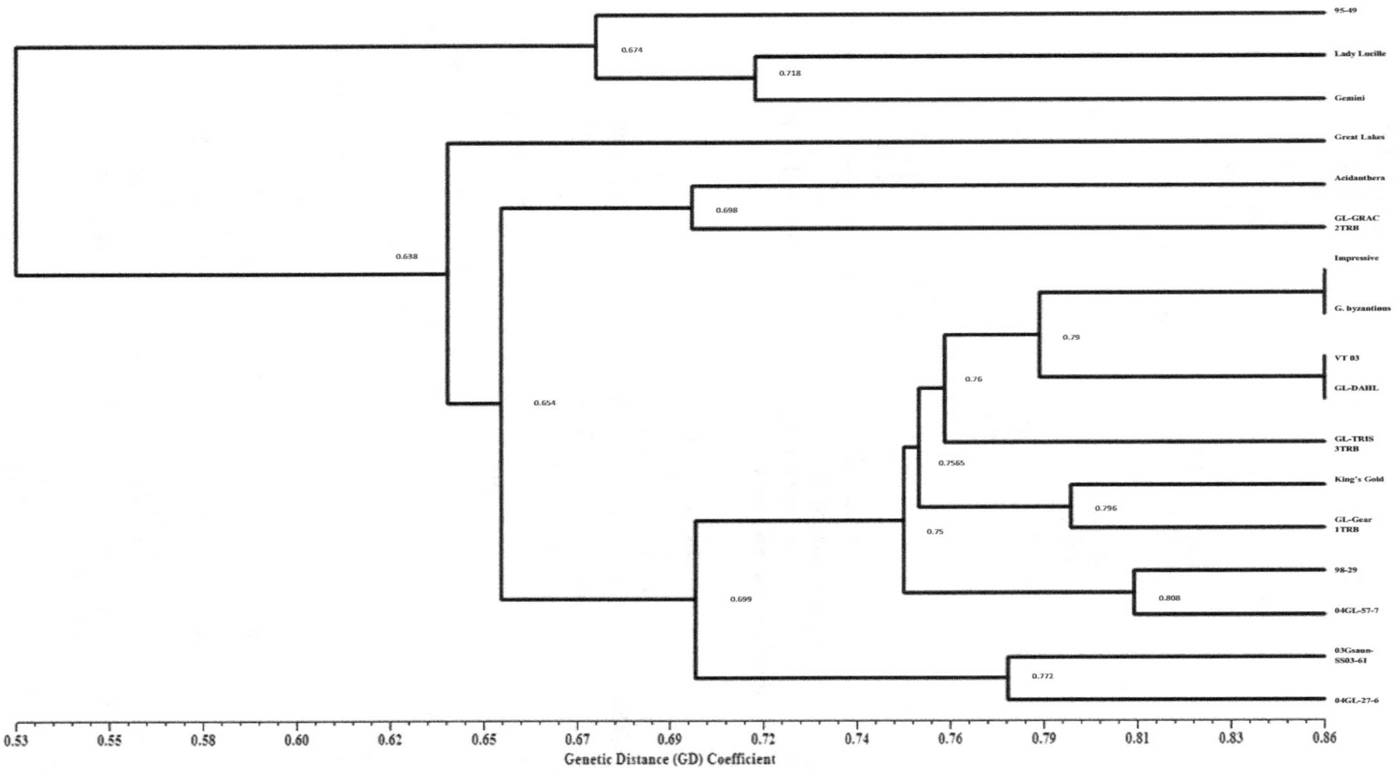

Fig. 2. Dendrograms of relatedness for wild *Gladiolus spp.* (*G. callianthus* [Code: Acidanthera], *G. byzantinus*, *G. carinatus* [1TRB, 2TRB, 3TRB], *G. dalenii* [GL-DAHL], *G. geardii* [GL-Gear], *G. gracilis* [GL-GRAC], *G. saundersii* [03 Gsaun-SS 03-61], *G. tristis* var. *concolor* [GL-TRIS]), five named cultivars (*G. nanus* 'Impressive', *G. xgrandiflorus* 'Gemini', 'Great Lakes', 'King's Gold', 'Lady Lucille'), and five numbered selections (*G. xgrandiflorus* 95-49, 98-29, VT03, 04GL-27-6, 04GL-57-7) derived from polymorphic, scorable bands for five ISSR primers (UBC 808, 810, 811, 814, 818) and based on UPGMA analysis using (Nei, 1972).

were very closely related and separated at GD=0.86, e.g., *G. gracilis* and *G. carinatus* 2 TRB; 'Impressive' and *G. byzantinus*; Sel'n. VT-03 and *G. dalenii*; *G. geardii* and *G. carinatus* 1TRB. Others separated at a lower level of genetic relatedness, i.e. Sel'n. 98-29 and 04GL-57-7 (GD=0.808), etc. Interestingly, Sel'n. VT-03 which has survived for >5 years in an unreplicated trial in USDA Z3 is closely related (GD=0.86) to *G. dahlenii*, introduced by Thornburn in 1908 (http://www.oldhousegardens.com/display.aspx?prod=SGL08), a source of cold hardiness.

### 3.3 Experiment 3: heritability of cold hardiness

Crossing groups, freezing temperatures, and their interaction were all highly significant ($p \leq 0.001$). In all crossing groups for -10°C, -12°C freezing temperatures, no living roots, root initials, apical meristems (shoots, shoot lengths) occurred (Fig. 3) and mean corm ratings for these temperatures (ranging from 1.1 to 1.2) barely exceeded the completely dead rating of 1.0 (Table 4). Thus, no transgressive segregants for corm ratings occurred with greater hardiness than the parents (Experiment 1, Table 2). In general, crossing groups involving at least one hardy parent (hardy x hardy, non-hardy x hardy) had significantly greater numbers and lengths of living roots and shoots than the non-hardy x non-hardy group.

Mean number of roots ranged from 0.0 (all three crossing groups at -10°C and -12°C) to 4.9 (non-hardy x hardy, -3°C) (Table 4). The significantly greatest number of roots occurred in hardy x hardy crosses at 0°C, -3°C (Fig. 3A) and non-hardy x hardy at -6°C (Fig. 3B). In all crossing groups and freezing temperatures, the number of roots was significantly lower than that found for the parents (Table 2). Average root lengths varied from 0.0 (all three crossing groups at -10°C and -12°C) to 6.89 (non-hardy x hardy, -3°C; Table 4) with the significantly greatest root lengths found in hardy x hardy at -3°C and non-hardy x hardy at -3°C, -6°C (Fig. 3). Root lengths, in some cases (6.89 for non-hardy x hardy at -3°C, Table 4), exceeded parental values (6.36 for Sel'n. 98-29 at -3°C, Table 2). Overall, roots in the progeny averaged longer lengths than the parents. Root number heritability ranged from $h^2 = 0.08$ (hardy x hardy) to $h^2 = 0.67$ (non-hardy x non-hardy) (Table 4). Root length heritability was both negative ($h^2 = -0.14$, non-hardy x non-hardy crosses) to positive ($h^2 = 0.37$) (Table 4).

Corm ratings of the hybrids ranged from 1.1 (hardy x hardy and non-hardy x non-hardy at -12°C; non-hardy x hardy at -10°C) to 2.9 (hardy x hardy, 0°C) (Table 4). Heritability of corm ratings was low, ranging from $h^2 = -0.04$ (hardy x hardy crosses) to $h^2 = 0.15$ (non-hardy x non-hardy) (Table 4). All corm ratings in all three crossing groups were significantly higher for freezing temperature treatmets of 0°C to -6°C than for lower temperatures (-10°C and -12°C) (Table 4). The parental values for 0°C to -6°C corm ratings (Table 2) were significantly higher than any of the progeny. These differences could be attributed to the significantly smaller corm size and stem tissue volume/density since the hybrids were only two-year-old corms (Fig. 3) and ~1/5 the size of the tested parental corms of commercial size (Fig. 1).

Fig. 3. Example freezing responses at 0°C, -3°C, -6°C, -10°C, and -12°C for gladiolus seedlings derived from A) Hardy x Hardy (HH), B) Hardy x Non-hardy (HN), and C) Non-hardy x Non-hardy (NN) parental crosses.

The average number of shoots among progenies ranged from 0.0 (all three crossing groups at -10°C and -12°C) to 1.1 (hardy x hardy at -6°C; non-hardy x hardy at 0°C, -3°C) (Table 4). In this case, shoot numbers >1.0 indicated transgressive segregation over the parents which all had 1.0 shoots/corm (Experiment 1). In all cases, a significantly greater number of shoots were found at the 0°C, -3°C, and -6°C temperatures for all crossing groups. Mean shoot lengths ranged from 0.0 (all three crossing groups at -10°C and -12°C) to 10.7 cm (non-hardy x hardy at -6°C, Table 4). Non-hardy x non-hardy crosses for 0°C, -3°C, and -6°C had significatly longer shoots than those with 0.0 cm lengths. Shoot lengths at 0°C, -3°C, and -6°C for both hardy x hardy and non-hardy x hardy groups were significantly longer than any other freezing temperatures within these groups or freezing temperatures for all non-hardy x non-hardy crosses (Table 4). When shoot lengths exceeded 0.0 cm for all crosses at 0°C, -3°C, and -6°C (Table 2), these were significantly longer than any parents (Table 2), indicating transgressive segregation for this trait. Heritability of shoot lengths remained low, however, ranging from $h^2$=-0.43 (hardy x hardy) to $h^2$=0.19 (non-hardy x hardy) (Table 4).

| Crossing group | Temp. Trmt. | Root No. | Root length (cm) | Corm damage | Shoot No. | Shoot Length (cm) | No. (%) plants survived |
|---|---|---|---|---|---|---|---|
| Hardy x Hardy | 0°C | 4.0±2.9cd | 5.7±4.8c | 2.9±1.7b | 1.0±0.2b | 6.4±6.6c | 17 (22.7%) |
| | -3°C | 4.0±2.4cd | 6.3±4.1d | 2.7±1.8b | 1.1±0.3b | 6.7±5.3c | 20 (26.7%) |
| | -6°C | 3.4±2.0c | 5.9±4.4c | 2.7±1.7b | 1.1±0.4b | 8.5±7.6d | 17 (22.7%) |
| | -10°C | 0.0±0.0a | 0.0±0.0a | 1.2±0.5a | 0.0±0.0a | 0.0±0.0a | 0 (0.0% |
| | -12°C | 0.0±0.0a | 0.0±0.0a | 1.1±0.4a | 0.0±0.0a | 0.0±0.0a | 0 (0.0%) |
| $h^2$ | | 0.08 | 0.29 | -0.04 | --- | -0.43 | --- |
| Non-hardy x Hardy | 0°C | 3.2±1.4bc | 4.4±2.9c | 2.3±1.6ab | 1.1±0.3b | 7.4±5.8c | 9 (12.0%) |
| | -3°C | 4.9±4.4cd | 6.9±4.3d | 2.7±1.8b | 1.1±0.3b | 7.5±9.5c | 16 (21.3%) |
| | -6°C | 3.5±1.7c | 6.3±6.8d | 2.0±1.5ab | 1.0±0.0b | 10.7±8.3e | 8 (10.7%)) |
| | -10°C | 0.0±0.0a | 0.0±0.0a | 1.1±0.4a | 0.0±0.0a | 0.0±0.0a | 0 (0.0% |
| | -12°C | 0.0±0.0a | 0.0±0.0a | 1.2±0.5a | 0.0±0.0a | 0.0±0.0a | 0 (0.0%) |
| $h^2$ | | 0.31 | 0.37 | 0.08 | --- | 0.19 | --- |
| Non-hardy x Non-hardy | 0°C | 2.1±1.5b | 2.6±2.5b | 2.3±1.6ab | 1.0±0.0b | 1.9±1.8ab | 4 (5.3%) |
| | -3°C | 2.8±1.6c | 5.7±4.4c | 2.0±1.5ab | 1.0±0.0b | 2.3±1.8b | 4 (5.3%) |
| | -6°C | 2.2±1.7b | 2.5±2.2b | 2.3±1.5ab | 1.0±0.0b | 2.6±1.8b | 7 (9.3%) |
| | -10°C | 0.0±0.0a | 0.0±0.0a | 1.2±0.5a | 0.0±0.0a | 0.0±0.0a | 0 (0.0% |
| | -12°C | 0.0±0.0a | 0.0±0.0a | 1.1±0.4a | 0.0±0.0a | 0.0±0.0a | 0 (0.0%) |
| $h^2$ | | 0.67 | -0.14 | 0.15 | --- | -0.15 | --- |

Table 4. Mean ±S.D. number of roots, root length (cm), corm damage rating, number of shoots, shoot length (cm), and number (%) of *Gladiolus* plants surviving within hardy x hardy, non-hardy x hardy, and non-hardy x non-hardy crossing groups after freezing to 0°C, -3°C, -6°C, -10°C, and -12°C and heritability ($h^2$) for each trait (except shoot number and plant survival) within crossing groups.

The total number of progeny with all living tissues after freezing ranged from zero (all three crossing groups at -10°C and -12°C) to 20 (26.7%; hardy x hardy at -3°C (Table 4). In general, a low range of hardy x hardy, non-hardy x hardy, and non-hardy x non-hardy progeny survived at 0°C, -3°C, and -6°C freezing temperatures (Table 4). Thus, while select progeny are hardy to -6°C, this is not within the minimum range required for herbaceous perennial survival in USDA Z4 and Z3 (-10°C and -12°C, respectively). Further breeding and testing will need to be done to determine whether or not the corms can be bred to survive lower temperatures.

## 4. Conclusion

A range in response was found among tested gladiolus genotypes for tissue damage after laboratory freezing (Experiment 1). 'King's Gold', for instance, was 'non-hardy' at 0°C to -10°C (dead roots/shoots). 'Great Lakes' was intermediate with living root/shoot tissues at 0, -3°C only whereas 'Lady Lucille' had living roots (0 to -6°C) and apical meristems (-3°C). $LT_{50}s$ = -10°C for stem tissues in most genotypes; all were severely damaged or dead at -12°C. The apical meristem is more sensitive to freezing than roots, which are likewise more sensitive than stem tissue (corms) to freezing. The genetic variation (ISSRs) for the tested genotypes ranged across a wide spectrum of the gladiolus genome (Experiment 2); no correlation with ability of tissues to survive cold temperatures was found, except Sel'n. VT-03 (USDA Z3) and G. dahlenii (GD=0.0.86). Two principle groups separated at GD=0.53 with only one monophyletic singleton ('Great Lakes'). Sel'n. 95-49 formed a distinct group with 'Lady Lucille' and 'Gemini' (GD=0.63). No transgressive hybrid segregants for corms occurred with greater hardiness than the parents (Experiment 3). In general, crosses with ≥1 hardy parent (hardy x hardy, non-hardy x hardy) had significantly greater numbers / lengths of living roots/shoots than non-hardy x non-hardy hybrids. In all crosses at -10°C, -12°C, no living roots, root initials, apical meristems occurred. Hybrids with ≥1 hardy parent had greater numbers / lengths of living roots and shoots than the non-hardy x non-hardy group. The highest number of roots occurred in hardy x hardy crosses (at 0°C, -3°C) and non-hardy x hardy (-6°C); root lengths, in some cases (6.89 for non-hardy x hardy at -3°C), exceeded parental values. Root number is barely heritable ($h^2$ = 0.08) for hardy x hardy hybrids but more so ($h^2$=0.67) with non-hardy parents. Root length had a wider range of heritability ($h^2$=-0.14 to 0.37). Heritability of corm ratings is likewise low ($h^2$=-0.04 to 0.15).

A significantly greater number of shoots in progeny (all crossing groups) than the parents were found at the 0°C, -3°C, and -6°C temperatures, although heritability remained low ($h^2$=-0.43 to 0.19). While select progeny are hardy to -6°C, this is not within the minimum range required for herbaceous perennial survival in USDA Z3-4. Further breeding and selection for increased cold tolerance would be required for gladiolus to reliably survive winter conditions in northern latitudes.

## 5. Acknowledgments

Support for this research was sponsored by the Minnesota Agricultural Experiment Station, the Minnesota Gladiolus Society, the Beatrice H. Anderson Memorial Flower Breeding Fund (University of Minnesota, Dept. of Horticultural Science), the University of Minnesota Undergraduate Research Opportunity Program (UROP), and the Institute of Horticultural Sciences, University of Agriculture, Faisalabad, Pakistan.

## 6. References

Anderson, N.O. & Ascher, P.D. (1996). Morphological and Biochemical Variability in Two-Species Congruity Backcross (CBC) *Phaseolus vulgaris x P. coccineus* Hybrids. *Bean Improvement Cooperative Annual Report,* Vol.39, (1996), pp.120-121, ISSN: 0084-7747

Anderson, N.O.; Younis, A. & Sun, Y. (2010). Intersimple Sequence Repeats Distinguish Genetic Differences in Easter Lily 'Nellie White' Clonal Ramets Within and Among Bulb Growers Over Years. *Journal of the American Society for Horticultural Science* Vol.135, No.5, (September 2010), pp.445-455, ISSN 0003-1062

Barnard, T.T. (1972). On Hybrids and Hybridization, In: *Gladiolus, a Revision of the South African Species,* G.J. Lewis, A.A. Obermeyer, & T.T. Barnard, (Eds.), *Journal of South African Botany, Supplement,* Vol.10, (1972), pp.304-310, ISSN: 0022- 4618

Bettaieb, T.; Mahmoud, M.; Ruiz de Galarreta, J.I. & Du Jardin, P. (2007). Relation Between the Low Temperature Stress and Catalase Activity in Gladiolus Somaclones (*Gladiolus grandiflorus* Hort.). *Scientia Horticulturae,* Vol.113, No.2, (June 2007), pp.49-51, ISSN: 0304-4238

Blum, A. (1988). *Plant Breeding for Stress Environments,* CRC Press, Inc., ISBN 0-8493-6388-8, Boca Raton, Florida

Cohen, A. & Barzilay, A. (1991). Miniature Gladiolus Cultivars Bred for Winter Flowering. *HortScience* Vol.26, No.2, (February 1991), pp.216-218, ISSN: 0018-5345

Doyle, J.J. & Doyle, J.L. (1987). A Rapid DNA Isolation Procedure for Small Quantities of Fresh Leaf Tissue. *Phytochemical Bulletin* Vol.19, (1987), pp.11-15, ISSN: 0898-3437

Duncan, G. (1982). *Gladiolus ochroleucus* – A Desirable Species for Pot Culture. *Veld & Flora* Vol.68, No.4, pp.112-113, ISSN: 0042-3203

Duncan, G. (1996). *Growing South African Bulbous Plants,* National Botanical Institute, ISBN 1-874907-15-3, Cape Town, S. Africa

Duncan, G. (2000). *Grow Bulbs: A Guide to the Species Cultivation, and Propagation of South African Bulbs,* National Botanical Institute, Kirstenbosch Gardening Series, ISBN 0-900048-53-0, Claremont, S. Africa

Fehr, W.R. (1987). *Principles of Cultivar Development, Vol.1. Theory and Technique,* MacMillan Publishing Co., ISBN 13: 978-0070203457, New York

Fuller, M.P. & Eagles, C.F. (1978). A Seedling Test for Cold Hardiness in *Lolium perenne* L. *Journal of Agricultural Science* Vol.91, No.1, (August 1978), pp.217-222, ISSN: 0021-8596

Goldblatt, P. (1996). *Gladiolus in Tropical Africa: Systematics, Biology & Evolution,* Timber Press, ISBN 10: 0-8819-233-38 / 0-88192-333-8, Portland, Oregon

Goldblatt, P. & Manning, J. (1998). *Gladiolus in Southern Africa,* Timber Press, ISBN-10: 1874950326, Portland, Oregon

Goldblatt, P.; Takei, M. & Razzaq, Z.A. (1993). Chromosome Cytology in Tropical African *Gladiolus* (Iridaceae). *Annals of the Missouri Botanical Garden* Vol.80, No.2, pp.461-470, ISSN: 0026-6493

Hancock, J.E. Jr. & Bringhurst, R.S. (1979). Ecological Differentiation in Perennial Octaploid Species of *Fragaria. American Journal of Botany* Vol.66, (1979), pp.367-375, ISSN 1537-2197

Kamo, K. (2008). Transgene Expression for *Gladiolus* Plants Grown Outdoors and in the Greenhouse. *Scientia Horticulturae* Vol.117, No. 3, (July 2008), pp.275-280, ISSN: 0304-4238

Kidd, M.M. (1996). *Cape Peninsula: South African Wild Flower Guide 3*, 4th Ed., Botanical Society of S. Africa, CTP Book Printers, ISBN: 13: 978-0620067461, Parow, S. Africa

Kim, D-C. & Anderson, N.O. (2006). Comparative Analysis of Laboratory Freezing Methods to Establish Cold Tolerance of Detached Rhizomes and Intact Crowns in Garden Chrysanthemums. *Scientia Horticulturae* Vol.109, No.4, (August 2006), pp.345-352, ISSN: 0304-4238

Knapp, S.J.; Stroup, W.W. & Ross, W.M. (1985). Exact Confidence Intervals for Heritability on a Progeny Mean Basis. *Crop Science* Vol.25, No.1, (January-February 1985), pp.192-194, ISSN: 0011-183X

Kumar, A.; Sood, A.; Palni, L.M.S. & Gupta, A.K. (1999). *In Vitro* Propagation of *Gladiolus hybridus* Hort.: Synergistic Effect of Heat Shock and Sucrose on Morphogenesis. *Plant Cell, Tissue and Organ Culture* Vol.57, (1999), pp.105-112, ISSN: 1573-5044

Manning, J.; Goldblatt, P., & Snijman, D. (2002). *The Color Encyclopedia of Cape Bulbs*, Timber Press, ISBN 9780881925470, Portland, Oregon

Nei, M. (1972). Genetic Distance Between Populations. *American Naturalist* Vol.106, No. 949, (May-June 1972), pp.283-292, ISSN: 0003-0147

Perry, L.P. & Herrick, T. (1996). Freezing Date and Duration Effects on Regrowth of Three Species of Container-Grown Herbaceous Perennials. *Journal of Environmental Horticulture* Vol.14, No.4, (December 1996), pp.214-216, ISSN: 0738-2898

Pietsch, G.M.; Anderson, N.O. & Li, P. (2009). Cold Tolerance and Short Day Acclimation in Perennial *Gaura coccinea* and *G. drummondii*. *Scientia Horticulturae* Vol.120, No.3, (May 2009), pp.418-425, ISSN: 0304-4238

Pomeroy, M.K. & Fowler, D.B. (1973). Use of Lethal Dose Temperature Estimates as Indices of Frost Tolerance for Wheat Cold Acclimated Under Natural and Controlled Environments. *Canadian Journal of Plant Science* Vol.53, No.3, (July 1973), pp.489-494, ISSN: 0008-4220

Roy, S.K.; Gangopadhyay, G.; Bandyopadhyay, T.; Modak, B.K.; Datta, S. & Mukheerjee, K. (2006). Enhancement of *In Vitro* Micro Corm Production in *Gladiolus* Using Alternative Matrix. *African Journal of Biotechnology* Vol.5, No.12, (June 2006), pp.1204-1209, ISSN 1684-5315

Russell, F.S. & Rao, T.R. (1940). On Habitat and Association of Species of Anopheline Larvae in South-Eastern Madras. *Journal of the Malaria Institute of India* Vol.3, (1940), pp.153-178, ISSN: 0368-2994

Sambrook, J.; Fritsch, E.F. & Maniatis, T. (1989). *Molecular Cloning: A Laboratory Manual*, 2nd ed., Cold Spring Harbor Laboratory Press, ISBN: 0879693096, Cold Spring Harbor, New York

Sinha, P. & Roy, S.K. (2002). Plant Regeneration Through *In Vitro* Cormel Formation from Callus Culture of *Gladious primulinus* Baker. *Plant Tissue Culture* Vol.12, No.2, (December 2002), pp.139-145, ISSN: 1817-3721

Takatsu, Y.; Manabe, T.; Kasumi, M.; Yamada, T.; Aoki, R.; Inoue, E.; Morinnaka, Y.; Marubashi, W. & Hayashi, M. (2002). Evaluation of Germplasm Collection in Wild *Gladiolus species* of Southern Africa. *Breeding Research* Vol.4, (February 2002), pp.87-94, ISSN: 1344-7629

Tcacenco, F.A.; Eagles, C.F. & Tyler, B.F. (1989). Evaluation of Winter Hardiness in Romanian Introductions of *Lolium perenne*. *Journal of Agricultural Science* Vol.112, No.2, (April 1989), pp.249-255, ISSN: 0021-8596

United States. Nautical Almanac Office. (1977). *Sunrise and Sunset Tables for Key Cities and Weather Stations of the U.S.* Gale Research, ISBN: 10-0810304643, Detroit, Michigan

University of Minnesota, Department of Soil, Water & Climate. (2005). *Minnesota Climatology Working Group, Historical Climate Data Retrieval,* Available from http://climate.umn.edu/

Waldron, B.L. (1997). *Breeding for Improved Winter Hardiness in Turf-Type Perennial Ryegrass (Lolium perenne L.).* University of Minnesota, Ph.D. Dissertation, St. Paul, Minnesota

Walker, M.D. ; Ingersoll, R.C. & Webber, P.J. (1995). Effects of Interannual Climate Variation on Phenology and Growth of Two Alpine Forbs. *Ecology* Vol.76, No.4, (June 1995), pp.1067-1083, ISSN: 0012-9658

Widmer, R.E. (1958). The Determination of Cold Resistance in the Garden Chrysanthemum and Its Relation to Winter Survival. *Proceedings of the American Society for Horticultural Science* Vol.71, (1958), pp.537-546, ISSN 0003-1062

Yamagishi M.; Hiromi A.; Michiharu N. & Akira N. (2002). PCR-Based Molecular Markers in Asiatic Hybrid Lily. *Scientia Horticulturae* Vol.96, No.1-4, (December 2002), pp.225-234, ISSN: 0304-4238

# 15

# Challenges, Opportunities and Recent Advances in Sugarcane Breeding

Katia C. Scortecci[1], Silvana Creste[2], Tercilio Calsa Jr.[3], Mauro A. Xavier[2],
Marcos G. A. Landell[2], Antonio Figueira[4] and Vagner A. Benedito[5*]

[1]*Department of Cell Biology and Genetics, Universidade*
*Federal do Rio Grande do Norte (UFRN), Natal, RN*
[2]*Centro de Cana – Instituto Agronômico de Campinas (IAC), Ribeirão Preto, SP*
[3]*Department of Genetics, Universidade Federal de Pernambuco (UFPE), Recife, PE*
[4]*Plant Breeding Laboratory , Centro de Energia Nuclear na Agricultura (CENA),*
*Universidade de São Paulo (USP), Piracicaba, SP*
[5]*Laboratory of Plant Functional Genetics, Genetics and Developmental Biology Program,*
*Plant & Soil Sciences Division, West Virginia University (WVU), Morgantown, WV*
[1,2,3,4]*Brazil*
[5]*USA*

## 1. Introduction

Sugarcane (*Saccharum* spp.) has a long history in line with European colonization and Imperialism. Its cultivation in the New World since the early Modern Period (XVI century) largely shaped social structures, national and international politics, global trading, slavery, culinary, and even disease proliferation (tooth decay, diabetes, obesity) with extensive significance to current history around the globe (cf. Hobhouse, 2005; Rogers, 2010). In 2009, 1,682 million metric tonnes (MT) of sugarcane were produced worldwide in a total area of 23.8 million hectares (ha) (close to the size of the United Kingdom, which occupies 24 million ha). Brazil is the largest sugarcane producer, contributing with 40% of the world production (700 MT in 2009), followed by India (285 MT), China (114 MT), Thailand (67 MT), Pakistan (50 MT), Colombia (38.5 MT), Australia (31 MT), Argentina (30 MT), United States (27.5 MT), Indonesia (26.5 MT) and the Philippines (23 MT) (http://faostat.fao.org). Sugarcane is also an economically important crop for the economies of Mexico and many countries in Central America, the Caribbean, Africa, and Southern Asia (**Fig. 1**).

Sugarcane is native to southeastern Asia, with its cultivation in India dating since before 5,000 years ago (Daniels & Daniels, 1975; Daniels & Roach, 1987). This species has C4 photosynthesis, resulting in a vigorous biomass accumulation under tropical conditions, but it also implies a less vigorous growth in temperate regions. In commercial settings, sugarcane is clonally propagated via stem cuttings, facilitating the preservation of cultivar genetic identity. The plant is semi-perennial, with a cycle ranging from ~12-18 months from planting to harvest at tropical conditions, and regrowth of ratoons (sugarcane stumps) that allow up to five harvests.

Sugarcane plantations are often criticized for occupying large extensions of fertile arable land that otherwise could be used for food production (the so-called "food or fuel" debate), for impacting the environment with deforestation and land degradation, monoculture, pollution (contamination of ground water via leaching and water bodies via run off of fertilizers, pesticides and molasses; pre-harvest burning and air pollution). It also relies heavily on low-paid seasonal jobs, and having many cases of labor abuses worldwide (child labor, slavery regimen, hazardous conditions, underpayment) (Martinelli & Filoso, 2008; Miranda, 2010; Uriarte et al., 2009). Nevertheless, it produces an alternative and renewable energy source with balanced carbon emission (Goldemberg et al., 2008).

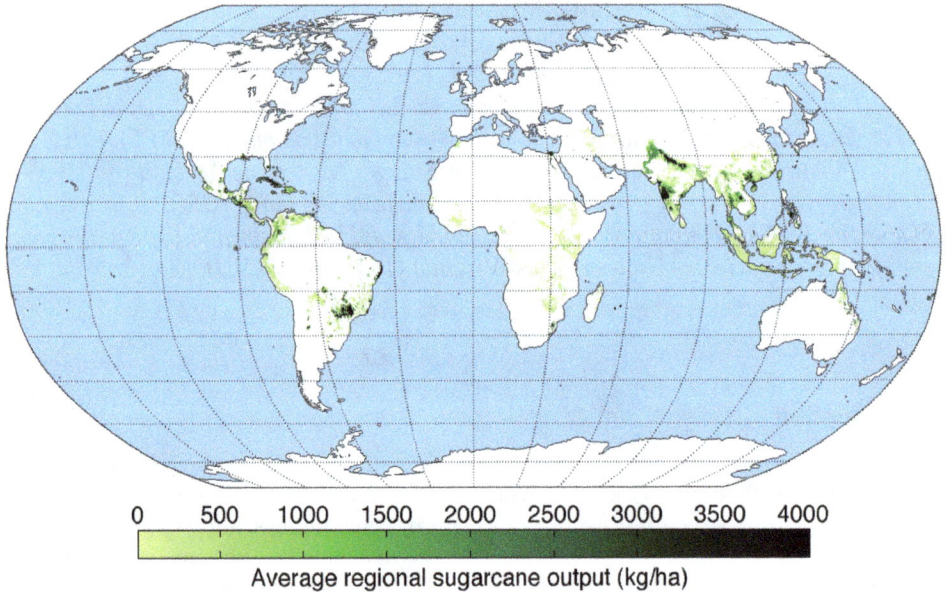

Average regional sugarcane output (kg/ha)

Fig. 1. Sugarcane production areas around the globe. Source: AndrewMT,
http://en.wikipedia.org/wiki/Sugarcane

The sustainability of ethanol production as a component of the energetic base of a country is highly dependent on how the source crop is managed, and this also depends on the genetic background of cultivars available to farmers. Modern agricultural practices depend heavily on high input of inorganic fertilizers. The answer for the huge challenges regarding energy production the world currently faces may lie on sugarcane. However, one has to consider the convoluted origin and genetics of sugarcane, the complex system required for a sustainable, productive and profitable cropping, the social and ecological aspects. The massive demand for a reliable source of energy can only be answered with sugarcane if sustainability and the social issues is resolved. Most of the technical issues related to sugarcane feedstock production for biofuel to meet the massive bioenergy demand can be solved genetically, via breeding programs that will have to look more seriously to production systems with lower input of resources (energy, water, fertilizers, pesticides). Efforts should not overlook traditional and modern breeding techniques, including biotechnological tools to achieve its goals in a timely manner.

Like no other contemporary crop, sugarcane is facing new paradigms and expected to at least partially solve one most important problems in present society: energy production. In this chapter we introduce the current status, potential and the new breeding challenges imposed to sugarcane crop, with a focus in Brazil as a model for sugarcane production. We further describe important traits for sugarcane production, delineate the breeding pipeline, and introduce recent advances in sugarcane genetics and genomics.

## 2. Current and potential utilization of sugarcane

Sugarcane is responsible for ~70% raw table sugar production worldwide (Contreras et al., 2009), with the remaining production coming from sugar beet in temperate countries. Sugarcane stores energy as the non-reducing disaccharide sucrose, which accumulates in large amounts in the vacuoles of parenchyma cells of stem tissues (up to 23% w/v, or 70 mM; Hawker, 1965). This organic compound is especially used for producing table sugar (via simple crystallization of sucrose from stem juice and further refining and clarification), ethanol and spirits such as rum and "cachaça" (via yeast fermentation of the stem juice and distillation). Moreover, chopped sugarcane stalks are widely used as cattle feed, especially during dry season when pastures are unavailable for grazing.

Sugarcane is considered a first-generation biofuel crop. In Brazil, most energy converted to ethanol biofuel is derived from the sucrose extracted from squeezing the stems and fermenting the juice. However, only a third of the plant's energy is extracted using this technology, with the remaining being stored in less readily available compounds, such as the cellulose deposited in plant cell walls. Today in refineries located in São Paulo state, Brazil, bagasse is burned to provide heat for distillation of the fermented must during ethanol purification and to co-generate electric power that is sold to the electric power grid. More efficient ways to use the cellulosic energy are being sought in order to increase ethanol yields at the refinery.

Plant cell walls are mainly composed by cellulose, hemicellulose and lignin. Cellulose is a polysaccharide consisting of linked [$\beta$ 1→6] glucose units, making around a third of the plant dry matter. The second generation of biofuels relies on the development of an efficient process of breaking down cellulose (thermally or enzymatically) into fermentable sugars from the bagasse and leaves. Current research goals in this regard are looking for: 1) plant varieties with improved constitution of cell wall contents to yield more cellulose and less lignin (Pandey et al., 2000; Ragauskas et al., 2006), 2) better methods for mechanic harvesting with low environmental impact, 3) use of the whole aerial plant parts as substrate for ethanol production, 4) use of high-pressure boilers to allow access to cellulose by the fermenting microorganisms and 5) genetically improved yeast strains with higher enzymatic capacity, including fermentation of pentoses. Special attention has also been given to novel product development from sugarcane industrial residues, or even from sucrose or ethanol, as a base for chemical synthesis of other organic molecules with higher market value such as xylitol and polyethylene glycol. In this context too, cellulose, hemicelluloses and lignin are acquiring increased relevance.

It is common practice (although not legal in many regions) to burn sugarcane fields to facilitate manual harvest. When fields are located near urban areas, this burning leads to air pollution and affect the population with smoke and carbonized particulates, which may lead to respiratory diseases and general population dissatisfaction. However, non-burned fields are quite hazardous to harvesters, as they have to cope with stinging stem trichomes, sharp leaf edges, and potential poisonous organisms, such as snakes, insects and spiders. By using mechanical harvesting, not only burning can be eliminated, but also along with the bagasse, the straw can be used in the refinery to either generate heat and electricity, or potentially be hydrolyzed to generate ethanol using second-generation technology, which should be more power efficient. Given the scale of ethanol production in Brazil, the energy stored in the straw and bagasse of sugarcane would be enough to generate up to ten gigawatt (GW) power, which is close to the 14 GW generated by the largest hydroelectric power plant in Brazil (Itaipu). This power plant supplies one fifth of the electricity consumed in Brazil and over 90% of the energy used in Paraguay (Lora & Andrade, 2009; Ripoli et al., 2000). The potential of biomass (bagasse and straw) energy generation in Brazil has been recently calculated to be over 25 GW (Azevedo & Galiana, 2009).

A major by-product of ethanol distillation is vinasse, a complex liquid residue derived from distillation of the fermentate that is rich in humic acids (soluble organic matter) and minerals (such as K and P). For every liter of ethanol distilled, 10-20 liters of vinasse is left over, which have been used as fertilizer in sugarcane fields, saving millions of liters of water and improving the physical conditions of the soil. Vinasse application to sugarcane fields allows the recycling of nutrients back to the soil. However, its application in the field requires careful procedures to avoid contamination of water tables (Amorim et al., 2011). Another potential destination of vinasse is biogas production (Gonçalves et al., 2009).

With a largely diversified energy basis, Brazil (with a population of 192 million, according to the 2010 census) is a worldwide model on how a large country can efficiently establish a renewable and diversified energy matrix. According to the Brazilian Ministry of Mines and Energy, the energy produced in 2008 in the country was 46% renewable, with 15% derived from hydroelectric, 16% from sugarcane, 12% from coal and wood, and 3% from others sources. In sugarcane refineries, 20% of the energy used in boilers comes from *bagasse*, the remaining fibrous matter after juice extraction from sugarcane stalks, whereas wood encompasses 7% (Cortez, 2010).

Land usage is an important economical and social issue to be considered during biofuel discussions. To meet the internal and global demand, Brazil needs to at least double its current ethanol production (Goldemberg, 2010; Waclawovsky et al., 2010). Sugarcane fields occupy 2% of agriculture land in Brazil (Goldemberg, 2010), and ethanol derived from sugarcane is currently responsible for ~50% of vehicle fuel in Brazil, where a biofuel program is in place since 1975 with the successful aim of relieving the nation's external dependence on oil (Leite, 2010). Yield has increased on average ~1.2% per year (considering a compound rate) in the last three decades in Brazil, while production increased almost five-fold (http://www.faostat.org), and sugarcane-based ethanol yield increased 3.8% per year between 1975 and 2004 (Goldemberg, 2008).

More recently, concerns on global warming and its association with human activities via $CO_2$ emissions and greenhouse effects, together with the foreseeable depletion of fossil fuel supply, launched a worldwide search for alternative energy sources that should be renewable and sustainable. During the mid-1990s, engines that accept both ethanol and gasoline (or a mixture) fueled the industry of flex fuel vehicles (i.e., with engine enabled to use any ratio of ethanol and gasoline blend) and is gaining market momentum in many countries with a worldwide increasing demand for ethanol production (Leite et al., 2009). The Brazilian Alcohol Program (Proalcool) helped to establish a massive bioethanol industry in the country along with the pioneer development of the ethanol-based internal combustion car engine (Grad, 2006), relieving the country of a heavy dependence on foreign petroleum and external market fluctuations. Brazil is the pioneer in flex fuel technology, and currently ~95% of the vehicles sold in Brazil are flex fuel. Blending ethanol into gasoline is also a current worldwide trend to relieve gas emissions derived from fossil fuels as well as to lower fuel costs. For comparison, net production costs of ethanol derived from several crops was estimated in 2004 (Henniges & Zeddies, 2004), with Brazilian sugarcane-derived ethanol for 15€/1,000L, while the same amount of ethanol was produced from U.S. corn for 24€, from German wheat for 48€ and from German sugar beet for 52€ per 1,000L. The net energy balance (i.e., energy derived from processing in relation to input energy to produce the fuel) of sugarcane ethanol is between 8.2 and 10-fold while it is ~1.3 for fossil fuels (Goldemberg, 2008; Hill et al., 2006), although these figures are controversial. The cost of sugarcane-derived ethanol in Brazil was calculated as US$30-35 per barrel of oil-equivalent (i.e., the amount of energy contained in a barrel of oil), whereas the corn-derived ethanol from U.S. was estimated as US$45-50 in the same period (Nass et al., 2007).

Other crops have more recently been target as biofuel crops, such as sugar beet (*Beta vulgares*), corn (*Zea mays*), switchgrass (*Panicum virgatum*), sweet sorghum (*Sorghum bicolor*), Brachipodium (*Brachypodium distachyon*), and Miscanthus (*Miscanthus giganteus*) (Goldemberg, 2008; Vega-Sánchez & Ronald, 2010). Oil-producing species (soybean, corn, sunflower, oil palm, algae, etc) are also considered alternative sources of sustainable biodiesel via transesterification reactions (i.e., the process of exchanging the organic group R″ of an ester with the organic group R′ of an alcohol). This carries higher power density than ethanol, and should be accounted in a diversified equation of energy matrix of a nation. However, sugarcane is considered the best biofuel crop in equatorial and tropical regions because of its yield and production costs (Goldemberg, 2008).

Together with expanding land occupation with sugarcane fields, improvement in crop and ethanol yields must be achieved. Crop yield in São Paulo, Brazil is expected to increase 12% over the next 10 years, with an 6.4% increase in total recoverable sugar, 6.2% in fermentation efficiency with improved yeast strains, and 2% in sugar extraction in the same period, adding to an overall increase of 29%, reaching 7,000-9,000 L ethanol per hectare (Goldemberg, 2008), with a perspective of reaching >14,000 L/ha in the next 20 years.

## 3. Limitations to production and breeding goals

### 3.1 Breeding programs

Sugarcane breeding programs play an essential role for the sugar-alcohol industry, as they are responsible for the development of cultivars, which consists of the major technological input for sugar and alcohol production. Sugarcane research and breeding programs have been very successful in Brazil, encompassing an agribusiness model of private and public funding integration. Fostered by government programs and focused research programs, São Paulo state is the major sugarcane producer in the country. Its production participation in Brazil rose from 20% in 1929, to 50% in 1970, and to 87% at present (Cortez, 2010). Breeding programs are partially responsible for this success, as they were able to deliver genotypes with superior traits, responding to new challenges that occasionally arise.

The basis of the germplasm for sugarcane breeding involves commercial varieties and *Saccharum*-related species. Because the major subtropical production regions in Brazil do not favor flowering, parental plants are cultivated in farms with more tropical conditions (such as in Bahia, Brazil) and used for crossing to obtain seeds. Segregating seeds are then distributed to diverse locations for field selection, which may occur twice a year. Plant selection is carried out in three phases, and new genotypes can be exchanged between institutions, especially the governmental ones. The best clones are identified, multiplied and tested in farms during the experimental phase. The release of a new variety may take 12-13 years, after being tested for genetic stability, performance, and agronomic traits (Gazaffi et al., 2010; Landell, 2003; Landell et al., 2010). The success of breeding program depends largely on the choice of parental genotypes, trait heritability, evaluation period, statistic model used, and clone evaluation (Gazaffi et al., 2010). Moreover, the breeder has to pay attention to environmental conditions. In variety development, all factors that may affect production, such as biotic and abiotic stresses (including future threats) have to be considered by breeders.

## 3.2 Production challenges

Until recently sugarcane had been bred mostly for high sucrose content, because sucrose was the main substrate to produce sugar and ethanol. Thus, there has not been a current distinction of sugarcane genotypes since both first-generation technology of ethanol and raw sugar production rely on extractable sucrose accumulation in the stalk. This is advantageous for the sugarcane industry, as processing factories are capable of producing both sugar and/or ethanol, the choice of which product to make is dictated by the demand for each commodity. With the advent of the second-generation ethanol that now relies on cellulose, sugarcane breeding programs will have to revisit their genetic pools. Traits of high cellulose content in sugarcane tissues will be introduced to deliver specialized genotypes that use metabolic energy towards accumulation of either sucrose or cellulose, thus diverging breeding programs to specific goals.

The challenge to increase biofuel production in the coming years (Lam et al., 2009; Melo & Poppe, 2010) can be achieved by expanding the cultivation area, increasing yield with new varieties and improving the whole production system from sugarcane management, harvesting, ethanol production and innovation on sugar and ethanol mills. Beyond the overall increase in carbohydrate yield (sucrose or total biomass), other traits are important to consider addressing the major challenges in production systems, such as better fitness against stresses classified as biotic (pests, diseases, weed competition) and abiotic (drought, salinity, cold, aluminum toxicity, poor and compacted soils), flowering, plant vigor (fast growth under limiting conditions), and plant architecture including height, stalk number, tillering, leaf angle (Carvalho, 2010; Creste et al., 2010).

## 3.3 Specific goals of breeding programs

### 3.3.1 Production goals

A key goal of sugarcane breeding programs is to increase sugar yield by increasing sugar production per area, which is closely associated with height, diameter and number of the stalk, along with sugar accumulation in the stalk. Sugar yields have been generally improved by increasing total biomass rather than directly increasing sugar concentration in stalks (Jackson, 2005). Regarding sugar accumulation in the cane, stalk diameter seems more important than length (Milligan et al., 1990). Other important traits to be considered are vigor and productivity of the ratoon (Aitken et al., 2008). Future varieties bred specifically for second-generation biofuels will be dissociated from high-sucrose yields, but instead will aim at total biomass production per area.

*Brix weight per stool* (BW) in sugarcane is a genetic trait derived from the combination of many components. This trait was deconvoluted to five genetic components using a sophisticate statistics tool. Some traits are additive while others present dominant effects. Additive effects can be useful for breeding in early generations of the program while dominant effects are more useful in hybrids (Liu et al., 2007). Increase of sugar yield (*tons of pol per hectare*, TPH) has always been a major goal of breeding programs. TPH Sugarcane clones are selected on the basis of TPH, estimated by the product of mass yield (*tons of sugarcane per hectare*, TSH), and *% Pol* (sucrose content) of the stem juice. Thus, TPH increment is of great economic interest, considering that costs of harvesting, transportation and grinding remain constant. Casu et al. (2005) proposed that sugar accumulation in stalks might be regulated by a network of genes that contribute to physiological process and abiotic stress tolerance. Papini-Terzi et al. (2009) observed an overlap between pathways related to sugar accumulation and drought stress. Iskandar et al. (2011) analyzed 51 genes related to abiotic stress and affecting sugar accumulation, and concluded that changes in gene expression as a response to water deficit involves different mechanisms, including genes related to biochemistry as well as development.

Another aspect to be considered in biomass production is maximization of radiation interception. It is proposed to potentially increase 10-15% production when the cultivar and plant density is combined with the appropriate planting date (either in planted or ratoon crops) (Singels et al., 2005), being key for crop management. Furthermore, as research on second-generation ethanol production is taking off, breeding programs are looking for traits that increase the lignocellulosic biomass. Another important aspect of plant anatomy to consider is the development of straight stalks in order to best adequate mechanical harvesting (Castro, 2010; Vega-Sánchez & Ronald, 2010).

### 3.3.2 Tolerance to abiotic stresses

Stress can be defined as any condition that hampers the expression of the full genetic potential of a living being. As sessile organisms, plants face many stressful conditions that required the evolutionary establishment of diverse developmental and physiological strategies to cope or avoid the stress condition. However, these strategies usually require metabolic energy that should rather be directed to useful production.

**Water deprivation or scarcity** is a major abiotic stress for sugarcane. Elucidating tolerance mechanisms would enable the development of cultivars more tolerant to drought, allowing cultivation in marginal areas, while assuring the sustainability and viability of the industry in such drought-prone areas. Plant irrigation is a good option for agriculture but it also increases salinity on soil. Besides, it corresponds to 65 % of global water demand and considering the expansion of cultivation to areas without fresh water, tolerance to drought will become increasingly important. Drought tolerance would also contribute to reduce irrigation and water use (Rocha et al., 2007). Even though sugarcane can survive long dry periods, it demands a fair amount of water for optimal yield, leading to the use of irrigation in many areas. Whereas irrigation of sugarcane plantations in Brazil is minimal, 60% of Australia fields and 40% of South African cultivation are irrigated (Innam-Bamber & Smith, 2005). However, the lack of genetic and molecular information about drought tolerance mechanisms and inheritance in sugarcane has limited the development of improved cultivars. There is a need to distinguish genes definitely associated with the response to water deficit, which hold an adaptive function to water deprivation and in stress environments. Genes associated with regulation of expression under water deficit or during the establishment of drought tolerance are potential candidates to evaluate differential expression between contrasting sugarcane genotypes. Studies conducted with rice, Arabidopsis and sugarcane have used microarray analyses (Seki et al., 2001, 2002; Yamaguchi-Shinozaki et al., 2002; Rabbani et al., 2003; Rocha et al., 2007), with subsequent validation through quantitative amplification of reversed transcripts (RT-qPCR). This is to further investigate differentially expressed genes at distinct moments during drought. Some studies have used these differentially expressed genes to obtain transformed plants more tolerant to water deficit (Kirch et al., 2005; Zheng et al., 2004). The identification of genes encoding structural proteins directly related with the establishment of drought tolerance could be useful to develop genetic markers to select tolerant and/or sensitive genotypes. This helps to obtain improved cultivars by direct manipulation (transgenic) or classical breeding. Microarrays data have showed a change in gene expression by drought conditions. In sugarcane 93 genes were differentially expressed, including orthologs of NAC and DREB transcription factors and the cysteine proteinase RD19A (Koizumi et al., 1993; Papini-Terzi et al., 2005; Tran et al., 2004; Yamaguchi & Shinozaki, 2006). Indeed, this knowledge led to the development of transgenic sugarcane more tolerant to drought (see section on Biotechnology).

### 3.3.3 Resistance to biotic stresses

Genetic resistance to pests and diseases is a *sine qua non* (i.e., indispensable and essential condition) in plant breeding. Pests and pathogens often conquer new territories and are well known to dynamically evolve towards breaking resistances, always posing new challenges. Indeed, biotic stresses are of special concern in sugarcane breeding programs, because they may cause great economical impact in plantations with susceptible cultivars.

Examples of biotic stresses to which sugarcane breeders, geneticists, pathologists, entomologists have been paying attention, depending on the location of the breeding program are fungal diseases such as rusts [especially the "brown rust" (*Puccinia melanocephala*) and the "orange rust" (*Puccinia kuehnii*), which recently invaded the American continent], as well as "smut" (*Ustilago scitaminea*). The main bacterial diseases are "ratoon stunting disease" (*Leifsonia xyli*) and "leaf scald" (*Xanthomonas albilineans*), and important viral diseases are "sugarcane mosaic virus" (SCMV) and "sugarcane yellow leaf virus" (SCYLV). Additional diseases with constrained proliferation or of potentially less significant economical impact comprise the fungal diseases "red rot" (*Glomerella tucumanensis*), "eye spot" (*Helminthosporium sacchari*), "pokka boeng" (*Fusarium moniloforme*), and "pineapple disease" (*Ceratocystis paradoxa*). Identification of genetic resistance for these diseases is important to allow incorporation of resistance traits as goals in breeding programs to reduce production threats (current and potential), as well as reduce fungicide spraying.

In the same context, insects are also potential threats to sugarcane production, either directly or as disease vectors. The main sugarcane pests include root froghopper (*Mahanarva fimbriolata*, Hemiptera: Cercopidae), the sugarcane weevil (*Sphenophorus levis*, Coleoptera: Curculionidade), longhorn beetle (*Migdolus fryanus*, Coleoptera: Cerambycidae), sugarcane borer (*Diatreae saccharallis*, Lepdoptera: Pyralidae), and the stem borer (*Telchin licus*, Lepdoptera: Castniidae). Whereas aphids are of little concern as pests *per se*, two species (*Melanaphis sacchari* and *Sipha flava*, Hemiptera: Aphididae) are SCYLV vectors. Sugarcane resistance against these insects is beneficial where the virus is a potential danger. Biological control of sugarcane pests by using natural enemies is a viable crop management technique in some cases, such as the fungus *Metarhizium anisopliae* that controls the root froghopper. However, the incorporation of genetic resistances against pests of economical or potential impact is indisputably the best option, when available.

### 3.3.4 Desirable developmental traits

**Flowering induction:** In sugarcane flowering is regulated by photoperiod (short day), temperature, humidity, plant age, and soil fertility. Flowering induction occurs when photoperiod decreases to more than 11.5-12.5 hours, depending on cultivar (Araldi et al., 2010; Moore & Nus, 1987). Panicle development and pollen fertility temperature is favored by ~28oC day/23oC night cycles (Clements & Awada, 1967). When sugarcane flowers, the plant stops growing, and sugars accumulated in the stalks are used for reproductive development, followed by plant senescence, as the plant's life cycle closes (Araldi et al., 2010). On the other hand, flowering is an important trait for breeding. Economic production of sugarcane in Equatorial areas is hampered by climatic conditions, as flowering may occur all year around, because photoperiod is always close to 12 h (Clements & Awasa, 1967). The location of sugarcane breeding stations is favored in these areas. *S. officinarum* shows a generally *low flowering index* in favorable conditions, which is used as an important source of this trait in breeding programs (Cheavegatti-Gianotto et al., 2011).

**Plant architecture:** it is composed mainly of three traits: tillering, stalk diameter and stalk height (Carvalho, 2010; Creste et al., 2010), as important aspects in determining biomass production. Leaf angle is also an important trait to consider, because shadowing of lower leaves by the upper ones leads to lower photosynthetic rates. More recently, a straight stalk development is sought in new cultivars to facilitate mechanic harvest. The genetics of these developmental traits are not well established in sugarcane and breeding programs. It must rely on performance field selection, making sugarcane germplasm evaluation a long-term effort.

**Ratooning capacity:** Ratoons can regrow after each harvest, although with decreased vigor, which allows on average five productive harvests before the necessary renovation of the field. This is a very important trait to take into consideration during the evaluation period of breeding programs, because poor ratooning capacity will compromise the longevity of the established plantation.

### 3.3.5 Important physiological traits to be considered

**Nitrogen use efficiency:** Nitrogen is one of the most expensive plant fertilizers, because the conversion of atmospheric nitrogen ($N_2$) to ammonia ($NH_3$) is produced commercially using the Haber-Bosch process, which is energetically demanding (Erisman et al., 2008). Sustainable production systems necessarily involve low demands of inorganic nitrogen. Interestingly, sugarcane is able to establish a mutualistic symbiosis with diazotrophic endophytes that reside in xylem cells (especially the bacteria *Glucoacetobacter diazotrophicus*; Cavalcante & Dobereiner, 1988) and in the rhizosphere (such as bacteria of the genera *Azotobacter, Azospirillum, Beijerinckia, Derxia, Enterobacter,* and *Erwinia*) (Thaweenut et al., 2011), which partially supply the plant's requirement of nitrogen, and contributes to minimize nitrogen fertilizer applications, at least under Brazilian conditions (Giller, 2001). Recently, *G. diazotrophicus* had its complete genome sequenced (Bertalan et al., 2009), which will help to understand the symbiotic interaction between this prokaryote and sugarcane at the genetic and molecular levels with perspectives of increasing symbiotic nitrogen supply to sugarcane plants. There are also indications that modern varieties can improve efficiency use of inorganic nitrogen via genetic breeding with focus on plant's physiology (Whan et al., 2010), which could be highly beneficial in breeding programs.

**Mycorrhizal association:** Another important symbiotic association involves mycorrhizae and roots, which improves the plant's nutrient uptake (notably phosphate, but also nitrogen and possibly water and other nutrients). Reis et al. (1999) analyzed the sugarcane rhizhosphere composition in three Brazilian regions and observed the presence of 14 distinct arbuscular mycorrhizae (AM) species, being *Glomus* the most representative, with the fungal diversity maintained when the field is not burned. Since this association is energetically demanding for the plant, which supplies the fungus with carbon skeletons in exchange for nutrients, the plant tightly controls this symbiosis. Low P soil conditions (e.g., ~20 mg kg$^{-1}$) tend to foster symbiotic infection rates (Takashasi, 2005). Crop rotation also benefits mycorrhizal association (Ambrosano et al., 2010) with observed 30% yield increase during three harvests along with increase of sugar content. The plant's genetic inheritance regarding mycorrhization is still elusive, but sugarcane geneticists and physiologists should put efforts in providing tools to enable breeding programs to create genotypes with higher mycorrhization potential, which would certainly have great impact on crop management costs, plant's fitness to adverse conditions and plantation sustainability.

**CO$_2$-enriched atmosphere:** sugarcane is a C$_4$ plant that evolved specific photosynthetic mechanism to fix CO$_2$. With global climatic changes, many questions have been raised about photosynthetic and water use efficiency in the environment where the CO$_2$ has increased. De Souza et al. (2008) analyzed Brazilian sugarcane varieties grown during 50 weeks under normal (360 ppm) and double CO$_2$ conditions (720 ppm). They observed that plants grown under double CO$_2$ increased photosynthesis by 30%, accumulated 40% more biomass and had higher water-use efficiency. Microarrays analysis of these plants indicated that 35 genes were differentially expressed on leaves: 14 genes were repressed and 22 genes were induced (De Souza et al., 2008). Moreover, Vu & Allen (2009) tested two varieties grown also in double CO$_2$ conditions (720 ppm) during 48 weeks and temperature increased by 1.5ºC or 6.0ºC. They observed that in the double CO$_2$ condition and temperature increase of 6.0ºC, there was an increase of 50% of leaf area, 26% increase in leaf dry weight, and 165% increase in stem juice volume, but the responses were variety specific. The data from both groups showed that sugarcane plants increase productivity under higher CO$_2$ and have better water-use efficiency.

## 4. Breeding strategies and current programs

### 4.1 Breeding methods

Hybridization (crossing) is the main procedure so far used for sugarcane to generate new genetic recombination events to further perform selection of superior genotypes, focusing on sugar, ethanol or biomass production. The genus *Saccharum* comprises mainly of six species (*S. officinarum, S. robustum, S. spontaneum, S. barberi, S. sinense*, and *S. edule*) that together with other closely related genera, such as *Erianthus, Miscanthus, Narenga* and *Sclerostachya*, constitute an inter-breeding group known as the 'Saccharum Complex' (Daniels et al., 1975), which represents the genetic variability pool available for sugarcane breeding.

All around the globe, diverse sugarcane breeding programs developed their own strategies. Briefly, conventional breeding is divided in tree steps: *(i)* parental selection, *(ii)* hybridization and, *(iii)* selection of superior genotypes. The criteria used for parental selection is based on parental value, defined on its potential to generate good progeny. Either biparental or poly-crosses can be used to generate segregating populations. The main advantage of biparental crosses is that the male and female parents are both known, whereas in polycrosses, the exact male parent of the progeny is not readily known, because several pollen sources are placed together to interbreed with only one female. In this case, if a superior genotype is selected, molecular markers can be used to identify the male parent. Although easier to perform, polycrosses are generally considered of lower quality, since it allows predicting only the *General Combining Ability* (GCA), whereas biparental crosses are more informative for predicting not only the GCA, but also the *Specific Combining Ability* (SCA) between the parents.

**Parental selection and pollination:** Crossing and seed production routines involve an inspection of the parental population to verify which genotypes are flowering (visible panicles). Florets are collected and pollen is quantified and tested for viability using iodine staining, to decide which genotype will be used as male, and a pollen fertility scale is made to decide the direction of crosses. Since sugarcane flowers are hermaphrodites, emasculation of the female parent is required to avoid pollen contamination. Heat treatment of the panicle (immersion in water at 50°C for 4.5 min) is used to eliminate pollen viability of the female parent. The best parental combination is determined using an algorithm that assesses existing database information of each breeding program, considering genetic distance, progeny performance in earlier crosses, and trait complementarity. Flowering stalks of the selected male parent are cut and labeled, transported to the crossing shed, and placed slightly above the female, as pollination occurs by gravity. The set is protected with a 'lantern' to avoid cross-pollination (**Fig. 2**). The stalks are kept in a nutritive solution, which is replaced frequently to preserve the stalks for about 25 days. After 14 days, the pollination process is complete, and the female stalks are set for another 7-10 days kept in nutrient solution. After the panicles bearing the seeds are dried in a controlled room, at 32°C and low humidity for 3 days, the seeds are ready for the next phase – the progeny selection.

**Progeny selection:** initially, seeds are germinated and the population is screened for traits of high heritability. This phase starts with thousands of plants and those selected are cloned to initiate proper (and recurrent) field experiments (with repetitions), which will be extended for 10-12 years until an elite genotype is released. In the first years, the experiments consist of many genotypes (each coming from a single seed) and little material (plants) from each genotype, thus requiring the adoption of experimental designs with small plots, little material from each genotype and a single location, with consequent implications of small experimental precision. As selection advances, the number of genotypes is decreased, allowing the increase of the replication number, plot size, and to include various experimental locations, especially at production environments, thus increasing statistical precision. At later stages, the clones are also evaluated for diverse harvesting time, if the local cropping management system includes more than one harvest a year. The process of clonal selection typically takes ~9 years to identify the superior genotype, and efforts have been made to increase selection efficiency to reduce the time required to develop a new cultivar. **Table 1** summarizes the many steps of a sugarcane selection process with the five basic steps of a sugarcane breeding program.

## 4.2 Selection methods

**Individual or bulk selection:** the success of a breeding program highly depends on the quality and quantity of genetic variability of the parental population, target trait heritability, and the genetic gain of the trait. According to Dudley & Moll (1969), the estimation of the genetic variance, genetic co-variance, heritability and the selection gain are fundamental in any plant breeding program, because they allow to answer basic breeding questions: 1) existence of sufficient genetic variability within the available germplasm to allow breeding for a trait of economical importance; 2) estimation of resource usage, including required time, experimental locations, and number of repetitions necessary to test the experimental material; 3) definition of the most efficient and fast method to generate an acceptable gain in the target trait; and 4) assess the method efficiency for simultaneous breeding of all traits being selected. During the seedling phase (P1), in which heritability coefficients of individual plants are low, the selection rate must be high. Therefore, selection of individual plants must be based only on traits with high heritability, such as sugar content measured by brix and disease resistance. Selection for traits of economical importance based on individual plants is usually more efficient when conducted based on family tests.

**Family selection:** for this approach, whole progenies are completely selected or rejected, according to its *mean phenotypic value*. Individual values are not considered. Family selection is preferred when the trait under selection presents low heritability, low environmental variation and large families. The efficiency of family selection is based on the deviation shown by environmental effects on each individual that tend to compensate one another. Thus, the mean phenotypic value of each family is close to the *mean genotypic value* (Falconer & Mackay, 1996). The number of individuals within a family is an important factor for family selection, because the larger the family, the highest the correlation between mean phenotypic and genotypic values of the family. As noted above, environmental effects are usually high in the first selection phase. For most of the traits of commercial importance, individual selection is ineffective, as ~80% of the variation is due to environmental factors. However, family selection for these traits might be efficient, since 75-80% of the phenotypic selection among families is due to genetic factors.

**Sequential selection:** Family selection efficiency can be enhanced by adding individual selection within the best families (also called *selection among and between families*). In this case, the selection criterion used within families is based on individual deviation from the mean

Fig. 2. Sugarcane crossing being carried out under *lanterns*. Source: Centro de Cana – IAC.

value of its corresponding family (Falconer & Mackay, 1996). Family selection is only more efficient than individual selection when the heritability based on family means is higher than the heritability of the individual plants. According to McRae et al. (1993) and to Cox et al. (1996), association of family selection with individual selection is more efficient than selection based only on families for sugarcane. Cox & Hogarth (1993) stated that family selection with repetition of each clone, followed by individual selection within the best family is the most efficient way to select sugarcane.

**Regional selection strategy:** Besides the methods described above, sugarcane breeding programs have adopted specific strategies for the development of varieties locally adapted to new environments, guiding the hybridization and selection processes as well as establishing regional experimental stations for selection (Landell & Bressiani, 2008). An important example is the *regional selection strategy* pioneered by the "Centro de Cana - Instituto Agronomico de Campinas", in Brazil, and now adopted by other Brazilian sugarcane breeding programs. This strategy includes a careful characterization of the production environment where seedling populations will be introduced, allowing breeders to isolate important environment factors, favoring the selection of regionally adapted genotypes. Thus, the environment mapping of typical sugarcane cultivation regions is an essential procedure to be considered during hybridization (parental choice) and selection phases. Precise information regarding genotype vs. environment interaction (G x E) is essential for breeders to define their initial objectives, for example, whether the aim is the development of varieties for a broad range of environments or for a specific environment (Borém, 1998).

### 4.3 Performance evaluation

*Complex interaction* is the most prominent type of interaction in sugarcane genotype selection, reflected by particular genotype responses to environmental variations. Varietal

| Phases | time | Evaluation |
|---|---|---|
| **Phase 1 (P1)** ⇒ *seedlings:* Individual plants with clumps spaced 0.6m x 1.5m. | 2 years | Phenotypic selection for stalk, number, diameter and height, sugar content, pests and disease tolerance of plants and ratoon canes. |
| **Phase 2 (P2)** ⇒ *clones:* Clones are spaced 2 (or 4) m) x 1.5m. | 2 years | Phenotypic selection for stalk number, diameter and height, sugar content, pests and disease tolerance of phase 1 ratoon canes. Technological evaluation and first harvest with quantification of yield (tons of cane per hectare, TCH). |
| **Phase 3 (P3)** ⇒ *local experiments:* One to two repetitions, with 30-60m long plots. | 1 year | Phenotypic selection of planted cane for yield components. |
| **Phase 4 (P4)** ⇒ *regional experiments:* Plots with 4-5 lines of 8-15m long spaced 1.1-1.5m from each other, with randomized block experimental design (2-4 repetitions in 3-5 locations). | 2 years | Technological evaluation and first cut with quantification of TPH (tons of pol per hectare); technological evaluation of second cut with TPH quantification. Clone selection for phase 5. Cultivation in different locations in the same region. |
| **Phase 5 (P5)** ⇒ *final characterization* | 4 years | Technological evaluation of first to fourth cut with TPH quantification and identification of superior genotype. |

Table 1. Basic steps of sugarcane selection process (Adapted from Landell & Bressiani, 2008).

adaptability and stability are other aspects that must be taken into account in breeding programs. *Adaptability* refers to the variety's capability to take advantage of environmental variations in a positive way. *Stability* refers to the variety's capability to show a predictable behavior due to environmental changes. There are two main types of stability: static and dynamic. Static stability occurs when a variety has a constant behavior, independently of changes in the environment, and does not show behavior deviations. It is also known as *biological stability*, and it is more correlated to traits less influenced by environment (qualitative traits), but also sucrose accumulation curve (maturation), and stalk color. On the other hand, dynamic stability, also designated as *agronomic stability*, is more correlated with quantitative traits. It is characterized when a specific genotype responds to environmental variation in a predictable way. This stability, if well estimated, consists of an important tool for varietal management. Therefore, a promising variety must show high yield and stability in different environmental conditions (Landell & Bressiani, 2008). Thus, for a good cultivar classification regarding its agronomic potential, it is necessary to associate knowledge about the production environment with individual performance. This way, a cultivar may be classified as: i) *stable*, when it shows reasonable response to most favorable growing conditions and an average response in non-favorable conditions; ii) *responsive*, which shows great responses in favorable growing conditions, but does not adapt to more restrictive environments; or iii) *rustic or low maintenance cultivar*, which, opposite to responsive cultivars, it adapts to more restrictive environments, but does not have top performance in favorable cultivation conditions.

Among the methods suggested to evaluate the genotypic performance, one of the most traditional methods is *analysis of experimental groups*. This method considers that a genotype with less variance is also the most stable. However, it is very common that low-variance genotypes also show low yield. Methods based on regression are still widely used, especially because they allow describing individual responses of different genotypes in a group of environments (for more information on selection, see Finlay & Wilkinsom, 1963; Eberhart & Russel, 1966; Verma et al., 1978; Duarte & Vencovsky, 1999).

*Regional selection* has the advantage to allow a better characterization of a new cultivar, regarding its performance on several yield environments. Studies on phenotypic stability enable one to summarize the huge amount of information obtained from an experimental network, characterizing the yield potential, adaptation to environmental conditions and stability of new cultivars (Raizer & Vencovsky, 1999). Therefore, new sugarcane cultivars are recommended for specific environments, in association to their specific agricultural management and harvesting period. This strategy allows the breeder to explore the genetic potential of new cultivars at its maximum. In addition to environmental adaptation, another essential aspect in regional selection is the relative importance of yield-related traits. For instance, the ability of a genotype to keep good tillering ability becomes more important in dry areas, which is also an indication of drought tolerance. Other major features for adapting to these particular conditions are ratooning ability (maintenance of stalk number during harvesting cycles) and absence of flowering. The study of production environments provides with a necessary support for identification of a superior genotype, allowing the adoption of suitable crop management strategies. Such strategies must gather heterogeneous environments, including the stratification of equivalent sub-regions, in which the interaction G x E is less significant. The *stratification*, or *agro-ecological zoning*, is a very useful procedure, yet having its efficacy restricted due to the occurrence of uncontrollable environmental factors, such as rain and variable thermal amplitude. Notwithstanding, the regional selection strategy has the advantage to enable the identification of a superior genotype in a much shorter period of time (~6-7 years, against 10-12 years as presented above for traditional selection).

## 5. Genetics & genomics, biotechnology & molecular biology

### 5.1 Sugarcane genetics – quite a convoluted system

Sugarcane, *Saccharum officinarum* ($2n$=70-140), also called "noble cane" due to the sweetness of its stalk juice, is a domesticated tropical, perennial grass species (Poaceae family; Andropogoneae tribe). Modern cultivars present a large range in chromosome number ($2n$=100-130) and a genome sequence of ~10 Gb originated from intricate interspecific hybridization, partial loss of chromosomes (aneuploidization), and polyploidization (8-10X) events. Notwithstanding, the basic sugarcane haplotype (X=10; 930 Mb) is remarkably small and syntenic to model grasses, such as sorghum.

In China and India, *S. officinarum* was crossed with *S. barberi* (India cane; $2n$=60-140) and *S. sinense* (China cane; $2n$=104-128) to generate hybrids, the latter thought to be already a hybrid between *S. officinarum* and *S. spontaneum* ($2n$=36-128). During the XIX century, crossings using the wild species *S. spontaneum* ($2n$=36-128) were carried out to improve sucrose yield and disease resistance (Roach, 1972, 1989). Thus, modern sugarcane cultivars correspond to introgression from the wild species *S. spontaneum* and *S. robustum* (2n=66-170) into the cultivated species *S. officinarum, S. sinense* and *S. barberi* (D'Hont et al., 2008, Grivet et al., 2006; Irvine, 1999). *S. edule* (*2n* = 60, 70, 80) is considered ornamental cultivated in New Guinea and Fiji Islands, with no contribution to modern cultivars. Portuguese introduced sugarcane to Brazil during the European colonization period (XV century) probably with hybrids between *S. officinarum* and *S. barberi* originated from India and Persia (Daniels & Daniels, 1975).

Cytogenetics of each species is largely controversial, given methodological difficulties to count such large chromosome number that are confined within a small cell nucleus, or even to establish reliable flow cytometry standards. Molecular genetics is now helping to better understanding the sugarcane origins, because its complex taxonomy has been established based solely on plant morphology and chromosome number. Molecular cytogenetics reveals that 15-25% of the sugarcane genome is derived from *S. spontaneum*, depending on the genotype. For example, the cultivar R570 has 10% of chromosomes inherited from *S. spontaneum*, 80% from *S. officinarum* and 10% as a result of recombined chromosomes (D'Hont et al., 2008). Sugarcane is closely related to maize and sorghum. In fact, sugarcane and sorghum shared the last common ancestor only about five million years ago (Paterson et al., 2004), suggesting that sorghum is a good model system to understand the more complex sugarcane biology, as it is diploid and has its genome sequence available (Wang et al., 2010; Paterson et al., 2009). Indeed, the basic sugarcane haplotype (930 Mb) is remarkably syntenic to model grasses, such as sorghum (730 Mb).

## 5.2 Molecular markers

Molecular markers have the potential to speed breeding up, and their main contribution in crops breeding relies on marker-assisted selection (MAS). There are many breeding challenges posed by sugarcane genetics, which consequently affect breeding programs, and so far these have benefit very little of molecular tools generated for sugarcane. The sugarcane genetic map has more than 1,100 molecular markers (considering diverse marker types) with a total map length of 2,600 cM and a marker density of 7.3 cM (Garcia et al., 2006), which is comparable to other crop species (Casu et al., 2005). Molecular markers have been useful to identify and map candidate genes in sugarcane breeding clones (Andru et al., 2011), from DNA regions (Tabasum et al., 2010) as well as from expressed RNA sequences (Wei et al, 2010). *Bulked segregant analysis* (BSA) coupled with molecular marker analysis of quantitative trait loci has been successful to develop genetic maps around resistance genes in sugarcane against diseases and pests (Asnaghi et al., 2004; Dussle et al., 2003). Molecular markers linked to yield were found in 27 regions of the sugarcane genome from a cross between the Australian variety Q165 and *S. officinarum*, whereas no significant correlations between stalk traits and sugar yield was found in the population analyzed (Aitken et al., 2008). Also, in a study with 40 sugarcane genotypes including *S. officinarum* and *S. barberi*, a high level of polymorphism was detected using 30 random amplification of polymorphic DNA (RAPD) markers, since more than a distinct allele could be identified by each marker (Tabasum et al., 2010). A promising large-scale tool may be the diversity array technology (DArT), which can cover the whole-genome to reveal hundreds of thousands of polymorphic markers in a single analysis via high-yield microarray platform (Wei et al., 2010).

However, sugarcane is a polyaneuploid species, and statistical segregation models have been developed to fit diploid organisms interpretations (Parida et al., 2010; Tabasum et al., 2010; Swapna et al., 2010). About 5% of publicly available sugarcane unigenes (i.e., assembled *expressed sequence tags* or ESTs) present single sequence repeats, and the frequency of perfect microsatellites is one marker for every 10.4 kb (Parida et al., 2010). Considering this, many polymorphic loci obtained during crossings cannot be properly analyzed, given the difficulties due to polyploid segregation (Garcia et al., 2006). In breeding of diploid species, molecular markers are significant for MAS through the use of single nucleotide polymorphisms (SNPs) for polygenic traits such as yield components and disease resistance. Thus, an important aspect of sugarcane breeding is the improvement of markers and statistical models to best fit the convoluted sugarcane genetics (Hotta et al., 2010). The polyploidy constitution of sugarcane makes it as the most difficult crop to apply MAS, that is a sugarcane breeders's dream. Finally, although many available papers report the identification of markers associated with qualitative and quantitative traits in sugarcane, it is noteworthy to mention that they have had very little impact in sugarcane breeding up to now.

## 5.3 Functional genomics – transcriptome, proteome and systems biology

The complex genome of the cultivated sugarcane is currently being sequenced (Sugarcane Genome Sequencing Initiative: http://sugarcanegenome.org), and more effort is supposed to accelerate the discovery of genes responsible for most desirable traits. It will allow the identification of regulatory regions, comparative studies of grasses chromosome, segments evolution and detection of intra and inter-cultivar allelic loci. Transcriptome efforts in sugarcane had a landmark in the late 1990s, when the large-scale cDNA libraries sequencing project SUCEST was set (Vettore et al., 2001, 2003), and from which almost 300,000 EST (ESTs: expressed sequence tags) were obtained, assembled into ~43,000 unique transcribed sequences, the closest picture of sugarcane transcriptional units. Most functional genomics projects performed in the 1990's focused on sucrose content, disease resistance and stress tolerance, and involved several techniques, such as EST characterization, microarray and SAGE analyses (Vettore et al., 2003; Papini-Terzi et al., 2005; Calsa Jr & Figueira 2007; Rocha et al., 2007; Menossi et al., 2008; Papini-Terzi et al., 2009; Waclawovsky et al., 2010; Iskandar et al., 2011). The post-genomic era comprises the use of this information into breeding programs, with the identified markers that reveal expression profile of genes in different environmental conditions (Moore, 2005; Waclawovsky et al., 2010; Khan et al., 2011). Although still rather incipient in this crop (Manners & Casu, 2011), proteomics is increasing its applicability in sugarcane. This was exemplified by recent reports on proteins extraction optimization (Amalraj et al., 2010) and identification of drought tolerance-associated peptides through comparative bi-dimensional electrophoresis followed by mass spectrometry (Ribeiro, 2010). Modern genomics offers the knowledge needed to assign a physiological function to a gene. However, the distance from genotype to phenotype still requires a more integrated approach from molecular data to sugarcane physiology and production, thus setting the basis for modeling the regulatory pathways that link genes, metabolites and physiological processes (Yin & Struik, 2010).

## 5.4 Bacterial artificial chromosome (BAC) library - a useful resource

Bacterial artificial chromosomes (BAC) clones are bacterial lines containing large chromosome fragments of a particular eukaryotic species of interest. BAC library is a collection of these clones that encompass the whole genome of a given species or genotype. These libraries are useful for physical genome sequencing as well as map-based cloning of specific loci or genes. Therefore, these clones are extremely helpful to associate molecular markers closely linked to traits of interest with genome sequences, facilitating identification of the actual genes responsible for the trait. Public BAC collections are available for many crop species, including the sugarcane BAC library at Clemson University Genomics Institute, North Carolina, USA. This library contains >100,000 clones of the R570 genotype, each clone with an average insert length of 130 kb, with a genome coverage of 4.5X and a probability of 98% of recovering any specific genomic sequence, considering a 3-Gb genome (Tomkins et al., 1999). The availability of BAC libraries of the parental species *S. officinarum* and *S. spontaneum* as well as a physic map would be extremely useful for the research community to foster sugarcane genetics.

An international effort involving Australia, Brazil, China, France, South Africa and the U.S. has been established to sequence the sugarcane genome. Given the complexity and variability of the sugarcane genome, this is a huge task, but it will be progressed more quickly with the next-generation Illumina or Solid sequencing, and the available sorghum genome. This will facilitate scaffolding efforts given the high synteny between these two species. The data will surely expand our understanding of the sugarcane genome on genomic organization, promoters, gene regulators and gene networks (Hotta et al., 2010), allowing for identification of major players on key agronomical traits.

## 5.5 Genetic transformation

The long time required for conventional breeding of sugarcane and its highly complex genome led to alternative complementary approaches to the to obtain novel or enhanced agronomic traits introduced in commercial hybrids. As mentioned, sugarcane breeding programs usually take 12-15 years to carry out, test and launch a new variety. The transgenic approach using candidate genes for targeted traits is an alternative to significantly shorten breeding time. Sugarcane is a recalcitrant species regarding genetic transformation and several parameters usually need optimization at the variety level to reach higher transformation efficiencies. Genetic transformation of sugarcane first relied on particle bombardment (biolistic) of cell suspension, embryogenic callus or meristem (Bower & Birch, 1992; Snyman et al., 2006). Efficiency of this method depends on callus formation and plant regeneration, which varies with genotype and culture conditions (Kaeppler et al., 2000). Later, a simpler protocol of genetic transformation of sugarcane using *Agrobacterium tumefaciens* was developed (Bower & Birch 1992; Arencibia, 1998; Brumbley et al. 2008). This approach is more efficient than biolistic for its higher stability on transgene expression, which derives from smaller number of transgene copies integrated into the genome (Dai et al. 2001). However, *Agrobacterium*-mediated transformation shows low efficiency and is highly genotype dependent. For this reason, several *in vitro* culture parameters have been pointed out as key factors to improve transformation, along with genotype screening, explant type and quality, selective agents, and *Agrobacterium* strains (Arencibia et al., 1998; Manickavasagam et al., 2004). For sugarcane, the most effective selectable marker is the neomycin phosphotransferase II (*nptII*) gene (for kanamycin resistance), commonly used to transform callus (Zhangsun et al., 2007; Joyce at al., 2010) that increases the efficiency of transgene integration and recovery of transgenic plants. Specific methodological parameters in transformation have not been widely investigated in sugarcane, and this is especially true for direct physical strategies such as biolistic bombardment. In the case of *Agrobacterium*-mediated transformation, efforts have also focused on *in vitro* regeneration capacity (Joyce et al., 2010).

Genes associated with sucrose content were identified and validated *in vivo* via genetic transformation, resulted in higher sucrose concentration in transgenic plants (Papini-Terzi et al., 2009). Another important application of genetic transformation is the development of resistance to pests and pathogens, including constructs against bacteria and viruses, as seen after biolistic transformation with the capsid gene of the leaf yellow virus (Arencibia et al., 1997, 1999; Falco et al., 2003; Ingelbrecht & Mirkov, 1999; Weng et al., 2010). Beyond the use of herbicide resistance genes (e.g., *bar* and *pat*) as selective markers, they also confer an attractive trait to reduce production costs (Manickavasagam et al, 2004). Increase in drought tolerance was correlated with proline accumulation in transgenic sugarcane (Molinari et al., 2007). In 2011, EMBRAPA, the Brazilian Company of Agricultural Research, announced the development of a drought-tolerant transgenic containing the transgene *DREB2A*. This gene codes for a transcription factor that increases the expression of several genes with function in tolerance against heat, drought and salinity.

Despite numerous studies on effective sugarcane transformation, none refers to plastid transformation, although it is a highly promising technology. There are several examples of agronomical/biotechnological applications of plastid transformation with enhanced biosafety (because the transgene is not transmitted via pollen), and higher transgene product yields in dicot species as cotton, soybean, lettuce, carrot, tomato and tobacco (Wang et al, 2009). *Chloroplast genetic transformation* is still very incipient in $C_3$ monocots rice (Lee et al., 2006) and wheat (Cui et al., 2011), and it has not been reported for $C_4$ grass species, as sugarcane. However, the avenue is widely open since sugarcane's plastid genome has been completely sequenced (Calsa Jr. et al., 2004), which enables recombination-based transformation, with huge potential for basic and applied research, especially in $C_4$ photosynthesis and *molecular pharming*. Moreover, the development of transgenic sugarcane for biodegradable polymers, as polyhydroxybutyrate (PHB) is another example of the potential usefulness of this grass, considering its high biomass production (Petrasovits et al., 2007).

Transgenic plants face release restrictions in many countries. The Brazilian Biosafety Commission (CTNBio) has approved more than 40 transgenic crop applications for sugarcane field experiments so far. Field trials are also been conducted in South Africa, Australia, and the U.S., but so far no transgenic variety has been commercially released in Brazil (Cheavegatti-Gianotto et al., 2011).

## 5.6 Gene promoters for sugarcane transformation

The regulation of gene expression involves DNA sequence upstream of the transcribed region and transcription factors that stabilizes RNA polymerases in these promoter regions to start transcription. The availability of useful promoter sequences in crop species enables molecular breeding to coordinate gene expression only in locations where it is necessary and only when it is necessary due to a fine control of place and time of expression. This minimizes pleiotropic effects of a transgene and saves cellular energy that otherwise would unnecessarily transcribe and translate a gene with spending metabolic energy. Although many promoters from heterologous species may function similarly in crop species, this opportunity is expected to diminish within evolutionarily more distant species. Thus, a repertoire of gene promoters that work efficiently and precisely regarding level, timing and location of expression is an important element of transgenic cultivar development. Due to the polyploidy nature of sugarcane, the great number of alleles in the same genotype makes promoter isolation difficult, because among the 8-10 alleles, it is difficult to point out which of these are effectively contributing to expression of interesting traits. A recent approach to isolate sugarcane promoters has been published (Damaj et al., 2010). This approach utilizes PCR of BACs. It is important, however, to bear in mind that each BAC holds only a single allele (haplotype) whereas alternative alleles (homoelogous loci) may contain regulatory sequences that diverge from each other.

Thus, despite the convoluted genetic system present in sugarcane, which largely limits the use of traditional genetic markers in breeding programs, it is becoming clear that molecular genetics and genomics will play important roles in sugarcane breeding programs, as transformation techniques become more efficient and more molecular tools (genes of interest, transformation vectors, specific promoters) become available.

## 6. Conclusion and perspectives

Traditionally, the main focus on sugarcane breeding had been on *sugar yield*. However, recently, a new sugarcane genotype concept is emerging, focusing on *biomass production* to enable better explore ethanol or energy production. Within this new concept, breeding programs must be reoriented to strengthen its efforts on the development of new cultivars that fit this new variety profile. For this, it is essential to quickly answer to question related to biometrics (stalk number, diameter, height) and processing (sucrose content, reducing sugars, fiber content). Surely, new germplasm resources must be explored by sugarcane breeding programs. The implementation of a parallel introgression program, aiming at broadening the genetic base of sugarcane cultivars for sugar content and/or biomass production, will definitively bring great contributions for increases on yield, ensuring a more sustainable cultivation of sugarcane. Gains on important traits, such as vigor (robustness), will contribute to biomass production and may be found within *S. spontaneum* accessions and related genera, such as *Miscanthus* and *Erianthus*.

New resources and tools are constantly been made available for sugarcane such as better understanding of its genome, genetics, physiology, molecular biology, new markers associated with traits of agronomical relevance and new analysis tools. Breeding programs should take advantage of these tools and incorporate in their selection pipelines to generate superior new cultivars that respond to current and future needs of the industry and the hope of the general society.

## 7. Acknowledgments

Article No. 3111 of the *West Virginia Agricultural and Forestry Experiment Station*, Morgantown. The authors acknowledge the following funding entities in Brazil: Conselho Nacional de Desenvolvimento Científico e Tecnológico (CNPq), Fundação de Amparo à Pesquisa do Estado de São Paulo (FAPESP), Coordenação de Aperfeiçoamento de Pessoal de Nível Superior (CAPES), Financiadora de Estudos e Projetos (FINEP), Fundação de Amparo à Ciência e Tecnologia do Estado de Pernambuco (FACEPE), Ministério de Ciência e Tecnologia (MCT), and Petrobras.

## 8. References

Aitken, K.S.; Hermann, S.; Karno, K.; Bonnett, G.D.; McIntyre, L.C. & Jackson, P.A. (2008). Genetic Control of Yield Related Stalk Traits in Sugarcane, *Theoretical and Applied Genetics*, Vol.117, No.7, (November 2008), pp. 1191-1203, ISSN 0040-5752

Amalraj, R.S.; Selvaraj, N.; Veluswamy, G.K.; Ramanujan, R.P.; Muthurajan, R.; Palaniyandi, M.; Agrawal, G.K.; Rakwal, R. & Viswanathan, R. (2010). Sugarcane Proteomics: Establishment of a Protein Extraction Method for 2-DE in Stalk Tissues and Initiation of Sugarcane Proteome Reference Map. *Electrophoresis*, Vol.31, No.12, (June 2010), pp. 1959-1974, ISSN 1522-2683.

Ambrosano, E.J.; Azcon, R.; Cantarella, H.; Ambrosano, G.M.B.; Schammass, E.A.; Muraoka, T.; Trivelin, P.C.O.; Rossi, F.; Guirado, N.; Ungaro, M.R.G. & Teramoto, J.R.S. (2010). Crop Rotation Biomass and Arbuscular Mycorrhizal Fungi Effects on Sugarcane Yield. *Scientia Agricola*, Vol.67, No.6, (December 2010), pp. 692-701, ISSN 0103-9016

Amorim, H.V; Lopes, M.L.; Oliveira, J.V.C.; Buckeridge, M.S.; Goldman, G.H. (2011). Scientific Challenges of Bioethanol Production in Brazil. *Appl. Microbiol. Biotechnol.*, DOI 10.1007/s00253-011-3437-6, (July 2011), ISSN 0175-7598

Andru, S.; Pan, Y.; Thongthawee, S.;Burner, D. M. & Kimbeng, C. A. (2011). Genetic Analysis of the Sugarcane (*Saccharum* spp.) Cultivar 'LCP 85-384'. Linkage Mapping Using AFLP, SSR, and TRAP markers. *Theoretical and Applied Genetics*, Vol.123, No.1, (June 2011), pp. 77-93, ISSN 0040-5752

Araldi, R.; Silva, F.M.L.; Ono, E.O. & Rodrigues, D. (2010). Flowering in Sugarcane. *Cienc. Rural*, Vol.40, No.3, (March 2010), pp. 694-702, ISSN 0103-8478

Arencibia A.; Vazquez, R.I.; Prieto, D.; Tellez, P.; Carmona, E.R.; Coego, A.; Hernandez, L.; Delariva, G.A. & Selmanhousein G. (1997). Transgenic Sugarcane Plants Resistant to Stem Borer Attack. *Mol Breeding*, Vol.3, No.4, (August 1997), pp. 247-255, ISSN 1380-3743

Arencibia, A.D.; Carmona, E.R.; Cornide, M.T.; Castiglione, S.; O'Reilly, J.; Chinea, A.; Oramas, P. & Sala, F. (1999). Somaclonal Variation in Insect-Resistant Transgenic Sugarcane (*Saccharum* hybrid) Produced by Cell Electroporation. *Transgenic Research*, Vol.8, No.5, (October 1999), pp. 349-360, ISSN 0962-8819

Arencibia, A.D.; Carmona, E.R.; Tellez, P.; Chan, M.T.; Yu, S.M.; Trujillo, L.E. & Oramas, P. (1998). An Efficient Protocol for Sugarcane (*Saccharum* spp. L.) Transformation Mediated by *Agrobacterium tumefaciens*. *Transgenic Research*, Vol.7, No.3, (May 1998), pp. 213–222, ISSN 1573-9368

Asnaghi, C.; Roques, D.; Ruffel, S.; Kaye, C.; Hoarau, J.Y.; Télismart, H.; Girard, J.C.; Raboin, L.M.; Risterucci, A.M.; Grivet, L. & D'Hont, A. (2004). Targeted Mapping of a Sugarcane Rust Resistance Gene (*Bru1*) Using Bulked Segregant Analysis and AFLP Markers. *Theoretical and Applied Genetics*, Vol.108, No.4, (February 2004), pp. 759-764, ISSN 0040-5752

Azevedo, J.M. & Galiana, F.D. (2009). The Sugarcane Ethanol Power Industry in Brazil: Obstacles, Success and Perspectives. *IEEE Electrical Power & Energy Conference 2009*, ISBN 9781424445080, Montreal, October 2009

Bertalan, M.; Albano, R.; Pádua, V. *et al.* (2009). Complete Genome Sequence of the Sugarcane Nitrogen-Fixing Endophyte *Glucoacetobacter diazotrophicus* Pal5. *BMC Genomics,* Vol.10, (September 2009), pp. 450, ISSN 1471-2164

Borém, A. (1998). *Melhoramento de plantas* (22 edition), Editora UFV, ISBN 85-7269-354-7 Viçosa

Bower, R. & Birch, R.G. (1992). Transgenic Sugarcane Plants Via Microprojectile Bombardment. *Plant Journal,* Vol.2, No.3, (May 1992), pp. 409–416, ISSN 0960-7412

Brumbley, S.M.; Snyman, S.J.; Gnanasambandam, A.; Joyce, P.; Hermann, S.R.; da Silva, J.A.G.; McQualter, R.B.; Wang, M.L.; Egan, B.T.; Patterson, A.H.; Albert, H.H. & Moore, P.H. (2008). Sugarcane. In: *Compendium of Transgenic Crop Plants: Transgenic Sugar, Tuber and Fiber Crops,* Kole C & Hall TC (Ed.), 1–58, ISBN 978-1-4051-6924-0, Blackwell, Oxford

Calsa Jr., T. & Figueira, A. (2007) Serial Analysis of Gene Expression in Sugarcane (*Saccharum* spp.) Leaves Revealed Alternative C4 Metabolism and Putative Antisense Transcripts. *Plant Molecular Biology,* Vol.63, No.6, pp. 745-762, ISSN 0735-9640

Calsa Jr., T.; Carraro, D.M.; Benatti, M.R.; Barbosa, A.C.; Kitajima, J.P. & Carrer, H. (2004) Structural Features and Transcript-editing Analysis of Sugarcane (*Saccharum officinarum* L.) Chloroplast Genome. *Current Genetics,* Vol.46, No.6, (December 2004), pp. 366-373, ISSN 0172-8083

Carvalho, L.C.C. (2010). Evolution of Sugarcane Industry in the State of São Paulo, In: *Sugar Cane Bioethanol: R&D for productivity and sustainability,* L.A.B. Cortez, (Ed.), 3-16, Blucher, ISBN: 9788521205302, São Paulo

Castro, O.M. (2010). The role of the State of São Paulo Research Centers on Bioenergy Technological Innovation, In: *Sugar Cane Bioethanol: R&D for productivity and sustainability,* L.A.B. Cortez (Ed.), 63-72, Blucher, ISBN: 9788521205302, São Paulo

Casu, R.; Manners, J.; Bonnett, G.; Jackson, P.; McIntyre, C.; Dunne, R.; Chapman, S.; Rae, A. & Grof, C. (2005). Genomics Approaches for the Identification of Genes Determining Important Traits in Sugarcane. *Field Crops Research,* Vol.92, No.2-3, (June 2005), pp. 137-147, ISSN 0378-4290

Cavalcante, V.A. & Dobereiner, J. (1988). A New Acid-tolerant Nitrogen-fixing Bacterium Associated with Sugarcane. *Plant and Soil,* Vol.108, No.1, (May 1988), pp. 23-31, ISSN 0032-079X

Cheavegatti-Gianotto, A.; de Abreu, H.M.C.; Arruda, P.; Bespalhok-Filho, J.C.; Burnquist, W.L.; Creste, S.; di Ciero, L.; Ferro, J.A.; Figueira, A.V.O.; Filgueiras, T.S.; Grossi-de-Sá, M.F.; Guzzo, E.C.; Hoffmann, H.P.; Landell, M.G.A.; Macedo, N.; Matsuoka, S.; Reinach, F.C.; Romano, E.; da Silva, W.J.; Silva-Filho, M.C. & Ulian, E.C. (2011). Sugarcane (Saccharum X officinarum): A Reference Study for the Regulation of Genetically Modified Cultivars in Brazil. *Tropical Plant Biol.,* Vol.4, (February 2001), pp. 62–89, ISSN 1935-9756

Clements, H.F. & Awada, M. (1967). Experiments on the Artificial Induction of Flowering in Sugarcane. *Proc. Int. Soc. Sugarcane. Technol.,* Vol.12, No.1, pp.795 – 812, 1967

Contreras, A.M.; Rosa, E.; Pérez, M.; van Langenhove, H. & Dewulf, J. (2009). Comparative Life Cycle Assessment of Four Alternatives for Using By-products of Cane Sugar Production. *Journal of Cleaner Production,* Vol.17, No.8, (May 2009), pp. 772-779, 0959-6526

Cortez, L.A.B. (2010). Introduction, In: *Sugar Cane Bioethanol: R&D for productivity and sustainability,* L.A.B. Cortez, (Ed.), 3-16, Blucher, ISBN: 9788521205302, São Paulo

Cox, M.C. & Hogarth, D.M. (1993). The Effectiveness of Family Selection in Early Stages of a Sugarcane Improvement Program. *Australian Plant Breeding Conference*, Vol. 10, pp. 53-54, ISBN 1864239573, Bundaberg, Brisbane: Watson Ferguson, April 1993

Cox, M.C.; Mcrae, T.A.; Bull, J.K. & Hogarth, D.M. (1996). Family Selection Improves the Efficiency and Effectiveness of a Sugarcane Improvement Program. In: *Sugarcane: research towards efficient and sustainable production*. J.R. Wilson; D.M. Hogarth; J.A. Campbell & A.L. Garside, (Ed.), 42-43, Brisbane: CSIRO Division of Tropical Crops and Pasture, ISBN 0643059415, Australia

Creste, S.; Pinto, L.R.; Xavier, M.A. & Landell, M.G.A. (2010). Sugarcane Breeding Method and Genetic Mapping, In: *Sugar Cane Bioethanol: R&D for productivity and sustainability*, L.A.B. Cortez (Ed.), 353-357, Blucher, ISBN: 9788521205302, São Paulo

Cui, C.; Song, F.; Tan, Y.; Zhou, X.; Zhao, W.; Ma, F.; Liu, Y.; Hussain, J.; Wang, Y.; Yang, G. & He, G. (2011). Stable Chloroplast Transformation of Immature Scutella and Inflorescences in Wheat (*Triticum aestivum* L.). *Acta Biochimica et Biophysica Sinica* (Shanghai), Vol.43, No.4, (April 2011), pp. 284-291, ISSN 1672-9145

D'Hont, A.; Souza, G.M.; Menossi, M.; Vincentz, M.; Van Sluys, M.A.; Glaszmann, J.C. & Ulian, E. (2008). Sugarcane: A major Source of Sweetness, Alcohol, and Bio-energy, In: *Genomics of Tropical Crop Plants*, P.H. Moore & R. Ming, (Ed.), 483-513, Springer, ISBN: 9781441924339, New York

Dai, S.H.; Zheng, P.; Marmey, P.; Zhang, S.P.; Tian, W.Z.; Chen, S.Y.; Beachy, R.N. & Fauquet, C. (2001). Comparative Analysis of Transgenic Rice Plants Obtained by *Agrobacterium* Mediated Transformation and Particle Bombardment. *Molecular Breeding*, Vol.7, No.1, (January 2001), pp. 25–33, ISSN 1380-3743

Damaj, M.B.; Beremand, P.D.; Buenrostro-Nava, M.T.; Ivy, J.; Kumpatla, S.P.; Jifon, J.; Beyene, G.; Yu, Q.; Thomas, T.L. & Mirkov, T.E. (2010). Isolating Promoters of Multigene Family Members from the Polyploid Sugarcane Genome by PCR-based Walking in BAC DNA. *Genome*, Vol.53, No.10, (October 2010), pp. 840-847, ISSN 0831-2796

Daniels, J. & Daniels C.A. (1975). Geographical, Historical and Cultural Aspects of the Origin of the Indian and Chinese Sugarcanes *S. barberi* and *S. sinense*. *Sugarcane Breeding Newsletters*, Vol. 36, pp. 4–23, ISSN 1028-1193

Daniels, J. & Roach, B.T. (1987). Taxonomy and Evolution, In: *Sugarcane Improvement through Breeding*, D. J. Heinz (Ed.), 7–84, Elsevier, ISBN: 0444427694, Amsterdam

Daniels, J.; Smith, P.; Paton, N. & Williams, C.A. (1975). The Origin of the Genus *Saccharum*. *Sugarcane Breeding Newsletter*, Vol. 36, 24-39, ISSN 1028-1193

de Souza, A.P.; Gaspar, M.; Da Silva, E.A.; Ulian, U.C.; Waclawovsky, A.J.; Nishiyama Jr., M.Y.; Santos, R.V.; Teixeira, M.M.; Souza, G.M. & Buckeridge, M.S. (2008). Elevated $CO_2$ Increases Photosynthesis, Biomass and Productivity, and Modifies Gene Expression in Sugarcane. *Plant, Cell and Environment*, Vol.31, No.8, (August 2008), pp. 1116-1127, ISSN 0140-7791

Duarte, J.B. & Vencovsky, R. (1999). Interação Genótipos x Ambientes: Uma Introdução à Análise AMMI. (9 edition), Sociedade Brasileira de Genética, ISBN 85-85572-79-5, Ribeirão Preto

Dudley, J.W. & Moll, R.H. (1969). Interpretation and Use of Estimates of Heritability and Genetic Variances in Plant Breeding. *Crop Science*, Vol.9, No.3, (July 1969), pp. 257-262, ISSN 0011-183X

Dussle, C.M.; Quint, M.; Melchinger, A.E.; Xu, M.L. & Lübberstedt, T. (2003). Saturation of Two Chromosome Regions Conferring Resistance to SCMV with SSR and AFLP Markers by Targeted BSA. *Theoretical and Applied Genetics*, Vol.106, No.3, (February 2003), pp. 485-493, ISSN 0040-5752

Eberhart, S.A. & Russel, W.A. (1966). Stability Parameters for Comparing Varieties. *Crop Science*, Vol.1, No.5, (July 1966), pp. 36-40, ISSN 0011-183X

Erisman, J.W.; Sutton, M.A.; Galloway, J.; Klimont, Z. & Winiwarter, W. (2008). How a Century of Ammonia Synthesis Changed the World. *Nature Geoscience*, Vol.1, (September 2008), pp. 636-639, ISSN *1752-0894*

Falco, M.C. & Silva-Filho, M.C. (2003). Expression of Soybean Proteinase Inhibitors in Transgenic Sugarcane Plants: Effects on Natural Defense Against *Diatraea saccharalis*. *Plant Physiology and Biochemistry*, Vol.41, No.8, (August 2003), pp 761-766, ISSN 0981-9428

Falconer, D.S. & Mackay, T.F.C. (1996). *Introduction to Quantitative Genetics* (4th ed.), Longman, ISBN 9780582243026, London

Finlay, K. W. & Wilkinson, G. N. (1963). The Analysis of Adaptation in a Plant Breeding Programme. *Australian Journal of Agricultural Research*, Vol. 14, No.6, (June 1963), pp. 742-754, ISSN 0004-9409

Garcia, A.A.; Kido, E.A.; Meza, A.N.; Souza, H.M.; Pinto, L.R.; Pastina, M.M.; Leite, C.S.; Silva, J.A.; Ulian, E.C.; Figueira, A. & Souza, A.P. (2006). Development of an Integrated Genetic Map of a Sugarcane (*Saccharum* spp.) Commercial Cross, Based on a Maximum-Likelihood Approach for Estimation of Linkage and Linkage Phases. *Theorical and Applied Genetics*, Vol.112, No.2, (January 2006), pp. 298-314, ISSN 0040-5752

Gazaffi, R.; Oliveira, K.M.; Souza, A.P. & Garcia, A.A.F. (2010). The Importance of the Germoplasm in Developing Agro-Energetic Profile Sugarcane Cultivars, In: *Sugar Cane Bioethanol: R&D for productivity and sustainability*, L.A.B. Cortez (Ed.), 333-343, Blucher, ISBN: 9788521205302, São Paulo

Giller, K. E. (2001). *Nitrogen Fixation in Tropical Cropping Systems* (2nd ed.), CABI Publishing, ISBN 0851994172, Oxon, U.K.

Goldemberg, J. (2008). The Brazilian Biofuels Industry. *Biotechnology for Biofuels*, Vol.1, No.6, (May 2008), pp. 1-7, ISSN 1754-6834

Goldemberg, J. (2010). The State of São Paulo Strategy for Fuel Ethanol, In: *Sugar Cane Bioethanol: R&D for productivity and sustainability*, L.A.B. Cortez (Ed.), 19-26, Blucher, ISBN: 9788521205302, São Paulo

Goldemberg, J.; Coelho, S.T. & Guardabassi, P. (2008). The Sustainability of Ethanol Production from Sugarcane. *Energy Policy*, Vol.36, (April 2008), pp. 1086-2097, ISSN 0301-4215

Gonçalves, H.M.; Borges, J.D. & da Silva, M.A.S. (2009). Heavy Metals and Sulphur Accumulation on Soil in Areas of Influence of Vinasse Channels of Fertilization. *Bioscience Journal*, Vol.25, No.6, (November-December 2009), pp. 66-74, ISSN 1516-3725

Grad, P. (2006). Biofuelling Brazil: An Overview of the Bioethanol Success Story in Brazil. *Refocus*, Vol.7, No.3, (May-June 2006), pp. 56-59, ISBN 1471-0846

Grivet, L.; Glaszmann, J.C. & D'Hont, A. (2006). Molecular Evidence for Sugarcane Evolution and Domestication, In: *Darwin's Harvest. New Approaches to the Origins,*

*Evolution and Conservation.* T. Moley; N. Zerega & H. Cross (Eds), 49-66, Columbia University Press, ISBN 9780231133166, West Sussex, UK

Hawker, J. S. (1965). The Sugar Content of Cell Walls and Intercellular Spaces in Surgarcane Stems and its Relation to Sugar Transport. *Australian Journal of Biological Sciences,* Vol.18, No.5, (May 1965), pp. 959-969, ISSN 0004-9417

Henniges, O. & Zeddies, J. (2004). Competiveness of Brazilian Ethanol in the EU. *FO Litch's World Ethanol and Biofuels Report,* Vol. 2, pp. 374-378, ISSN 1478-5765

Hill, J.; Nelson, E.; Tilman, D.; Polasky, S. & Tiffany, D. (2006). Environmental, Economic, and Energetic Costs and Benefits of Biodiesel and Ethanol Biofuels. *Proceeding of the National Academy of Sciences U.S.A.,* Vol.103, No.30, (July 2006), pp. 11206-11210, ISSN *0027-8424*

Hobhouse, H. (2005). *Seeds of Change: six plants that transformed mankind.* Counterpoint, ISBN 978-1593760496, Berkeley, CA

Hotta, C.T.; Lembke, C.G.; Domingues, D.S.; Ochoa, E.A.; Cruz, G.M.Q.; Melotto-Passarin, D.; Marconi, T.G.; Santos, M.O.; Mollinari, M.; Margarido, G.R.A.; Crivellari, A.C.; Santos, W.D.; Souza, A.P.; Hoshino, A.A.; Carrer, H.; Souza, A.P.; Garcia, A.A.F.; Buckeridge, M.S.; Menossi, M.; Van Sluys, M.A. & Souza, G.M. (2010). The Biotecnological Roadmap for Sugarcane Improvement. *Tropical Plant Biology,* Vol.3, No.2, (April 2010), pp. 75-87, ISSN 1935-9756

Ingelbrecht, I.L.; Irvine, J.E. & Mirkov, T.E. (1999). Posttranscriptional Gene Silencing in Transgenic Sugarcane. Dissection of Homology-Dependent Virus Resistance in a Monocot That Has a Complex Polyploid Genome. *Plant Physiology, Vol.119, No.4,* (April 1999), pp. 1187-1198, ISSN 0032-0889

Innan-Bamber, N.G. & Smith, D.M. (2005). Water Relations in Sugarcane and Response to Water Deficits. *Field Crops Research,* Vol.92, No.2-3, (June 2005), pp. 185-202, ISSN 0378-4290

Irvine, J.E. (1999). *Saccharum* Species as Horticultural Classes. *Theoretical and Applied Genetics,* Vol.98, No.2, (February 1999), pp. 186-194, ISSN 0040-5752

Iskandar, H.M.; Casu, R.E.; Fletcher, A.T.; Schmidt, S.; Xu, J.; Maclean, D.J.; Manners, J.M. & Bonnett, G.D. (2011). Identification of Drought Response Genes and a Study of Their Expression During Sucrose Accumulation and Water Deficit in Sugarcane Culms. *BMC Plant Biology,* Vol.11, (January 2011), pp. 12, ISSN 1471-2229

Jackson, P.A. (2005). Breeding for Improved Sugar Content in Sugarcane. *Field Crops Research,* Vol.92, No.2-3, (June 2005), pp. 277-290, ISSN 0378-4290

Joyce, P.; Kuwahata, M.; Turner, N. & Lakshmanan, P. (2010). Selection System and Co-Cultivation Medium are Important Determinants of *Agrobacterium* Mediated Transformation of Sugarcane. *Plant Cell Reports,* Vol.29, No.2, (February 2010), pp. 173–183, ISSN 0721-7714

Kaeppler, S.M.; Kaeppler, H.F. & Rhee, Y. (2000). Epigenetic Aspects of Somaclonal Variation in Plants. *Plant Molecular Biology,* Vol.43, No.2-3, (June 2000), pp. 179-188, ISSN 0735-9640

Khan, M. S.; Yadav,S.; Srivastava,S.; Swapna, M.; Chandra, A. & Singh, R. K. (2011). Development and Utilization of Conserved Intron Scanning Marker in Sugarcane. *Australian Journal of Botany,* Vol.59, No.1, (February 2011), pp. 38-45, ISSN *0067-1924*

Kirch, H.H.; Schlingensiepen, S.; Kotchoni, S.; Sunkar, R. & Bartels, D. (2005). Detailed Expression Analysis of Selected Genes of the *Aldehyde Dehydrogenase* (ALDH) Gene

Superfamily in *Arabidopsis thaliana*. *Plant Mol Biol.*, Vol.57, No.3, (February 2005), pp. 315-332, ISSN 0735-9640

Koizumi, M.; Yamaguchi-Shinozaki, K., Tsuji, H. & Shinozaki, K. (1993). Structure and Expression of Two Genes That Encode Distinct Drought-Inducible Cysteine Proteinases in *Arabidopsis thaliana. Gene*, Vol.129, No.2, (July 1993), pp. 175-182, ISSN 0378-1119

Lam, E.; Shine, J.; Silva, J. da; Lawton, M.; Bonoss, S.; Calvino, M.; Carrer, H.; Silva-Filho, M.; Glynn, N.; Helsels, Z.; Ma, J.; Richard, E.; Souza, G.M. & Ming, R. (2009). Improving Sugarcane for Biofuel: Engineering for an Even Better Feedstock. *GCB Bioenergy*, Vol.1, (July 2009), pp. 251-255, ISSN 1557-1693

Landell, M. (2003). ProCana – O Programa Cana-de-açúcar do Instituto Agronômico. *O Agronomico*, Vol.55, No.1, pp. 5-8, ISSN 0365-2726

Landell, M.G.A. & Bressiani, J.A. (2008). Melhoramento Genético, Caracterização e Manejo Varietal. In: *Cana-de-açúcar*. Dinardo-Miranda, L.L.; Vasconcelos, A.C.M.; Landell, M.G.A. (Eds.), 101 – 155, Instituto Agronômico, ISBN 9788585564179, Campinas

Landell, M.G.A.; Creste, S.; Pinto, L.R.; Xavier, M.A. & Bressiani, J.A. (2010). The Strategy For Regional Selection For The Development of Energy Sugarcane Cultivar Profile. In: *Sugar Cane Bioethanol: R&D for productivity and sustainability* L.A.B. Cortez (Ed.), 345-352, Blucher, ISBN: 9788521205302, São Paulo

Lee, S.M.; Kang, K.; Chung, H.; Yoo, S.H.; Xu, X.M.; Lee, S.B.; Cheong, J.J.; Daniell, H. & Kim, M. (2006). Plastid Transformation in the Monocotyledonous Cereal Crop, Rice (*Oryza sativa*) and Transmission of Transgenes to Their Progeny. *Molecular Cells*, Vol.21, No.3, (June 2006), pp. 401-410, ISSN 1097-2765

Leite, R.C.C. (2010). The Brazilian Strategy for Bioethanol, In: *Sugar Cane Bioethanol: R&D for productivity and sustainability*, L.A.B. Cortez, (Ed.), 17-18, Blucher, ISBN 9788521205302, São Paulo

Leite, R.C.C.; Leal, M.R.L.V.; Cortez, L.A.B.; Griffin, W.M. & Scandiffio, M.I.G. (2009). Can Brazil Replace 5% of the 2025 Gasoline World Demand With Ethanol? *Energy*, Vol.34, pp. 655-661, ISSN 0360-5442

Liu, G.F.; Zhou, H.K.; Zhu, Z.H.; Xu, H.M. & Yang, J. (2007). Genetic Analysis for Brix Weight per Stool and Its Component Traits in Sugarcane (*Saccharum officinarum*). *Journal of Zheijang University Science B*, Vol.8, No.12, (December 2007), pp. 860-866, ISSN 1673-1581

Lora, E.S. & Andrade, R.V. (2009). Biomass As Energy Source in Brazil. *Renewable and Sustainable Energy Reviews*, Vol.13, No.4, (May 2009), pp. 777-788, ISSN 1364-0321

Manickavasagam, M.; Ganapathi, A.; Anbazhagan, V.R.; Sudhakar, B.; Selvaraj, N.; Vasudevan, A. & Kasthurirengan, S. (2004). *Agrobacterium* Mediated Genetic Transformation and Development of Herbicide-Resistant Sugarcane (*Saccharum* species hybrids) Using Axillary Buds. *Plant Cell Reports*, Vol.23, No.3, (September 2004), pp. 134–143, ISSN 0721-7714

Manners, J.M & Casu, R.E. (2011) Transcriptome Analysis and Functional Genomics of Sugarcane. *Tropical Plant Biology*, Vol.4, No.1, (March 2011), pp. 9–21, ISSN 1935-9756

Martinelli, L.A. & Filoso, L. (2008). Expansion of Sugarcane Ethanol Production in Brazil: Environmental and Social Challenges. *Ecological Applications*, Vol.18, No.4, (April 2008), pp. 885-898, ISSN 1051-0761

McRae, T.A.; Hogath, D.M.; Foreman, J.W. & Braithwaite, M.J. (1993). Selection of Sugarcane Seedling Families in Burdekin District, *Australian Plant Breeding Conference*, Vol. 1, pp. 77-82, ISBN 0646134361, Gold Coast, Australia, April 1993

Melo, L.C.P. & Poppe, M.K. (2010). Challenges in Research, Development and Innovation in Biofuels in Brazil, In: *Sugar Cane Bioethanol: R&D for productivity and sustainability* L.A.B. Cortez (Ed.), 27-34, Blucher, ISBN: 9788521205302, São Paulo

Menossi, M.; Silva-Filho M.C.; Vincentz, M.; Van-Sluys, M.A. & Souza, G.M. (2008). Sugarcane Functional Genomics: Gene Discovery for Agronomic Trait Development. *International Journal of Plant Genomics*, Vol.2008, (December 2007), pp. 458732, ISSN 16875370

Milligan, S.B.; Gravois, K.A.; Bischoff, K.P. & Martin, F.A. (1990). Crop Effects on Genetic Relationships Among Sugarcane Traits. *Crop Science*, Vol.30, No.4, (April 2004), pp. 927-931, ISSN 0011-183X

Miranda, E.E. (2010). Environmental (Local and Global) Impact and Energy Issues on Sugarcane Expansion and Land Occupation in the São Paulo State, In: *Sugar Cane Bioethanol: R&D for productivity and sustainability* L.A.B. Cortez (Ed.), 41-52, Blucher, ISBN: 9788521205302, São Paulo

Molinari, H.B.C.; Marur, C.J.; Daros, E.; Campos, M.K.F.; Carvalho, J.F.R.P.; Bespalhok-Filho, J.C.; Pereira, L.F.P. & Vieira, L.G.E. (2007). Evaluation of the Stress-Inducible Production of 6-proline in Transgenic Sugarcane (*Saccharum* spp.): Osmotic Adjustment, Chlorophyll Fluorescence and Oxidative Stress. *Physiologia Plantarum*, Vol.130, No.2, (June 2007), pp. 218-229, ISSN 0031-9317

Moore, P. H. (2005). Integration of Sucrose Accumulation Processes Across Hierarchical Scales: Towards Developing an Understanding of the Gene-to-Crop Continuum. *Field Crops Research*, Vol.92, pp. 119–135, ISSN 0378-4290

Moore, P.H. & Nuss, K.J. (1987). Flowering and Flower Synchronization. In: *Sugarcane improvement through breeding*, Heinz, D.J. (Ed), 273-311, Elsevier, ISBN 0444427694, Amsterdam.

Nass, L.L.; Pereira, P.A.A., & Ellis, D. (2007). Biofuels in Brazil: An overview. *Crop Science*, Vol.47, (November-December 2007), pp. 2228-2237, ISSN 0931-2250

Pandey, A.; Soccol, C.R.; Nigam, P. & Soccol, V.T. (2000). Biotecnological Potential of Agro-Industrial Residues. I: Sugarcane Bagasse. *Bioresource Tecnhology*, Vol.74, No.1, (August 2000), pp. 69-80, ISSN 0960-8524

Papini-Terzi, F. S.; Rocha, F. R.; Vêncio, R. Z. N.; Felix, J. M.; Branco, D. S.; Waclawovsky, A. J.; Del Bem, L. E. V.; Lembke, C. G.; Costa, M. D. L.; Nishiyama Jr, M. Y.; Vicentini, R.; Vincentz, M. G. A.; Ulian, E. C.; Menossi, M. & Souza, G. M. (2009). Sugarcane Genes Associated with Sucrose Contente. *BMC Genomics*, Vol.10, No.120, (March 2009), pp. 120, ISSN 1471-2164

Papini-Terzi, F.S.; Rocha, F.R.; Vêncio, R.Z.; Oliveira, K.C.; Felix, J. de M.; Vicentini, R.; Rocha, C. de S.; Simões, A.C.; Ulian, E.C.; di Mauro, S.M.; da Silva, A.M.; Pereira, C.A.; Menossi, M. & Souza, G.M. (2005). Transcription Profiling of Signal Transduction-Related Genes in Sugarcane Tissues. *DNA Research*, Vol.12, No.1, (January 2005), pp. 27-38, ISSN 1340-2838

Parida, S.K.; Pandit, A.; Gaikwad, K.; Sharma, T.R.; Srivastava, P.S.; Singh, N.K. & Mohapatra, T. (2010). Functionally Relevant Microsatellites in Sugarcane Unigenes. *BMC Plant Biology*, Vol.10, (November 2010), pp. 251, ISSN 1471-2229

Paterson, A.H., Bowers, J.E. & Chapman, B.A. (2004) Ancient Polyploidization Predating Divergence of the Cereals, And Its Consequences For Comparative Genomics. *Proc. Natl. Acad. Sci. USA* Vol.101, No.26, (June 2004), pp. 9903–9908, ISSN *0027-8424*

Paterson, A.H.; Bowers, J.E. *et al.* (2009). The *Sorghum bicolor* Genome and the Diversification of Grasses. *Nature,* Vol.457, (January 2009), pp. 551-556, ISSN 0028-0836

Petrasovits, L.A.; Purnell, M.P.; Nielsen, L.K. & Brumbley, S.M. (2007). Production of Polyhydroxybutyrate in Sugarcane. *Plant Biotechnology Journal,* Vol.5, No.1, (January 2007), pp. 162-172, ISSN 1467-7644

Rabbani, M.A.; Maruyama, K.; Abe, H.; Khan, M.A.; Katsura, K.; Ito, Y.; Yoshiwara, K.; Seki, M.; Shinozaki K. & Yamaguchi-Shinozaki, K. (2003). Monitoring Expression Profiles of Rice (*Oryza sativa* L.) Genes Under Cold, Drought and High-Salinity Stresses, and ABA Application Using Both cDNA Microarray and RNA Gel Blot Analyses. *Plant Physiology,* Vol.133, No.4, (December 2003), pp. 1755–1767, ISSN 0032-0889

Ragauskas, A.J.; Williams, C.K.; Davison, B.H.; Britovsek, G.; Cairney, J.; Eckert, C.K.; Frederick Jr., W.J.; Hallett, J.P.; Leak, D.J.; Liotta, C.L.; Mielenz, J.R.; Murphey, R.; Templer, R. & Tschaplinski, T. (2006). The Path Forward for Biofuels and Biomaterials. *Science,* Vol.311, (January 2006), pp. 484-489, ISSN 0036-8075

Raizer, A.J. & Vencovsky, R. (1999). Estabilidade Fenotípica de Novas Variedades de Cana-de-Açúcar para o Estado de São Paulo. *Pesquisa Agropecuária Brasileira,* Vol.34, No.12, pp. 2241-2246, ISSN 0100-204X

Reis, V.M.; de Paula, M.A. & Döbereiner, J. (1999) Ocorrência de Micorrizas Arbusculares e da Bactéria Diazotrófica *Acetobacter diazotrophicus* em Cana-de-açúcar. *Pesquisa Agropecuária Brasileir,* Vol.34, No.7, (Julho 1999), pp. 1933-1941, ISSN 0100-204X

Ribeiro, I.L.A.C. (2010) *Proteômica de cana-de-açúcar em condição de estresse hídrico.* Dissertation. Universidade Federal de Pernambuco, Recife, Brasil. 127 p.

Ripoli, T.C.C.; Molina Jr., W.F. & Ripoli, M.L.C. (2000). Energy Potential of Sugarcane Biomass in Brazil. *Scientia Agricola,* Vol.57, No.4, (October-November 2000), pp. 677-681, ISSN 0103-9016

Roach, B.T. (1972). Nobilisation of sugarcane. *Proc. Int. Soc. Sugarcane. Technol.,* Vol.14, pp. 206–216

Roach, B.T. (1989). Origin and Improvement of the Genetic Base of Sugarcane. XX *Proc. Int. Soc. Sugarcane. Technol.,* Vol. 11, pp. 35-47, São Paulo, 1989

Rocha, F.R.; Papini-Terzi, F.S.; Nishiyama Jr., M.Y.; Vêncio, R.Z.; Vicentini, R.; Duarte, R.D.; de Rosa Jr., V.E.; Vinagre, F.; Barsalobres, C.; Medeiros, A.H.; Rodrigues, F.A.; Ulian, E.C.; Zingaretti, S.M.; Galbiatti, J.A.; Almeida, R.S.; Figueira, A.V.; Hemerly, A.S.; Silva-Filho, M.C.; Menossi, M. & Souza, G.M. (2007). Signal Transduction Rlated Responses to Phytohormones and Environmental Challenges in Sugarcane. *BMC Genomics,* Vol.8, (March 2007), pp. 71, ISSN 1471-2164

Rogers, T.D. (2010). *The Deepest Wounds: A Labor and Environmental History of Sugar in Northeast Brazil.* University of North Carolina Press, ISBN 978-0807871676, Chapel Hill, NC

Seki, M.; Ishida, J.; Narusaka, M.; Fujita, M.; Nanjo, T.; Umezawa, T.; Kamiya, A.; Nakajima, M.; Enju, A.; Sakurai, T.; Satou, M.; Akiyama, K.; Yamaguchi-Shinozaki, K.; Carninci, P.; Kawai, J.; Hayashizaki, Y. & Shinozaki, K. (2002). Monitoring the expression pattern of around 7,000 Arabidopsis genes under ABA treatments using a full-length cDNA microarray. *Funct Integr Genomics,* Vol.2, No.6, (November 2002), pp. 282-291, ISSN 1438-793X

Seki, M.; Narusaka M.; Abe, H.; Kasuga M.; Yamaguchi-Shinozaki, K.; Carninci, P.; Hayashizaki, Y. & Shinozaki, K. (2001). Monitoring the Expression Pattern of 1300 Arabidopsis Genes Under Drought and Cold Stresses by Using a Full-Length cDNA Microarray. *The Plant Cell*, Vol.13, No.1, (January 2001), pp. 61–72, ISSN 1040-4651

Singels, A.; Smit, M.A.; Redshaw, K.A. & Donaldson, R.A. (2005). The Effect of Crop Start Date, Crop Class and Cultivar on Sugarcane Canopy Development and Radiation Interception. *Field Crops Research*, Vol.92, No.2-3, (June 2005), pp. 249-260, ISSN 0378-4290

Snyman, S.J.; Meyer, G.M.; Richards, J.M.; Haricharan N.; Ramgareeb, S. & Huckett, B.I. (2006). Refining The Application of Direct Embryogenesis In Sugarcane: Effect of the Developmental Phase of Leaf Disc Explants and the Timing of DNA Transfer on Transformation Efficiency. *Plant Cell Rep*, Vol.25, No.10, (October 2006), pp. 1016-1023, ISSN 0721-7714

Swapna, M.; Sivaraju, K.; Sharma, R. K.; Singh, N. K. & Mohapatra, T. (2010). Single-Strand Conformational Polymorphism of EST-SSRs: a Potential Tool for Diversity Analysis and Varietal Identification in Sugarcane. *Plant Molecular Biology Reports*, Vol.5, pp. 254-263, ISSN 0735-9640

Tabasum, S.; Khan, F.A.; Nawaz, S.; Iqbal, M.Z. & Saeed, A. (2010). DNA Profiling of Sugarcane Genotypes Using Randomly Amplified Polymorphic DNA. *Genetics and Molecular Research*, Vol.9, No.1, pp. 471-483, ISSN 1676-5680

Thaweenut, N.; Hachisuba, Y.; Ando, S.; Yanagisawa, S. & Yoneyama, T. (2011). Two Seasons Study of *nifH* Gene Expression and Nitrogen Fixation by Fiazotrophic Fndophytes in Sugarcane (*Saccharum* spp. hybrids): Expression of *nifH* Genes Similar to Those of Rhizobia. *Plant and Soil*, Vol.338, No.1-2, (January 2011), pp. 435-449, ISBN 0032-079X

Tomkins, J.P.; Yu, Y.; Miller-Smith, H.; Frisch, D.A.; Woo, S.S. & Wing, R.A. (1999). A Bacterial Artificial Chromosome Library for Sugarcane. *Theoretical and Applied Genetics*, Vol.99, No.3-4, (August 1999), pp. 419-424, ISSN 0040-5752

Tran, L.S.P.; Nakashima, K.; Sakuma, Y.; Simpson, S.D.; Fujita, Y.; Maruyama, K.; Fujita, M.; Seki, M.; Shinozaki, K. & Yamaguchi-Shinozaki, K. (2004). Isolation and Functional Analysis of Arabidopsis Stress-Inducible NAC Transcription Factors That Bind to a Drought-Responsive Cis-Element in The Early Responsive to Dehydration Stress 1 Promoter. *The Plant Cell*, Vol.16, No.9, (September 2009), pp. 2481–2498, ISSN 1040-4651

Uriarte, M.; Yackulic, C.B.; Cooper, T.; Flynn, D.; Cortes, M.; Crk, T.; Cullman, G.; McGinty, M. & Sircely, J. (2009). Expansion of Sugarcane Production in São Paulo, Brazil: Implications For Fire Occurrence and Respiratory Health. *Agriculture Ecosystems & Environment*, Vol.132, No.1-2, (Julho 2009), pp. 48-56, ISSN 0167-8809

Vega-Sánchez, M.E. & Ronald P.C. (2010). Genetic and Biotechnological Approaches for Biofuel Crop Improvement. *Current Opinion in Biotechnology*, Vol.21, No.2, (February 2010), pp. 218-224, ISSN 0958-1669

Verma, M.M.; Chahal, G.S. & Murty, B.R. (1978). Limitations of Conventional Regression Analysis: a Proposed Modifications. *Theoretical and Applied Genetics*, Vol. 53, No.2, (March 1978), pp. 89-91, ISSN 0040-5752

Vettore, A.L.; da Silva, F.R.; Kemper, E.L. & Arruda, P. (2001). The Libraries That Made SUCEST. *Genetic Molecular Biology*, Vol.24, No.1-4, pp. 1–7, ISSN 1415-4757

Vettore, A.L.; da Silva, F.R.; Kemper, E.L.; *et al.* (2003). Analysis and Functional Annotation of an Expressed Sequence Tag Collection for Tropical Crop Sugarcane. *Genome Research*, Vol.13, No.12, (November 2003), pp. 2725-2735, ISSN 1088-9051

Vu, J.C.V. & Allen-Jr., L.H. (2009). Growth at Elevated $CO_2$ Delays the Adverse Effects of Drought Stress on Leaf Photosynthesis of the $C_4$ Sugarcane. *Journal of Plant Physiology*, Vol.166, No. 2, (January 2009), pp. 107-116, ISSN 0176-1617

Waclawovsky, A.J.; Sato, P.M.; Lembke, C.G.; Moore, P.H. & Souza, G.M. (2010). Sugarcane for Bioenergy Production: An Assessment of Yield and Regulation of Sucrose Content. *Plant Biotechnology Journal*, Vol.8, No.3, (April 2010), pp. 263-276, ISSN 1467-7644

Wang, H.H.; Yin, W.B. & Hu, Z.M. (2009). Advances in Chloroplast Engineering. *Journal of Genetics & Genomics*, Vol.36, No.7, (July 2009), pp. 387-398, ISSN 1673-8527

Wang, J.; Roe, B.; Macmil, S.; Yu, Q.; Jan, E.; Murray, J.E.; Tang, G.; Chen, C.; Najar, F.; Wiley, G.; Bower, J.; Van Sluys, M.A.; Rokhsar, D.S.; Hudson, M.E.; Moose, S.P.; Paterson, A.H. & Ming, R. (2010). Microcollinearity Between Autopolyploid Sugarcane and Diploid Sorghum Genomes. *BMC Genomics*, Vol.11, (April 2010), pp. 261, ISSN 1471-2164

Wei, X.; Jackson, P.A.; Hermann, S.; Kilian, A.; Heller-Uszynska, K. & Deomano, E. (2010). Simultaneously Accounting for Population Structure, Genotype by Environment Interaction, and Spatial Variation in Marker-Trait Associations in Sugarcane. *Genome*, Vol.53, No.11, (November 2010), pp. 973-981, ISSN 0831-2796

Weng, L.X.; Deng, H.H.; Xu, J.L.; Li, Q.; Zhang, Y.Q.; Jiang, Z.D.; Li, Q.W., Chen, .JW. & Zhang, L.H. (2010). Transgenic Sugarcane Plants Expressing High Levels of Modified *Cry1Ac* Provide Effective Control Against Stem Borers in Field Trials. *Transgenic Research*, Vol.20, No.4 (November, 2010), pp. 1-14, ISSN 1573-9368

Whan, A.; Robinson, N.; Lakshmanan, P.; Schmidt, S. & Aitken, K. (2010). A Quantitative Genetics Approach to Nitrogen Use Efficiency in Sugarcane. *Plant Functional Genetics*, Vol.37, No.5, (April 2010), pp. 448-454, ISSN 1445-4408

Yamaguchi-Shinozaki, K. & Shinozaki, K. (2006). Transcriptional Regulatory Networks in Cellular Responses and Tolerance to Dehydration and Cold Stresses. *Annu. Rev. Plant Biol.*, Vol.57, pp. 781-803, ISSN 1040-2519

Yamaguchi-Shinozki, K.; Kasuga, M.; Liu, Q.; Nakashima, K.; Sakuma, Y.; Abe, H.; Shinwari, Z.K.; Seki, M. & Shinozaki, K. (2002). Biological mechanisms of drought stress response. *JIRCAS Working Report*, pp. 1-8

Yin, X. & Struik, P.C. (2010). Modelling the Crop: From System Dynamics to Systems Biology. *Journal of Experimental Botany*, Vol.61, No.8, (January 2010), pp. 2171–2183, ISSN 0022-0957

Zhangsun, D.; Luo, S.; Chen, R. & Tang, K. (2007). Improved *Agrobacterium* Mediated Genetic Transformation of GNA Transgenic Sugarcane. *Biologia*, Vol.62, No.4, (August 2007), pp. 386–393, ISSN 0006-3088

Zheng, J.; Zhao, J.; Tao, Y.; Wang, J.; Liu, Y.; Fu, J.; Jin, Y.; Gao, P.; Zhang, J.; Bai, Y. & Wang, G. (2004). Isolation and analysis of water stress induced genes in maize seedlings by subtractive PCR and cDNA macroarray. *Plant Mol. Biol.*, Vol.55, No.6, (August 2004), pp. 807-823, ISSN 0735-9640

# Permissions

The contributors of this book come from diverse backgrounds, making this book a truly international effort. This book will bring forth new frontiers with its revolutionizing research information and detailed analysis of the nascent developments around the world.

We would like to thank Ibrokhim Y. Abdurakhmonov, for lending his expertise to make the book truly unique. He has played a crucial role in the development of this book. Without his invaluable contribution this book wouldn't have been possible. He has made vital efforts to compile up to date information on the varied aspects of this subject to make this book a valuable addition to the collection of many professionals and students.

This book was conceptualized with the vision of imparting up-to-date information and advanced data in this field. To ensure the same, a matchless editorial board was set up. Every individual on the board went through rigorous rounds of assessment to prove their worth. After which they invested a large part of their time researching and compiling the most relevant data for our readers. Conferences and sessions were held from time to time between the editorial board and the contributing authors to present the data in the most comprehensible form. The editorial team has worked tirelessly to provide valuable and valid information to help people across the globe.

Every chapter published in this book has been scrutinized by our experts. Their significance has been extensively debated. The topics covered herein carry significant findings which will fuel the growth of the discipline. They may even be implemented as practical applications or may be referred to as a beginning point for another development. Chapters in this book were first published by InTech; hereby published with permission under the Creative Commons Attribution License or equivalent.

The editorial board has been involved in producing this book since its inception. They have spent rigorous hours researching and exploring the diverse topics which have resulted in the successful publishing of this book. They have passed on their knowledge of decades through this book. To expedite this challenging task, the publisher supported the team at every step. A small team of assistant editors was also appointed to further simplify the editing procedure and attain best results for the readers.

Our editorial team has been hand-picked from every corner of the world. Their multi-ethnicity adds dynamic inputs to the discussions which result in innovative outcomes. These outcomes are then further discussed with the researchers and contributors who give their valuable feedback and opinion regarding the same. The feedback is then collaborated with the researches and they are edited in a comprehensive manner to aid the understanding of the subject.

Apart from the editorial board, the designing team has also invested a significant amount of their time in understanding the subject and creating the most relevant covers. They scrutinized every image to scout for the most suitable representation of the subject and create an appropriate cover for the book.

The publishing team has been involved in this book since its early stages. They were actively engaged in every process, be it collecting the data, connecting with the contributors or procuring relevant information. The team has been an ardent support to the editorial, designing and production team. Their endless efforts to recruit the best for this project, has resulted in the accomplishment of this book. They are a veteran in the field of academics and their pool of knowledge is as vast as their experience in printing. Their expertise and guidance has proved useful at every step. Their uncompromising quality standards have made this book an exceptional effort. Their encouragement from time to time has been an inspiration for everyone.

The publisher and the editorial board hope that this book will prove to be a valuable piece of knowledge for researchers, students, practitioners and scholars across the globe.

# List of Contributors

**Sven B. Andersen**
University of Copenhagen, Dept. Agronomy and Ecology, Denmark

**Patu Khate Zeliang and Arunava Pattanayak**
Division of Plant Breeding, ICAR Research Complex for NEH Region, India

**Jiankang Wang**
Institute of Crop Science and CIMMYT China, Chinese Academy of Agricultural Sciences (CAAS), China

**Jana Murovec and Borut Bohanec**
University of Ljubljana, Biotechnical Faculty, Slovenia

**Sukumar Saha**
United States Department of Agriculture-Agriculture Research Service, Crop Science Research Laboratory, USA

**A. Dewitte**
KATHO Catholic University College of Southwest Flanders, Department of Health Care and Biotechnology, Belgium

**K. Van Laere and J. Van Huylenbroeck**
Institute for Agricultural and Fisheries Research (ILVO), Plant Sciences Unit, Belgium

**Siva P. Kumpatla, Ramesh Buyyarapu and Jafar A. Mammadov**
Department of Trait Genetics and Technologies, Dow Agro Sciences LLC, USA

**Ibrokhim Y. Abdurakhmonov**
Center of Genomic Technologies, Institute of Genetics and Plant Experimental Biology, Academy of Sciences of Uzbekistan, Uzbekistan

**Katarzyna Mikolajczyk, Iwona Bartkowiak-Broda, Wieslawa Poplawska, Stanislaw Spasibionek and Agnieszka Dobrzycka**
Plant Breeding and Acclimatization Institute – National Research Institute, Research Division in Poznan, Poland

**Miroslawa Dabert**
Molecular Biology Techniques Laboratory, Faculty of Biology, Adam Mickiewicz University in Poznan, Poland

**José Luis Piña-Escutia, Luis Miguel Vázquez-García and Amaury Martín Arzate-Fernández**
Centro de Investigación y Estudios Avanzados en Fitomejoramiento, Facultad de Ciencias Agrícolas, Universidad Autónoma del Estado de México, México

**Santelmo Vasconcelos, Alberto V. C. Onofre, Ana Maria Benko-Iseppon and Ana Christina Brasileiro-Vidal**
Laboratory of Plant Genetics and Biotechnology, Federal University of Pernambuco, Brazil

**Máira Milani**
Embrapa Algodão, Campina Grande, Brazil

**Sónia Gomes, Henrique Guedes-Pinto and Paula Martins-Lopes**
Institute of Biotechnology and Bioengineering, Centre of Genomics and Biotechnology - University of Trás-os-Montes and Alto Douro (IBB/CGB-UTAD), Portugal

**Pilar Prieto**
Department of Mejora Genética Vegetal, Instituto de Agricultura Sostenible (CSIC), Spain

**Teresa Carvalho**
National Station of Plant Breeding, Department of Oliviculture, Portugal

**S. Hossain**
Department of Primary Industries, Australia

**G.P. Kadkol**
NSW Department of Primary Industries, Tamworth Agricultural Institute, Australia

**R. Raman and H. Raman**
EH Graham Centre for Agricultural Innovation, an alliance between NSW Department of Primary, Industries and Charles Sturt University, Wagga Wagga Agricultural Institute, Australia

**P.A. Salisbury**
Department of Primary Industries, Australia
Department of Agriculture and Food Systems, Melbourne School of Land and Environment, The University of Melbourne, Australia

**Elsa Gonçalves and Antero Martins**
Instituto Superior de Agronomia/Technical University of Lisbon, Portugal

**Neil O. Anderson**
University of Minnesota, USA

**Janelle Frick**
University of Minnesota, USA
University of Arkansas, USA

**Adnan Younis**
University of Minnesota, USA
University of Agriculture, Pakistan

**Christopher Currey**
University of Minnesota, USA
Purdue University, USA

**Katia C. Scortecci**
Department of Cell Biology and Genetics, Universidad Federal do Rio Grande do Norte (UFRN), Natal, RN, Brazil

**Silvana Creste, Mauro A. Xavier and Marcos G. A. Landell**
Centro de Cana – Instituto Agronômico de Campinas (IAC), Ribeirão Preto, SP, Brazil

**Tercilio Calsa Jr.**
Department of Genetics, Universidad Federal de Pernambuco (UFPE), Recife, PE, Brazil

**Antonio Figueira**
Plant Breeding Laboratory, Centro de Energia Nuclear na Agricultura (CENA), Universidad de São Paulo (USP), Piracicaba, SP, Brazil

**Vagner A. Benedito**
Laboratory of Plant Functional Genetics, Genetics and Developmental Biology Program, Plant & Soil Sciences Division, West Virginia University (WVU), Morgantown, WV, USA